Revised Student Solutions Manual

# Intermediate Algebra for College Students

Revised Student Solutions Manual

**Lea Campbell**
*Lamar University Port Arthur*

THIRD EDITION

# Intermediate Algebra for College Students

**Allen R. Angel**
*Monroe Community College*

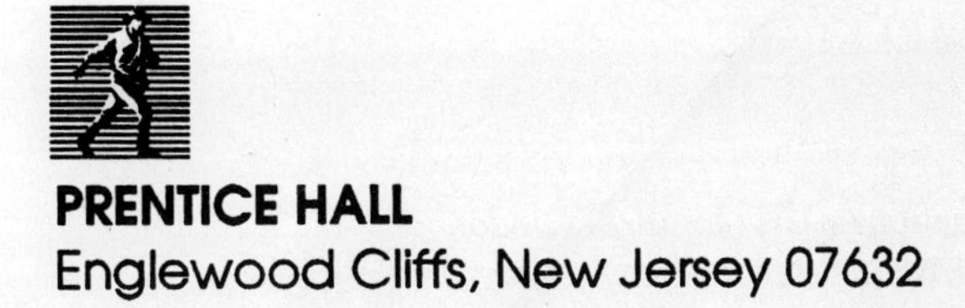

**PRENTICE HALL**
Englewood Cliffs, New Jersey 07632

Editorial production/supervision: *Shirin Khan*
Prepress Buyer: *Paula Massenaro*
Manufacturing buyer: *Lori Bulwin*
Acquisitions Editor: *Priscilla McGeehon*

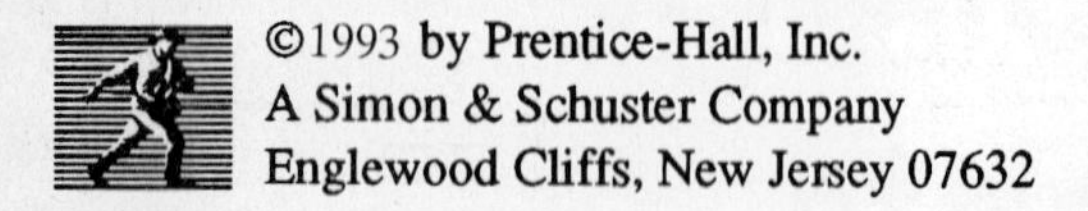


A Simon & Schuster Company
Englewood Cliffs, New Jersey 07632



Printed in the United States of America

10 9 8 7 6 5 4 3

ISBN 0-13-482662-0

Prentice-Hall International (UK) Limited, *London*
Prentice-Hall of Australia Pty. Limited, *Sydney*
Prentice-Hall Canada Inc. *Toronto*
Prentice-Hall Hispanoamericana, S.A., *Mexico*
Prentice-Hall of India Private Limited, *New Delhi*
Prentice-Hall of Japan, Inc., *Tokyo*
Simon & Schuster Asia Pte. Ltd., *Singapore*
Editora Prentice-Hall do Brasil, Ltda., *Rio de Janeiro*

# CHAPTER 1

## EXERCISE SET 1.1

9. a) Do the homework.
   b) Write down any questions you have as you do the homework. Star any problems you had difficulty with and ask your instructor about those problems.
   c) Prior to class, preview new material.
   d) After the new material is presented, read the corresponding section of the text.

11. a) Prepare for the exam a little bit each day to eliminate the need to cram the night before the exam.
    b) Review class notes and homework assignments.
    c) Study the formulas, definitions, and procedures given in the text.
    d) Read the Common Student Error boxes and Helpful Hint boxes.
    e) Read the summary at the end of each chapter.
    f) Work the review exercises at the end of each chapter. If you have difficulty, restudy those sections. If you still have trouble, seek help.
    g) Work the practice chapter test.

## EXERCISE SET 1.2

1. A = {4,5,6,7}
3. C = {6,8,10}
5. E = {0,1,2,3,4,5,6}
7. G = { }
9. I = {-4,-3,-2,-1...}
11. K = { }
13. $\notin$
15. $\in$
17. $\notin$
19. $\notin$
21. $\subseteq$
23. $\not\subseteq$
25. $\not\subseteq$
27. $\not\subseteq$
29. $\not\subseteq$
31. $\subseteq$
33. $\subseteq$
35. $\not\subseteq$
37. $\subseteq$
39. $\not\subseteq$
41. $\not\subseteq$
43. $\subseteq$
45. True
47. False
49. True
51. False
53. True
55. True
57. True
59. False
61. True
63. 4
65. -6,4,0
67. $\sqrt{7}$, $\sqrt{5}$,
69. 2,4
71. 2,4,-5.33,9/2,-100,-7, 4.7
73. 2,4,-5.33,9/2, $\sqrt{7}$ $\sqrt{2}$ -100, -7, 4.7
75. $A \cup B$ = { 1,2,3,4}  $A \cap B$ = {2,3}
77. $A \cup B$ = {-1,-2,-4,-5,-6}
    $A \cap B$ = {-2,-4}
79. $A \cup B$ = {0,1,2,3}  $A \cap B$ = { }
81. $A \cup B$ = {2,4,6,8...}
    $A \cap B$ = { 2,4,6}
83. $A \cup B$ = 0,1,2,3,4,5,6,7,8,}
    $A \cap B$ = { }
85. $A \cup B$ = {1,2,3,4...}
    $A \cap B$ = {2,4,6,8}
87. The set of odd natural numbers greater than or equal to 5.
89. The set of lower case letters in the English alphabet.
91. The set of states in the United States.
93. a) Set B is the set of all x such that x is one of the last five capital letters in the English alphabet.
    b) B = {V,W,X,Y,Z}

EXERCISE SET 1.3

1. Commutative property of addition
3. Distributive property
5. Associative property of addition
7. Commutative property of addition
9. Commutative property of multiplication
11. Identity property of addition
13. Commutative property of addition
15. Commutative property of addition
17. Commutative property of addition
19. Distributive property
21. Double negative property
23. Identity property of multiplication
25. Inverse property of addition
27. Inverse property of addition
29. Inverse property of addition
31. Identity property of multiplication
33. Inverse property of multiplication
35. Double negative property
37. Multiplication property of 0
39. Double negative property
41. $3 + x$
43. $x + (2 + 3)$
45. $x$
47. $x$
49. $x$
51. 0
53. $1x + 1y$ or $x + y$
55. 3
57. 0
59. $-4$ is the additive inverse and $1/4$ is the multiplicative inverse.
61. 3 is the additive inverse and $-1/3$ is the multiplicative inverse.
63. $-2/3$ is the additive inverse and $3/2$ is the multiplicative inverse.
65. 6 is the additive inverse and $-1/6$ is the multiplicative inverse.
67. $3/7$ is the additive inverse and $-7/3$ is the multiplicative inverse.
69. $a + b = b + a$
71. $a(b + c) = ab + ac$
73. True
75. a) 3, 4, −2, 0
    b) 3, 4, −2, 5/6, 0
    c) $\sqrt{3}$
    d) 3, 4, −2, 5/6, $\sqrt{3}$, 0

EXERCISE SET 1.4

1. >
3. >
5. >
7. <
9. <
11. >
13. <
15. >
17. >
19. <
21. >
23. >
25. 6
27. 4
29. 2
31. $\left|\frac{-1}{2}\right| = \frac{1}{2}$
33. $|0| = 0$
35. $|45| = 45$
37. $|-13.84| = 13.84$
39. $-|7| = -(7) = -7$
41. $-|-3| = -(3) = -3$
43. $-\left|\frac{5}{9}\right| = \frac{-5}{9}$

45. $|6| = |-6|$
47. $|-9| > |3|$
49. $|-10| > -5$
51. $|-3| > |-2|$
53. $|-20| > -|24|$
55. $-|4| > -|8|$
57. $6 < |-12|$
59. $|19| < |-25|$
61. $-1, 2, |3|, |-5|, 6$
63. ~~$-2, 4, |-6|, -8$~~ $-3$
65. $-3, |0|, |-5|, |7|, |-12|$
67. $|-9|, 12, 24, |36|, |-45|$
69. $-|-6|, -4, -2, 6, |-8|$
71. $-|2.9|, -2.4, -2.1, -2, |-2.8|$
73. $-2, 1/3, |-1/2|, |3/5|, |-3/4|$

75. The absolute value of 6 is 6.
77. $|4| = 4$ and $|-4| = 4$ so two numbers whose absolute value is 4 are 4 and $-4$.
79. $|6 + 5| = 11$ and $|6 - 17| = 11$ so two numbers when added to 6 result in an absolute value of 11 are 5 and $-17$.
81. All real numbers
83. $a \leq 0$
85. All numbers whose absolute value is 5 are 5 and $-5$.
87. The absolute value of 7 is greater than 4.

89. Negative 4 is equal to the additive inverse of the absolute value of 4.
91. a) If a represents any real real number then

$$|a| = \begin{cases} a \text{ if } a \geq 0 \\ -a \text{ if } a < 0 \end{cases}$$

93. The question can be rewritten as "Does $|a| - |b|$ always equal a positive number. The answer is no and to show this we can use a counter example. Let $a = 3$ and $b = 9$.
$|3| - |9| = 3 - 9 = -6$
95. Commutative property of addition
97. Associative property of addition

EXERCISE SET 1.5

1. $4 + (-3) = 1$
3. $12 + (-2) = 10$
5. $-3 + 8 = 5$
7. $-9 + 17 = 8$
9. $-16 - (-5) = -16 + 5 = -11$
11. $35 - (-4) = 35 + 4$
$= 39$
13. $-9.5 - (-3.72) = -9.5 + 3.72$
$= -5.78$

15. $\dfrac{-3}{8} - \left[\dfrac{-5}{7}\right] = \dfrac{-3}{8} + \dfrac{5}{7}$

$= \dfrac{19}{56}$

17. $4 + 6 - 3 = 10 - 3$
$= 7$
19. $6.23 - 4.5 - (-9.67)$
$= 6.23 - 4.5 + 9.67$
$= 1.73 + 9.67$
$= 11.4$

21. $-6 - 4 - \dfrac{1}{2} = -10 - \dfrac{1}{2}$

$= \dfrac{-20}{2} - \dfrac{1}{2} = \dfrac{-21}{2}$

23. $-3 + (4 - 9) + 3 = -3 + (-5) + 3$
$= -8 + 3$
$= -5$

25. $-(-4 + 2) + (-6 + 3) + 2$
$= 2 + (-3) + 2$
$= 1$

27. $|4| - |3| + |1| = 4 - 3 + 1$
$= 1 + 1$
$= 2$

29. $3 - |-8| - 5 = 3 - 8 - 5$
$= -5 - 5$
$= -10$

31. $|6 - 9| - 5 = |-3| - 5$
$= 3 - 5$
$= -2$

33. $-|-3| - |7| + (6 + |-2|)$
$= -|-3| - |7| + (6 + 2)$
$= -3 - 7 + 8$
$= -10 + 8$
$= -2$

35. $-4(12) = -48$

37. $-4\left(\frac{-5}{16}\right) = \frac{-4}{1}\cdot\left(\frac{-5}{16}\right)$

$= \frac{5}{4}$

39. $(-1)(-1)(-1)(2)(-3)$
$= 1(-1)(2)(-3)$
$= -1(2)(-3)$
$= -2(-3)$
$= 6$

41. $-6 \div (\ 2) = -3$

43. $-3 \div (-3) = 1$

45. $36 \div \left(\frac{-1}{4}\right) = 36\cdot\left(\frac{-4}{1}\right)$

$= -144$

47. $\frac{-5}{9} \div \left(\frac{-5}{9}\right) = \frac{-5}{9}\cdot\left(\frac{-9}{5}\right)$

$= 1$

49. $-3|8| = -3(8) = -24$

51. $\left|\frac{3}{5}\right| \left|\frac{-10.}{6}\right| = \frac{3}{5}\cdot\frac{10}{6}$

$= \frac{2}{2} = 1$

53. $\left|\frac{3}{8}\right| \div (-2) = \frac{3}{8} \div \left(\frac{-2}{1}\right)$

$= \frac{3}{8}\cdot\left(\frac{-1}{2}\right)$

$= \frac{-3}{16}$

55. $\frac{-5}{9} \div |-5| = \frac{-5}{9} \div \frac{5}{1}$

$= \frac{-5}{9}\cdot\frac{1}{5} = \frac{-1}{9}$

57. $5 - 7 = 5 + (-7) = -2$

59. $-64 \div 4 = \frac{-64}{1}\cdot\frac{1}{4} = -16$

61. $\frac{-3}{5} - \left(\frac{-5}{9}\right) = \frac{-3}{5} + \left(\frac{5}{9}\right)$

$= \frac{-27}{45} + \left(\frac{25}{45}\right)$

$= \frac{-2}{45}$

63. $3 - (-4) + 6 - 3 = 3 + 4 + 6 - 3$
$= 7 + 6 - 3$
$= 13 - 3$
$= 10$

65. $(4)(-1)(6)(-2)(-2) = -4(6)(-2)(-2)$
$= -24(-2)(-2)$
$= 48(-2)$
$= -96$

67. $-6 - 6 - (6 + 6) - 3 = -6 - 6 - (12) - 3$
$= -12 - 12 - 3$
$= -24 - 3 = -27$

69. $-|4|\cdot\left|\frac{-1}{2}\right| = \frac{-4}{1}\cdot\frac{1}{2}$

$= \frac{-4}{2} = -2$

71. $|-1| \div \frac{5}{12} = 1 \div \frac{5}{12}$

$= \frac{1}{1}\cdot\frac{12}{5} = \frac{12}{5}$

73. $(-|3| + |5|) - (6 - |-9|)$
$= (-3 + 5) - (6 - 9)$
$= 2 - (-3)$
$= 2 + 3$
$= 5$

75. $4 - |8| + (4 - 6) - |12|$
$= 4 - |8| + (-2) - |12|$
$= 4 - 8 - 2 - 12$
$= -4 - 2 - 12$
$= -6 - 12$
$= -18$

77. $(|-4| - 3) - (3 \cdot |-5|)$
$= (4 - 3) - (3)(5)$
$= 1 - 15$
$= -14$

79. $(25 - |36|)(-6 - 5)$
$= (25 - 36)(-6 - 5)$
$= (-11)(-11)$
$= 121$

81. After the submarine's first move, its position can be represented by - 350 ft since it is below sea level. It's second move can be represented by + 180 feet which the submarine is moving upward.
$-350 \text{ ft} + 180 \text{ ft} = -170 \text{ ft}$
The final position is 170 ft below sea level.

85. a) The \$350,000 advance is deducted (subtracted) from the record royalties of \$267,000.
$\$267{,}000 - \$350{,}000 = -\$83{,}000$
Bon Jovi still owes the company \$83,000.

85. b) In this question, the advance of \$350,000 is deducted from royalties of \$1,400,000.
$\$1{,}400{,}000 - \$350{,}000 = \$1{,}050{,}000$
Since the answer is positive, Bon Jovi has repaid the advance of \$350,000 and the record company still owes him an additional \$1,050,000.

87. True

89. False
Possible Counter Example:
Consider the positive numbers 5 and 6. The difference between 5 and 6 is $5 - 6 = -1$

91. True

93. True

95. False
Possible Counter Example:
Consider the numbers $-5$ and 15.
$-5 + 15 = 10$

97. False
Possible Counter Example:
Consider the numbers $-3$ and $-7$.
Their difference is represented by $-3 - (-7) = -3 + 7 = 4$

99. True

105. Identity Property of Multiplication

107. $|-4| > -3$ since the absolute value of $-4$ is 4 and $4 > -3$

JUST FOR FUN

1. Begin by grouping the 100 numbers into pairs of 2 resulting in 50 pairs.
$(1-2) + (3-4) + (5-6) + \ldots + (99-100)$
Notice the the difference in each set of parentheses is $-1$ so when simplified we have
$(-1) + (-1) + (-1) + \ldots + (-1)$
Since we had 50 pairs and $50(-1)$ the sum of the numbers in the problem is $-50$.

3.
$$\frac{(1) \cdot |-2| \cdot (-3) \cdot |4| \cdot (-5)}{|-1| \cdot (-2) \cdot |-3| \cdot (4) \cdot |-5|}$$
$$= \frac{(1)(2)(-3)(4)(-5)}{(1)(-2)(3)(4)(5)}$$
$$= \frac{120}{-120} = -1$$

EXERCISE SET 1.6

1. $3^2 = 3 \cdot 3 = 9$

3. $5^2 = 5 \cdot 5 = 25$

5. $(-2)^4 = (-2)(-2)(-2)(-2) = 16$

Since the negative sign is inside the parentheses, $-2$ is multiplied by itself 4 times.

7. $(-3)^4 = (-3)(-3)(-3)(-3) = 81$

Since the negative sign is inside the parentheses, $-3$ is multiplied by itself 4 times.

9. $-2^5 = -2 \cdot 2 \cdot 2 \cdot 2 \cdot 2 = -32$

There are no parentheses around the negative sign so only the 2 is repeatedly multiplied 5 times.

13. $(0.3)^2 = (0.3) \cdot (0.3) = 0.09$

11. $\left[\frac{2}{3}\right]^4 = \frac{2}{3} \cdot \frac{2}{3} \cdot \frac{2}{3} \cdot \frac{2}{3} = \frac{16}{81}$

15. $(0.2)^3 = (0.2)(0.2)(0.2) = .008$

17. $6^0 = 1$

19. $4x^0 = 4\left[x^0\right] = 4(1) = 4$

Notice that only the x is raised to the zero power. If the 4 was to also be raised to the zero power, the original problem would have been expressed as

$$(4x)^0 = 1$$

21. $-3y^0 = -3\left[y^0\right] = (-3)(1) = -3$

Only the y is to be raised to the zero power.

23. $-7^0 = -1$

Only the 7 is to be raised to the zero power. If $-7$ was to be raised to the zero power the original problem would have been written as

$$(-7)^0 = 1$$

25. $\sqrt{16} = 4$ since $4(4) = 16$

27. $\sqrt{64} = 8$ since $8(8) = 64$

29. $\sqrt{\frac{25}{36}} = \frac{5}{6}$ since $\frac{5}{6} \cdot \frac{5}{6} = \frac{25}{36}$

31. $\sqrt{\frac{225}{81}} = \frac{15}{9} = \frac{5}{3}$

since $\frac{15}{9} \cdot \frac{15}{9} = \frac{225}{81}$

33. $\sqrt{.04} = 0.2$ since $(0.2) \cdot (0.2) = 0.04$

35. $\sqrt{0.25} = 0.5$ since $(0.5) \cdot (0.5) = 0.25$

37. $\sqrt[3]{64} = 4$ since $4(4)(4) = 64$

39. $\sqrt[3]{-8} = -2$ since

$(-2)(-2)(-2) = -8$

41. $\sqrt[3]{-64} = -4$ since

$(-4)(-4)(-4) = -64$

43. $\sqrt[4]{1} = 1$ since

$1 \cdot 1 \cdot 1 \cdot 1 = 1$

45. $\sqrt[3]{125} = 5$ since

$5 \cdot 5 \cdot 5 = 125$

47. $\sqrt[3]{-216} = -6$ since

$(-6) \cdot (-6) \cdot (-6) = -216$

49. $\sqrt[3]{\frac{1}{64}} = \frac{1}{4}$ since

$\frac{1}{4} \cdot \frac{1}{4} \cdot \frac{1}{4} = \frac{1}{64}$

51. $\sqrt[3]{.001} = 0.1$ since

$(0.1)(0.1)(0.1) = .001$

53. $2^5 = 2 \cdot 2 \cdot 2 \cdot 2 \cdot 2 = 32$

55. $\sqrt[3]{27} = 3$ since $3(3)(3) = 27$

57. $\sqrt{\frac{4}{9}} = \frac{2}{3}$ since

$\frac{2}{3} \cdot \frac{2}{3} = \frac{4}{9}$

59. $-3x^0 = -3\left[x^0\right]$

$= -3(1) = -3$

Only the x is to be raised to the zero power.

61. $\sqrt[3]{-125} = -5$ since

$(-5) \cdot (-5) \cdot (-5) = -125$

63. $\left[\frac{-1}{4}\right]^4 = \left[\frac{-1}{4}\right]\left[\frac{-1}{4}\right]\left[\frac{-1}{4}\right]\left[\frac{-1}{4}\right]$

$= \frac{1}{256}$

65. Replace x by 3 in the expressions

$x^2$ and $-x^2$

a) $(3)^2 = 9$

b) $-3^2 = -9$

67. Replace x by 1 in the expressions

$x^2$ and $-x^2$

a) $1^2 = 1$

b) $-1^2 = -1$

69. Replace x by -1 in the expressions

$x^2$ and $-x^2$

a) $(-1)^2 = (-1)(-1) = 1$

b) $-(-1)^2 = -(-1)(-1) = -1$

71. Replace x by 1/3 in the expressions

$x^2$ and $-x^2$

a) $\left[\frac{1}{3}\right]^2 = \frac{1}{3}\cdot\frac{1}{3} = \frac{1}{9}$

b) $-\left[\frac{1}{3}\right]^2 = -\left[\frac{1}{3}\cdot\frac{1}{3}\right] = \frac{-1}{9}$

73. Replace x by 3 in the expressions

$x^3$ and $-x^3$

a) $3^3 = 3\cdot3\cdot3 = 27$

b) $-3^3 = -3\cdot3\cdot3 \quad -27$

75. Replace x by -3 in the expressions

$x^3$ and $-x^3$

a) $(-3)^3 = (-3)(-3)(-3)$
$= -27$

b) $-(-3)^3 = -(-3)(-3)(-3)$
$= 27$

77. Replace x by -2 in the expressions

$x^3$ and $-x^3$

a) $(-2)^3 = (-2)(-2)(-2)$

$= -8$

b)

$-(-2)^3 = -(-2)(-2)(-2)$

$= 8$

79. Replace x by 2/3 in the expressions

$x^3$ and $-x^3$

a) $\left[\frac{2}{3}\right]^3 = \frac{2}{3}\cdot\frac{2}{3}\cdot\frac{2}{3}$

$= \frac{8}{27}$

b) $-\left[\frac{2}{3}\right]^3 = -\left[\frac{2}{3}\right]\cdot\frac{2}{3}\cdot\frac{2}{3}$

$= \frac{-8}{27}$

81. $4^2 + 3^2 - 2^2 = 16 + 9 - 4$
$= 21$

83. $2^3 + 3^2 + (-4)^2 = 8 + 9 + 16$
$= 33$

85. $(3 - 2)^3 + (2 - 3)^3 = (1)^3 + (-1)^3$
$= 1 + (-1)$
$= 0$

87. $-2^2 - (2)^3 + 1^0 + (-2)^3 =$
$= -4 - (8) + 1 + (-8)$
$= -12 + 1 - 8$
$= -11 - 8$
$= -19$

89. $|-3| + 4^2 - |-2| - 3^0 =$

$= 3 + 4^2 - (2) - 3^0$
$= 3 + 16 - 2 - 1$
$= 19 - 2 - 1$
$= 17 - 1$
$= 16$

91. $|5| - |3| - |-8| - 6 =$
$= 5 - (3) - (8) - 6$
$= -12$

93. $(0.2)^2 - (1.6)^2 - (3.2)^2 =$
$= .04 - 2.56 - 10.24$
$= -2.52 - 10.24$
$= -12.76$

95. $$\left[\frac{-1}{2}\right]^3 - \left[\frac{1}{3}\right]^2 - \left[\frac{-2}{3}\right]^2 =$$

$$= \frac{-1}{8} - \frac{1}{9} - \frac{4}{9} =$$

$$= \frac{-9}{72} - \frac{8}{72} - \frac{32}{72}$$

$$= -\frac{49}{72}$$

97. The solution to the square root of n is the number that when multiplied by itself gives the number under the radical. For example,

$\sqrt{9} = 3$ because $3 \cdot 3 = 9$.

$\sqrt{-4}$ is not a real number because there is no real number that can be multiplied by itself to give $-4$

$\sqrt{-4}$ does not equal 2 because $2 \cdot 2$ does not equal $-4$.

97 cont.

$\sqrt{-4}$ does not equal $-2$ because $(-2)*(-2)$ does not equal $-4$ either.

99. A negative number used as a factor an odd number of times will always result in a negative answer.

101. $|a| = a$ only if a is equal to or greater than zero. To see this, choose a positive and a negative value for a. Suppose $a = 5$. The $|5| = 5$ We see that $|a|$ does equal a. Now let $a = -3$. Then $|-3| = 3$. In this case $|a|$ does not equal a.

103. $4 - (-6) + 3 - 7 =$

$= 4 + 6 + 3 - 7$

$= 10 + 3 - 7$

$= 13 - 7$

$= 6$

EXERCISE SET 1.7

1. $6 + 4 \quad 5 = 6 + 20$
$= 26$

3. $2 + 3 \cdot 4^2 = 2 + 3(16)$
$= 2 + 48 = 50$

5. $$6 \div 2 + 5 \cdot \frac{3}{4} = 3 + \frac{15}{4}$$

$$= \frac{12}{4} + \frac{15}{4} = \frac{27}{4}$$

7. $$24 \cdot 2 \div \frac{1}{3} \div 6 = 48 \div \frac{1}{3} \div 6$$

$$= 48 \cdot \frac{3}{1} \div 6$$

$$= 144 \div 6$$

$$= 24$$

9. $\left[\sqrt{4} - 3\right]\cdot(5-1)^3 = (2-3)(4)^3$
$= (-1)(64)$
$= -64$

11. $\frac{3}{4} \div \frac{5}{6} + \frac{1}{2}\cdot\frac{9}{4} = \frac{9}{10} + \frac{9}{8}$

$= \frac{36}{40} + \frac{45}{40}$

$= \frac{81}{40}$

13.

$2(1 - (4\cdot 5)) + 6^3$

$= 2(1 - 20) + 6^3$

$= 2(-19) + 6^3$
$= 2(-19) + 216$
$= -38 + 216$
$= 178$

15. $\left[3^2 - 1\right] \div (3 + 1)$

$= (9 - 1) \div (3 + 1)$

$= 8 \div 4 = 2$

17. $3\left[(4 + 6)^2 - \sqrt[3]{8}\right]$

$= 3\left[10^2 - \sqrt[3]{8}\right]$

$= 3(100 - 2)$

$= 3(98) = 294$

19.

$\left[(3(14 \div 7))^2 - 2\right]^2$

$= \left[(3(2))^2 - 2\right]^2$

$= \left[(6)^2 - 2\right]^2$

19. cont.

$= (36 - 2)^2$

$= (34)^2 = 1156$

21.

$3(6 - ((25 \div 5) - 2))^3$

$= 3(6 - (5 - 2))^3$

$= 3(6 - 3)^3$

$= 3\cdot(3)^3$

$= 3\cdot(27) = 81$

23. $\dfrac{\left[\frac{1}{6}\right] - 4 \div 2}{8 - 3 + 6}$

$= \dfrac{\frac{1}{6} - 2}{5 + 6}$

$= \dfrac{\frac{1}{6} - \frac{12}{6}}{11}$

$= \dfrac{\frac{-11}{6}}{\frac{11}{1}}$

$= \left[\frac{-11}{6}\right]\cdot\frac{1}{11} = \frac{-1}{6}$

25.

$$\frac{\frac{1}{2}\cdot\frac{1}{3} \div 4 - 2}{3^2 - 4\cdot 2 + 3}$$

$$= \frac{\frac{1}{6} \div 4 - 2}{9 - 8 + 3}$$

$$= \frac{\frac{1}{6}\cdot\frac{1}{4} - 2}{9 - 8 + 3}$$

$$= \frac{\frac{1}{24} - 2}{9 - 8 + 3}$$

$$= \frac{\frac{1}{24} - \frac{48}{24}}{1 + 3}$$

$$= \frac{\frac{-47}{24}}{\frac{4}{1}}$$

$$= \frac{-47}{24}\cdot\frac{1}{4} = \frac{-47}{96}$$

27.

$$\frac{4 - (2 + 3)^2 - 6}{4(3 - 2) - 3^2}$$

$$= \frac{4 - 5^2 - 6}{4(1) - 3^2}$$

$$= \frac{4 - 25 - 6}{4(1) - 9}$$

$$= \frac{-21 - 6}{-5}$$

$$= \frac{-27}{-5} = \frac{27}{5}$$

29.

$$\frac{2(-3) + 4\cdot 5 - 3^2}{5 + \left[\sqrt{4}\right]\left[2^2 - 1\right]}$$

$$= \frac{2(-3) + 4\cdot 5 - 3^2}{5 + \left[\sqrt{4}\right](4 - 1)}$$

$$= \frac{-6 + 20 - 9}{5 + 2(3)}$$

$$= \frac{-6 + 20 - 9}{5 + 6}$$

$$= \frac{14 - 9}{11}$$

$$= \frac{5}{11}$$

31. $$\frac{8 - 4 \div 2 \cdot 3 - 4}{5^2 - 3^2 \cdot 2 - 6}$$

$$= \frac{8 - 4 \div 2 \cdot 3 - 4}{25 - 9 \cdot 2 - 6}$$

$$= \frac{8 - 2 \cdot 3 - 4}{25 - 18 - 6}$$

$$= \frac{8 - 6 - 4}{25 - 18 - 6}$$

$$= \frac{2 - 4}{7 - 6}$$

$$= \frac{-2}{1} = -2$$

33. $$-2 \cdot \left| -3 - \frac{2}{3} \right| + 4$$

$$= -2 \cdot \left| \frac{-9}{3} - \frac{2}{3} \right| + 4$$

$$= -2 \cdot \left| \frac{-11}{3} \right| + 4$$

$$= -2 \cdot \left[ \frac{11}{3} \right] + 4$$

$$= \frac{-22}{3} + 4$$

$$= \frac{-22}{3} + \frac{12}{3}$$

$$= \frac{-10}{3}$$

35. $$3 \cdot |4 - 6| - 2 \cdot |-4 - 2| + 3^2$$

$$= 3 \cdot |-2| - 2 \cdot |-6| + 3^2$$

$$= 3(2) - 2(6) + 3^2$$

$$= 3(2) - 2(6) + 9$$

$$= 6 - 12 + 9 = 3$$

37. $$12 - 15 \div |5| + 2(|4| - 2)^2$$

$$= 12 - 15 \div |5| + 2(4 - 2)^2$$

$$= 12 - 15 \div |5| + 2(2)^2$$

$$= 12 - 15 \div |5| + 2(4)$$

$$= 12 - 15 \div 5 + 9$$

$$= 12 - 3 + 8 = 17$$

39. $$\frac{6 - 2 \cdot |9 - 4| + 8}{4 - |-4| + 4^2 \div 2^2}$$

$$= \frac{6 - 2 \cdot |5| + 8}{4 - |-4| + 16 \div 4}$$

$$= \frac{6 - 2(5) + 8}{4 - 4 + 16 \div 4}$$

$$= \frac{6 - 10 + 8}{4 - 4 + 4}$$

$$= \frac{-4 + 8}{0 + 4}$$

$$= \frac{4}{4} = 1$$

41.

$$\frac{24 - 5 - 4^2}{|8| - 4 \div 2(3)} + \frac{4 - (-3)^2 - |4|}{3^2 - 4 \cdot 3 + |-7|}$$

$$= \frac{24 - 5 - 4^2}{8 - 4 \div 2(3)} + \frac{4 - (-3)^2 - 4}{3^2 - 4 \cdot 3 + 7}$$

$$= \frac{24 - 5 - 16}{8 - 4 \div 2(3)} + \frac{4 - 9 - 4}{9 - 4 \cdot 3 + 7}$$

$$= \frac{24 - 5 - 16}{8 - 2(3)} + \frac{4 - 9 - 4}{9 - 12 + 7}$$

$$= \frac{19 - 16}{8 - 6} + \frac{-5 - 4}{-3 + 7}$$

$$= \frac{3}{2} + \frac{-9}{4}$$

$$= \frac{6}{4} - \frac{9}{4} = \frac{-3}{4}$$

43. Substitute 1 for each x in the expression.

$$\begin{aligned} -3x^2 - 4 &= -3(1)^2 - 4 \\ &= -3(1) - 4 \\ &= -3 - 4 \\ &= -7 \end{aligned}$$

45. Substitute 3 for each x in the expression.

$$\begin{aligned} 5x^2 - 2x + 5 &= 5(3)^2 - 2(3) + 5 \\ &= 5(9)-2(3) + 5 \\ &= 45 - 6 + 5 \\ &= 39 + 5 \\ &= 44 \end{aligned}$$

47. Substitute 1/4 for each x in the expression.

$$\begin{aligned} 3(x - 2)^2 &= 3\left[\left[\frac{1}{4}\right] - 2\right]^2 \\ &= 3\left[\frac{1}{4} - \frac{8}{4}\right]^2 \\ &= 3\left[\frac{-7}{4}\right]^2 \\ &= 3\left[\frac{49}{16}\right] \\ &= \frac{147}{16} \end{aligned}$$

49. Substitute 1 for each x in the expression.

$$\begin{aligned} 4(x - 3)(x + 4) &= 4(1 - 3)(1 + 4) \\ &= 4(-2)(5) \\ &= -8(5) \\ &= -40 \end{aligned}$$

51. Substitute 2 for each x and for each y in the expression.

$$\begin{aligned} -6x + 3y &= -6(2) + 3(4) \\ &= -12 + 12 \\ &= 0 \end{aligned}$$

53. Substitute 2 for each x and −3 for each y in the expression.

$$\begin{aligned} &4(x + y)^2 + 4x - 3y \\ &= 4(2 + (-3))^2 + 4(2) - 3(-3) \\ &= 4(-1)^2 + 4(2) - 3(-3) \\ &= 4(1) + 4(2) - 3(-3) \\ &= 4 + 8 + 9 = 21 \end{aligned}$$

55. Substitue 4 for each a and −1 for each b in the expression

$3(a + b)^2 + 4(a + b) - 6$

$= 3(4 + (-1))^2 + 4(4 + (-1)) - 6$

$= 3(3)^2 + 4(3) - 6$

$= 3(9) + 4(3) - 6$

$= 27 + 12 - 6$

$= 39 - 6 = 33$

57. Substitute 2 for each x and 3 for each y in the expression.

$x^3 \cdot y^2 - 6xy + 3x$

$= 2^3 \cdot 3^2 - 6(2)(3) + 3(2)$

$= 8(9) - 6(2)(3) + 3(2)$

$= 72 - 36 + 6$

$= 42$

59. Substitute 2 for each x and −3 for each y in the expression.

$\frac{1}{2} \cdot \left[x^2 + y^2 - 2xy\right]$

$= \frac{1}{2} \cdot \left[2^2 + (-3)^2 - 2(2)(-3)\right]$

$= \frac{1}{2} \cdot (4 + 9 - 2(2)(-3))$

$= \frac{1}{2} \cdot (4 + 9 + 12)$

$= \frac{1}{2} \cdot (25)$

$= \frac{25}{2}$

61. Substitue 2 for each x and −1 for each y in the expression.

$x^2 \cdot y^4 - y^3 + 3(x + y)$

$= 2^2 \cdot (-1)^4 - (-1)^3 + 3(2 + (-1))$

$= 2^2 \cdot (-1)^4 - (-1)^3 + 3(1)$

$= 4(1) - (-1) + 3$

$= 4 + 1 + 3$

$= 8$

63. Substitue 0 for each x and 2 for each y in the expression.

$\frac{x^2}{25} + \frac{y^2}{9} = \frac{0^2}{25} + \frac{2^2}{9}$

$= \frac{0}{25} + \frac{4}{9}$

$= 0 + \frac{4}{9}$

$= \frac{4}{9}$

65. Substitue 3 for a, 5 for b, and −12 for c in each expression.

a) $b^2 - 4ac = 5^2 - 4(3)(-12)$

$= 25 - 4(3)(-12)$

$= 25 - 12(-12)$

$= 25 + 144$

$= 169$

65 cont.

b)

$$\sqrt{b^2 - 4ac} = \sqrt{5^2 - 4(3)(-12)}$$

$$= \sqrt{25 - 4(3)(-12)}$$

$$= \sqrt{25 - 12(-12)}$$

$$= \sqrt{25 + 144}$$

$$= \sqrt{169} = 13$$

c)

$$-b + \sqrt{b^2 - 4ac} = -5 + \sqrt{5^2 - 4(3)(-12)}$$

$$= -5 + \sqrt{25 - 4(3)(-12)}$$

$$= -5 + \sqrt{25 + 144}$$

$$= -5 + \sqrt{169}$$

$$= -5 + 13$$

$$= 8$$

d)

$$\frac{-b + \sqrt{b^2 - 4ac}}{2a} = \frac{-5 + \sqrt{5^2 - 4(3)(-12)}}{2(3)}$$

$$= \frac{-5 + \sqrt{25 - 4(3)(-12)}}{6}$$

$$= \frac{-5 + \sqrt{25 + 144}}{6}$$

$$= \frac{-5 + \sqrt{169}}{6} = \frac{8}{6} = \frac{4}{3}$$

e)

$$\frac{-b - \sqrt{b^2 - 4ac}}{2a} = \frac{-5 - \sqrt{5^2 - 4(3)(-12)}}{2(3)}$$

$$= \frac{-5 - \sqrt{25 + 144}}{6}$$

$$= \frac{-5 - \sqrt{169}}{6}$$

$$= \frac{-5 - 13}{6}$$

$$= \frac{-18}{6} = -3$$

67. Substitute 3 for each x in the expression.

$$(3x + 6)^2 = (3 \cdot 3 + 6)^2$$

$$= (9 + 6)^2$$

$$= 15^2$$

$$= 225$$

69 Substitute 3 for each x in the expression.

$$6(6 + 3x) - 9 = 6(6 + 3 \cdot 3) - 9$$

$$= 6(6 + 9) - 9$$

$$= 6(15) - 9$$

$$= 90 - 9$$

$$= 81$$

71. Substitute 5 for each x and 2 for each y in the expression.

$$\left[\frac{3+x}{2y}\right]^2 - 3 = \left[\frac{3+5}{2(2)}\right]^2 - 3$$

$$= \left[\frac{8}{4}\right]^2 - 3$$

$$= 2^2 - 3$$

$$= 4 - 3$$

$$= 1$$

73 b)

$$\frac{5 - 18 \div (3)^2}{4 - 3\cdot 2} = \frac{5 - 18 \div 9}{4 - 3\cdot 2}$$

$$= \frac{5 - 2}{4 - 6}$$

$$= -\left[\frac{3}{2}\right]$$

75.

$$16 \div 2^2 + 6\cdot 4 - 24 \div 6$$

$$= 16 \div 4 + 6(4) - 24 \div 6$$

$$= 4 + 24 - 4$$

$$= 28 - 4$$

$$= 24$$

77. Associative property of addition

79. $8 - (-4) + (7 - 5) - 10$

$$= 8 + 4 + 2 - 10$$

$$= 12 + 2 - 10$$

$$= 14 - 10$$

$$= 4$$

JUST FOR FUN

1.

$$\left[(3 \div 6)^2 + 4\right]^2 + 3\cdot 4 \div 12 \div 3$$

$$= \left[\left[\frac{1}{2}\right]^2 + 4\right]^2 + 3(4) \div 12 \div 3$$

$$= \left[\frac{1}{4} + 4\right]^2 + 3(4) \div 12 \div 3$$

$$= \left[\frac{17}{4}\right]^2 + 3(4) \div 12 \div 3$$

1 cont.

$$= \frac{289}{16} + 12 \div 12 \div 3$$

$$= \frac{289}{16} + 1 \div 3$$

$$= \frac{289}{16} + \frac{1}{3}$$

$$= \frac{867}{48} + \frac{16}{48}$$

$$= \frac{883}{48}$$

3. Substitute 2 for each x and 3 for each y in the expression.

$$\frac{2x + 4 - y\cdot\left[2 + \frac{3}{x}\right]}{\frac{y-2}{6} + \frac{3x^2}{4}}$$

$$= \frac{2(2) + 4 - (3)\cdot\left[2 + \frac{3}{2}\right]}{\frac{(3)-2}{6} + \frac{3(2)^2}{4}}$$

$$= \frac{2(2) + 4 - (3)\cdot\left[\frac{7}{2}\right]}{\frac{(3)-2}{6} + \frac{3(2)^2}{4}}$$

$$= \frac{4 + 4 - \frac{21}{2}}{\frac{(3)-2}{6} + \frac{3(4)}{4}}$$

3 cont.

$$= \frac{8 - \frac{21}{2}}{\frac{1}{6} + 3}$$

$$= \frac{\frac{16}{2} - \frac{21}{2}}{\frac{1}{6} + \frac{18}{6}}$$

$$= \frac{\frac{-5}{2}}{\frac{19}{6}}$$

$$= \frac{-5}{2}\cdot\frac{6}{19}$$

$$= \frac{-15}{19}$$

REVIEW EXERCISES

1. $A = \{3,4,5,6\}$

3. $0 \in \{0,1,2,3\}$

5. $\{5\} \notin \{4,5,6\}$

7. $\{3\} \subseteq \{1, 2, 3\}$

9. $5 \nsubseteq \{3,4,5,6\}$

11. $N \subseteq W$

13. $I \subseteq Q$

15. $H \subseteq$ Reals

17. 4,6

19. −3,4−6,0

21. $\sqrt{5}, \sqrt{3}$

23. True

25. False

27. True

29. $A \cup B = \{2,3,4,5,6,7,8,9\}$
$A \cap B = \{ \}$

31. $A \cup B = \{3,4,5,6,9,10,11,12\}$
$A \cap B = \{9, 10\}$

33. Commutative property of addition

35. Commutative property of multiplication

37. Identity property of addition

39. Identity property of multiplication

41. Multiplication property of zero

43. Identity proper;ty of multiplication

45. Inverse property of multiplication

47. Identity property of multiplication

49. $3x + 15$

51. $x \cdot 3$

53. $4x - 4y + 20$

55. a

57. 1

59. $3 > 2$

61. $-2 < 3$

63. $-8 < 0$

65. $1.06 < 1.6$

67. $|3| = 3$

69. $|4| < |6|$

71. $13 > |-5|$

73. $|-2/3| > 3/5$

75. $-5, -2, 4, |7|$

77. $-2, 3, |-5|, |-7|$

79. $-|-3|, -4, 5, 6$

81.
$$4 - 2 + 3 - \frac{3}{5} = 2 + 3 - \frac{3}{5}$$
$$= 5 - \frac{3}{5}$$
$$= \frac{25}{5} - \frac{3}{5}$$
$$= \frac{22}{5}$$

83.
$$-4 \cdot |6| - 3(-4) = -4(6) - 3(-4)$$
$$= -24 + 12$$
$$= -12$$

85.
$$3 \cdot |-2| - (4 - 3) + 2(-3)$$
$$= 3 \cdot |-2| - 1 + 2(-3)$$
$$= 3(2) - 1 + 2(-3)$$
$$= 6 - 1 - 6$$
$$= 5 - 6$$
$$= -1$$

87.
$$(6 - 9) \div (9 - 6) = (-3) \div 3$$
$$= -1$$

89.
$$|6 - 3| \div 3 + 4 \cdot 8 - 12$$
$$= |3| \div 3 + 4(8) - 12$$
$$= 3 \div 3 + 4(8) - 12$$
$$= 1 + 32 - 12$$
$$= 33 - 12$$
$$= 21$$

91.
$$3^2 - 6 \cdot 9 + 4 \div 2^2 - 3$$
$$= 9 - 6(9) + 4 \div 4 - 3$$
$$= 9 - 54 + 1 - 3$$
$$= -45 + 1 - 3$$
$$= -44 - 3$$
$$= -47$$

93.

$$4^2 - \left[2 - 3^2\right]^2 + 4^3$$

$$= 4^2 - (2 - 9)^2 + 4^3$$

$$= 4^2 - (-7)^2 + 4^3$$
$$= 16 - 49 + 64$$
$$= -33 + 64$$
$$= 31$$

95.

$$\left[\left[(9 \div 3)^2 - 1\right]^2 \div 8\right]^3$$

$$= \left[\left[3^2 - 1\right]^2 \div 8\right]^3$$

$$= \left[(9 - 1)^2 \div 8\right]^3$$

$$= \left[8^2 \div 8\right]^3$$

$$= (64 \div 8)^3$$

$$= 8^3 = 512$$

97.

$$\frac{-(4 - 6)^2 - 3(-2) + |-6|}{18 - 9 \div 3 \cdot 5}$$

$$= \frac{-(-2)^2 - 3(-2) + |-6|}{18 - 9 \div 3 \cdot 5}$$

$$= \frac{-4 - 3(-2) + 6}{18 - 9 \div 3 \cdot 5}$$

$$= \frac{-4 + 6 + 6}{18 - 3 \cdot 5}$$

$$= \frac{2 + 6}{18 - 15} = \frac{8}{3}$$

99.

$$\frac{9 \cdot |3 - 5| - 5 \cdot |4| \div 10}{-3 \cdot 5 - 2 \cdot 4 \div 2}$$

$$= \frac{9 \cdot |-2| - 5 \cdot |4| \div 10}{-3 \cdot 5 - 2 \cdot 4 \div 2}$$

$$= \frac{9(2) - 5(4) \div 10}{-3(5) - 2 \cdot 4 \div 2}$$

$$= \frac{18 - 20 \div 10}{-15 - 8 \div 2}$$

$$= \frac{18 - 2}{-15 - 4}$$

$$= \frac{16}{-19} = -\left[\frac{16}{19}\right]$$

101. Substitute −2 for each x in the expression.

$$(x - 2)^2 + 3x$$

$$= ((-2) - 2)^2 + 3(-2)$$

$$= (-4)^2 + 3(-2)$$

$$= 16 - 6$$

$$= 10$$

103. Substitute 1 for each x and 3 for each y in the expression.

$$-3x^2 \cdot y + 6xy^2 - 2xy$$

$$= -3(1)^2 \cdot (3) + 6(1)(3)^2 - 2(1)(3)$$

$$= -3(1)(3) + 6(1)9 - 2(1)(3)$$
$$= -9 + 54 - 6$$
$$= 45 - 6$$
$$= 39$$

105. Substitute −1 for each x and −3 for each y in the expression.

$$4(x - 3) + 5(y - 3) - 4$$
$$= 4((-1)-3)+5(-3 -3) - 4$$
$$= 4(-4) + 5(-6) - 4$$
$$= -16 - 30 - 4 = -50$$

107. Substitute −2 for each x and 3 for each y in the expression.

$$-x^2 \cdot y - 6xy^2 + 4y^3$$
$$= -(-2)^2 \cdot 3 - 6(-2)3^2 + 4(3)^3$$
$$= -4(3) - 6(-2)(9) + 4(27)$$
$$= -12 + 12(9) + 108$$
$$= -12 + 108 + 108$$
$$= 204$$

PRACTICE TEST

1. $A = \{6,7,8,9....\}$

2. $3 \notin \{1,2,3,4\}$

3. $\{5\} \subseteq \{1,2,3,4,5\}$

4. True

5. False
   Zero is a whole number that is not a natural number.

6. True

7. $\frac{-3}{5}, 2, -4, 0, \frac{19}{12}, 2.57, -1.92$

8. $\frac{-3}{5}, 2, -4, 0, \frac{19}{12}, 2.57, \sqrt{8}, \sqrt{2}, -1.92$

9. $A \cup B = \{5,7,8,9,10,11,14\}$
   $A \cap B = \{8,10\}$

10. $A \cup B = \{1,3,5,7,9...\}$
    $A \cap B = \{3,5,7,9,11\}$

11. $-4 < |-9|$

12. $|-3| > -|5|$

13. $-|4|, -2, |3|, 6$

14. Distributive property

15. Associative property of addition

16. Commutative property of addition

17. Inverse property of multiplication

18. Identity property of addtion

19. 
$$(4 - (6 - 3(4 - 5)))^2 \div (-5)$$
$$= (4 - (6 - 3(-1)))^2 \div (-5)$$
$$= (4 - (6 + 3))^2 \div (-5)$$
$$= (4 - 9)^2 \div (-5)$$
$$= (-5)^2 \div (-5)$$
$$= 25 \div (-5)$$
$$= -5$$

20. 
$$5^2 + 16 \div 4 - 3 \cdot 2$$
$$= 25 + 16 \div 4 - 3 \cdot 2$$
$$= 25 + 4 - 6$$
$$= 29 - 6$$
$$= 23$$

21. 
$$\frac{-3 \cdot |4 - 8| \div 2 + 4}{\sqrt{36} + 18 \div 3^2}$$
$$= \frac{-3 \cdot |-4| \div 2 + 4}{\sqrt{36} + 18 \div 9}$$
$$= \frac{-3(4) \div 2 + 4}{6 + 18 \div 9}$$
$$= \frac{-12 \div 2 + 4}{6 + 18 \div 9} = \frac{-2}{8} = \frac{-1}{4}$$

22.

$$\frac{-6^2 + 3\cdot(4 - |6|) \div 6}{4 - (-3) + 12 \div 4\cdot 5}$$

$$= \frac{-6^2 + 3(4 - 6) \div 6}{4 + 3 + 12 \div 4\cdot 5}$$

$$= \frac{-6^2 + 3(-2) \div 6}{4 + 3 + 3\cdot 5}$$

$$= \frac{-36 + 3(-2) \div 6}{4 + 3 + 15}$$

$$= \frac{-36 - 6 \div 6}{7 + 15}$$

$$= \frac{-36 - 1}{22}$$

$$= \frac{-37}{22}$$

23.

$$\frac{(4 - (2 - 5))^2 + 6 \div 2\cdot 5}{|4 - 6| + |-6| \div 2}$$

$$= \frac{(4 - (-3))^2 + 6 \div 2\cdot 5}{|-2| + |-6| \div 2}$$

$$= \frac{7^2 + 6 \div 2\cdot 5}{|-2| + |-6| \div 2}$$

$$= \frac{49 + 6 \div 2\cdot 5}{2 + 6 \div 2}$$

$$= \frac{49 + 3\cdot 5}{2 + 3}$$

$$= \frac{49 + 15}{5} = \frac{64}{5}$$

24. Substitute 2 for each x and 3 for each y in the expression.

$$-x^2 + 2xy + y^2 = -(2)^2 + 2(2)(3) + 3^2$$

$$= -4 + (2)(2)(3) + 9$$

$$= -4 + 12 + 9$$

$$= 8 + 9$$

$$= 17$$

25. Substitute 2 for each x and − 3 for each y in the expression.

$$(x - 5)^2 + 2xy^2 - 6$$

$$= ((2) - 5)^2 + 2(2)(-3)^2 - 6$$

$$= (-3)^2 + 2(2)(-3)^2 - 6$$

$$= 9 + 2(2)(9) - 6$$

$$= 9 + 4(9) - 6$$

$$= 9 + 36 - 6$$

$$= 45 - 6$$

$$= 39$$

# CHAPTER 2

## EXERCISE SET 2.1

1. Reflexive property
3. Symmetric property
5. Transitive property
7. Reflexive property
9. Addititition property
11. Multiplication property
13. Multiplication property
15. Addition property
17. Multiplication property
19. Symmetric property

21. $4x$ is a first degree term since $4x^1 = 4x$

23. $3xy$ is a second degree term since $3xy = 3x^1 \cdot y^1$ and the sum of the exponents is 2.

25. $\frac{1}{2} \cdot x^4 \cdot y$ is a fifth degree term since $\frac{1}{2} \cdot x^4 \cdot y = \frac{1}{2} \cdot x^4 \cdot y^1$ and the sum of the exponents is 5.

27. The degree of $-3$ is zero since $-3 = -3x^0$

29. $3x^4 \cdot y^6 \cdot z^3$ is a thirteenth degree term since the sum of the exponents on the variables is 13.

31. $3x^5 \cdot y^6 \cdot z$ is a twelth degree term since $3x^5 \cdot y^6 \cdot z = 3x^5 \cdot y^6 \cdot z^1$ and the sum of the exponents on the variables is 12.

33. $8x + 7 + 7x - 12 = 8x + 7x + 7 - 12$
$= 15x - 5$

35. $5x^2 - 3x + 2x - 5 = 5x^2 - x - 5$

37. $-4x^2 - 3x - 5x + 7 = -4x^2 - 8x + 7$

39. $6y^2 + 6xy + 3$ cannot be simplified further.

41. $xy + 3xy + y^2 - 2$
$= y^2 + 1xy + 3xy - 2$
$= y^2 + 4xy - 2$

43. $4(x + 3) - 7(2x - 5)$
$= 4x + 12 - 14x + 35$
$= -10x + 47$

45. $3\left[x + \frac{1}{2}\right] - \frac{1}{3} \cdot x + 5$
$= 3x + \frac{3}{2} - \frac{1}{3} \cdot x + 5$
$= \frac{9}{3} \cdot x - \frac{1}{3} \cdot x + \frac{3}{2} + \frac{10}{2}$
$= \frac{8}{3} \cdot x + \frac{13}{2}$

47. $4 - \left[6(3x + 2) - x\right] + 4$
$= 4 - \left[18x + 12 - x\right] + 4$
$= 4 - \left[17x + 12\right] + 4$
$= 4 - 17x - 12 + 4$
$= 17x - 4$

49. $4x - \left[3x - (5x - 4y)\right] + y$
$= 4x - \left[3x - 5x + 4y\right] + y$
$= 4x - \left[-2x + 4y\right] + y$
$= 4x + 2x - 4y + y$
$= 6x - 3y$

51.

$$2x + 3 = 5$$
$$2x + 3 - 3 = 5 - 3$$
$$2 \cdot x = 2$$
$$\frac{2x}{2} = \frac{2}{2}$$
$$x = 1$$

53.

$$4x + 3 = -12$$
$$4x + 3 - 3 = -12 - 3$$
$$4 \cdot x = -15$$
$$\frac{4x}{4} = \frac{-15}{4}$$
$$x = \frac{-15}{4}$$

55.

$$\frac{-x}{4} = 8$$
$$4\left[\frac{-x}{4}\right] = 4(8)$$
$$-x = 32$$
$$-1(-x) = -1(32)$$
$$x = -32$$

57.

$$\frac{10}{3} = x + 6$$
$$\frac{10}{3} - 6 = x + 6 - 6$$
$$\frac{-8}{3} = x$$

59.

$$3.2(x - 1.6) = 5.88 + .2x$$
$$3.2x - 5.12 = 5.88 + .2x$$
$$3.2x - 5.12 + 5.12 = 5.88 + .2x + 5.12$$
$$3.2x = 11 + .2x$$
$$3.2x - .2x = 11 + .2x - .2x$$
$$3 \cdot x = 11$$
$$x = \frac{11}{3} = 3.67$$

61.

$$\frac{3x - 9}{3} = \frac{2x - 6}{6}$$

Cross multiply:
$$6(3x - 9) = 3(2x - 6)$$
$$18x - 54 = 6x - 18$$
$$18x - 54 + 54 = 6x - 18 + 54$$
$$18x = 6x + 36$$
$$18x - 6x = 6x - 6x + 36$$
$$12x = 36$$
$$\frac{12x}{12} = \frac{36}{x}$$
$$x = 3$$

63.

$$\frac{\frac{1}{4}x - 2}{5} = \left[\frac{3(x - 2)}{4}\right]$$
$$4\left[\frac{1}{4} \cdot x - 2\right] = 5\left[3(x - 2)\right]$$
$$x - 8 = 5(3x - 6)$$
$$x - 8 = 15x - 30$$
$$x - 8 + 30 = 15x - 30 + 30$$
$$x + 22 = 15x$$
$$x - x + 22 = 15x - x$$
$$22 = 14x$$
$$\frac{22}{14} = \frac{14x}{14}$$
$$\frac{22}{14} = \frac{11}{7} = x$$

65.

$$\frac{1.5x}{5} = \frac{x - 4.2}{8}$$
$$8(1.5 \cdot x) = 5(x - 4.2)$$
$$12x = 5x - 21$$
$$12x - 5x = 5x - 5x - 21$$
$$7 \cdot x = -21$$
$$7 \cdot \frac{x}{7} = \frac{-21}{7}$$
$$x = -3$$

67.

$$-(x - 4) + 3x = -12$$
$$-x + 4 + 3x = -12$$
$$2x + 4 = -12$$
$$2x + 4 - 4 = -12 - 4$$
$$2 \cdot x = -16$$
$$\frac{2x}{2} = \frac{-16}{2}$$
$$x = -8$$

69.

$$\frac{1}{2} \cdot (x - 4) = 8$$

Multiply both sides of the equation by 2.

$$(2) \cdot \frac{1}{2} \cdot (x - 4) = 8(2)$$
$$x - 4 = 16$$
$$x - 4 + 4 = 16 + 4$$
$$x = 20$$

71.

$$6 - 2(x - 3) + 5 = 15$$
$$6 - 2x + 6 + 5 = 15$$
$$-2x + 17 = 15$$
$$-2x + 17 - 17 = 15 - 17$$
$$-2 \cdot x = -2$$
$$-2 \cdot \frac{x}{-2} = \frac{-2}{-2}$$
$$x = 1$$

73.

$$\frac{2x - 5}{3} = -5$$
$$\frac{2x - 5}{3} = \frac{-5}{1}$$
$$1(2x - 5) = 3(-5)$$
$$2x - 5 = -15$$
$$2x - 5 + 5 = -15 + 5$$
$$2 \cdot x = -10$$
$$2 \cdot \frac{x}{2} = \frac{-10}{2}$$
$$x = -5$$

75.

$$\frac{4 - 3x}{2} + x = 4$$

Multiply every term by 2.

$$2 \cdot \frac{4 - 3x}{2} + 2(x) = 2(4)$$
$$4 - 3x + 2x = 8$$
$$4 - x = 8$$
$$4 - 4 - x = 8 - 4$$
$$-x = 4$$
$$-x(-1) = 4(-1)$$
$$x = -4$$

77.

$$\frac{-3}{5} \cdot (15 - 2x) = -3$$

Multiply both sides by -5/3.

$$\frac{-5}{3} \cdot \frac{-3}{5} \cdot (15 - 2x) = -3 \cdot \frac{-5}{3}$$
$$15 - 2x = 5$$
$$15 - 15 - 2x = 5 - 15$$
$$-2 \cdot x = -10$$
$$-2 \cdot \frac{x}{-2} = \frac{-10}{-2}$$
$$x = 5$$

79.

$$-4.2(3.2x - 4) = 2.56x$$
$$-13.44x + 16.8 = 2.56x$$
$$-13.44x + 13.44x + 16.8 = 2.56x + 13.44x$$
$$16.8 = 16x$$
$$\frac{16.8}{16} = \frac{16x}{16.8}$$
$$1.05 = x$$

81.

$$4x - 5(x + 3) = 2x - 3$$
$$4x - 5x - 15 = 2x - 3$$
$$-x - 15 = 2x - 3$$
$$-x + x - 15 = 2x + x - 3$$
$$-15 = 3x - 3$$
$$-15 + 3 = 3x - 3 + 3$$
$$-12 = 3x$$
$$\frac{-12}{3} = \frac{3x}{3}$$
$$-4 = x$$

83. $\frac{7x}{5} = 2x + 3$

Multiply each term in the equation by 5.

$$5 \cdot \frac{7x}{5} = (5) \cdot 2x + (5) \cdot 3$$

$$7x = 10x + 15$$
$$7x - 10x = 10x - 10x + 15$$
$$-3x = 15$$
$$\frac{-3x}{-3} = \frac{15}{-3}$$
$$x = -5$$

85.
$$2(x - 3) + 2x = 4x - 5$$
$$2x - 6 + 2x = 4x - 5$$
$$4x - 6 = 4x - 5$$
$$4x - 4x - 6 = 4x - 4x - 5$$
$$-6 = -5$$

Solving the equation results in a false statement. Therefore, the equation is inconsistent and has no solution.

87. $\frac{x - 25}{3} = 2x - \frac{2}{5}$

Multiply each term by the least common multiple of 3 and 5 which is 15.

$$15 \cdot \frac{x - 25}{3} = 15 \cdot 2x - 15 \cdot \frac{2}{5}$$

$$5 \cdot (x - 25) = 30x - 6$$
$$5x - 125 = 30x - 6$$
$$5x - 5x - 125 = 30x - 5x - 6$$
$$-125 = 25x - 6$$
$$-125 + 6 = 25x - 6 + 6$$
$$-119 = 25x$$
$$\frac{-119}{25} = \frac{25x}{25}$$
$$\frac{-119}{25} = x$$

89.
$$4 \cdot (2 - 3x) = -\left[6x - (8 - 6x)\right]$$
$$8 - 12x = -6x + 8 - 6x$$
$$8 - 12x = -12x + 8$$
$$8 - 12x + 12x = -12x + 12x + 8$$
$$8 = 8$$

Solving the equation results in an identity. The solution is all real numbers.

91.
$$\frac{1}{2} \cdot (2x + 1) = \frac{1}{4} \cdot (x - 4)$$

Multiply both sides of the equation by the least common multiple of 2 and 4 which is 4.

$$\left[4 \cdot \frac{1}{2}\right] \cdot (2x + 1) = 4 \cdot \frac{1}{4} \cdot (x - 4)$$

$$2(2x + 1) = x - 4$$
$$4x + 2 = x - 4$$
$$4x - x + 2 = x - x - 4$$
$$3x + 2 = -4$$
$$3x + 2 - 2 = -4 - 2$$
$$3x = -6$$
$$\frac{3x}{3} = \frac{-6}{3}$$
$$x = -2$$

93.
$$4(x - (5x - 2)) = 2(x - 3)$$
$$4(x - 5x + 2) = 2(x - 3$$
$$4(-4x + 2) = 2x - 6$$
$$-16x + 8 = 2x - 6$$
$$-16x + 16x + 8 = 2x + 16x - 6$$
$$8 = 18x - 6$$
$$8 + 6 = 18x - 6 + 6$$
$$14 = 18x$$
$$\frac{14}{18} = \frac{18x}{18}$$
$$\frac{7}{9} = x$$

95. $$\frac{2}{3}\cdot(x-4)=\frac{2}{3}\cdot(4-x)$$

Multiply both sides of the equation by 3.

$$3\cdot\frac{2}{3}\cdot(x-4)=3\cdot\frac{2}{3}\cdot(4-x)$$

$$2(x-4)=2(4-x)$$
$$2x-8=8-2x$$
$$2x-8=8-2x$$
$$2x+2x-8=8-2x+2x$$
$$4x-8=8$$
$$4x-8+8=8+8$$
$$4x=16$$
$$\frac{4x}{4}=\frac{16}{4}$$
$$x=4$$

97. $$\frac{3x}{4}-2=6+\frac{x}{3}$$

Multiply every term in the equation by the least common multiple of 4 and 3 which is 12.

$$12\cdot\frac{3x}{4}-2(12)=6(12)+12\cdot\frac{x}{3}$$

$$9x-24=72+4x$$
$$9x-4x-24=72+4x-4x$$
$$5x-24=72$$
$$5x-24+24=72+24$$
$$5x=96$$
$$\frac{5x}{5}=\frac{96}{5}$$
$$x=\frac{96}{5}$$

99. $$\frac{x+1}{4}=\frac{x-4}{2}-\frac{2x-3}{4}$$

Multiply each term by the least common multiple of 2 and 4 which is 4.

$$4\cdot\frac{x+1}{4}=4\cdot\frac{x-4}{2}-4\cdot\frac{2x-3}{4}$$

99. cont.

$$x+1=2\cdot(x-4)-(2x-3)$$
$$x+1=2x-8-2x+3$$
$$x+1=-5$$
$$x+1-1=-5-1$$
$$x=-6$$

101. b)

$$2x-\frac{2}{5}=\frac{2}{3}\cdot(x+5)$$

Multiply by the least common multiple of 5 and 3 which is 15.

$$(15)\cdot 2x-15\cdot\frac{2}{5}=15\cdot\frac{2}{3}\cdot(x+5)$$

$$30x-6=10(x+5)$$
$$30x-6=10x+50$$
$$30x-10x-6=10x-10x+50$$
$$20x-6=50$$
$$20x-6+6=50+6$$
$$20x=56$$
$$\frac{20x}{20}=\frac{56}{20}$$
$$x=\frac{56}{20}=\frac{14}{5}$$

103. An identity is an equation that is true for all real numbers. If at any point while solving an equation you realize both sides of the equation are identical, the equation is an identity.

105. An inconsistent equation is an equation with no solution. An equation is inconsistent if attempting to solve the equation results in a false statement such as $0 = 1$ or $5 = -5$.

107. There are an infinite number of answers to this question. An equivalent equation can be found by multiplying both sides of the equation by a real number such as 4.

$$x=4$$
$$4x=4(4)$$
$$4x=16$$

Another equivalent equation can be found by adding the same number such as 2, to both sides of the

107. cont.
equation.

$$x = 4$$
$$x + 2 = 4 + 2$$
$$x + 2 = 6$$

A third equivalent equation could be found by subtracting the same number from both sides.

$$x = 4$$
$$x - 3 = 4 - 3$$
$$x - 3 = 1$$

$x = 4$ is equivalent to $4x = 16$, $x + 2 = 6$, and $x - 3 = 1$.

109. If a represents any real number, then

$$|a| = \begin{cases} a \text{ if } a \geq 0 \\ -a \text{ if } a < 0 \end{cases}$$

111.

$$\left[\frac{-3}{4}\right]^3 = \left[\frac{-3}{4}\right]\left[\frac{-3}{5}\right]\left[\frac{-3}{5}\right]$$
$$= \frac{-27}{64}$$

JUST FOR FUN

1. $$\frac{-3}{5}\cdot(x + 2) - \frac{4}{3}\cdot(2x - 3) + 4 = \frac{1}{2}\cdot(x + 4) - 6x + 3(5 - x)$$

Multiply through by the least common multiplie of 5,3,and 2 which is 30.

$$30\cdot\frac{-3}{5}\cdot(x + 2) - 30\cdot\frac{4}{3}\cdot(2x - 3) + 30(4) = 30\cdot\frac{1}{2}\cdot(x + 4) - 30(6x) + 30(3)(5 - x)$$

$$-18\cdot(x + 2) - 40(2x - 3) + 120 = 15(x + 4) - 180x + 90(5 - x)$$

$$-18x - 36 - 80x + 120 + 120 = 15x + 60 - 180x + 450 - 90x$$

$$-98x + 204 = -255x + 510$$

$$-98x + 255x + 204 = -255x + 255x + 510$$

$$157x + 204 = 510$$

$$157x + 204 - 204 = 510 - 204$$

$$157x = 306$$

$$\frac{157x}{157} = \frac{306}{157}$$

$$x = \frac{306}{157}$$

3. $$\frac{x}{3} + \frac{x-2}{4} + \frac{2x-3}{5} = \frac{x-3}{6} + 4(x-7) - x + 2$$

Multiply through by the least common multiple of 3,4,5,and 6 which is 60.

$$60\cdot\frac{x}{3} + 60\cdot\frac{x-2}{4} + 60\cdot\frac{2x-3}{5} = 60\cdot\frac{x-3}{6} + 60\cdot 4(x-7) - 60x + 60(2)$$

$$20x + 15(x-2) + 12(2x-3) = 10(x-3) + 240(x-7) - 60x + 120$$

$$20x + 15x - 30 + 24x - 36 = 10x - 30 + 240x - 1680 - 60x + 120$$

$$59x - 66 = 190x - 1590$$

$$59x - 59x - 66 = 190x - 59x - 1590$$

$$-66 = 131x - 1590$$

$$-66 + 1590 = 131x - 1590 + 1590$$

$$1524 = 131x$$

$$\frac{1524}{131} = \frac{131x}{131}$$

$$\frac{1524}{131} = x$$

EXERCISE SET 2.2

1. $P = 2L + 2w$
$= 2(15) + 2(6)$
$= 30 + 12$
$= 42$

3. $$A = \frac{1}{2}\cdot h\left[b_1 + b_2\right]$$

$$= \frac{1}{2}\cdot 10(20 + 30)$$

$$= \frac{1}{2}\cdot 10(50) = 250$$

5. $c = a + by_D$

$= 500 + 10(60)$
$= 500 + 600$
$= 1100$

7. $E = a_1 p_1 + a_2 p_2 + a_3 p_3$

$= 10(.2) + 100(.3) + 1000(.5)$

$= 2 + 30 + 500$

$= 532$

9.

$$K = \frac{F - 32}{1.8} + 273.1$$

$$= \frac{100 - 32}{1.8} + 273.1$$

$$= \frac{68}{1.8} + 273.1$$

$$= 37.78 + 273.1 = 310.88$$

11.

$$m = \frac{y_2 - y_1}{x_2 - x_1}$$

$$= \frac{4 - (-3)}{-2 - (-6)}$$

$$= \frac{4 + 3}{-2 + 6}$$

$$= \frac{7}{4}$$

13.

$$z = \frac{\overline{x} - \mu}{\frac{\sigma}{\sqrt{n}}}$$

$$= \frac{80 - 70}{\frac{15}{\sqrt{25}}}$$

$$= \frac{10}{\left[\frac{15}{5}\right]}$$

$$= \frac{10}{3} = 3.333$$

15.

$$d = \sqrt{\left[x_2 - x_1\right]^2 + \left[y_2 - y_1\right]^2}$$

$$= \sqrt{(5 - (-3))^2 + (-6 - 3)^2}$$

$$= \sqrt{8^2 + (-9)^2}$$

$$= \sqrt{64 + 81}$$

$$= \sqrt{145} = 12.042$$

17.

$$R_T = \frac{R_1 \cdot R_2}{R_1 + R_2}$$

$$= \frac{100(200)}{100 + 200}$$

$$= \frac{20000}{300} = 66.667$$

19.

$$x = \frac{-b + \sqrt{b^2 - 4ac}}{2a}$$

$$= \frac{-(-5) + \sqrt{(-5)^2 - 4(2)(-12)}}{2(2)}$$

$$= \frac{5 + \sqrt{25 + 96}}{4}$$

$$= \frac{5 + \sqrt{121}}{4}$$

$$= \frac{5 + 11}{4}$$

$$= 4$$

21.
$$\begin{aligned} R &= O + (V - D)r \\ &= 500 + (200 - 12)4 \\ &= 500 + (188)4 \\ &= 500 + 752 \\ &= 1252 \end{aligned}$$

23.
$$\begin{aligned} S &= \pi r^2 + \pi r s \\ &= 3.14(3)^2 + 3.14(3)(4) \\ &= 3.14(9) + 3.14(12) \\ &= 28.26 + 37.68 \\ &= 65.94 \end{aligned}$$

25.
$$Y = \frac{\left[\frac{F}{A}\right]}{\left[\frac{\Delta l}{l_o}\right]} = \frac{\left[\frac{20}{2}\right]}{\left[\frac{5}{10}\right]} = \frac{20}{2}\cdot\frac{10}{5} = 20$$

27.
$$P_c = \frac{P_1 - P_2}{\frac{P_1 + P_2}{2}} = \frac{800 - 600}{\frac{800 + 600}{2}} = \frac{200}{\left[\frac{1400}{2}\right]} = \frac{200}{700} = 0.286$$

29.
$$\bar{x}_w = \frac{w_1 \cdot x_1 + w_2 \cdot x_2 + w_3 \cdot x_3}{w_1 + w_2 + w_3} = \frac{4(60) + 6(80) + 10(96)}{4 + 6 + 10} = \frac{240 + 480 + 960}{20} = \frac{1680}{20} = 84$$

31.
$$A = P\left[1 + \frac{r}{n}\right]^{nt} = 100\left[1 + \frac{.06}{1}\right]^{1(3)} = 100(1.06)^3 = 100\cdot(1.191016) = 119.102$$

33.
$$\begin{aligned} 2x + y &= 3 \\ 2x - 2x + y &= 3 - 2x \\ y &= -2x + 3 \end{aligned}$$

35.
$$\begin{aligned} 2x + 3y &= 6 \\ 2x - 2x + 3y &= 6 - 2x \\ 3y &= -2x + 6 \\ \frac{3y}{3} &= \frac{-2x + 6}{3} \\ y &= \frac{-2x + 6}{3} \end{aligned}$$

37.
$$\begin{aligned} x - y &= 8 \\ x - x - y &= -x + 8 \\ -y &= -x + 8 \\ -y\cdot(-1) &= -x\cdot(-1) + 8 \\ y &= x - 8 \end{aligned}$$

39.
$$2x - 4y = 6$$
$$2x - 2x - 4y = 6 - 2x$$
$$-4y = -2x + 6$$
$$\frac{-4y}{-4} = \frac{-2x + 6}{-4}$$
$$y = \frac{-2(x - 3)}{-4}$$
$$y = \frac{x - 3}{2}$$

41.
$$2y = 8x - 3$$
$$\frac{2y}{2} = \frac{8x - 3}{2}$$
$$y = \frac{8x - 3}{2}$$

43.
$$\frac{3}{5}\cdot x + \frac{1}{3}\cdot y = 1$$

Multiply each term by the least common multiple of 5 and 3 which is 15.

$$15\cdot\frac{3}{5}\cdot x + 15\cdot\frac{1}{3}\cdot y = 15(1)$$
$$9x + 5y = 15$$
$$9x - 9x + 5y = 15 - 9x$$
$$5y = -9x - 15$$
$$\frac{5y}{5} = \frac{-9x - 15}{5}$$
$$y = \frac{-9x - 15}{5}$$

45.
$$2(x + 3y) = 4(x - y) + 5$$
$$2x + 6y = 4x - 4y + 5$$
$$2x + 6y + 4y = 4x + 4y - 4y + 5$$
$$2x + 10y = 4x + 5$$
$$2x - 2x + 10y = 4x - 2x + 5$$
$$10y = 2x + 5$$
$$\frac{10y}{10} = \frac{2x + 5}{10}$$
$$y = \frac{2x + 5}{10}$$

47.
$$\frac{1}{5}\cdot(x - 2y) = \frac{3}{4}\cdot(y + 2) + 3$$

Multiply through by the least common multiple of 5 and 4 which is 20.

$$20\,\frac{1}{5}\,(x - 2y) = 20\cdot\frac{3}{4}\cdot(y + 2) + 20(3)$$
$$4(x - 2y) = 5(3)(y + 2) + 60$$
$$4x - 8y = 15y + 30 + 60$$
$$4x - 8y = 15y + 90$$
$$4x - 8y + 8y = 15y + 8y + 90$$
$$4x = 23y + 90$$
$$4x - 90 = 23y + 90 - 90$$
$$4x - 90 = 23y$$
$$\frac{4x - 90}{23} = \frac{23y}{23}$$
$$\frac{4x - 90}{23} = y$$

49. $d = rt$ Solve for r.

$$\frac{d}{t} = \frac{rt}{t}$$
$$\frac{d}{t} = r$$

51.
$$A = \frac{1}{2}\cdot bh$$ Solve for b.

Multiply both sides of the equation by 2.

$$2\cdot A = 2\cdot\frac{1}{2}\cdot bh$$
$$2\cdot A = bh$$
$$\frac{2A}{h} = \frac{bh}{h}$$
$$\frac{2A}{h} = b$$

53. $i = prt$ Solve for $t$.

$$\frac{i}{pr} = \frac{prt}{pr}$$

$$\frac{i}{pr} = t$$

55. $P = 2L + 2w$ Solve for $L$.

$$P - 2w = 2L + 2w - 2w$$
$$P - 2w = 2L$$

$$\frac{P - 2w}{2} = \frac{2L}{2}$$

$$\frac{P - 2w}{2} = L$$

57. $V = \frac{1}{3} \cdot Bh$ Solve for $h$

Multiply both sides by 3.

$$3 \cdot V = 3 \cdot \frac{1}{3} \cdot Bh$$

$$3 \cdot V = Bh$$

$$\frac{3V}{B} = \frac{Bh}{B}$$

$$\frac{3V}{B} = h$$

59. $V = \pi r^2 \cdot h$ Solve for $h$.

$$\frac{V}{\pi r^2} = \frac{\pi r^2 \cdot h}{\pi r^2}$$

$$\frac{V}{\pi r^2} = h$$

61. $z = \frac{\bar{x} - \mu}{\sigma}$ Solve for $\sigma$.

Multiply both sides by $\sigma$.

$$\sigma \cdot z = \sigma \cdot \frac{\bar{x} - \mu}{\sigma}$$

$$\sigma \cdot \frac{z}{z} = \frac{\bar{x} - \mu}{z}$$

$$\sigma = \frac{\bar{x} - \mu}{z}$$

63. $P = I^2 \cdot R$ Solve for $R$.

$$\frac{P}{I^2} = \frac{I^2 \cdot R}{I^2}$$

$$\frac{P}{I^2} = R$$

65. $A = \frac{1}{2} \cdot h\left[b_1 + b_2\right]$ Solve for $h$.

Multiply both sides of the equation by 2.

$$2 \cdot A = 2 \cdot \frac{1}{2} \cdot h\left[b_1 + b_2\right]$$

$$2 \cdot A = h\left[b_1 + b_2\right]$$

$$\frac{2A}{b_1 + b_2} = \frac{h\left[b_1 + b_2\right]}{b_1 + b_2}$$

$$\frac{2A}{b_1 + b_2} = h$$

69. $y - y_1 = m\left[x - x_1\right]$

Solve for m.

$$\frac{y - y_1}{x - x_1} = \frac{m\left[x - x_1\right]}{x - x_1}$$

$$\frac{y - y_1}{x - x_1} = m$$

71. $S = \frac{n}{2}\cdot(f + \ell)$ Solve for n.

Multiply both sides of the equatio by 2 to clear the fraction.

$$2\cdot S = 2\cdot\frac{n}{2}\cdot(f + \ell)$$

$$2\cdot S = n\cdot(f + \ell)$$

$$\frac{2S}{f + \ell} = \frac{n(f + \ell)}{f + \ell}$$

$$\frac{2S}{f + \ell} = n$$

73. $C = \frac{5}{9}\cdot(F - 32)$ Solve for F.

Multiply both sides of the equation by 9/5 to clear the fraction.

$$\frac{9}{5}\cdot C = \left[\frac{9}{5}\right]\cdot\frac{5}{9}\cdot(F - 32)$$

$$\frac{9}{5}\cdot C = F - 32$$

$$\frac{9}{5}\cdot C + 32 = F - 32 + 32$$

$$\frac{9}{5}\cdot C + 32 = F$$

75. $y = \frac{kx}{z}$ Solve for z.

Multiply both sides of the equation by z to clear the fraction.

$$z\cdot y = z\cdot\frac{kx}{z}$$

$$z\cdot y = kx$$

$$\frac{zy}{y} = \frac{kx}{y}$$

$$z = \frac{kx}{y}$$

77. $F = \frac{km_1\cdot m_2}{d_2^2}$ Solve for $m_1$

Multiply both sides of the equation by $d^2$

to clear the fraction.

$$d^2\cdot F = d^2\cdot\frac{km_1\cdot m_2}{d^2}$$

$$d_2^2\cdot F = km_1\cdot m_2$$

$$\frac{d^2\cdot F}{km_2} = \frac{km_1\cdot m_2}{km_2}$$

$$\frac{d^2\cdot F}{km_2} = \left[m_1\right]$$

79.

$$\mu = \frac{0.5cV^2}{Ad} \quad \text{Solve for A.}$$

Multiply both sides of the equation by Ad to clear the fraction and simplify.

$$Ad\cdot\mu = Ad\cdot\frac{0.5cV^2}{Ad}$$

$$Ad\cdot\mu = \left[0.5cV^2\right]$$

Divide both sides of the equation by d$\mu$ to isolate A. Simplify.

$$\frac{Ad\mu}{d\mu} = \frac{0.5cV^2}{d\mu}$$

$$A = \frac{0.5cV^2}{d\mu}$$

81.

$$-(5-8)^2 + |5-8| - 4^2$$

$$= -(-3)^2 + |-3| - 4^2$$

$$= -9 + 3 - 16$$

$$= -6 - 16$$

$$= -22$$

83. Substitute 2 for each x and 3 for each y in the expression.

$$6x^2 - 3xy + y^2 = 6(2)^2 - 3(2)(3) + 3^2$$

$$= 6(4) - 3(2)(3) + 3^2$$

$$= 24 - 18 + 9$$

$$= 6 + 9$$

$$= 15$$

## EXERCISE SET 2.3

1. Let x = one number
   6x = other number

$$x + 6x = 56$$
$$7x = 56$$
$$\frac{7x}{7} = \frac{56}{7}$$
$$x = 8$$
$$6x = 48$$

The two numbers are 8 and 48.

3. Let x = first integer
   x + 1 = next consecutive integer

$$x + (x + 1) = 51$$
$$2x + 1 = 51$$
$$2x = 50$$
$$\frac{2x}{2} = \frac{50}{2}$$
$$x = 25$$
$$x + 1 = 26$$

The two numbers are 25 and 26.

5. Let x = the number.

$$2x - 8 = 38$$
$$2x = 46$$
$$\frac{2x}{2} = \frac{46}{2}$$
$$x = 23$$

The number is 23.

7. Let x = the distance one train travels.
   3x = the distance traveled by the other train.

$$x + 3x = 48$$
$$4x = 48$$
$$\frac{4x}{4} = \frac{48}{4}$$
$$x = 12$$
$$3x = 36$$

One train traveled 12 miles and the other traveled 36 miles.

9. Let $x$ = the first number
$x + 2$ = the second even number
$x + 4$ = the third even number

$$x + (x + 2) + (x + 4) = 66$$
$$3x + 6 = 66$$
$$3x = 60$$
$$\frac{3x}{3} = \frac{60}{3}$$
$$x = 20$$
$$x + 2 = 22$$
$$x + 4 = 24$$

The three consecutive even integers are 20,22, and 24.

11. Let $x$ = the number.

$$10 - \frac{3}{5} \cdot x = 4$$
$$5 \cdot 10 - 5 \cdot \frac{3}{5} \cdot x = 5(4)$$
$$50 - 3x = 20$$
$$50 - 50 - 3x = 20 - 50$$
$$-3x = -30$$
$$\frac{-3x}{-3} = \frac{-30}{-3}$$
$$x = 10$$

13. Let $x$ = smaller number
$4x + 2$ = larger number

$$5x = \frac{1}{2} \cdot (4x + 2) + 2$$

Multiply through by 2 to clear the fraction.

$$2(5x) = 2 \cdot \frac{1}{2} \cdot (4x + 2) + 2(2)$$
$$10x = 4x + 2 + 4$$
$$10x = 4x + 6$$
$$6x = 6$$
$$\frac{6x}{6} = \frac{6}{6}$$
$$x = 1$$
$$4x + 2 = 6$$

The smaller number is 1 and the larger is 6.

15. Let $x$ = base angle
$2x$ = third angle

$$x + x + 2x = 180$$
$$4x = 180$$
$$\frac{4x}{4} = \frac{180}{4}$$
$$x = 45$$
$$2x = 90$$

The two base angles are 45 degrees each and the third angle is 90 degrees.

17. Let $x$ = the length of one of the equal sides.
$x + 15$ = the length of the third side.

$$x + x + (x + 15) = 45$$
$$3x + 15 = 45$$
$$3x = 30$$
$$x = 10$$

The sides are 10", 10", and 25".

21. Let $x$ = the number of rides.

$$1.25x = 30$$
$$\frac{1.25x}{1.25} = \frac{30}{1.25}$$
$$x = 24$$

Ron would have to ride the bus more than 24 times per month.

23. Let $x$ = the number of years.

$$5200 + 300x = 8800$$
$$300x = 3600$$
$$\frac{300x}{300} = \frac{3600}{3600}$$
$$x = 12$$

It will take 12 years for the population to reach 8800.

23. Let $x$ = the number of years

$5200 + 300x = 8800$

$300x = 3600$

$$\frac{300x}{300} = \frac{3600}{300}$$

$x = 12$

It will take 12 years for the population to reach 8800.

25. Let $x$ = the number of miles.

$35 + .2x = 80$

$.2x = 45$

$$\frac{.2x}{.2} = \frac{45}{.2}$$

$x = 225$

Kendra can drive 225 miles.

27. Let $x$ = sales.

$240 + .12x = 540$

$.12x = 300$

$$\frac{.12x}{.12} = \frac{300}{.12}$$

$x = 2500$

Bridget must have $2500 in sales to make a weekly salary of $540.

29. Let $x$ = the cost of the car.

$x + .04x = 10000$

$1.04x = 10000$

$$\frac{1.04x}{1.04} = \frac{10000}{1.04}$$

$x = 9615.38$

The maximum cost of the car is $9615.38.

31. Let $x$ = the price of a 1 way ticket.

$499 = x + .07x$

$499 = 1.07x$

$$\frac{499}{1.07} = \frac{1.07}{1.07}x$$

$466.36 = x$

The price of a one way ticket should be $466.36.

33. Let $x$ = the price of the dinner.

$x + .07x + .15x = 9.25$

$1.22x = 9.25$

$$\frac{1.22x}{1.22} = \frac{9.25}{1.22}$$

$x = 7.58$

The maximum amount Mrs. Englers can spend for the dinner is $7.58.

35. Let $x$ = the number of times the center is used.

$150 + 6x = 510$

$6x = 360$

$$\frac{6x}{6} = \frac{360}{6}$$

$x = 60$

Ms. Smith would need to use the center 60 times for the cost of the plans to be equal.

37. Let $w$ = width

$2w + 2$ = length

Using the paremeter formula
$P = 2L + 2w$,

$40 = 2(2w + 2) + 2w$

$40 = 4w + 4 + 2w$

$40 = 6w + 4$

$36 = 6w$

$$\frac{36}{6} = \frac{6w}{6}$$

$6 = w$

$14 = 2w + 2$

The length would be 14 ft and the width 6 ft.

39. Let L = length

$\frac{1}{2} \cdot L + 1 = \text{width}$

Using the paremeter formula P = 2L + 2w

$$20 = 2\left[\frac{1}{2} \cdot L + 1\right] + 2L$$

$$20 = L + 2 + 2L$$
$$20 = 3L + 2$$
$$18 = 3L$$
$$6 = L$$

$$4 = \frac{1}{2} \cdot L + 1$$

The length would be 6 meters and the width 4 meters.

41. Let L = the length of the shelves.
L + 3 = the heigth of the bookshelves.

$$2(L + 3) + 4L = 30$$
$$2L + 6 + 4L = 30$$
$$6L + 6 = 30$$
$$6L = 24$$
$$L = 4$$
$$L + 3 = 7$$

The heigth of the bookshelves 7 feet and each shelf would be 4 feet.

43. Let x = the original price of the calculator.

$$x - .1x - 5 = 49$$
$$.90x - 5 = 49$$
$$.90x = 54$$
$$\frac{.90x}{.90} = \frac{54}{.90}$$
$$x = 60$$

The original price of the calculator was $60.

45. Let x = the original price of the camera.

$$x - \frac{1}{4} \cdot x - 10 = 290$$

Multiply through by 4 to eliminate the fraction.

$$4 \cdot x - 4 \cdot \frac{1}{4} \cdot x - (4)10 = 290(4)$$

$$4x - x - 40 = 1160$$
$$3x - 40 = 1160$$
$$3x = 1200$$
$$\frac{3x}{3} = \frac{1200}{3}$$
$$x = 400$$

The original price of the camera was $400.

47. Let x = the area of the smallest region.
2x = the area of the second region.
3x − 4 = the area of the third region.

$$x + 2x + (3x - 4) = 512$$
$$6x - 4 = 512$$
$$6x = 516$$
$$x = 86$$
$$2x = 172$$
$$3x - 4 = 254$$

The areas of the 3 regions are 86 acres, 172 acres, and 254 acres.

49.a) Let x = the score of the fifth test.

$$\frac{87 + 93 + 97 + 96 + x}{5} = 90$$

c)

$$\frac{87 + 93 + 97 + 96 + x}{5} = 90$$

$$\frac{373 + x}{5} = 90$$

$$5 \cdot \frac{373 + x}{5} = 5 \cdot 90$$

$$373 + x = 450$$
$$x = 77$$

Paula needs to make a 77.

53.

$$\frac{\left[(2((5-3)-4))^{2}\right] \div (-8)}{-|8-5| - 4^{2}}$$

$$\frac{(2(2-4))^{2} \div (-8)}{-|-3| - 16}$$

$$\frac{(2(-2))^{2} \div (-8)}{-3 - 16}$$

$$\frac{(-4)^{2} \div (-8)}{-19}$$

$$\frac{16 \div (-8)}{-19}$$

$$\frac{-2}{-19} = \frac{2}{19}$$

55. $\frac{1}{5}\cdot x + \frac{2}{3} = \frac{5}{4}\cdot x$

Multiply through by the least common multiple of 5,3,and 4 which is 60.

$$60\cdot\frac{1}{5}\cdot x + 60\cdot\frac{2}{3} = 60\cdot\frac{5}{4}\cdot x$$

$$12x + 40 = 75x$$

$$40 = 63x$$

$$\frac{40}{63} = x$$

JUST FOR FUN

1. Let x = the original stock value.

   $x + 5\%x$ = the value of the stock on Tuesday.

$$x + (5\%)x - 5\%(x + 5\%(x)) = 59.85$$
$$x + .05x - .05(x + .05x) = 59.85$$
$$x + .05x - .05x - .0025x = 59.85$$
$$.9975x = 59.85$$
$$x = 60$$

The original stock price was $60.00

3. Let n = the number.

$$\frac{2n + 33 - 13}{2} - n = 10$$

$$\frac{2n + 20}{2} - n = 10$$

Multiply through by 2 to clear the fraction.

$$2\cdot\left[\frac{2n + 20}{2}\right] - 2n = 2\cdot 10$$

$$2n + 20 - 2n = 20$$

$$20 = 20$$

This is an example of an equation which is an identity. The solution is all real numbers.

EXERCISE SET 2.4

1. Let r = the average speed of the stage coach.
   $d = rt$
   $1.2 = r(.3)$
   $4 = r$
   The average speed is 4 mph.

3. a) Let r = the average speed.
   $d = rt$
   $10.5 = r(5.5)$
   $1.91 = r$
   The average speed down is 1.91 mph.
   b) $d = rt$
   $8 = r(4.5)$
   $1.78 = r$ = avg. speed up

5. Let t = the time takes to fill the truck.
   Amount = rate x time
   $29{,}500 = 1800\ t$
   $16.39 = t$
   It will take 16.39 min. to fill the truck.

7. Let t = the time needed for the coral to gain 24 ounces. Before we plug values into the equation Amount = rate x time we need to be sure that all units are consistent. Let's convert to ounces. There are 16oz in 1 lb so

$$\frac{3\text{lb}}{\text{year}} = \frac{3(16)\text{oz}}{\text{year}} = \frac{48\text{oz}}{\text{year}}$$

   Amount = rate x time
   $24 = 48(t)$
   $.5 = t$
   It will take 1/2 a year or 6 months for the coral to gain 24 ounces.

9. a) Let t = the number of days for the diapers to reach the moon.
   Amount = rate x time
   $239000 = 9167.1(t)$
   $26.07 = t$
   It would take 26.07 days.
   b) Let A = the distance the diapers would reach.
   Amount = rate x time
   $A = 9167.1(365)$
   $A = 3{,}345{,}991.5$ miles

11. a) Let t = the time for the briefcase to reach the end of the walkway.
   $d = rt$
   $275 = 120t$
   $2.29 = t$
   It would take 2.29 min. for the briefcase to reach the end of the walkway.

   b) Let t = the time needed to walk the length of the sidewalk.
   $d = rt$
   $275 = 150t$
   $1.83 = t$
   It would take 1.83 min. to walk the length of the sidewalk.

   c) Let t = the time for a person to walk the length of the sidewalk. Since both the sidewalk is moving and the person is also walking, the rate of the sidewalk is added to the rate the person is walking to get the overall rate the person is moving.
   $r = 120 + 150 = 270$

   $d = rt$
   $275 = 270t$
   $1.02 = t$
   It will take the person 1.02 min. to walk the length of the sidewalk.

   d) 1.83 = the time needed to walk the distance beside the walkway.
   1.02 = the time needed to walk the distance on the walkway.
   $1.83 - 1.02 = .81$
   The time saved by walking on the walkway is .81 min.

13. Let t = the time each plane is in the air.

| Plane | distance | = rate | x time |
|---|---|---|---|
| 1 | 550 t | 550 | t |
| 2 | 650 t | 650 | t |

   The total distance the planes will fly is 3000 miles. Therefore,
   $550t + 650t = 3000$
   $1200t = 3000$
   $t = 2.5$
   Each plane will fly 2.5 hours.

15. Let $r$ = the rate of the second car.

| Car | distance = | rate x | time |
|---|---|---|---|
| 1 | 180 | 60 | 3 |
| 2 | 3r | r | 3 |

The total distance the cars travel is 330 miles. Therefore,

$$180 + 3r = 330$$
$$3r = 150$$
$$r = 50$$

The rate of the second car is 50 mph.

17. Let $t$ = the time needed for the trains to meet.

| Train | distance = | rate x | time |
|---|---|---|---|
| 1 | 60t | 60 | t |
| 2 | 70t | 70 | t |

The total distance the trains must travel is 520 miles. Therefore,

$$60t + 70t = 520$$
$$130t = 520$$
$$t = 4$$

The trains will meet in 4 hours.

19. Let $r$ = the rate the freight train is traveling.

| train | distance = | rate x | time |
|---|---|---|---|
| freight | 4.2r | r | 3+1.2= 4.2 |
| passenger | 3(r+20) | r + 20 | 3 |

When the passenger train overtakes the freight train they will have traveled the same distance. Therefore,

$$4.2r = 3(r + 20)$$
$$4.2r = 3r + 60$$
$$1.2r = 60$$
$$r = 50$$
$$r + 20 = 70$$

The rate of the freight is 50 mph the the rate of the passenger train is 70 mph.

21. a) Let $r$ = the rate of the jogger.
$t = 2$ since both had been traveling for 2 hours.

| | distance = | rate x | time |
|---|---|---|---|
| jogger | 2r | r | 2 |
| cyclist | 8r | 4r | 2 |

After 2 hours the distance between the two was 18 miles.

21 a) cont.
Therefore,

$$8r - 2r = 18$$
$$6r = 18$$
$$r = 3$$
$$4r = 12$$

The rate of the cyclist was 12 mph.

b) The distance the cyclist rode is $8r$. Since $r = 3$, the distance is $8(3) = 24$ miles.

23. a) Let $t$ = the time needed to reach the bottom of the canyon. $16 - t$ = the time needed to hike back up.

| trip | distance | = rate x | time |
|---|---|---|---|
| down | 2.6t | 2.6 | t |
| up | 1.2(16−t) | 1.2 | 16 − t |

Assuming the distance down equals the distance hiked back up,

$$2.6t = 1.2(16 - t)$$
$$2.6t = 19.2 - 1.2t$$
$$3.8t = 19.2$$
$$t = 5.05$$

McDonald took 5.05 hours to hike down.

b) The total distance traveled can be found by doubling the distance down. Remember we are assuming the distance down equals the distance up.

$$2(2.6t) = 2(2.6)(5.05)$$
$$= 26.26$$

The total distance traveled is 26.26 miles.

25. Let $x$ be the number of hours the smaller machine operates. Then $x + 2$ is the number of hours the larger machine operates.

$400x$ is the number of boxes packed by the smaller machine and $600(x+2)$ is the number of boxes packed by the larger machine.

Since the sum must be 15,000 boxes,

$$600(x + 2) + 400x = 15000$$
$$600x + 1200 + 400x = 15000$$
$$1000x + 1200 = 15000$$
$$1000x = 13800$$
$$x = 13.8$$

The smaller machine is on for 13.8 hours.

27. Let $x$ = the amount invested at 9%.
$11000 - x$ = the amount invested at 10%

| Investment | Amount Invest. | x Rate | = Interest |
|---|---|---|---|
| #1 | x | .09 | .09x |
| #2 | 11000-x | .1 | .1(11000-x) |

The total amount of interest is $1050. Therefore,

$$.09x + .1(11000 - x) = 1050$$
$$.09x + 1100 - .1x = 1050$$
$$-.01x + 1100 = 1050$$
$$-0.1x = -50$$
$$x = 5000$$
$$11000 - x = 6000$$

$5000 is invested at 9% and $6000 is invested at 10%.

29. Let $x$ = the number of shares invested in General Motors.
$5x$ = the number of shares Reebok

| Company | # of shares | x price per share | = cost of shares |
|---|---|---|---|
| G.M. | x | 45 | 45x |
| Reebok | 5x | 14 | 5x(14) |

The total amount to be invested is $9000. Therefore,

$$45x + 5x(14) = 9000$$
$$45x + 70x = 9000$$
$$115x = 9000$$
$$x = 78.26$$

Since stock can only be purchased in whole units, $x$ is rounded to 78 shares.

$$5x = 5(78) = 390$$

Bob Davis can purchase 78 shares of General Motors and 390 shares of Reebok.

b) $78(45) + 390(14)$
$= 3510 + 5460$
$= 8970$

The purchase of stock would cost $8970. This would leave $30.00.

31. Let $d$ = the number of dimes
$33 - d$ = the number of quaters.

| Coin | Face value | x number | = value |
|---|---|---|---|
| dime | .10 | d | .1d |
| quarter | .25 | .25(33-d) | .25(33-d) |

The total value of the coins is $4.50. Therefore,

$$.10d + .25(33 - d) = 4.50$$
$$.10d + 8.25 - .25d = 4.50$$
$$-.15d + 8.25 = 4.50$$
$$-.15d = -3.75$$
$$d = 25$$
$$33 - d = 8$$

The number of dimes is 25 and the number of quaters is 8.

33. Let $x$ = the amount of $6.20 coffee.

| Coffee | Cost per lb. | x # of lbs | Total value |
|---|---|---|---|
| 1 | 6.20 | x | 6.20x |
| 2 | 5.80 | 18 | 5.80·18=104.4 |
| Mixture | 6.10 | (x+18) | 6.10(x+18) |

$$6.20x + 104.4 = 6.10(x + 18)$$
$$6.20x + 104.4 = 6.10x + 109.80$$
$$.1x + 104.4 = 109.80$$
$$.1x = 5.4$$
$$x = 54$$

Larry needs to mix in 54 lbs. of the $6.20 per lb. coffee.

35. Let $x$ = the ounces of pure vinegar.
Note: pure vinegar has a strength of 100% = 1

| Strength of solution | x # of oz. | = Amount |
|---|---|---|
| 100% = 1 | x | 1x |
| 10% = .1 | 40 | 4 |
| Mixture 25% = .25 | (40 + x) | .25(40+x) |

$$x + 4 = .25(40 + x)$$
$$x + 4 = 10 + .25x$$
$$.75x + 4 = 10$$
$$.75x = 6$$
$$x = 8$$

The cook needs to add 8oz of pure vinegar.

37. Let $x$ = the milliliters of 20% solution.
$12 - x$ = the milliliters of 50% solution.

| Strength of solution | x # of milliliters | = Amount |
|---|---|---|
| .2 | x | .2x |
| .5 | (12−x) | .5(12−x) |
| Mixture .3 | 12 | .3(12) = 3.6 |

$$\begin{aligned} .2x + .5(12-x) &= 3.6 \\ .2x + 6 - .5x &= 3.6 \\ -.3x + 6 &= 3.6 \\ -.3x &= -2.4 \\ x &= 8 \\ 12 - x &= 4 \end{aligned}$$

The 30% mixture will require 8 milliliters of 20% solution and 4 milliliters of 50% solution.

39. Let $x$ = the germination rate of the second seed.

| Seed type | Germination Rate | x Quantity of seed | = Amount |
|---|---|---|---|
| low | .76 | 16 | .76x16=12.16 |
| high | x | 12 | 12x |
| Mix | .82 | 28 | .82(28)= 22.96 |

$$\begin{aligned} 12.16 + 12x &= 22.96 \\ 12x &= 10.8 \\ x &= .9 = 90\% \end{aligned}$$

The germination rate of the high quality seed is 90%.

41. Let $x$ = Ms. Clar's portion of the income tax deduction
$6400 - x$ = Mr. Clar's portion.
Note: Amount of taxable income = income − deduction.

| | Taxable | = Income | − deduction |
|---|---|---|---|
| Ms. | 32450−x | 32450 | x |
| Mr. | 28200 − (6400−x) | 28200 | 6400 − x |

Since the smallest amount of tax owed occurs when the husband's and wife's income equals,

$$\begin{aligned} 32450 - x &= 28200 - (6400 - x) \\ 32450 - x &= 28200 - 6400 + x \\ 32450 &= 21800 + 2x \\ 10650 &= 2x \\ 5325 &= x \\ 6400 - x &= 1075 \end{aligned}$$

Ms. Clar should claim $5325 in deductions and Mr. Clar should claim $1075.

43. Let $r$ = the rate one hiker hikes.
$r + .4$ = the rate of the second hiker.

| hiker | distance | = rate | x time |
|---|---|---|---|
| 1 | 2r | r | 2 |
| 2 | 2(r+.4) | (r+.4) | 2 |

The total distance the hikers travel is 8.2 miles. Therefore,

$$\begin{aligned} 2r + 2(r + .4) &= 8.2 \\ 2r + 2r + .8 &= 8.2 \\ 4r + .8 &= 8.2 \\ 4r &= 7.4 \\ r &= 1.85 \\ r + .4 &= 2.25 \end{aligned}$$

The rate of the first hiker is 1.85 mph and the rate of the second is 2.25 mph.

45. a) Let $x$ = the number of US Steel shares
$3x$ = the number of Sears shares

| Stock | # of shares | x price per share | = cost of shares |
|---|---|---|---|
| US Steel | x | 23 | 23x |
| Sears | 3x | 34 | 3x(34)=102x |

The total amount to be invested is $6000. Therefore,

$$\begin{aligned} 23x + 102x &= 6000 \\ 125x &= 6000 \\ x &= 48 \\ 3x &= 144 \end{aligned}$$

The Bernhams should purchase 48 shares of US Steel and 144 shares of Sears stock.

b) The total cost of the shares is
$23(48) + 34(144)$
$= 1104 + 4896$
$= 6000$
There would not be any money left over.

47. Let t = the time the pumps will be working to remove the water.

| pump | amount of water pumped | = rate | x time |
|---|---|---|---|
| 1 | 10t | 10 | t |
| 2 | 20t | 20 | t |

The total amount of water to be pumped is 15000 gallons. Therefore,

$$10t + 20t = 15000$$
$$30t = 15000$$
$$t = 500$$

It wuld take 500 min. or 8 1/3 hours to empty the pool.

$$\frac{500\text{min}}{60\text{min}} = 8\ \frac{1}{3}\ \text{hours}$$

49. Let r = the interest rate on one account.
A total of $4000 is invested. If $2500 is invested in the 9% account, this leaves $1500 to invest in the second.

| Amount Invested | x Rate | = Interest |
|---|---|---|
| 2500 | .09 | 225 |
| 1500 | r | 1500r |

The total interest earned is $315. Therefore,

$$225 + 1500r = 315$$
$$1500r = 90$$
$$r = .06$$
$$r = 6\%$$

The interest rate of the second account is 6%.

51. a) Let t = the flying time to Tallahassee.
11.2 - t = the flying time for the return trip.

| Trip | Distance | = rate | x time |
|---|---|---|---|
| to | 300t | 300 | t |
| from | 220(11.2-t) | 220 | (11.2-t) |

Assuming the distance on both legs of the journey is the same,

$$300t = 220(11.2 - t)$$
$$300t = 2464 - 220t$$
$$520t = 2464$$
$$t = 4.74$$

The trip will take 4.74 hours.

51 b) cont.
The distance between the two airports is given by 300t.

$$300t = 300(4.74)$$
$$= 1422\ \text{miles}$$

53. Let x = the number of hours worked at $6 per hour.
18 - x = the number of hours worked at $6.50 per hour.

| Job | Hours worked | x Rate of pay | = Total pay |
|---|---|---|---|
| 1 | x | 6 | 6x |
| 2 | 18-x | 6.50 | 6.5(18-x) |

The total pay for the week was $114. Therefore,

$$6x + 6.5(18 - x) = 114$$
$$6x + 117 - 6.50x = 114$$
$$-.5x + 117 = 114$$
$$-.5x = -3$$
$$x = 6$$
$$18 - x = 12$$

Diedre worked 6 hours at $6 per hour and 12 hours at $6.50 per hour.

55 . a) Let t = the time to travel to Goat Island.
8 - x = the time needed to return.

| Trip | Distance = | rate | x time |
|---|---|---|---|
| to | 20t | 20 | t |
| from | 12(8-t) | 12 | (8-t) |

Assuming the distance both to and from the island is the same,

$$20t = 12(8 - t)$$
$$20t = 96 - 12t$$
$$32t = 96$$
$$t = 3$$

The trip to the island took 3 hours.

b) The expression 20t gives the distance to the island. Therefore, 2(20t) gives the round trip distance.

$$2(20t) = 2(20)(3) = 120$$

The total distance traveled is 120 nautical miles.

57. Let $x$ = the percentage of milk the mixture will contain.
Before writing an equation it is necessary to have consistent units. Express both liquids in ounces.
1 pt. = 16 oz
1 gal.= 64 oz

| % of milk | x | # of oz | = | Amount |
|---|---|---|---|---|
| 3% | | 16 | | 3%(16)=.48 |
| 7% | | 64 | | 7%(64)=4.48 |
| Mixture x | | (16+64)=80 | | 80x |

$$.48 + 4.48 = 80x$$
$$4.96 = 80x$$
$$.062 = x$$
$$6.2\% = x$$

The mixture contains 6.2% milk.

59. Let $x$ = the amount of 80% solution.
$128 - x$ = the amount of water.
Note: Water has no methal alcohol so its percentage will be zero.

| % of methal alcohol | x | # of oz. | = | amount |
|---|---|---|---|---|
| 80% = .8 | | x | | .8x |
| 0 | | (128-x) | | 0(128-x) |
| Mixture 6% = .06 | | 128 | | .06(128)=7.68 |

$$.8x + 0(128 - x) = 7.68$$
$$.8x = 7.68$$
$$x = 9.6$$
$$128 - x = 118.4$$

Philip should mix 9.6 oz. of methal alcohol solution with 118.4 oz. of water to make the windshield washer solution.

61. Let $x$ = the amount of 1.5% butterfat milk.

| % of butter-fat | x | # of quarts | = | Amount |
|---|---|---|---|---|
| 5% = .05 | | 400 | | .05(400)=20 |
| 1.5%=.015 | | x | | .015x |
| Mixture 2% = .02 | | 400+x | | .02(400+x) |

61 cont.
$$20 + .015x = .02(400 + x)$$
$$20 + .015x = 8 + .02x$$
$$20 = .005x + 8$$
$$12 = .005x$$
$$2400 = x$$

2400 qts. of 1.5% butterfat milk should be added to produce milk containing 2% butterfat.

63. Let $t$ = the amount of time each machine works.

| machine | rate | x | time | = | amount of cartons produced |
|---|---|---|---|---|---|
| old | 50 | | t | | 50t |
| new | 70 | | t | | 70t |

Since the old machine has already produced 200 cartons, the total number of cartons it has made is $200 + 50t$. We are looking for the length of time needed for the new machine to produce the same amount of cartons as the old machine. Therefore,
$$70t = 50t + 200$$
$$20t = 200$$
$$t = 10$$

It will take the new machine 10 min. to produce the same number of cartons as the old has produced.

65. Mike can measure the amount of time the mower needs to mow a certain distance. Using the formula $d = rt$, and plugging in the values for $d$ and $t$, he can solve for $r$ which is the average speed the mower will travel.

69.
$$0.6x + .22 = .4(x - 2.3)$$
$$0.6x + .22 = .4x - .92$$
$$0.2x + .22 = -.92$$
$$0.2x = -1.14$$
$$x = -5.7$$

71. $\frac{3}{5}\cdot(x - 2) = \frac{2}{7}\cdot(2x + 3y)$

Solve for y.

Clear the fractions by multiplying each side of the equation by the least common multiple of 5 and 7 which is 35.

$$35\cdot\frac{3}{5}\cdot(x - 2) = 35\cdot\frac{2}{7}\cdot(2x + 3y)$$

$$21(x - 2) = 10(2x + 3y)$$
$$21x - 42 = 20x + 30y$$
$$21x - 20x - 42 = 20x - 20x + 30y$$
$$x - 42 = 30y$$

$$\frac{x - 42}{30} = \frac{30y}{30}$$

$$\frac{x - 42}{30} = y$$

JUST FOR FUN

1. a) Car A has completed half the race or 250 laps. Since each lap is 1 mile, the distance Car A has traveled is 250 miles.

$$d = r \times t$$
$$250 = 125(t)$$
$$2 = t$$

The cars have been racing 2 hours.

Let r = the average speed of Car B.

Car B is exactly 6.2 laps or 6.2 miles behind Car A. Therefore, Car B has traveled a distance of
250 - 6.2 = 243.8 miles.

$$d = r \times t$$
$$243.8 = r(2)$$
$$121.9 = r$$

The rate of Car B is 121.9mph.

1."Just for Fun" cont.

b) Another way to word this portion of the problem is "How long will it take Car B to catch Car A?". To catch Car A, Car B must travel the 6.2 miles it is behing.

Let t = the time Car B needs to travel 6.2 miles.

$$d = r \times t$$
$$6.2 = 121.9t$$
$$.0509 = t$$

The time needed for Car B to catch Car A is .0509 hours or 183.2 seconds.
(.0509 x 3600 seconds per hr. = 183.2 seconds)

3. First, determine the number of stitches needed to make the afghan. Since the instructions indicate 4 stitches = 1 inch and 6 rows = 1 inch, convert the dimensions of the afghan to inches.

4 ft. wide = 4(12inches)
= 48 inches wide
6 ft. long = 6(12inches)
= 72 inches long.

There are 4 stitches per inch in the width so
48(4) = 192 stitches.
There are 6 rows per inch in length so
72(6) = 432 rows.

The total number of stitches =
432 x 192 = 82,944 stitches.

The rate Lester knits is 8 stitches per minute.

Amount of stitches = rate x time
82,944 = 8t
10,368 = t
It will take Lester 10368 minutes or 10368/60 = 172.8 hours to knit the aphgan.

EXERCISE SET 2.5

1. a) <━━━━━━━━o--------->
-3

b) $(-\infty, -3)$
c) $\{x \mid x < -3\}$

3. a) <-----------●━━━━━━━>
5

b) $[5, \infty)$
c) $\{x \mid x \geq 5\}$

5. a) <----●━━━━━━o--------->
2 12/5

b) $[2, 12/5)$
c) $\{x \mid 2 \leq x < 12/5\}$

7. a) <----o━━━━━━●--------->
-6 -4

b) $(-6, -4]$
c) $\{x \mid -6 < x \leq -4\}$

9. a) <----●━━━━━━●--------->
-4 5

b) $[-4, 5]$
c) $\{x \mid -4 \leq x \leq 5\}$

11.
$$x + 3 < 8$$
$$x < 5$$

13.
$$2x > 12$$
$$\frac{2x}{2} > \frac{12}{2}$$
$$x > 6$$

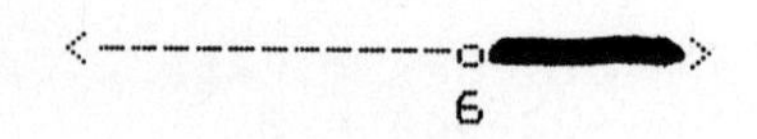

15.
$$2x + 3 > 4$$
$$2x > 1$$
$$x > 1/2$$

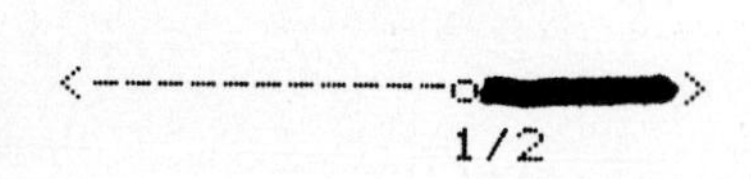

17.
$$4x + 5 > 3x$$
$$x + 5 > 0$$
$$x > -5$$

<-----------o━━━━━━>
-5

19.
$$4x + 3 \leq -2x + 9$$
$$6x + 3 \leq 9$$
$$6x \leq 6$$
$$x \leq 1$$

21.
$$14 > 3a - 6$$
$$20 > 3a$$
$$20/3 > a$$

<━━━━━━━━o------->
20/3

23.
$$-(x - 3) + 4 \leq -2x + 5$$
$$-x + 3 + 4 \leq -2x + 5$$
$$-x + 7 \leq -2x + 5$$
$$-x + 2x + 7 \leq -2x + 2x + 5$$
$$x + 7 \leq 5$$
$$x \leq -2$$

<━━━━━━━━●------->
-2

25.
$$2y - 6y + 10 \leq 2(-2y + 3)$$
$$-4y + 10 \leq -4y + 6$$
$$-4y + 4y + 10 \leq -4y + 4y + 6$$
$$10 \leq 6$$

Since 10 is never less than or equal to 6, there is no solution.

27.
$$(w - 5) \leq \frac{3}{4}(2w + 6)$$

Multiply both sides of the inequality by 4 to clear the fractions.

$$4 \cdot (w - 5) \leq 4 \cdot \frac{3}{4}(2w + 6)$$

$$4 \cdot (w - 5) \leq 3 \cdot (2w + 6)$$
$$4w - 20 \leq 6w + 18$$
$$4w - 4w - 20 \leq 6w - 4w + 18$$
$$-20 \leq 2w + 18$$
$$-20 - 18 \leq 2w + 18 - 18$$
$$-38 \leq 2w$$
$$-19 \leq w$$
$$\text{or} \quad w \geq -19$$

<-----------●━━━━━━>
-19

$[-19, \infty)$

29. $4 + \frac{3x}{2} < 6$

Multiply by 2 to clear the fraction.

$$2 \cdot 4 + 2 \cdot \frac{3x}{2} < 2 \cdot 6$$

$$8 + 3x < 12$$
$$3x < 4$$
$$x < \frac{4}{3}$$
$(-\infty, 4/3)$

31. $\frac{5 - 6y}{3} \le 1 - 2y$

Multiply by 3 to clear the fraction.

$$3 \cdot \frac{5 - 6y}{3} \le 3 \cdot 1 - 3 \cdot 2y$$

$$5 - 6y \le 3 - 6y$$

$$5 \le 3$$

Since 5 is never less than or equal to 3, there is no solution.

33. $x + 1 < 3(x + 2) - 2x$
$x + 1 < 3x + 6 - 2x$
$x + 1 < x + 6$
$1 < 6$

Since 1 is always less than 6, the solution is the set of all real numbers.

$(-\infty, \infty)$

35. $4 < x + 3 < 9$
$4 - 3 < x + 3 - 3 < 9 - 3$
$1 < x < 6$
$(1, 6)$

37. $-3 < 5x \le 8$

$$\frac{-3}{5} < \frac{5x}{5} \le \frac{8}{5}$$

$$\frac{-3}{5} < x \le \frac{8}{5}$$

$(-3/5, 8/5]$

39. $4 \le 2x - 3 < 7$
$4 + 3 \le 2x - 3 + 3 < 7 + 3$
$7 \le 2x < 10$

$$\frac{7}{2} \le \frac{2x}{2} < \frac{10}{2}$$

$$\frac{7}{2} \le x < 5$$

$[7/2, 5)$

41. $\frac{1}{2} < 3x + 4 < 6$

Multiply through by 2 to clear the fraction.

$$2 \cdot \frac{1}{2} < 2 \cdot 3 \cdot x + 2 \cdot 4 < 2 \cdot 6$$

$1 < 6x + 8 < 12$
$1 - 8 < 6x + 8 - 8 < 12 - 8$
$-7 < 6x < 4$

$$\frac{-7}{6} < \frac{6x}{6} < \frac{4}{6}$$

$$\frac{-7}{6} < x < \frac{2}{3}$$

$(-7/6, 2/3)$

43. $-6 < \frac{-2x - 3}{4} \le 8$

Multiply through by 4 to clear the fraction.

$$4 \cdot (-6) < 4 \cdot \frac{-2x - 3}{4} \le 4 \cdot 8$$

$-24 < -2x - 3 \le 32$
$-24 + 3 < -2x - 3 + 3 \le 32 + 3$
$-21 < -2x \le 35$

$$\frac{-21}{-2} > \frac{-2x}{-2} \ge \frac{35}{-2}$$

$$\frac{21}{2} > x \ge \frac{-35}{2}$$

$$\frac{-35}{2} \le x < \frac{21}{2}$$

$[-35/2, 21/2)$

45. $$-12 \le \frac{4 - 3x}{-5} < 2$$

Multiply by -5 to clear the fraction.

$$-12\cdot(-5) \ge -5\cdot\frac{4 - 3x}{-5} > (-5)2$$

$$60 \ge 4 - 3x > -10$$
$$60 - 4 \ge 4 - 4 - 3x > -10 - 4$$
$$56 \ge -3x > -14$$

$$\frac{56}{-3} \le \frac{-3x}{-3} < \frac{-14}{-3}$$

$$\frac{-56}{3} \le x < \frac{14}{3}$$

$$\{x| \quad \frac{-56}{3} \le x < \frac{14}{3} \quad \}$$

47. $$6 \le -3(2x - 4) < 12$$
$$6 \le -6x + 12 < 12$$
$$6 - 12 \le -6x + 12 - 12 < 12 - 12$$
$$-6 \le -6x < 0$$

$$\frac{-6}{-6} \ge \frac{-6x}{-6} > \frac{0}{-6}$$

$$1 \ge x > 0$$

$$0 < x \le 1$$

$$\{x| \quad 0 < x \le 1 \quad \}$$

49. $$-15 < \frac{3(x - 2)}{5} \le 0$$

Multiply through by 5 to clear the fraction.

$$-15\cdot 5 < 5\cdot\frac{3(x - 2)}{5} \le 5\cdot 0$$

$$-75 < 3(x - 2) \le 0$$
$$-75 < 3x - 6 \le 0$$
$$-75 + 6 < 3x - 6 + 6 \le 0 + 6$$
$$-69 < 3x \le 6$$

$$\frac{-69}{3} < \frac{3x}{3} \le \frac{6}{3}$$

$$-23 < x \le 2$$

$$\{x| \quad -23 < x \le 2 \quad \}$$

51. $$1 < \frac{4 - 6x}{2} < 5$$

Multiply by 2 to clear the fraction.

$$2\cdot 1 < 2\cdot\frac{4 - 6x}{2} < 2\cdot 5$$

$$2 < 4 - 6x < 10$$
$$2 - 4 < 4 - 4 - 6x < 10 - 4$$
$$-2 < -6x < 6$$

$$\frac{-2}{-6} > \frac{-6x}{-6} > \frac{6}{-6}$$

$$\frac{1}{3} > x > -1$$

$$-1 < x < \frac{1}{3}$$

$$\{x| \quad -1 < x < \frac{1}{3} \quad \}$$

53. $x<4$ and $x>2$

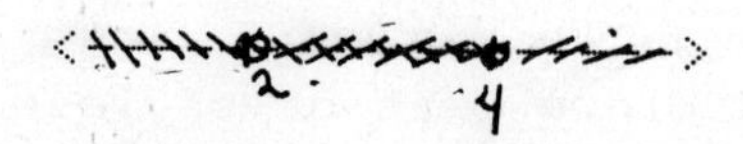


The solution occurs where the regions overlap.
$\{x|\ 2 < x < 4\}$

55. $x < 2$ and $x > 4$

2 4

Since there is no overlap the solution is the empty set.

57. $x + 2 < 3$ and $x + 1 > -2$
$x < 1$ and $x > -3$

-3 1

The solution occurs where the regions overlap.
$\{x|\ -3 < x < 1\}$

59. $5x - 3 \le 7$ or $-x + 3 < -5$
$5x \le 10$ or $-x < -8$
$x \le 2$ or $x > 8$

The solution is any area that is shaded.
$\{x \mid x \le 2 \text{ or } x > 8\}$

61. $3x - 6 \le 4$ or $2x - 3 < 5$
$3x \le 10$ or $2x < 8$
$x \le \frac{10}{3}$ or $x < 4$

The solution is all shaded areas.
$(-\infty, 4)$

63. $4x + 5 \ge 5$ and $3x - 4 \le 2$
$4x \ge 0$ and $3x \le 6$
$x \ge 0$ and $x \le 2$

The solution occurs where regions overlap.
$[0,2]$

65. $5x - 3 > 10$ and $4 - 3x < -2$
$5x > 13$ and $-3x < -6$
$x > \frac{13}{5}$ and $x > 2$

The solution occurs where the regions overlap.
$(13/5, \infty)$

67. $4 - x < -2$ or $3x - 1 < -1$
$-x < -6$ or $3x < 0$
$x > 6$ or $x < 0$

The solution is all shaded regions.
$(-\infty, 0) \cup (6, \infty)$

69. Let x = the maximum number of boxes.
$80(x) \le 900$
$x \le 11.25$

The maximum number of boxes is 11.

71. Let x = the maximum length of time.

$4.25 + .48x \le 9.5$
$.48x \le 5.25$
$x \le 10.9375$

The maximum length of time the customer can talk is 10 minutes plus the first three additional minutes for a total of 13 minutes.

73. To make a profit, Miriam's cost must be less than the revenue she earns.
cost < revenue
Cost = 10,025 + 1.09x and Revenue = 6.42x. Therefore,
$10{,}025 + 1.09x < 6.42x$
$10{,}025 < 5.33x$
$1880.86 < x$

Mariam must sell a minimum of 1881 books to make a profit.

75. The cost of the bulk mail must be less than the regular mailing rate. Therefore, let x = the number of pieces of mail and
"$60 + .084x < .167x$
$60 < .083x$
$722.89 < x$
The organization must mail 723 or more pieces for the bulk permit to be profitable.

77. Ray's average must be greater than or equal to 90 for him to receive an A. Let x = the grade Ray needs to make on the fifth exam.

$$90 \le \frac{90 + 87 + 96 + 95 + x}{5}$$

$$90(5) \le 5 \cdot \frac{90 + 87 + 96 + 95 + x}{5}$$

$$450 \le 90 + 87 + 96 + 95 + x$$
$$450 \le 368 + x$$
$$82 \le x$$

Ray must make an 82 or better on the fifth exam to receive an A.

79. Let x = the value of the third pollutant.

$$\frac{2.7 + 3.42 + x}{3} < 3.2$$

$$\frac{6.12 + x}{3} < 3.2$$

$$3 \cdot \frac{6.12 + x}{3} < (3)3.2$$

$$6.12 + x < 9.6$$
$$x < 3.48$$

Any value less than 3.48 parts per million is considered "clean" air.

81. Let x = the value of the third reading.

$$7.2 < \frac{7.48 + 7.85 + x}{3} < 7.8$$

$$7.2 < \frac{15.33 + x}{3} < 7.8$$

$$7.2(3) < 3 \cdot \frac{15.33 + x}{3} < 3(7.8)$$

$$21.6 < 15.33 + x < 23.4$$
$$6.27 < x < 8.07$$

Any value between 6.27 and 8.07 would result in a normal pH reading.

83. a) Clear the parenthesies.

$$x + 5 < 3x - 8 \le 2(x + 7)$$
$$x + 5 < 3x - 8 \le 2x + 14$$

b) Rewrite the inequality as

$$x + 5 < 3x - 8$$

and

$$3x - 8 \le 2x + 14$$

c) Solve each inequality.

$$x + 5 < 3x - 8$$
$$5 < 2x - 8$$
$$13 < 2x$$
$$\frac{13}{2} < x$$

and

$$3x - 8 \le 2x + 14$$
$$x - 8 \le 14$$
$$x \le 22$$

d) Graph each solution.

13/2 22

e) The region where the shading overlaps is the solution.
(13/2, 22]

85. a) 4

b) 4, 0

c) $-3,\ 4\ \ \frac{5}{2}, 0, \frac{-29}{80}$

d) $-3,\ 4\ \ \frac{5}{2}, \sqrt{7}, 0, \frac{-29}{80}$

87. Commutative property of addition.

JUST FOR FUN

Let x = the final exam grade. If the final counts 1/3 then the remaining test grades must count 2/3 of Russell's grade.

$$80 \le \frac{2}{3}\cdot\frac{82 + 90 + 74 + 76 + 68}{5} + \frac{1}{3}\cdot(x) < 90$$

$$80\cdot(3) \le 3\cdot\frac{2}{3}\cdot\frac{82 + 90 + 74 + 76 + 68}{5} + 3\cdot\frac{1}{3}\cdot(x) < 3(90)$$

$$240 \le \frac{2(82 + 90 + 74 + 76 + 68)}{5} + x < 270$$

$$240 < \frac{2(390)}{5} + x < 270$$

$$240 < 2(78) + x < 270$$
$$240 < 156 + x < 270$$
$$84 \le x < 114$$

Since the maximum grade on the final is 100,
$$84 \le x < 100$$

EXERCISE SET 2.6

1. $|x| = 5$
Either $x = 5$ or $x = -5$.
$\{-5, 5\}$

3. $|x| = 12$
Either $x = 12$ or $x = -12$.
$\{-12, 12\}$

5. $|x| = -2$
The absolute value of a number is never negative so there is no solution.
$\{\ \}$

7. $|x + 5| = 7$
Either
$x + 5 = 7$ or $x + 5 = -7$.
$x = 2$ or $x = -12$
$\{-12, 2\}$

9. $|2w + 4| = 6$
Either
$2w + 4 = 6$ or $2w + 4 = -6$
$2w = 2$ or $2w = -10$
$w = 1$ or $w = -5$
$\{-5, 1\}$

11. $|5 - 3x| = \frac{1}{2}$
Either
$$5 - 3x = \frac{1}{2}$$
$$2\cdot 5 - 2\cdot 3x = 2\cdot\frac{1}{2}$$
$$10 - 6x = 1$$
$$-6x = -9$$
$$\frac{-6x}{-6} = \frac{-9}{-6}$$
$$x = \frac{3}{2}$$
or
$$5 - 3x = -\left[\frac{1}{2}\right]$$
$$2\cdot 5 - 2\cdot 3x = 2\cdot\frac{-1}{2}$$
$$10 - 6x = -1$$
$$-6x = -11$$
$$x = \frac{11}{6}$$
$$\left\{\frac{3}{2}, \frac{11}{6}\right\}$$

13. $|4(x - 2)| = 18$

Either

$$4(x - 2) = 18 \quad \text{or} \quad 4(x - 2) = -18$$
$$4x - 8 = 18 \qquad 4x - 8 = -18$$
$$4x = 26 \qquad 4x = -10$$
$$x = \frac{26}{4} \qquad x = \frac{-10}{4}$$
$$x = \frac{13}{2} \qquad x = \frac{-5}{2}$$

$$\left\{ \frac{-5}{2}, \frac{13}{2} \right\}$$

15. $$\left|\frac{3z + 5}{6}\right| - 3 = 6$$

$$\left|\frac{3z + 5}{6}\right| - 3 + 3 = 6 + 3$$

$$\left|\frac{3z + 5}{6}\right| = 9$$

Either

$$\frac{3z + 5}{6} = 9 \quad \text{or} \quad \frac{3z + 5}{6} = -9$$

$$6 \cdot \frac{3z + 5}{6} = 6 \cdot 9 \qquad 6 \cdot \frac{3z + 5}{6} = 6 \cdot (-9)$$

$$3z + 5 = 54 \qquad 3z + 5 = -54$$
$$3z = 49 \qquad 3z = -59$$
$$z = \frac{49}{3} \qquad z = \frac{-59}{3}$$

$$\left\{ \frac{-59}{3}, \frac{49}{3} \right\}$$

17. $$\left|\frac{5x - 3}{2}\right| + 2 = 6$$

$$\left|\frac{5x - 3}{2}\right| = 4$$

Either

$$\left[\frac{5x - 3}{2}\right] = 4$$

17. cont.

$$2 \cdot \left[\frac{5x - 3}{2}\right] = 2 \cdot (4)$$

$$5x - 3 = 8$$
$$5x = 11$$
$$x = \frac{11}{5}$$

or

$$2 \cdot \left[\frac{5x - 3}{2}\right] = 2 \cdot (-4)$$

$$5x - 3 = -8$$
$$5x = -5$$
$$x = -1$$
$$\left\{ -1, \frac{11}{5} \right\}$$

19. $$|y| \le 5$$

$$-5 \le y \le 5$$

$$\{y \mid -5 \le y \le 5\}$$

21. $$|x - 7| \le 9$$

$$-9 \le x - 7 \le 9$$
$$-9 + 7 \le x - 7 + 7 \le 9 + 7$$
$$-2 \le x \le 16$$

$$\{x \mid -2 \le x \le 16\}$$

23. $$|3z - 5| \le 5$$
$$-5 \le 3z - 5 \le 5$$
$$-5 + 5 \le 3z - 5 + 5 \le 5 + 5$$
$$0 \le 3z \le 10$$

$$\frac{0}{3} \le \frac{3z}{3} \le \frac{10}{3}$$

$$0 \le z \le \frac{10}{3}$$

$$\left\{z \mid 0 \le z \le \frac{10}{3}\right\}$$

25. $$|2x + 3| - 5 \le 10$$
$$|2x + 3| - 5 + 5 \le 10 + 5$$
$$|2x + 3| \le 15$$
$$-15 \le 2x + 3 \le 15$$
$$-15 - 3 \le 2x + 3 - 3 \le 15 - 3$$
$$-18 \le 2x \le 12$$
$$-18 \le 2x \le 12$$
$$-9 \le x \le 6$$

$$\{x \mid -9 \le x \le 6\}$$

27. $$|x - 5| \leq \frac{1}{2}$$

$$\frac{-1}{2} \leq x - 5 \leq \frac{1}{2}$$

$$2 \cdot \frac{-1}{2} \leq 2(x - 5) \leq 2 \cdot \frac{1}{2}$$

$$1 \leq 2x - 10 \leq 1$$
$$-1 + 10 \leq 2x - 10 + 10 \leq 1 + 10$$
$$9 \leq 2x \leq 11$$

$$\frac{9}{2} \leq \frac{2x}{2} \leq \frac{11}{2}$$

$$\frac{9}{2} \leq x \leq \frac{11}{2}$$

$$\{x \mid \frac{9}{2} \leq x \leq \frac{11}{2}\}$$

29. $$|2x - 6| + 5 \leq 2$$
$$|2x - 6| \leq -3$$

The absolute value of a negative number is never negative. Therefore, there does not exist a value of x such that
$$|2x - 6| + 5 \leq 2$$

31. $$\left|5 - \frac{3x}{4}\right| < 8$$

$$-8 < 5 - \frac{3x}{4} < 8$$

$$4 \cdot (-8) < 4 \cdot 5 - 4 \cdot \frac{3x}{4} < 4 \cdot 8$$
$$-32 < 20 - 3x < 32$$
$$-52 < -3x < 12$$

$$\frac{-52}{-3} > \frac{-3x}{-3} > \frac{12}{-3}$$

$$\frac{52}{3} > x > -4 \quad \text{or}$$

$$-4 < x < \frac{52}{3}$$

$$\{x \mid -4 < x < \frac{52}{3}\}$$

33. $$|x| > 3$$
Either
$$x > 3 \text{ or } x < -3$$
$$\{x \mid x > 3 \text{ or } x < -3\}$$

35. $$|x + 4| > 5$$
Either
$$x + 4 > 5 \text{ or } x + 4 < -5$$
$$x > 1 \quad \text{or} \quad x < -9$$
$$\{x \mid x > 1 \text{ or } x < -9\}$$

37. $$|3x + 1| > 4$$
Either
$$3x + 1 > 4 \text{ or } 3x + 1 < -4$$
$$3x > 3 \text{ or } \quad 3x < -5$$
$$x > 1 \text{ or } \quad x < \frac{-5}{3}$$

$$\{x \mid x > 1 \text{ or } x < \frac{-5}{3}\}$$

39. $$\left|\frac{6 + 2z}{3}\right| > 2$$

Either
$$\frac{6 + 2z}{3} > 2 \quad \text{or} \quad \frac{6 + 2z}{3} < -2$$

$$3 \cdot \frac{6 + 2z}{3} > 3 \cdot 2 \qquad 3 \cdot \frac{6 + 2z}{3} < 3 \cdot (-2)$$

$$6 + 2z > 6 \quad \text{or} \quad 6 + 2z < -6$$
$$2z > 0 \quad \text{or} \quad 2z < -12$$
$$z > 0 \quad \text{or} \quad z < -6$$

$$\{z \mid z > 0 \text{ or } z < -6\}$$

41. $$|4x - 3| + 2 > 7$$
$$|4x - 3| > 5$$
Either
$$4x - 3 > 5 \text{ or } 4x - 3 < -5$$
$$4x > 8 \text{ or } \quad 4x < -2$$
$$x > 2 \text{ or } \quad x < -1/2$$

$$\{x \mid x > 2 \text{ or } x < -1/2\}$$

43. $$\left|\frac{2x - 4}{3}\right| > -5$$

The absolute value of a number is always positive and therefore, always greater than −5. The solution is all real numbers.

45. $\left|\frac{x}{2} + 4\right| \geq 5$

Either

$\frac{x}{2} + 4 \geq 5$ or $\frac{x}{2} + 4 \leq -5$ □

$2 \cdot \frac{x}{2} + 2 \cdot 4 \geq 2 \cdot 5$ $\quad 2 \cdot \frac{x}{2} + 2 \cdot 4 \leq 2 \cdot (-5)$ □

$x + 8 \geq 10$ or $x + 8 \leq -10$
$x \geq 2$ or $x \leq -18$
$\{x \mid x \geq 2$ or $x \leq -18$ □ $\}$

47. $|w| = 7$
$w = 7$ or $w = -7$
$\{-7, 7\}$

49. $|x - 3| < 5$

$-5 < x - 3 < 5$
$-2 < x < 8$
$\{x \mid -2 < x < 8\}$

51. $|x + 5| > 9$
$x + 5 > 9$ or $x + 5 < -9$
$x > 4$ or $x < -14$
$\{x \mid x < -14$ or $x > 4\}$

53. $|2y + 4| < 1$

$-1 < 2y + 4 < 1$
$-5 < 2y < -3$

$\frac{-5}{2} < y < \frac{-3}{2}$

$\{y \mid \frac{-5}{2} < y < \frac{-3}{2}\}$

55. $|4x + 2| = 9$
$4x + 2 = 9$ or $4x + 2 = -9$
$4x = 7$ or $4x = -11$
$x = \frac{7}{4}$ or $x = \frac{-11}{4}$
$\{\frac{-11}{4}, \frac{7}{4}\}$

57. $|5 + 2x| \geq 3$
$5 + 2x \geq 3$ or $5 + 2x \leq -3$
$2x \geq -2$ or $2x \leq -8$
$x \geq -1$ or $x \leq -4$
$\{x \mid x \geq -1$ or $x \leq -4$ □ $\}$

59. $|4 + 3x| \leq 9$
$-9 \leq 4 + 3x \leq 9$
$-13 \leq 3x \leq 5$

$\frac{-13}{3} \leq x \leq \frac{5}{3}$

$\{x \mid \frac{-13}{3} \leq x \leq \frac{5}{3}\}$

61. $|3x - 5| + 4 = 2$

$|3x - 5| = -2$

The absolute value of a number is never negative. Therefore, there does not exist a value of x such that $|3x-5| + 4 = 2$.

63. $\left|\frac{3x - 2}{4}\right| - 5 = 1$

$\left|\frac{3x - 2}{4}\right| = 6$

Either

$\frac{3x - 2}{4} = 6$ or $\frac{3x - 2}{4} = -6$

$4 \cdot \frac{3x - 2}{4} = 4(6)$ $\quad 4 \cdot \frac{3x - 2}{4} = 4(-6)$

$3x - 2 = 24$ or $3x - 2 = -24$
$3x = 26$ or $3x = -22$
$x = \frac{26}{3}$ $\quad x = \frac{-22}{3}$
$\{\frac{-22}{3}, \frac{26}{3}\}$

65. $\left|\frac{w + 4}{3}\right| < 4$

$-4 < \frac{w + 4}{3} < 4$

$-4 \cdot 3 < 3 \cdot \frac{w + 4}{3} < 3 \cdot 4$
$-12 < w + 4 < 12$
$-16 < w < 8$

$\{x \mid -16 < w < 8\}$

67. $$\left|\frac{3x - 2}{4}\right| + 5 \ge 5$$

$$\left|\frac{3x - 2}{4}\right| \ge 0$$

The absolute value of a number is always greater than or equal to zero so the solution is all real numbers.

69. $$|2 - 3x| - 4 \ge -2$$

$$|2 - 3x| \ge 2$$

$$2 - 3x \ge 2 \quad \text{or} \quad 2 - 3x \le -2$$
$$-3x \ge 0 \quad \text{or} \quad -3x \le -4$$
$$x \le 0 \quad \text{or} \quad x \ge \frac{4}{3}$$

$$\{x \mid x \le 0 \quad \text{or} \quad x \ge \frac{4}{3}\}.$$

71. $$\left|2\left[\frac{3 - x}{5}\right]\right| < \frac{9}{5}$$

$$\frac{-9}{5} < \frac{2(3 - x)}{5} < \frac{9}{5}$$

$$5 \cdot \frac{-9}{5} < 5 \cdot \frac{2(3 - x)}{5} < 5 \cdot \frac{9}{5}$$

$$-9 < 2(3 - x) < 9$$
$$-9 < 6 - 2x < 9$$
$$-15 < -2x < 3$$

$$\frac{15}{2} > x > \frac{-3}{2}$$

$$\frac{-3}{2} < x < \frac{15}{2}$$

$$\{x \mid \frac{-3}{2} < x < \frac{15}{2}\}$$

73. Given the rule, "If $|x| = a$ and $a > 0$, the $x = a$ or $x = -a$. , we know that either $ax + b = c$ or $ax + b = -c$. Solve each linear equation for $x$.

$$ax+b = c \quad \text{or} \quad ax+b = -c$$
$$ax+b-b= c-b \quad \text{or} \quad ax+b-b= -c-b$$
$$ax= c-b \quad \text{or} \quad ax= -c-b$$

$$\frac{ax}{a} = \frac{c - b}{a} \quad \text{or} \quad \frac{ax}{a} = \frac{-c - b}{a}$$

$$x = \frac{c - b}{a} \quad \text{or} \quad x = \frac{-c - b}{a}$$

The solution set is:

$$\{\frac{-c - b}{a}, \frac{c - b}{a}\}$$

75. Given the rule, "If $|x| > a$ and $a > 0$, then $x < -a$ or $x > a$." we know that $ax + b < -c$ or $ax + b > c$. Solve each linear inequality for $x$.

$$ax+b-b<-c-b \quad \text{or} \quad ax+b-b>c-b$$
$$ax<-c-b \quad \text{or} \quad ax>c-b$$

$$\frac{ax}{a} < \frac{-c - b}{a} \quad \text{or} \quad \frac{ax}{a} > \frac{c - b}{a}$$

$$x < \frac{-c - b}{a} \quad \text{or} \quad x > \frac{c - b}{a}$$

The solution set is:

$$\{x \mid x < \frac{-c - b}{a} \quad \text{or} \quad x > \frac{c - b}{a}\}$$

77. $$\frac{1}{3} + \frac{1}{4} - \frac{2}{5} \cdot \left[\frac{1}{3}\right]^2$$

$$= \frac{1}{3} + \frac{1}{4} - \frac{2}{5} \cdot \left[\frac{1}{9}\right]$$

$$= \frac{1}{3} + \frac{1}{4} \cdot \frac{5}{2} \cdot \frac{1}{9}$$

$$= \frac{1}{3} + \frac{5}{72}$$

$$= \frac{24}{72} + \frac{5}{72} = \frac{29}{72}$$

79. Let $x$ = the time needed to make the trip.
$1.5-x$ = the time needed to make the return trip.

| Trip | Distance | = rate | x time |
|---|---|---|---|
| 1st trip | $2x$ | 2 | $x$ |
| return | $1.6(1.5-x)$ | 1.6 | $(1.5-x)$ |

Assuming the distance traveled on both trips is the same:

$$2x = 1.6(1.5 - x)$$
$$2x = 2.4 - 1.6x$$
$$3.6x = 2.4$$
$$x = .667$$

The time necessary to make the trip is .667 hours. Therefore, the distance across the lake is $2x = 2(.667) = 1.33$ miles.

JUST FOR FUN

1. $|x - 3| = |3 - x|$

The values inside the absolute value bars are either the same or differ only in sign. Therefore,

$x - 3 = 3 - x$ or $x - 3 = -(3 - x)$
$2x - 3 = 3$ or $x - 3 = -3 + x$
$2x = 6$ or $-3 = -3$
$x = 3$

Since $-3$ always equal $-3$, the second equation is an identity. The solution set is all real numbers.

3. $|x + 6| = |2x - 3|$

The values inside the absolute value bars are either the same or differ only in sign. Therefore,

$x + 6 = 2x - 3$ or $x + 6 = -(2x-3)$
$6 = x - 3$ or $x + 6 = -2x+3$
$9 = x$ or $3x + 6 = 3$
$3x = -3$
$x = -1$

$\{-1, 9\}$

REVIEW EXERCISES

1. tenth

3. Since $-4xyz^5 = -4x^1y^1z^5$ and $1 + 1 + 5 = 7$, the expression is a seventh degree term.

5. $x^2 + 2xy + 6x^2 - 4$

$= 7x^2 + 2xy - 4$

7. $2\left[-(x - y) + 3x\right] - 5y + 6$

$= 2\left[-x + y + 3x\right] - 5y + 6$

$= 2(2x + y) - 5y + 6$

$= 4x + 2y - 5y + 6$

$= 4x - 3y + 6$

9. $3(x + 2) - 6 = 4(x - 5)$
$3x + 6 - 6 = 4x - 20$
$3x = 4x - 20$
$-x = -20$
$x = 20$

11. $-6 - 2x = \frac{1}{2}(4x + 12)$

$-6 - 2x = \frac{4}{2}x + \frac{12}{2}$

$-6 - 2x = 2x + 6$
$-6 = 4x + 6$
$-12 = 4x$
$-3 = x$

13. $3x - 4 = 6x + 4 - 3x$
$3x - 4 = 3x + 4$
$-4 = 4$

Since $-4$ never equals 4, the equation is inconsistant and there is no solution.

15. $P = \frac{nRT}{V}$ $\quad n = 10,\ R = 100$, $T = 4$ and $V = 20$

$$P = \frac{10(100)(4)}{20}$$

$$P = 200$$

17. $L = \frac{1}{2} \cdot at^2 + V_o \cdot t + h_o$

$a = -32,\quad V_o = 60$

$h_o = 120,\quad t = 2$

$$L = \frac{1}{2} \cdot (-32)(2)^2 + 60(2) + 120$$

$$L = \frac{1}{2} \cdot (-32)(4) + 60(2) + 120$$

$$L = -16(4) + 120 + 120$$
$$L = 176$$

19. $A = lw$ □ Solve for $l$

$$\frac{A}{w} = \frac{lw}{w}$$

$$\frac{A}{w} = l$$

21. $P = 2L + 2w$ Solve for w.

$$P - 2L = 2L - 2L + 2w$$
$$P - 2L = 2w$$

$$\frac{P - 2L}{2} = \frac{2w}{2}$$

$$\frac{P - 2L}{2} = w$$

23. $y = mx + b$ Solve for m.

$$y - b = mx + b - b$$
$$y - b = mx$$

$$\frac{y - b}{x} = \frac{mx}{x}$$

$$\frac{y - b}{x} = m$$

25. $P_1 \cdot V_1 = P_2 \cdot V_2$ Solve for $V_2$

$$\frac{P_1 \cdot V_1}{p_2} = \frac{P_2 \cdot V_2}{P_2}$$

$$\frac{P_1 \cdot V_1}{p_2} = V_2$$

27. $k = 2(d + L)$ Solve for L.

$$k = 2d + 2L$$
$$k - 2d = 2d - 2d + 2L$$
$$k - 2d = 2L$$

$$\frac{k - 2d}{2} = \frac{2L}{2}$$

$$\frac{k - 2d}{2} = L$$

29. $A = \frac{1}{2} \cdot h \cdot \left[b_1 + b_2\right]$

Solve for $b_1$

$$2 \cdot A = 2 \cdot \frac{1}{2} \cdot h \cdot \left[b_1 + b_2\right]$$

$$2A = h\left[b_1 + b_2\right]$$

$$2A = hb_1 + hb_2$$

$$2A - hb_2 = hb_1 + hb_2 - hb_2$$

$$2A - hb_2 = hb_1$$

$$\frac{2A - hb_2}{h} = \frac{hb_1}{h}$$

$$\frac{2A - hb_2}{h} = b_1$$

31. Let x = the unknown number.

$$x + 4x = 80$$
$$5x = 80$$
$$x = 16$$

33. Let x = the unknown number

$4x + 12 = 32$

$4x = 20$

$x = 5$

35. Let x = the unknown number

$1x - .6x = 20$

$.4x = 20$

$x = 50$

37. Let x = the unknown number

$1x - .1x = 180$

$.9x = 180$

$x = 200$

39. Let r = the hourly inspection rate.

Amount = rate x time

$245 = r(8)$

$30.6 = r$

Tanya's inspection rate is 30.6 rolls per hour.

41. Let x = the time needed for the trains to travel a combined distance of 400 miles.

| Train | Distance | = | rate | x | time |
|---|---|---|---|---|---|
| #1 | 60x | | 60 | | x |
| #2 | 90x | | 90 | | x |

The combined distance coved by the two trains is 400 miles. Therefor,

$60x + 90x = 400$

$150x = 400$

$x = 2\ 2/3$

The time needed for the trains to cover 400 miles is 2 2/3 hours.

43. Let x = the amount of $6 per pound coffee.

40 - x = the amount of $6.80 per pound coffee.

| Coffee | Cost Per Lbs | x | Number of Lbs. | = | Total Value |
|---|---|---|---|---|---|
| #1 | 6 | | x | | 6x |
| #2 | 6.80 | | (40-x) | | 6.80(40-x) |
| Mixture | 6.50 | | 40 | | 6.50(40) |

$6x + 6.80(40-x) = 6.50(40)$

$6x + 272 - 6.80x = 260$

$-.8x + 272 = 260$

$-.8x = -12$

$x = 15$

$40 - x = 25$

The mixture should contain 15 pounds of the $6 per pound coffee and 25 pounds of the $6.80 per pound coffee.

45. a) Let x = the amount of time jogging.

4 - x = the amount of time walking.

| | Distance | = | rate | x | time |
|---|---|---|---|---|---|
| Jogging | 7.2x | | 7.2 | | x |
| Walking | 2.4(4-x) | | 2.4 | | (4-x) |

Assuming the distance walked equals the distance jogged,

$7.2x = 2.4(4-x)$

$7.2x = 9.6 - 2.4x$

$9.6x = 9.6$

$x = 1$

$4 - x = 4 - 1 = 3$

Nicolle jogged 1 hour and walked 3 hours.

b) The total distance she traveled is

$7.2x + 2.4(4-x)$

$= 7.2(1) + 2.4(4 - 1)$

$= 7.2 + 9.6 - 2.4 = 14.4$

Nicolle traveled a total distance of 14.4 miles.

47. Let x = the rate of the smaller hose.

1.5x = the rate of the larger hose.

| Hose | Amount | = | Rate | x | Time |
|---|---|---|---|---|---|
| Larger | 1.5x(5) | | 1.5x | | 5 |
| Smaller | 3x | | x | | 3 |

The amount the larger hose pumps plus the amount the smaller hose pumps is 3150 gallons. Therefore,

$1.5x(5) + 3x = 3150$

$7.5x + 3x = 3150$

$10.5x = 3150$

$x = 300$

$1.5x = 450$

The rate of the smaller hose is 300 gallons per minute and the rate of the larger is 450 gallons per minute.

49. Let $x$ = the amount of 20% solution.

| Strength of dye | x Number of oz. | = Amount of dye |
|---|---|---|
| 6% | 10 | .06(10) |
| 20% | x | .20x |
| Mixture 12% | 10 + x | .12(10+x) |

$$
\begin{aligned}
.06(10) + .20x &= .12(10 + x) \\
.6 + .20x &= 1.2 + .12x \\
.08x + .6 &= 1.2 \\
.08x &= .6 \\
x &= 7.5
\end{aligned}
$$

7.5 ounces of 20% dye needs to be mixed with the 6% solution.

51. Let $x$ = the number of visits.

$$
\begin{aligned}
40 + 1(x) &= 25 + 4(x) \\
40 + x &= 25 + 4x \\
40 &= 25 + 3x \\
15 &= 3x \\
5 &= x
\end{aligned}
$$

Mike would have to make more than 5 visits for the first plan to be more advantageous.

53. $x - 3 \geq 4$

$x \geq 7$

<------●━━━━━━━>
7

55. $2x + 4 > 9$

$2x > 5$

$x > \frac{5}{2}$

<------○━━━━━━━>
$\frac{5}{2}$

57. $\frac{4x + 3}{5} > -3$

$5 \cdot \frac{4x + 3}{5} > 5 \cdot (-3)$

$4x + 3 > -15$

$4x > -18$

$x > -18/4$

$x > -9/2$

<------○━━━━━━━>
$\frac{-9}{2}$

59. $-4(x - 2) \leq 6x + 4$

$-4x + 8 \leq 6x + 4$

$8 \leq 10x + 4$

$4 \leq 10x$

$\frac{4}{10} \leq \frac{10x}{10}$

$\frac{2}{5} \leq x$ or $x \geq \frac{2}{5}$

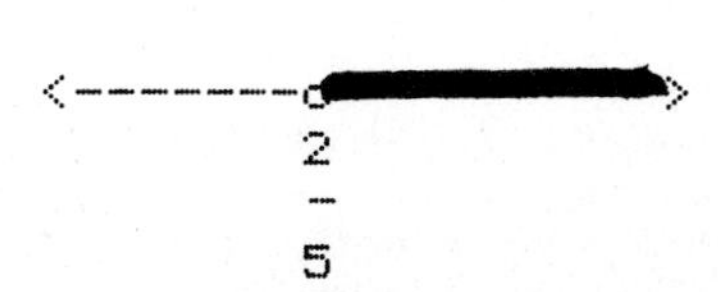

61. Let $x$ = the number of boxes. Since the maximum load is 1525 pounds, the total weight of the passengers and boxes must be less than or equal to 1525 pounds.

$$
\begin{aligned}
468 + 80x &\leq 1525 \\
80x &\leq 1057 \\
x &\leq 13.2125
\end{aligned}
$$

The maximum number of boxes the plane could accomidate is 13.

63. Let $x$ = the number of weeks needed to lose 27 pounds.

$3 + \frac{3}{2} \cdot x \geq 27$

$2 \cdot 3 + 2 \cdot \frac{3}{2} \cdot x \geq 2 \cdot 27$

$$
\begin{aligned}
6 + 3x &\geq 54 \\
3x &\geq 48 \\
x &\geq 16
\end{aligned}
$$

The amount of time needed is 16 weeks plus the first initial week for a total of 17 weeks.

65. $2 \leq x + 5 < 8$

$-3 \leq x < 3$

$[-3,3)$

67. $-12 < 6 - 3x < -2$

$-18 < -3x < -8$

$\frac{-18}{-3} > \frac{-3x}{-3} > \frac{-8}{-3}$

$6 > x > \frac{8}{3}$ or $\frac{8}{3} < x < 6$

$(-8/3,6)$

69.
$$-8 < \frac{4 - 2x}{3} < 0$$

$$3\cdot(-8) < 3\cdot\frac{4 - 2x}{3} < 3\cdot 0$$

$$-24 < 4 - 2x < 0$$
$$-28 < -2x < -4$$

$$\frac{-28}{-2} > \frac{-2x}{-2} > \frac{-4}{-2}$$

$14 > x > 2$ or $2 < x < 14$
$(2,14)$

71. $x < 3$ and $2x - 4 > -10$
$x < 3$ and $2x > -6$
$x < 3$ and $x > -3$

$-3 \quad 3$

The solution set is found where the shaded regions overlap.
$\{x \mid -3 < x < 3\}$

73. $3x + 5 > 2$ or $6 - x < 1$
$3x > -3$ or $-x < -5$
$x > -1$ or $x > 5$

$-1 \quad 5$

The solution includes all the shaded regions.

$\{x \mid x > -1\}$

75. $4x - 5 < 11$ and $-3x - 4 \geq 8$
$4x < 16$ and $-3x \geq 12$
$x < 4$ and $x \leq -4$

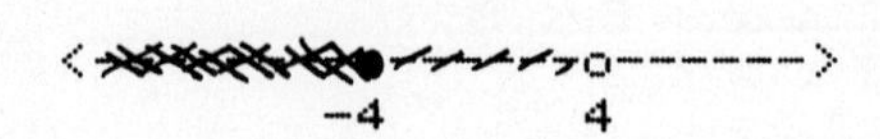


The solution set is found where the shaded regions overlap.
$\{x \mid x \leq -4\}$

77. $|x| = 4$
$x = 4$ or $x = -4$
$\{-4, 4\}$

79. $|x| \geq 4$

$x \geq 4$ or $x \leq -4$
$\{x \mid x \geq 4$ or $x \leq -4\}$

81. $|x - 2| \geq 5$
$x - 2 \leq -5$ or $x - 2 \geq 5$
$x \leq -3$ or $x \geq 7$

$\{x \mid x \leq -3$ or $x \geq 7\}$

83. $|3 - 2x| < 7$
$-7 < 3 - 2x < 7$
$-10 < -2x < 4$

$$\frac{-10}{-2} > \frac{-2x}{-2} > \frac{4}{-2}$$

$5 > x > -2$ or $-2 < x < 5$

$\{x \mid -2 < x < 5\}$

85.
$$\left|\frac{x - 4}{3}\right| < 6$$

$$-6 < \frac{x - 4}{3} < 6$$

$$3\cdot(-6) < 3\cdot\frac{x - 4}{3} < 3\cdot 6$$

$$-18 < x - 4 < 18$$
$$-14 < x < 22$$

$\{x \mid -14 < x < 22\}$

87. $|4(2 - x)| > 5$

$4(2 - x) < -5$ or $4(2 - x) > 5$
$8 - 4x < -5$ or $8 - 4x > 5$
$-4x < -13$ or $-4x > -3$

$$x > \frac{13}{4} \quad \text{or} \quad x < \frac{3}{4}$$

$$\{x \mid x > \frac{13}{4} \quad \text{or} \quad x < \frac{3}{4}\}$$

89.
$$\frac{5x - 3}{2} > 6$$

$$2\cdot\frac{5x - 3}{2} > 2\cdot 6$$

$$5x - 3 > 12$$
$$5x > 15$$
$$x > 3$$
$(3, \infty)$

91. $|x + 6| < -1$
Since the absolute value of a number is always greater than or equal to 0, there does not exist an x such that $|x + 6|$ will be less than −1.

93.
$$-6 \le \frac{3 - 2x}{4} < 5$$
$$4 \cdot (-6) \le 4 \cdot \frac{3 - 2x}{4} < 4 \cdot 5$$
$$-24 \le 3 - 2x < 20$$
$$-27 \le -2x < 17$$
$$\frac{-27}{-2} \ge \frac{-2x}{-2} > \frac{17}{-2}$$
$$\frac{27}{2} \ge x > \frac{-17}{2}$$
or
$$\frac{-17}{2} < x \le \frac{27}{2}$$
$(-17/2, 27/2]$

95.
$$|3(x + 2)| \le \frac{9}{2}$$
$$\frac{-9}{2} \le 3(x + 2) \le \frac{9}{2}$$
$$2 \cdot \frac{-9}{2} \le 2 \cdot 3(x + 2) \le 2 \cdot \frac{9}{2}$$
$$-9 \le 6x + 12 \le 9$$
$$-21 \le 6x \le -3$$
$$\frac{-21}{6} \le \frac{6x}{6} \le \frac{-3}{6}$$
$$\frac{-7}{2} \le x \le \frac{-1}{2}$$
$[-7/2, -1/2]$

97.
$$|x + 3| - 2 < 7$$
$$|x + 3| < 9$$
$$-9 < x + 3 < 9$$
$$-12 < x < 6$$
$(-12, 6)$

99.
$$\left|\frac{x - 4}{2}\right| - 3 > 5$$
$$\left|\frac{x - 4}{2}\right| > 8$$
$$\frac{x - 4}{2} < -8 \quad \text{or} \quad \frac{x - 4}{2} > 8$$
$$2 \cdot \frac{x - 4}{2} < 2 \cdot (-8) \qquad 2 \cdot \frac{x - 4}{2} > 2 \cdot 8$$
$$x - 4 < -16 \quad \text{or} \quad x - 4 > 16$$
$$x < -12 \quad \text{or} \quad x > 20$$
$(-\infty, -12) \cup (20, \infty)$

## PRACTICE TEST

1. Since $-6xy^2 \cdot z^3 = -6x^1 \cdot y^2 \cdot z^3$

   and $1 + 2 + 3 = 6$, $-6xy^2 \cdot z^3$ is a sixth degree term.

2. $$3(x - 2) = 4(4 - x) + 5$$
   $$3x - 6 = 16 - 4x + 5$$
   $$3x - 6 = 21 - 4x$$
   $$7x - 6 = 21$$
   $$7x = 27$$
   $$x = \frac{27}{7}$$

3. $$\frac{3}{5} - \frac{x}{2} = 4$$

   Multiply each term by the Least Common Multiple of 5 and 2 which is 10.

   $$10 \cdot \frac{3}{5} - 10 \cdot \frac{x}{2} = 10 \cdot 4$$
   $$6 - 5x = 40$$
   $$-5x = 34$$
   $$x = \frac{-34}{5}$$

4. $$\frac{3x}{4} - 1 = 5 + \frac{2x - 1}{3}$$

   Multiply each term by the Least Common Multiple of 4 and 3 which is 12.

   $$12 \cdot \frac{3x}{4} - 12 \cdot 1 = 12 \cdot 5 + 12 \cdot \frac{2x - 1}{3}$$
   $$9x - 12 = 60 + 4(2x - 1)$$
   $$9x - 12 = 60 + 8x - 4$$
   $$9x - 12 = 56 + 8x$$
   $$x - 12 = 56$$
   $$x = 68$$

5. $$S_n = \frac{\left[a_1\right]\left[1 - r^n\right]}{1 - r}$$

   $a_1 = 3$ $r = 1/3$, $n = 3$


5. $$S_n = \frac{\left[a_1\right]\left[1 - r^n\right]}{1 - r}$$

   $a_1 = 3$ $r = 1/3$, $n = 3$


$$S_3 = \frac{(3)\left[1 - \left[\frac{1}{3}\right]^3\right]}{1 - \frac{1}{3}}$$

$$= \frac{3\left[1 - \frac{1}{27}\right]}{\frac{2}{3}}$$

$$= \frac{3\left[\frac{26}{27}\right]}{\frac{2}{3}}$$

$$= \frac{\frac{26}{9}}{\frac{2}{3}}$$

$$= \frac{26}{9} \cdot \frac{3}{2}$$

$$= \frac{13}{3}$$

6. $C = \dfrac{a - 3b}{2}$ Solve for b.

$$C \cdot 2 = 2 \cdot \frac{a - 3b}{2}$$

$$2C = a - 3b$$
$$2C - a = a - a - 3b$$
$$2C - a = -3b$$

$$\frac{2C - a}{-3} = \frac{-3b}{-3}$$

$$\frac{a - 2C}{3} = b$$

7. $A = \dfrac{1}{2} \cdot h \left[b_1 + b_2\right]$

Solve for $b_2$

$$2 \cdot A = 2 \cdot \frac{1}{2} \cdot h \left[b_1 + b_2\right]$$

$$2 \cdot A = h \left[b_1 + b_2\right]$$

$$2 \cdot A = hb_1 + hb_2$$

$$2 \cdot A - hb_1 = hb_1 - hb_1 + hb_2$$

$$2 \cdot A - hb_1 = hb_2$$

$$\frac{2 \cdot A - hb_1}{h} = \frac{hb_2}{h}$$

$$\frac{2 \cdot A - hb_1}{h} = b_2$$

8. Let x = the first integer.
x + 1 = the next consecutive integer.

$$x + (x + 1) = 47$$
$$2x + 1 = 47$$
$$2x = 46$$
$$x = 23$$
$$x + 1 = 24$$

The two consecutive integers are 23 and 24.

9. Let x = the number of miles.

$$35 + .15x = 65$$
$$.15x = 30$$
$$x = 200$$

Valerie can drive 200 miles on $65.

10. Let x = the distance between the joggers.

| Jogger | Distance | = rate | x time |
|---|---|---|---|
| Homer | 4(1.25) | 4 | 1.25 |
| Frances | 5.25(1.25) | 5.25 | 1.25 |

The distance between the joggers is the distance Homer runs plus the distance Frances runs. Therefore,

$$x = 4(1.25) + 5.25(1.25)$$
$$x = 5 + 6.56$$
$$x = 11.56$$

The runners will be 11.56 miles apart in 1 1/4 hours.

11. Let x = the liters of 12% salt solution.

| Strength of solution | x Number of liters | = Amount of solution |
|---|---|---|
| 12% | x | .12x |
| 25% | 10 | .25(10) |
| Mixture 20% | x + 10 | .2(x+10) |

$$.12x + .25(10) = .2(x + 10)$$
$$.12x + 2.5 = .2x + 2$$
$$2.5 = .08x + 2$$
$$.5 = .08x$$
$$6.25 = x$$

6.25 liters of 12% salt solution must be added to the 25% salt solution.

12. Let x = the amount invested at 8%.
12000 − x = the amount invested at 7%.

| Amount Invested | x Rate | = Interest |
|---|---|---|
| x | .08 | .08x |
| 12000 − x | .07 | .07(12000−x) |

The total interest earned off the two accounts is $910. Therefore

$$.08x + .07(12000-x) = 910$$
$$.08x + 840 - .07x = 910$$
$$.01x + 840 = 910$$
$$.01x = 70$$
$$x = 7000$$
$$12000 - x = 5000$$

$7000 was invested at 8% and $5000 was invested at 7%.

13. $\dfrac{6 - 2x}{5} \ge -12$

$5 \cdot \dfrac{6 - 2x}{5} \ge 5 \cdot (-12)$

$6 - 2x \ge -60$

$-2x \ge -66$

$\dfrac{-2x}{-2} \le \dfrac{-66}{-2}$

$x \le 33$

14. $-4 < \dfrac{x + 4}{2} < 8$

$2 \cdot (-4) < 2 \cdot \dfrac{x + 4}{2} < 2 \cdot 8$

$-8 < x + 4 < 16$

$-12 < x < 12$

$(-12, 12)$

15. $|x - 4| = 5$

$x - 4 = -5$ or $x - 4 = 5$

$x = -1$ or $x = 9$

$\{-1, 9\}$

16. $|2x - 3| + 1 > 6$

$|2x - 3| > 5$

$2x - 3 < -5$ or $2x - 3 > 5$

$2x < -2$ or $2x > 8$

$x < -1$ or $x > 4$

$\{x | x < -1 \text{ or } x > 4\}$

17. $\left|\dfrac{2x - 3}{4}\right| \le \dfrac{1}{2}$

$\dfrac{-1}{2} \le \dfrac{2x - 3}{4} \le \dfrac{1}{2}$

$4 \cdot \dfrac{-1}{2} \le 4 \cdot \dfrac{2x - 3}{4} \le 4 \cdot \dfrac{1}{2}$

$-2 \le 2x - 3 \le 2$

$1 \le 2x \le 5$

$\dfrac{1}{2} \le \dfrac{2x}{2} \le \dfrac{5}{2}$

$\dfrac{1}{2} \le x \le \dfrac{5}{2}$

$\{x | \dfrac{1}{2} \le x \le \dfrac{5}{2}\}$

CUMULATIVE REVIEW TEST

1. $A \cup B = \{1,2,3,4,5,6,7,9,10,12\}$
   $A \cap B = \{4,6,9,12\}$

2. a) Commutative Property of Addition
   b) Associative Property of Multiplication
   c) Distributive Property

3. $-|-3| = -(3) = -3$ and $|-5| = 5$
   Therefore $-|-3| < |-5|$ since $-3 < 5$.

4. $4 - |-3| - (6 + |-3|)^2$

$= 4 - |-3| - (6 + 3)^2$

4. cont.

$= 4 - |-3| - 9^2$

$= 4 - 3 - 81$

$= 1 - 81 = -80$

5. $-4^2 + (-3)^2 - 2^3 + (-2)^0$

$= -16 + 9 - 8 + 1$

$= -7 - 8 + 1$

$= -15 + 1$

$= -14$

6. Substitute − 3 for each x and −2 for each y in the expression.

$$x^3 - xy^2 + y^3 = (-3)^3 - (-3)(-2)^2 + (-2)^3$$
$$= -27 - 6 + 4$$
$$= -33 + 4$$
$$= -29$$

7.

$$\frac{8 - \sqrt[3]{27} \cdot 3 \div 9}{|-5| - (5 - (12 \div 4))^2}$$

$$= \frac{8 - (3) \cdot 3 \div 9}{5 - (5 - 3)^2}$$

$$= \frac{8 - 9 \div 9}{5 - 2^2}$$

$$= \frac{8 - 1}{5 - 4}$$

$$= \frac{7}{1} = 7$$

8.

$$3x - 4 = -2(x - 3) - 9$$
$$3x - 4 = -2x + 6 - 9$$
$$3x - 4 = -2x - 3$$
$$5x - 4 = -3$$
$$5x = 1$$
$$x = \frac{1}{5}$$

9.

$$1.2(x - 3) = 2.4x - 4.98$$
$$1.2x - 3.6 = 2.4x - 4.98$$
$$-3.6 = 1.2x - 4.98$$
$$1.38 = 1.2x$$
$$1.15 = x$$

10.

$$\frac{x}{4} - 5 = 3x - \frac{1}{3}$$

Multiply each term by the least common multiple of 4 and 3 which is 12.

$$12 \cdot \frac{x}{4} - 12 \cdot 5 = 12 \cdot 3x - 12 \cdot \frac{1}{3}$$

$$3x - 60 = 36x - 4$$
$$-60 = 33x - 4$$
$$-56 = 33x$$
$$\frac{-56}{33} = x$$

11.

$$\frac{\frac{1}{4} \cdot x + 2}{3} = \frac{x - 4}{4}$$

Cross multiply.

$$4 \cdot \left[\frac{1}{4} \cdot x + 2\right] = 3 \cdot (x - 4)$$

$$x + 8 = 3x - 12$$
$$8 = 2x - 12$$
$$20 = 2x$$
$$10 = x$$

12. A conditional linear equation has only one solution.
Example:

$$4x + 7 = 9$$
$$4x = 2$$
$$x = 1/2$$

An identity is an equation with an infinite number of solution.
Example:

$$4x + 12 = 4(x + 3)$$
$$4x + 12 = 4x + 12$$
$$12 = 12$$

Since 12 always = 12, the solution set is all real numbers.

An inconsistent linear equation has no solution.
Example:

$$3x + 6 = 3(x + 7)$$
$$3x + 6 = 3x + 21$$
$$6 = 21$$

Since 6 never equals 21, here is no value of x such that $3x + 6 = 3(x + 7)$. The solution is the empty set.

13. Substitue 3 for a, −8 for b and −3 for c in the expression.

$$x = \frac{-b + \sqrt{b^2 - 4ac}}{2a}$$

$$= \frac{-(-8) + \sqrt{(-8)^2 - 4(3)(-3)}}{2(3)}$$

$$= \frac{8 + \sqrt{100}}{6} = \frac{8 + 10}{6} = 3$$

14. $I = p + prt$ Solve for t.

$$I - p = p - p + prt$$
$$I - p = prt$$

$$\frac{I - p}{pr} = \frac{prt}{pr}$$

$$\frac{I - p}{pr} = t$$

15.
$$-4 < \frac{5x - 2}{3} < 2$$

$$3 \cdot (-4) < 3 \cdot \frac{5x - 2}{3} < 3 \cdot 2$$

$$-12 < 5x - 2 < 6$$
$$-10 < 5x < 8$$

$$\frac{-10}{5} < \frac{5x}{5} < \frac{8}{5}$$

$$-2 < x < \frac{8}{5}$$

a) <--------o●●●●●●o-------> with open circles at $-2$ and $\frac{8}{5}$

b) $\{x \mid -2 < x < 8/5\}$

c) $(-2, 8/5)$

16. $|4z + 8| = 12$

$4z + 8 = -12$ or $4z + 8 = 12$
$4z = -20$ or $4z = 4$
$z = -5$ $z = 1$

$\{-5, 1\}$

17. $|2x - 4| - 6 \geq 18$

$|2x - 4| \geq 24$

$2x - 4 \leq -24$ or $2x - 4 \geq 24$
$2x \leq -20$ or $2x \geq 28$
$x \leq -10$ or $x \geq 14$

$\{x \mid x \leq -10$ or $x \geq 14\}$

18. Let x = the original price.

$$x - .2x = 1800$$
$$.8x = 1800$$
$$x = 2250$$

The original price was $2250.

19. Let x = the speed of the car traveling east.
x + 10 = the speed of the car moving west.

| Car | Distance | = | Rate | x Time |
|---|---|---|---|---|
| East | 3x | | x | 3 |
| West | 3(x+10) | | x + 10 | 3 |

The combined distance the cars traveled is 270 miles. Therefore,

$$3x + 3(x + 10) = 270$$
$$3x + 3x + 30 = 270$$
$$6x + 30 = 270$$
$$6x = 240$$
$$x = 40$$
$$x + 10 = 50$$

The speed of the east bound car is 40 mph and the speed of the west bound car is 50 mph.

20. Let x = the number of liters of 20% salt solution.
2 − x = the liters of 50% salt solution.

| Strength | x Number of | = Amount of |
|---|---|---|
| 20% | X | .2x |
| 50% | (2−x) | .5(2−x) |
| Mixture 30% | 2 | .3(2) |

$$.2x + .5(2-x) = .3(2)$$
$$.2x + 1 - .5x = .6$$
$$1 - .3x = .6$$
$$-.3x = -.4$$
$$x = 1.33 \text{ or } 1\ 1/3$$
$$2 - x = .67 \text{ or } 2/3$$

1 1/3 liters of the 20% solution should be mixed with 2/3 of the 50% salt solution.

CHAPTER 3

SECTION 3.1

1. A(3,1), B(-6,0), C(2,-4), D(-2,-4) E(0,3), F(-8,1), G(3/2,-1)

3.

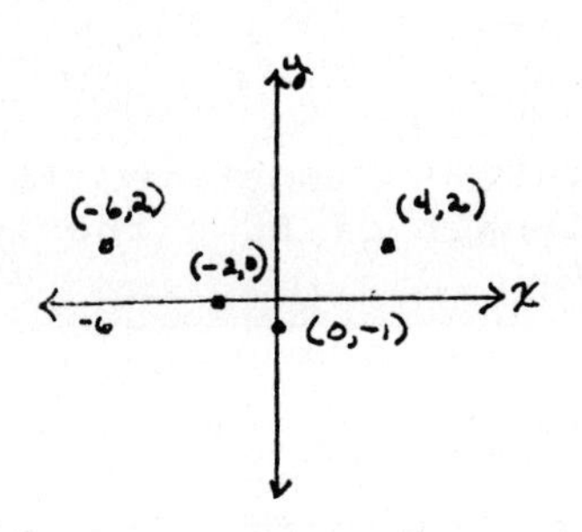


5.

$$d = \sqrt{(2 - 2)^2 + (-2 - (-5))^2}$$

$$d = \sqrt{(0)^2 + (3)^2}$$

$$d = \sqrt{0 + 9}$$

$$d = \sqrt{9} = 3$$

7.

$$d = \sqrt{(-4 - 5)^2 + (3 - 3)^2}$$

$$d = \sqrt{(-9)^2 + (0)^2}$$

$$d = \sqrt{81} = 9$$

9.

$$d = \sqrt{(1 - (-3))^2 + (4 - 1)^2}$$

$$d = \sqrt{(4)^2 + (3)^2}$$

$$d = \sqrt{16 + 9}$$

$$d = \sqrt{25} = 5$$

11.

$$d = \sqrt{(-3 - 6)^2 + (-5 - (-2))^2}$$

$$d = \sqrt{(-9)^2 + (-3)^2}$$

$$d = \sqrt{81 + (9)}$$

$$d = \sqrt{90} \approx 9.49$$

13.

$$d = \sqrt{(0 - 5)^2 + (6 - (-1))^2}$$

$$d = \sqrt{(-5)^2 + (7)^2}$$

$$d = \sqrt{25 + 49}$$

$$d = \sqrt{74} \approx 8.60$$

15.

$$d = \sqrt{(-1.6 - (-4.3))^2 + (3.5 - (-1.7))^2}$$

$$d = \sqrt{2.7^2 + 5.2^2}$$

$$d = \sqrt{7.29 + 27.04}$$

$$d = \sqrt{34.33} \approx 5.86$$

17.

$$d = \sqrt{\left[\frac{3}{4} - \left[\frac{-1}{2}\right]\right]^2 + (2 - 6)^2}$$

$$d = \sqrt{\left[\frac{5}{4}\right]^2 + (-4)^2}$$

$$d = \sqrt{\frac{25}{16} + \frac{256}{16}}$$

$$d = \sqrt{\frac{281}{16}} \approx 4.19$$

19.
Midpoint $= \left[\dfrac{5 + (-1)}{2}, \dfrac{2 + 4}{2}\right]$

$= \left[\dfrac{4}{2}, \dfrac{6}{2}\right]$

$= (2,3)$

21.
Midpoint $= \left[\dfrac{-5 + 5}{2}, \dfrac{3 + (-3)}{2}\right]$

$= (0,0)$

23.
Midpoint $= \left[\dfrac{-2 + (-6)}{2}, \dfrac{-8 + (-2)}{2}\right]$

$= \left[\dfrac{-8}{2}, \dfrac{-10}{2}\right]$

$= (-4,-5)$

25.
Midpoint $= \left[\dfrac{1 + (-8)}{2}, \dfrac{-6 + (-4)}{2}\right]$

$= \left[\dfrac{-7}{2}, \dfrac{-10}{2}\right]$

$= \left[\dfrac{-7}{2}, -5\right]$

27.
Midpoint $= \left[\dfrac{-9.62 + 3.52}{2}, \dfrac{12.58 + 6.57}{2}\right]$

$= \left[\dfrac{-6.1}{2}, \dfrac{19.15}{2}\right]$

$= (-3.05, 9.575)$

29.
Midpoint $= \left[\dfrac{\frac{5}{2} + 2}{2}, \dfrac{3 + \frac{9}{2}}{2}\right]$

$= \left[\dfrac{\frac{9}{2}}{2}, \dfrac{\frac{15}{2}}{2}\right]$

29. cont.

$= \left[\left[\dfrac{9}{2}\cdot\dfrac{1}{2}\right], \left[\dfrac{15}{2}\cdot\dfrac{1}{2}\right]\right]$

$= \left[\dfrac{9}{4}, \dfrac{15}{4}\right]$

31. Perimeter = length of AB + the length of BC + the length of CA.

$= \sqrt{(7 - 7)^2 + (7 - 1)^2}$

$+ \sqrt{(7 - (-1))^2 + (1 - 1)^2}$

$+ \sqrt{(-1 - 7)^2 + (1 - 7)^2}$

$= \sqrt{36} + \sqrt{64} + \sqrt{100}$

$= 6 + 8 + 10 = 24$

33. Perimeter = length of AB + the length of BC + the length of CA

$= \sqrt{(0 - 2)^2 + (-4 - 3)^2}$

$+ \sqrt{(2 - 4)^2 + (3 - 6)^2}$

$+ \sqrt{(4 - 0)^2 + (6 - (-4))^2}$

$= \sqrt{53} + \sqrt{13} + \sqrt{116} \approx 21.66$

35. The square of any non-zero number is positive and the square root of a positive number is positive.

37. Let $x$ = the number of miles traveled.

$30 + .14x = 16 + .24x$
$30 = 16 + .1x$
$14 = .1x$
$140 = x$

You would have to drive 140 miles for the cost of renting from Hertz to equal the cost of renting from National.

39. $|3x + 2| > 5$

$$3x + 2 < -5 \text{ or } 3x + 2 > 5$$
$$3x < -7 \text{ or } 3x > 3$$
$$x < -7/3 \text{ or } x > 1$$

$\{x \mid x < -7/3 \text{ or } x > 1\}$

## EXERCISE SET 3.2

1. $y = 4$

| x | y |
|---|---|
| 0 | 4 |
| 1 | 4 |
| 2 | 4 |

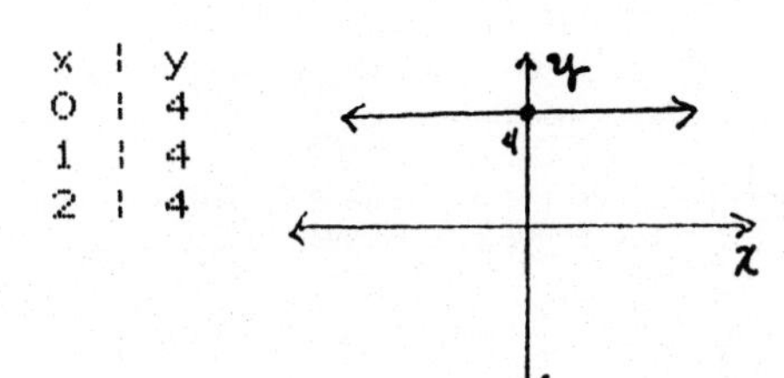


3. $x = -2$

| x | y |
|---|---|
| -2 | 0 |
| -2 | 1 |
| -2 | 2 |

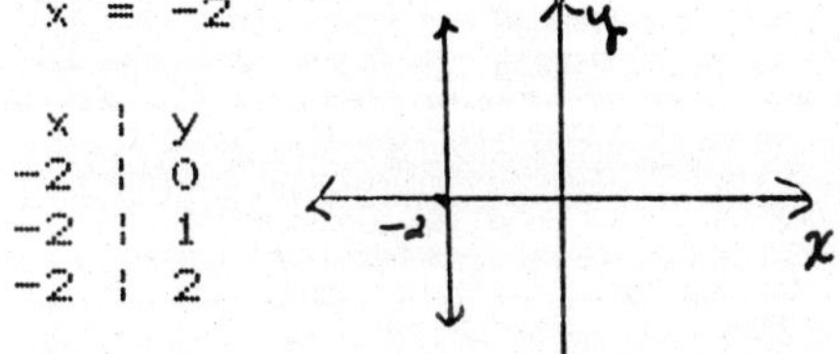


5. $y = 4x - 2$

| x value | y value |
|---|---|
| 0 | $y = 4(0) - 2 = -2$ |
| 1 | $y = 4(1) - 2 = 2$ |
| 2 | $y = 4(2) - 2 = 6$ |

| x | y |
|---|---|
| 0 | -2 |
| 1 | 2 |
| 2 | 6 |

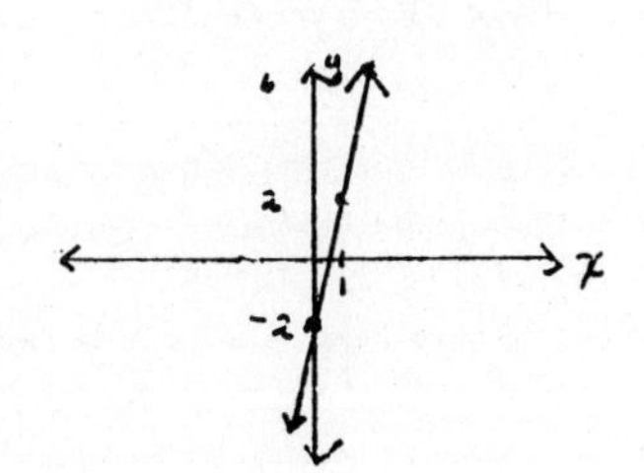


7. $y = x + 2$

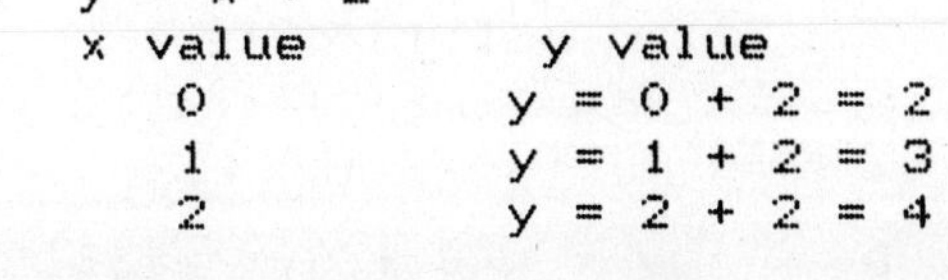

| x value | y value |
|---|---|
| 0 | $y = 0 + 2 = 2$ |
| 1 | $y = 1 + 2 = 3$ |
| 2 | $y = 2 + 2 = 4$ |

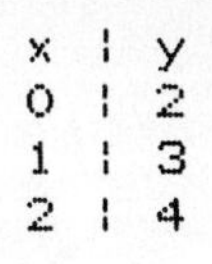

| x | y |
|---|---|
| 0 | 2 |
| 1 | 3 |
| 2 | 4 |

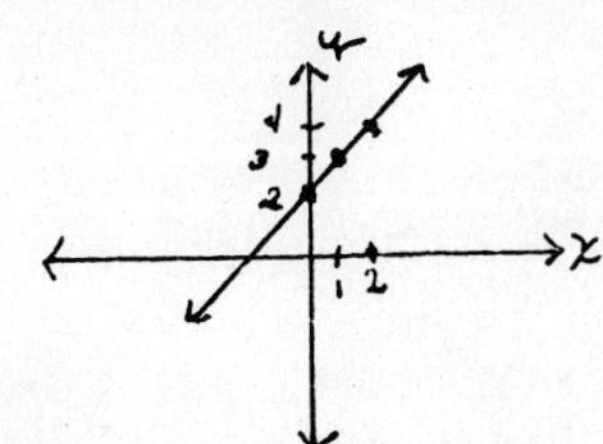


9. $y = \frac{-1}{2} \cdot x + 5$

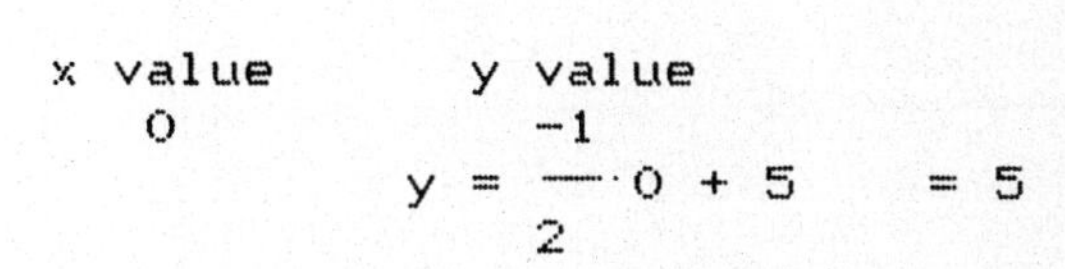

| x value | y value |
|---|---|
| 0 | $y = \frac{-1}{2} \cdot 0 + 5 = 5$ |
| 2 | $y = \frac{-1}{2} \cdot 2 + 5 = 4$ |
| 4 | $y = \frac{-1}{2} \cdot 4 + 5 = 3$ |

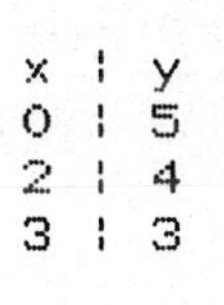

| x | y |
|---|---|
| 0 | 5 |
| 2 | 4 |
| 3 | 3 |

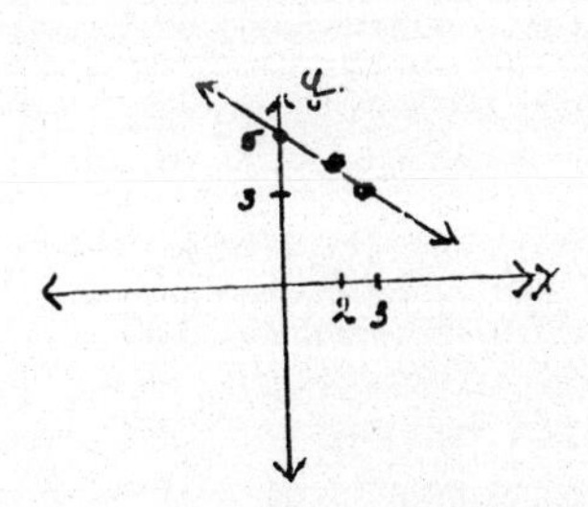


11. $6x - 2y = 4$

$$-2y = -6x + 4$$
$$y = 3x - 2$$

| x value | y value |
|---|---|
| 0 | $y = 3(0) - 2 = -2$ |
| 1 | $y = 3(1) - 2 = 1$ |
| 2 | $y = 3(2) - 2 = 4$ |

| x | y |
|---|---|
| 0 | -2 |
| 1 | 1 |
| 2 | 4 |

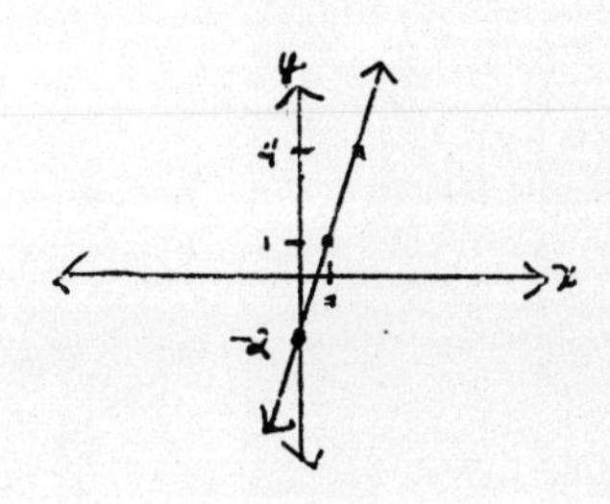

13. $5x - 2y = 8$
$-2y = -5x + 8$

$$y = \frac{5}{2}x - 4$$

| x value | y value |
|---|---|
| 0 | $y = \frac{5}{2}0 - 4 = -4$ |
| 2 | $y = \frac{5}{2}2 - 4 = 1$ |
| 4 | $y = \frac{5}{2}4 - 4 = 6$ |

| x | y |
|---|---|
| 0 | -4 |
| 2 | 1 |
| 4 | 6 |

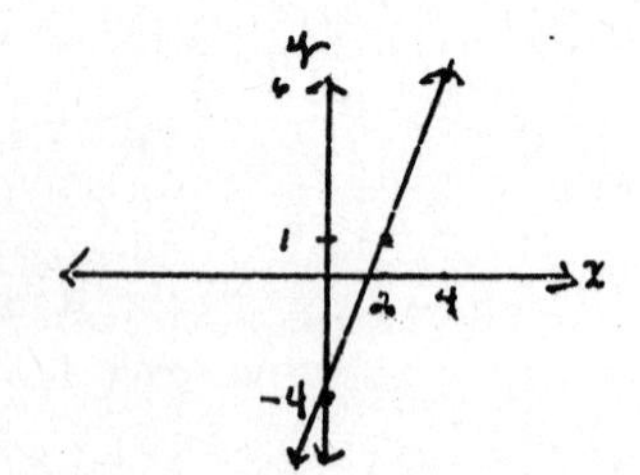

15. $6x + 5y = 30$
$5y = -6x + 30$

$$y = \frac{-6}{5}x + 6$$

| x value | y value |
|---|---|
| -5 | $y = \frac{-6}{5}(-5) + 6 = 12$ |
| 0 | $y = \frac{-6}{5}(0) + 6 = 6$ |
| 5 | $y = \frac{-6}{5}(5) + 6 = 0$ |

| x | y |
|---|---|
| -5 | 12 |
| 0 | 6 |
| 5 | 0 |

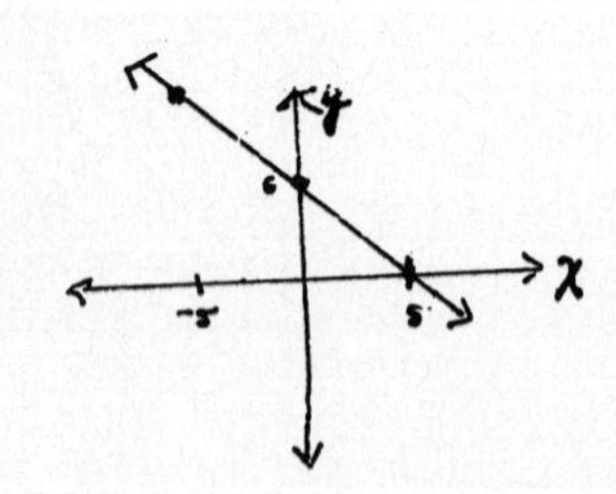

17. $-6x - y = -7$
$-y = 6x - 7$
$y = -6x + 7$

| x value | y value |
|---|---|
| 0 | $y = -6(0) + 7 = 7$ |
| 1 | $y = -6(1) + 7 = 1$ |
| 2 | $y = -6(2) + 7 = -5$ |

| x | y |
|---|---|
| 0 | 7 |
| 1 | 1 |
| 2 | -5 |

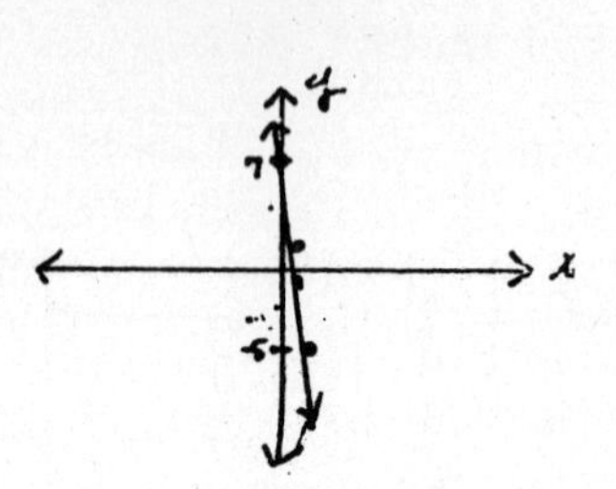

19. $y = 20x + 40$

| x value | y value |
|---|---|
| -2 | $y = 20(-2) + 40 = 0$ |
| -1 | $y = 20(-1) + 40 = 20$ |
| 0 | $y = 20(0) + 40 = 40$ |

| x | y |
|---|---|
| -2 | 0 |
| -1 | 20 |
| 0 | 40 |

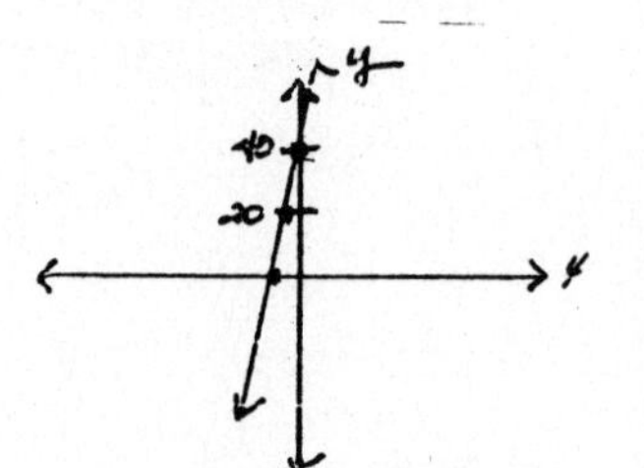

21. $-2x + 5y = 15$
$5y = 2x + 15$

$$y = \frac{2}{5}x + 3$$

| x value | y value |
|---|---|
| -5 | $y = \frac{2}{5}(-5) + 3 = 1$ |
| 0 | $y = \frac{2}{5}0 + 3 = 3$ |
| 5 | $y = \frac{2}{5}(5) + 3 = 5$ |

| x | y |
|---|---|
| -5 | 1 |
| 0 | 3 |
| 5 | 5 |

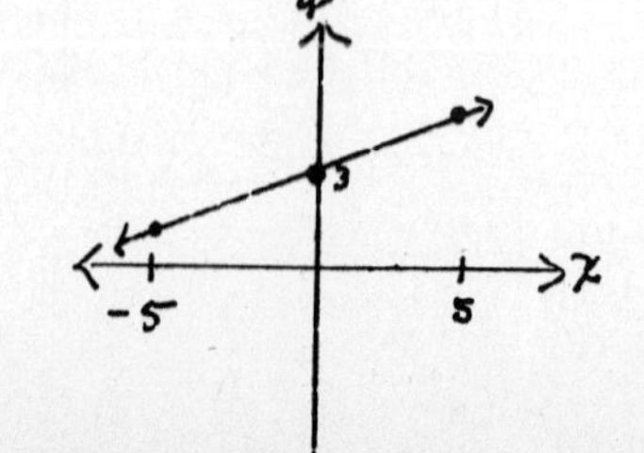

23. $-4x - 3y = -12$

$-3y = 4x - 12$

$y = \frac{-4}{3}x + 4$

| x value | y value | |
|---|---|---|
| -3 | $y = \frac{-4}{3}(-3) + 4$ | $=8$ |
| 0 | $y = \frac{-4}{3}(0) + 4$ | $= 4$ |
| 3 | $y = \frac{-4}{3}(3) + 4$ | $= 0$ |

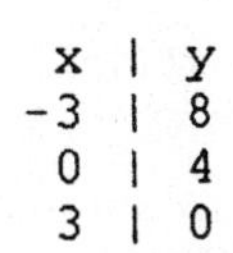

| x | y |
|---|---|
| -3 | 8 |
| 0 | 4 |
| 3 | 0 |

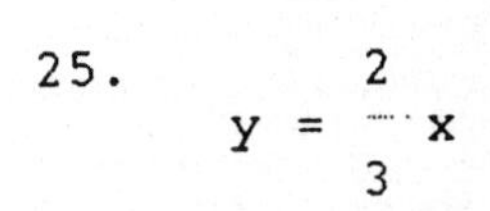

25. $y = \frac{2}{3}x$

| x value | y value | |
|---|---|---|
| -3 | $y = \frac{2}{3}(-3)$ | $= -2$ |
| 0 | $y = \frac{2}{3}0$ | $=0$ |
| 3 | $y = \frac{2}{3}3$ | $= 2$ |

| x | y |
|---|---|
| -3 | -2 |
| 0 | 0 |
| 3 | 2 |

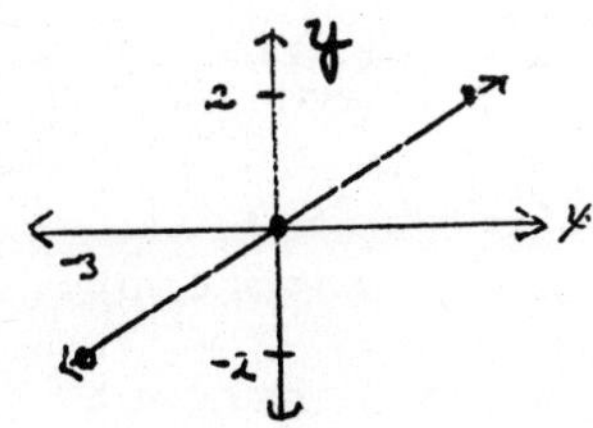

27. $y = \frac{1}{2}x + 4$

| x value | y value | |
|---|---|---|
| 0 | $y = \frac{1}{2}0 + 4$ | $= 4$ |
| 2 | $y = \frac{1}{2}2 + 4$ | $= 5$ |
| 4 | $y = \frac{1}{2}4 + 4$ | $= 6$ |

27. cont.

| x | y |
|---|---|
| 0 | 4 |
| 2 | 5 |
| 4 | 6 |

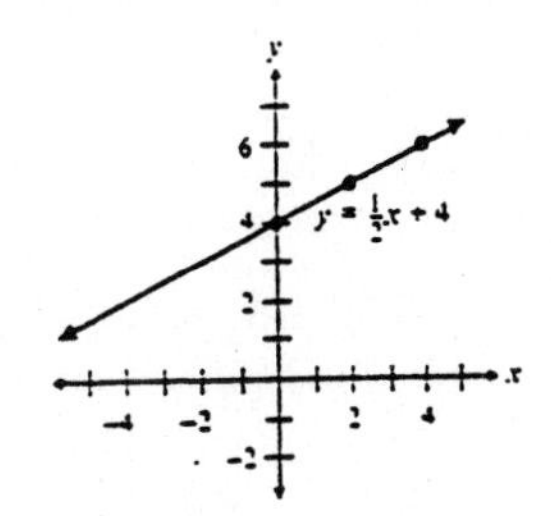


29. $x - \frac{2}{3}y = -2$

$3 \cdot x - 3 \cdot \frac{2}{3}y = -2(3)$

$3x - 2y = -6$

$-2y = -3x - 6$

$y = \frac{3}{2}x + 3$

| x-value | y-value | |
|---|---|---|
| -2 | $y = \frac{3}{2}(-2) + 3$ | $= 0$ |
| 0 | $y = \frac{3}{2}0 + 3$ | $= 3$ |
| 2 | $y = \frac{3}{2}2 + 3$ | $= 6$ |

| x | y |
|---|---|
| -2 | 0 |
| 0 | 3 |
| 2 | 6 |

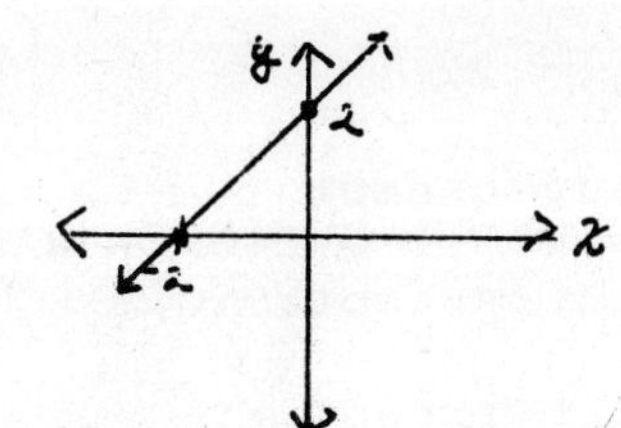

31. $\frac{2}{3}x + \frac{1}{2}y = \frac{3}{2}$

$6 \cdot \frac{2}{3}x + 6 \cdot \frac{1}{2}y = 6 \cdot \frac{3}{2}$

$4x + 3y = 9$
$3y = -4x + 9$
$y = \frac{-4}{3}x + 3$

x value, y value

-3 $y = \frac{-4}{3}x + 3 = 7$

0 $y = \frac{-4}{3}x + 3 = 3$

3 $y = \frac{-4}{3}x + 3 = -1$

| x | y |
|---|---|
| -3 | 7 |
| 0 | 3 |
| 3 | -1 |

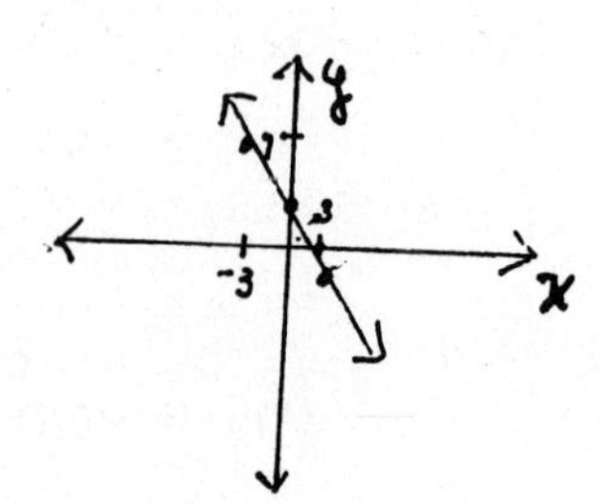


33. $y = 8x + 4$

y intercept
$x = 0$ $y = 8(0) + 4 = 4$
y intercept: (0,4)

x intercept
$y = 0$ $0 = 8x + 4$
$-4 = 8x$
$-1/2 = x$
x intercept: (-1/2,0)

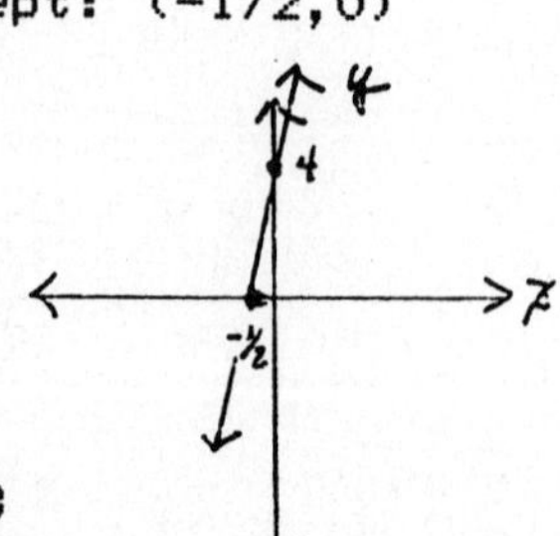


35. $y = 2x + 3$

y intercept
$x = 0$ $y = 2(0) + 3 = 3$
y intercept: (0,3)

x intercept
$y = 0$ $0 = 2x + 3$
$-3 = 2x$
$-3/2 = x$
x intercept: (-3/2,0)

35. cont.

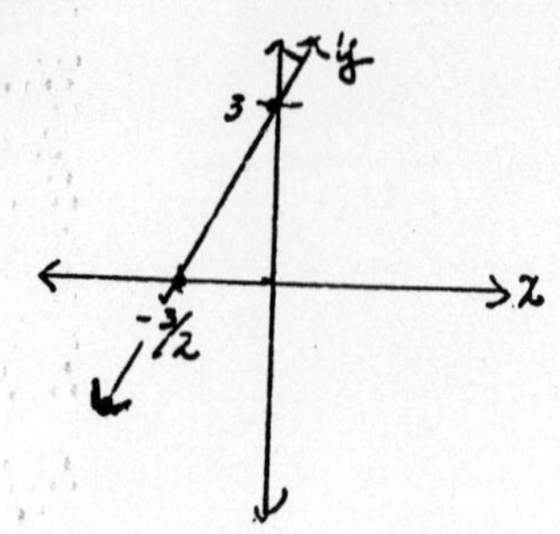


37. $y = -6x + 5$

y intercept
$x = 0$ $y = -6(0) + 5 = 5$
y intercept: (0,5)

x intercept
$y = 0$ $0 = -6x + 5$
$-5 = -6x$
$5/6 = x$
x intercept: (5/6,0)

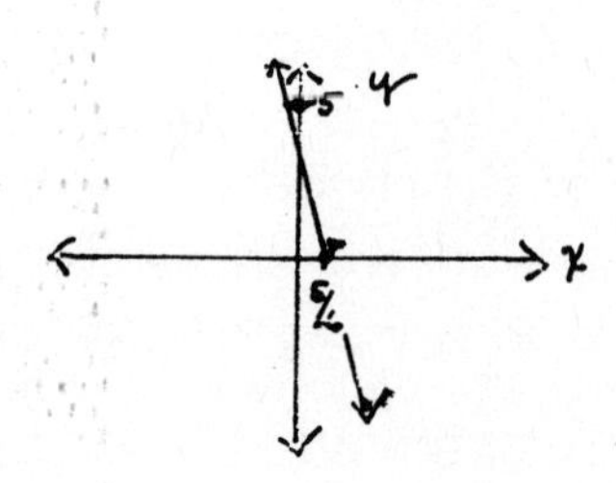


39. $4y + 3x = 12$

y intercept
$x = 0$ $4y + 3(0) = 12$
$4y = 12$
$y = 3$
y intercept: (0,3)

x intercept
$y = 0$ $4(0) + 3x = 12$
$3x = 12$
$x = 4$
x intercept: (4,0)

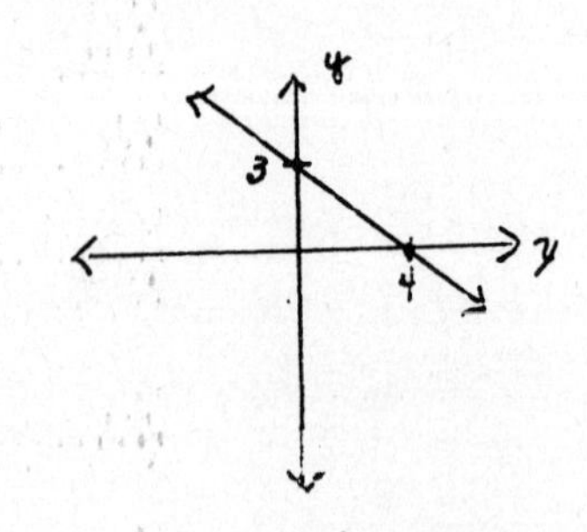

41. $4x = 3y - 9$

y intercept

$x = 0 \quad 4(0) = 3y - 9$

$0 = 3y - 9$

$9 = 3y$

$3 = y$

y intercept: $(0,3)$

x intercept

$y = 0 \quad 4x = 3(0) - 9$

$4x = -9$

$x = -9/4$

x intercept: $(-9/4,0)$

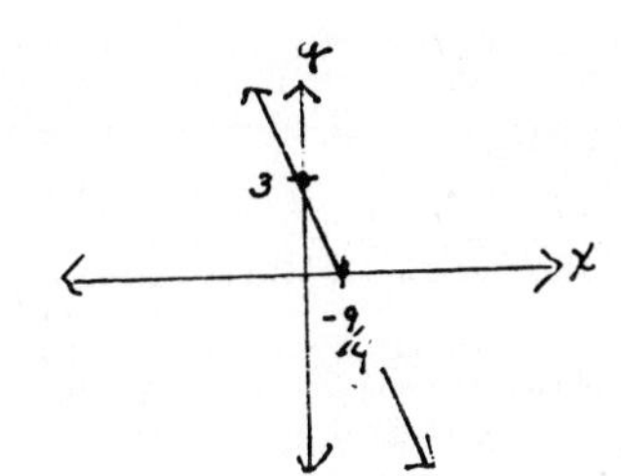

43. $\frac{1}{2}x + 2y = 4$

y intercept

$x = 0 \quad \frac{1}{2}\cdot 0 + 2y = 4$

$2y = 4$

$y = 2$

y intercept: $(0,2)$

x intercept

$y = 0 \quad \frac{1}{2}x + 2(0) = 4$

$\frac{1}{2}x = 4$

$2\cdot\frac{1}{2}x = 2\cdot 4$

$x = 8$

x intercept: $(8,0)$

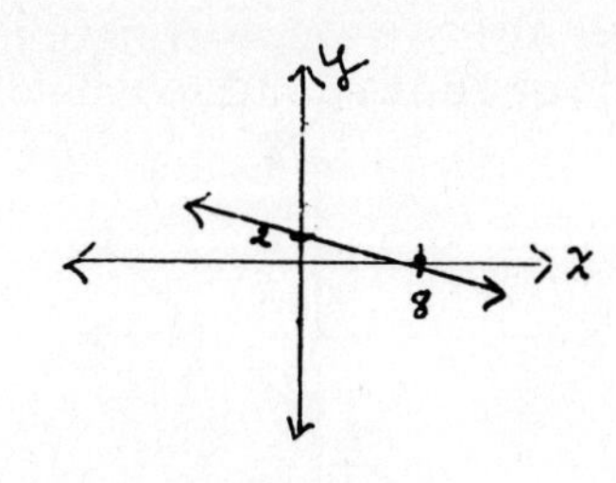

45. $6x - 12y = 24$

y intercept

$x = 0 \quad 6(0) - 12y = 24$

$-12y = 24$

$y = -2$

y intercept: $(0,-2)$

x intercept

$y = 0 \quad 6x - 12(0) = 24$

$6x = 24$

$x = 4$

x intercept $(4,0)$

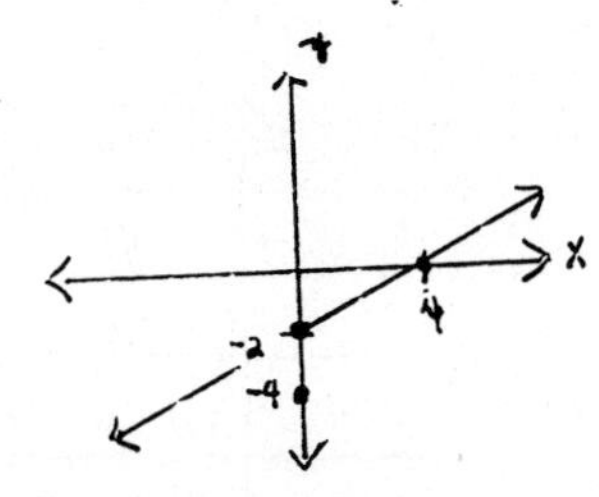

47. $-16y = 4x + 96$

y intercept

$x = 0 \quad -16y = 4(0) + 96$

$-16y = 96$

$y = -6$

y intercept: $(0,-6)$

x intercept

$y = 0 \quad -16(0) = 4x + 96$

$0 = 4x + 96$

$-96 = 4x$

$-24 = x$

x intercept: $(-24,0)$

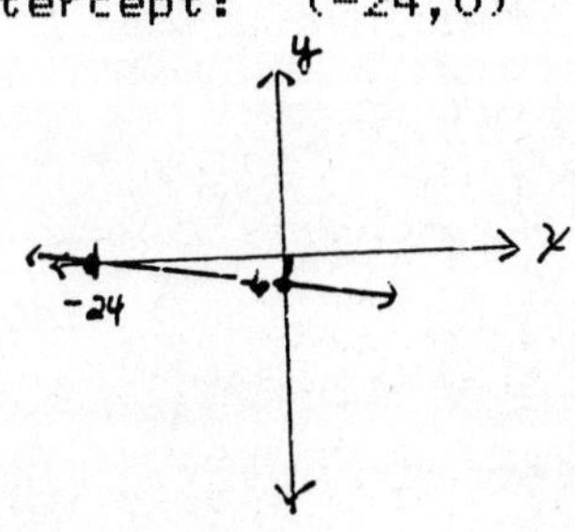

49. $30y + x = 45$

y intercept

$x = 0 \quad 30y + 0 = 45$

$30y = 45$

$y = 45/30 = 3/2$

y intercept: $(0,3/2)$

x intercept

$y = 0 \quad 30(0) + x = 45$

$x = 45$

x intercept: $(45,0)$

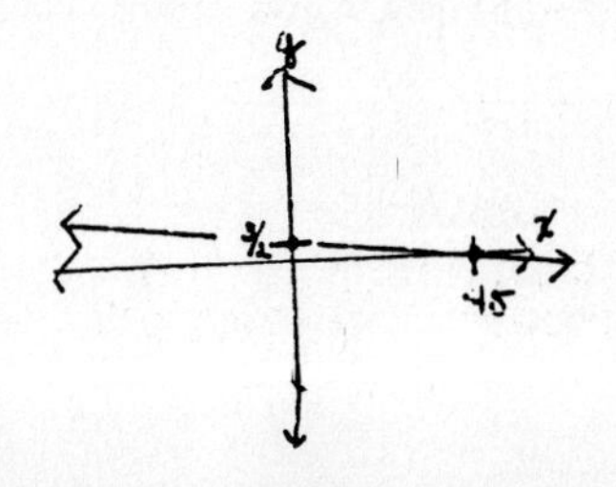

51. $40x + 6y = 40$

y intercept

$x = 0 \quad 40(0) + 6y = 40$

$y = 40/6$

$y = 20/3$

y intercept: (0, 20/3)

x intercept

$y = 0 \quad 40x + 6(0) = 40$

$40x = 40$

$x = 1$

x intercept: (1,0)

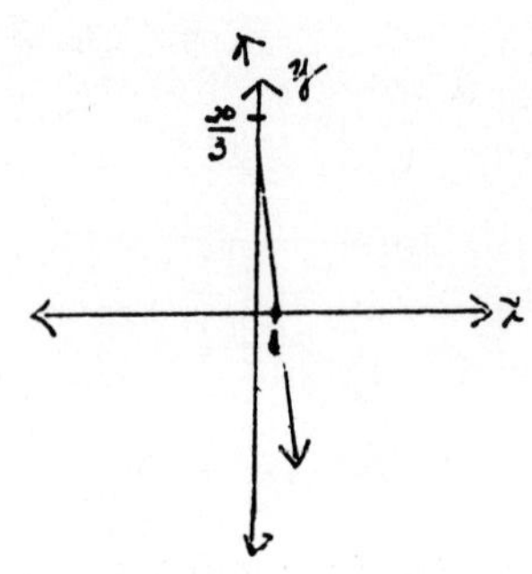

53. $\frac{1}{3}\cdot x + \frac{1}{4}\cdot y = 12$

y intercept

$x = 0 \quad \frac{1}{3}\cdot 0 + \frac{1}{4}\cdot y = 12$

$\frac{1}{4}\cdot y = 12$

$4\cdot\frac{1}{4}\cdot y = 4\cdot 12$

$y = 48$

y intercept: (0,48)

x intercept

$y = 0 \quad \frac{1}{3}\cdot x + \frac{1}{4}\cdot 0 = 12$

$\frac{1}{3}\cdot x = 12$

$3\cdot\frac{1}{3}\cdot x = 3\cdot 12$

$x = 36$

x intercept: (36,0)

53. cont.

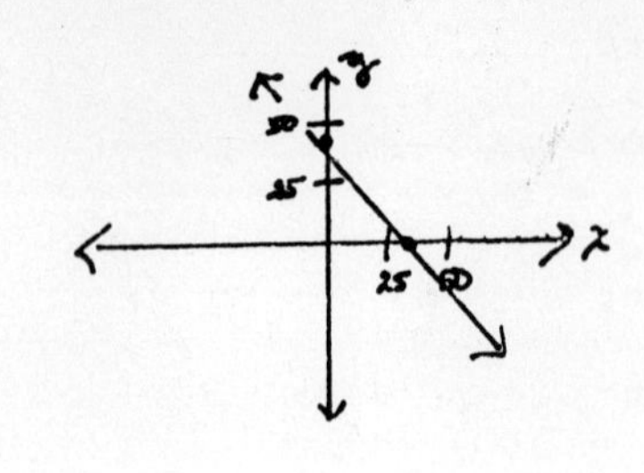

55. $\frac{1}{2}\cdot y = \frac{3}{8}\cdot x - \frac{3}{4}$

y intercept

$x = 0 \quad \frac{1}{2}\cdot y = \frac{3}{8}\cdot x - \frac{3}{4}$

$\frac{1}{2}\cdot y = \frac{3}{8}\cdot 0 - \frac{3}{4}$

$\frac{1}{2}\cdot y = \frac{-3}{4}$

$y = \frac{-3}{2}$

y intercept: (0,-3/2)

x intercept

$y = 0 \quad \frac{1}{2}\cdot 0 = \frac{3}{8}\cdot x - \frac{3}{4}$

$0 = \frac{3}{8}\cdot x - \frac{3}{4}$

$\frac{3}{4} = \frac{3}{8}\cdot x$

$\left[\frac{8}{3}\right]\cdot\frac{3}{4} = \left[\frac{8}{3}\right]\cdot\frac{3}{8}\cdot x$

$2 = x$

x intercept: (2,0)

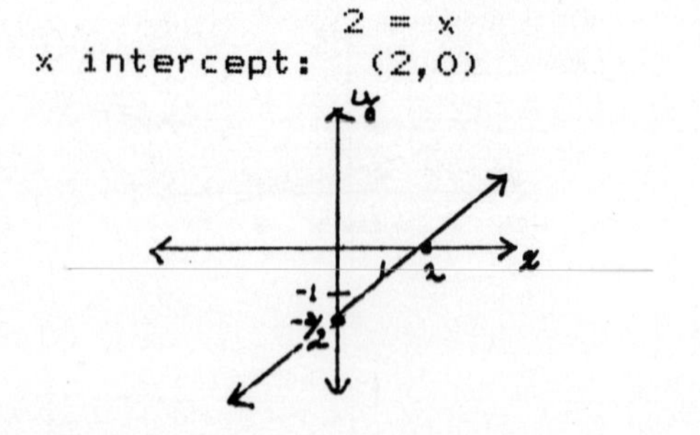

57. $r = 50$, $d = 50t$

| t values | d values |
|---|---|
| 1 | $d = 50(1) = 50$ |
| 2 | $d = 50(2) = 100$ |
| 3 | $d = 50(3) = 150$ |

| t | d |
|---|---|
| 1 | 50 |
| 2 | 100 |
| 3 | 150 |

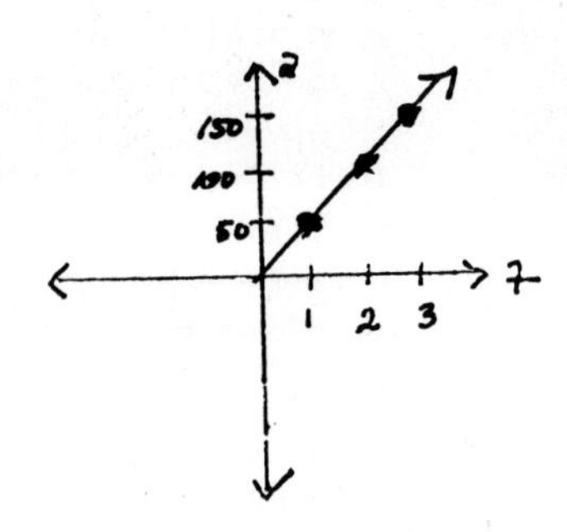


59. $P = 60x - 80,000$

a)

| x values | P values |
|---|---|
| 0 | $P = 60(0) - 80000 = -80000$ |
| 2500 | $P = 60(2500) - 80000 = 70000$ |
| 5000 | $P = 60(5000) - 80000 = 220000$ |

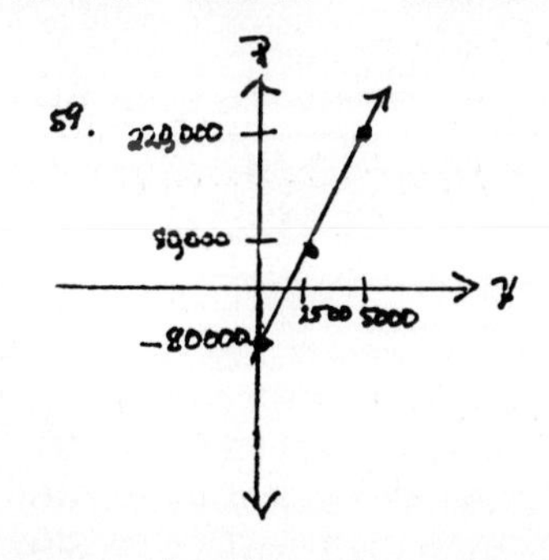


b) For the company to break even, profit would be zero. Therefore,

$0 = 60x - 80000$

$80000 = 60x$

$1333.33 = x$

(Round to the nearest hundred)
Approximately 1300 bicylcles would need to be sold for the company to break even.

c) $P = 150,000$. Therefore

$150,000 = 60x - 80,000$

$230,000 = 60x$

$3833 = x$

The company would need to sell approximately 3800 to make $150,000 profit. (Round to the nearest hundred)

61. a) $C = 50 + .12m$
where $C$ = cost and
$m$ = miles

61. cont.

b)

| m values | C values |
|---|---|
| 0 | $C = 50 + .12(0) = 50$ |
| 100 | $C = 50 + .12(100) = 62$ |
| 200 | $C = 50 + .12(200) = 74$ |

| m | C |
|---|---|
| 0 | 50 |
| 100 | 62 |
| 200 | 74 |

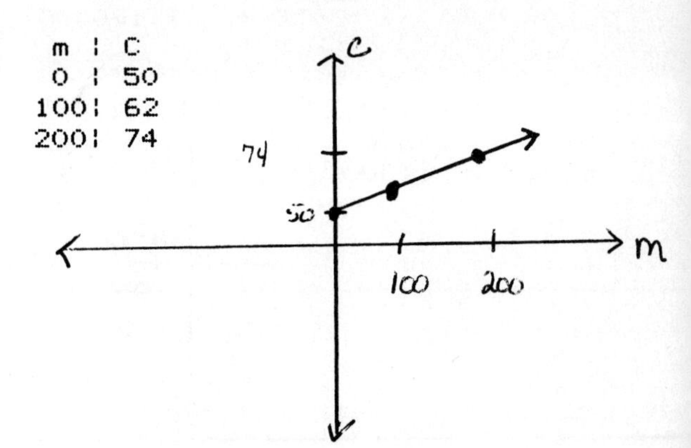


c) If the weekly cost is $70 Jack would have driven about 170 miles.

d) If the weekly cost is $60, Jack would have driven about 85 miles.

63. a) $s = 150 + .01x$
$s$ = weekly salary
$x$ = sales

b)

| x values | s values |
|---|---|
| 0 | $s = 150 + .01(0) = 150$ |
| 50,000 | $s = 150 + .01(50000) = 650$ |
| 100,000 | $s = 150 + .01(100000) = 1150$ |

| x | y |
|---|---|
| 0 | 150 |
| 50000 | 650 |
| 100000 | 1150 |

c) Using the graph, Ms. Tocci would earn about $950 per week if she sold one $80,000 house per week.

65. Step 1: Solve the equation for y.
Step 2: Choose 3 arbitrary values for x.
Step 3: Substitute each x value into the equation and calculate the cooresponding value of y.
Step 4: Plot the 3 ordered pairs and connect the points.

67.
$$3x - 2 = \frac{1}{3}\cdot(3x - 3)$$
$$3\cdot 3x - 3\cdot 2 = 3\cdot\frac{1}{3}\cdot(3x - 3)$$
$$9x - 6 = 3x - 3$$
$$6x - 6 = -3 \quad \square$$
$$6x = 3 \quad \square$$
$$x = \frac{1}{2} \quad \square$$

69
$$\frac{3}{5}\cdot(x - 3) > \frac{1}{4}\cdot(3 - x)$$
$$20\cdot\frac{3}{5}\cdot(x - 3) > 20\cdot\frac{1}{4}\cdot(3 - x)$$
$$12(x - 3) > 5(3 - x)$$
$$12x - 36 > 15 - 5x$$
$$17x - 36 > 15$$
$$17x > 51$$
$$x > 3$$

a) <---------o━━━━━━━━> (open circle at 3)

b) $(3, \infty)$

c) $\{x \mid x > 3\}$

JUST FOR FUN

1. $y = |x|$

$$y = |x| = \begin{cases} x & x \geq 0 \\ -x & x < 0 \end{cases}$$

$x \geq 0$ □

| x values | y values |
|---|---|
| 0 | $y = \|0\| = 0$ |
| 1 | $y = \|1\| = 1$ |
| 2 | $y = \|2\| = 2$ |

$x < 0$ □

| x values | y values |
|---|---|
| -1 | $y = \|-1\| = 1$ |
| -2 | $y = \|-2\| = 2$ |
| -3 | $y = \|-3\| = 3$ |

1. cont.
Ordered pairs:
(-3,3) (0,0)
(-2,2) (1,1)
(-1,1) (2,2)

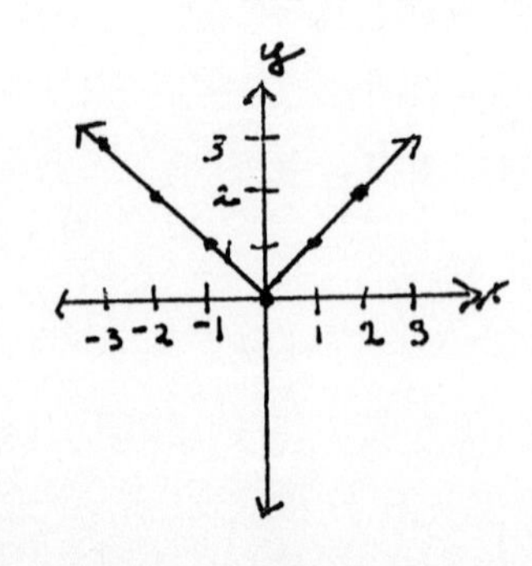

3.

$y = |x - 2|$

$$= \begin{cases} x-2, & x-2 \geq 0 \text{ or } x \geq 2 \\ -(x-2), & x-2 < 0 \text{ or } x < 2 \end{cases}$$

| x values | y values |
|---|---|
| x < 2 | |
| 0 | y = \|0-2\| = 2 |
| 1 | y = \|1-2\| = 1 |
| x ≥ 2 | |
| 2 | y = \|2-2\| = 0 |
| 3 | y = \|3-2\| = 1 |
| 4 | y = \|4-2\| = 2 |

| x | y |
|---|---|
| 0 | 2 |
| 1 | 1 |
| 2 | 0 |
| 3 | 1 |
| 4 | 2 |

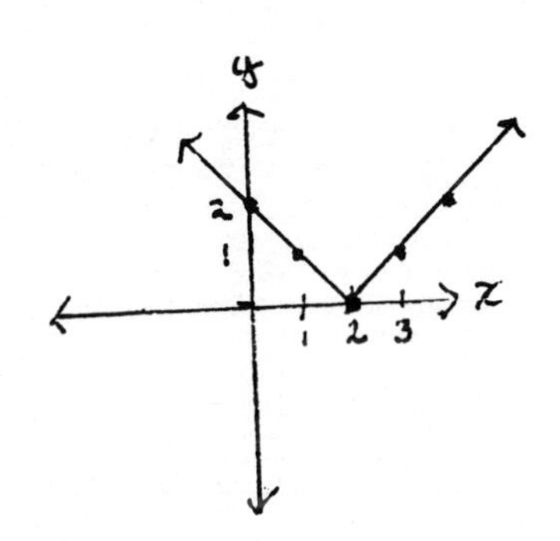


5.

$$y = \begin{cases} x + 3, & x > 4 \\ 3x - 5, & x \leq 4 \end{cases}$$

| x values | y values |
|---|---|
| x ≤ 4 | |
| 2 | y = 3(2)-5 = 1 |
| 3 | y = 3(3)-5 = 4 |
| 4 | y = 3(4)-5 = 7 |
| x > 4 | |
| 5 | y = 5 + 3 = 8 |
| 6 | y = 6 + 3 = 9 |

| x | y |
|---|---|
| 2 | 1 |
| 3 | 4 |
| 4 | 7 |
| 5 | 8 |
| 6 | 9 |

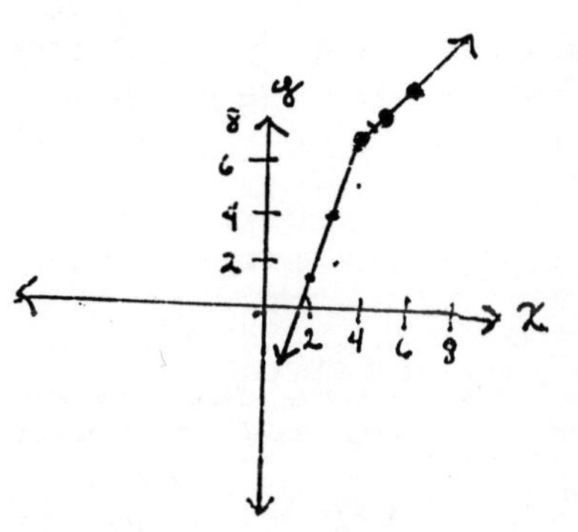


EXERCISE SET 3.3

1.
$$m = \frac{5 - (-3)}{1 - 2}$$
$$= \left[\frac{8}{-1}\right] = -8$$

3.
$$m = \frac{2 - 4}{5 - 1}$$
$$= \left[\frac{-2}{4}\right] = \frac{-1}{2}$$

5.
$$m = \frac{4 - 3}{-1 - 0}$$
$$= \left[\frac{1}{-1}\right] = -1$$

7.
$$m = \frac{2 - (-1)}{4 - 4}$$
$$= \frac{3}{0}$$

Undefined slope.

9.
$$m = \frac{4 - (-6)}{-3 - (-1)}$$
$$= \left[\frac{10}{-2}\right] = -5$$

11.
$$m = \frac{5 - 5}{2 - (-1)}$$
$$= \frac{0}{3} = 0$$

13.
$$m = \frac{-4 - (-3)}{2 - (-5)}$$

$$= -\left[\frac{1}{7}\right]$$

15.
$$m = \frac{0 - (-2)}{2 - (-4)}$$

$$= \left[\frac{2}{6}\right] = \frac{1}{3}$$

17. The ordered pairs (-2,1) and (3,2) are shown.

$$m = \frac{2 - 1}{3 - (-2)}$$

$$= \frac{1}{5}$$

19. The ordered pairs (-4,3) and (2,3) are shown.

$$m = \frac{3 - 3}{2 - (-4)}$$

$$= \frac{0}{6} = 0$$

Since the slope is zero, this is a horizontal line.

21. The line is vertical so the slope is undefined.

23. The ordered pairs (0,-1) and (2,1) are shown.

$$m = \frac{1 - (-1)}{2 - 0}$$

$$= \frac{2}{2} = 1$$

25.
$$m_1 = \frac{4 - 8}{0 - 2} = \frac{-4}{-2} = 2$$

$$m_2 = \frac{-1 - 5}{0 - 3} = \frac{-6}{-3} = 2$$

25. cont.

Since the slopes are equal, the slopes are parallel.

27.
$$m_1 = \frac{2 - (-2)}{3 - (-1)} = \left[\frac{4}{4}\right] = 1$$

$$m_2 = \frac{0 - (-1)}{2 - 3} = \frac{1}{-1} = -1$$

Since -1 is the negative reciprocal of 1, the lines are perpendicular.

29.
$$m_1 = \frac{4 - 3}{3 - (-2)} = \frac{1}{5}$$

$$m_2 = \frac{-3 - (-1)}{0 - 2} = \frac{-2}{-2} = 1$$

The lines are neither parallel nor perpendicular.

31.
$$m_1 = \frac{2 - (-2)}{0 - 6} = \frac{4}{-6} = \frac{-2}{3}$$

$$m_2 = \frac{0 - 3}{4 - 6} = \frac{3}{2}$$

Since 3/2 is the negative reciprocal of -2/3, the lines are perpendicular.

33.
$$m_1 = \frac{5 - (-1)}{1 - (-2)} = \frac{6}{3} = 2$$

$$m_2 = \frac{-2 - 2}{1 - 3} = \frac{-4}{-2} = 2$$

Since the slopes are equal, the lines are parallel.

35. $$\frac{a-4}{6-3} = 1$$

$$\frac{a-4}{3} = 1$$

$$3\cdot\frac{a-4}{3} = 3\cdot 1$$

$$a - 4 = 3$$
$$a = 7$$

37. $$\frac{b-(-4)}{5-2} = 2$$

$$\frac{b+4}{3} = 2$$

$$3\cdot\frac{b+4}{3} = 3\cdot 2$$

$$b + 4 = 6$$
$$b = 2$$

39. $$\frac{-3-c}{2-3} = -1$$

$$\frac{-3-c}{-1} = -1$$

$$(-1)\cdot\frac{-3-c}{-1} = -1(-1)$$

$$-3 - c = 1$$
$$-c = 4$$
$$c = -4$$

41. $$\frac{2-(-4)}{x-3} = 2$$

$$\frac{6}{x-3} = 2$$

Cross Multiply

$$6(1) = 2(x - 3)$$
$$6 = 2x - 6$$
$$12 = 2x$$
$$6 = x$$

43. $$\frac{5-3}{3-x} = \frac{2}{3}$$

$$\frac{2}{3-x} = \frac{2}{3}$$

43. cont.

Cross Multiply

$$2\cdot(3) = 2(3 - x)$$
$$6 = 6 - 2x$$
$$0 = -2x$$
$$0 = x$$

45. Identify two points on the line. Using the formula

$$m = \frac{y_2 - y_1}{x_2 - x_1}$$

find the change in y over the change in x.

47. A negative slope means the line is falling as you follow the graph from left to right.

49. For a vertical line, the change in x is zero so the denominator of the slope equation is zero. Since we cannot divide by zero, we say the slope is undefined.

51. Let x equal the first odd integer, x + 2 equal the next and x + 4 equal the third.

$$x + (x + 2) + (x + 4) = 27$$
$$3x + 6 = 27$$
$$3x = 21$$
$$x = 7$$
$$x + 2 = 9$$
$$x + 4 = 11$$

The three consecutive odd integers are 7, 9, and 11.

53. Using the rule, "If $|x| < a$, then $-a < x < a$.", rewrite $|x - a| < b$ as

$$-b < x - a < b$$

Add a to each of the three parts of the inequality and simplify.

$$-b + a < x - a + a < b + a$$
$$a - b < x < a + b$$

$$\{x \mid a - b < x < a + b\}$$

JUST FOR FUN

1a) Since there are 91 steps and each step has a height of 14.2", the total vertical distance is 14.2(91) = 1292.2".

b) The width of each step is 6.4". The total horizontal distance would be 6.4(91) = 582.4

c) To find the slope of the steps, draw the stairs on a the x-y coordinate plane. The point where the stairs begin can be labeled as (582.4,0) The point where the stairs end can be labeled as (0,1292.2)

$$|m| = \left|\frac{0 - 1292.2}{582.4 - 0}\right| = \frac{1292.2}{582.4} = 2.219$$

EXERCISE SET 3.4

1. The slope is positive and y changes 1 unit for each unit change of x, so $m = 1$. The line crosses the y axis at 2, so the y intercept is 2. For $m = 1$ and $b = 2$, the equation of the line is $y = x + 2$.

3. The slope is negative and y changes 15 units when x changes 10 units, so

$$m = \frac{-15}{10} = \frac{-3}{2}$$

The line crosses the y axis at 15, so $b = 15$.

$$y = \frac{-3}{2}\cdot x + 15$$

5. The equation of a horizontal line through $(0,-4)$ is $y = -4$.

7. $m = -2 = -2/1$, y-intercept = 3
(Graph on pg. 591 of text.)

9. $m = 3 = 3/1$, y-intercept = 1/2
(Graph on pg. 591 of text.)

11. x-intercept = −3, slope is undefined
Since the slope is undefined, the line is a vertical line and crosses the x-axis at −3
(Graph on pg. 591 of text.)

13. $y = -x + 2 = -1x + 2$
$m = -1$ and $b = 2$
(Graph on pg. 591 of text.)

15. Write the equation in slope-intercept form by solving for y.

$$20x - 30y = 60$$
$$-30y = -20x + 60$$
$$y = \frac{2}{3}\cdot x - 2$$

$m = \frac{2}{3}$ $\quad b = -2$

(Graph on pg. 591 of text.)

17. Write the equation in slope-intercept form by solving for y.

$$20y = 50x + 40$$
$$y = \frac{5}{2}\cdot x + 2$$ □

$m = \frac{5}{2}$ $\quad b = 2$

(Graph on pg. 591 of text.)

19. Since $m = 2$ for both lines, the lines are parallel.

21. Write each equation in slope-intercept form.

$4x + 2y = 8$ $\qquad 8x = 4 - 4y$
$2y = -4x + 8$ $\qquad 8x - 4 = -4y$
$y = -2x + 4$ $\qquad -2x + 1 = y$
$m = -2$ $\qquad m = -2$

Since $m = -2$ for both lines, the lines are parallel.

23. Write each equation in slope-intercept form.

$2x + 5y = 10$ $\qquad -x + 3y = 9$
$5y = -2x + 10$ $\qquad 3y = x + 9$
$y = \frac{-2}{5}\cdot x + 2$ $\qquad y = \frac{1}{3}\cdot x + 3$
$m = \frac{-2}{5}$ $\qquad m = \frac{1}{3}$

Since the slopes are neither equal nor negative reciprocals, the lines are neither parallel nor perpendicular.

25. Write $-3y = 6x + 9$ in slope intercept form. $y = 1/2(x) - 6$ is already in the proper form.

$y = \frac{1}{2}\cdot x - 6$ $\qquad -3y = 6x + 9$
$\qquad\qquad y = -2x - 3$
$m = \frac{1}{2}$ □ $\qquad m = -2$

The slopes are negative reciprocals so the lines are perpendicular.

27. Write $x = -2y - 4$ in slope-intercept form. $y = 2x - 6$ is already in proper form.

$y = 2x - 6$ $\qquad x = -2y - 4$
$m = 2$ $\qquad x + 4 = -2y$
$\qquad \frac{-1}{2}\cdot x - 2 = y$
$\qquad m = \frac{-1}{2}$

The slopes are negative reciprocals so the lines are perpendicular.

29. Write each equation in slope-intercept form.

$2x + y - 6 = 0$ $\qquad 6x + 3y = 12$
$y = -2x + 6$ $\qquad 3y = -6x + 12$
$m = -2$ $\qquad y = -2x + 4$
$\qquad m = -2$

The slopes are equal so the lines are parallel.

31. Write each equation in slope-intercept form.

$-6x + 4y = 6$ $\qquad$ $3y = -2x + 12$

$4y = 6x + 6$ $\qquad$ $y = \frac{-2}{3}x + 4$

$y = \frac{3}{2}x + \frac{3}{2}$ $\qquad$ $m = \frac{-2}{3}$ ▫

$m = \frac{3}{2}$ ▫

The slopes are negative reciprocals so the lines are prependicular.

33. $m = 4 \quad (x,y) = (2,3)$

Using the point-slope form of a line:

$$y - 3 = 4(x - 2)$$

Write the equation in slope-intercept form by solving for y.

$$y - 3 = 4x - 8$$
$$y = 4x - 5$$

35. $m = -1 \quad (x,y) = (6,0)$

Using the point-slope form of a line:

$$y - 0 = -1(x - 6)$$

Write the equation in slope-intercept form by solving for y.

$$y = -x + 6$$

37. $m = \frac{-2}{3} \qquad (x,y) = (-1,-2)$

Using the point-slope form of a line:

$$y - (-2) = \frac{-2}{3}(x - (-1))$$

Write the equation in slope-intercept form by solving for y.

$$y + 2 = \frac{-2}{3}(x + 1)$$
$$y + 2 = \frac{-2}{3}x - \frac{2}{3}$$
$$y = \frac{-2}{3}x - \frac{8}{3}$$

39. Before we can use the point-slope formula, we must determine the slope of the line.

$$m = \frac{6 - 1}{4 - (-2)} = \frac{5}{6}$$

Using the point-slope formula and one of the given points:

$$y - 6 = \frac{5}{6}(x - 4)$$

39. cont.

$$y - 6 = \frac{5}{6}x - \frac{10}{3}$$
$$y = \frac{5}{6}x + \frac{8}{3}$$

41. Before using the point-slope formula, we must compute the slope.

$$m = \frac{3 - 2}{6 - 5} = 1$$

Using the point-slope formula and one of the given points:

$y - 3 = 1(x - 6)$

To write the equation in slope-intercept form solve for y.

$$y - 3 = x - 6$$
$$y = x - 3$$

43. Before using the point-slope formula, we must compute the slope.

$$m = \frac{0 - 4}{1 - (-2)} = \frac{-4}{3}$$

Using the point-slope formula and one of the given points:

$$y - 0 = \frac{-4}{3}(x - 1)$$
$$y = \frac{-4}{3}x + \frac{4}{3}$$

45. To write the equation, we need to first determine the slope. The slope of $y = 2x + 4$ is 2 and since the lines are parallel, the slope of the line we wish to write the equation for is also 2. Using the point slope formula and the given point on the line:

$$y - 4 = 2(x - 1)$$

To write the equation in slope intercept form, solve for y.

$$y - 4 = 2x - 2$$
$$y = 2x + 2$$

47. To write the equation, we need to determine the slope. Since the line is parallel to $2x + 3y - 9 = 0$ the lines will have the same slope. Find the slope of $2x + 3y - 9 = 0$ by writing the equation in slope-intercept form:

$$2x + 3y - 9 = 0$$
$$3y = -2x + 9$$
$$y = \frac{-2}{3}x + 3$$

47. cont.

$$m = \frac{-2}{3}$$

Using the point-slope formula and a slope of -2/3, the equation of the line is:

$$y - 3 = \frac{-2}{3}\cdot\left[x - \frac{1}{2}\right]$$

Write the equation in standard form.

$$y - 3 = \frac{-2}{3}\cdot x + \frac{1}{3}$$

$$3y - 9 = -2x + 1$$

$$3y + 2x = 10$$

49. To write the equation we need to determine the slope. The slope of y = 2/3(x) - 5 is 2/3 and since the line we wish to write the equation for is parallel, its slope is also 2/3. Using the point-slope formula:

$$y - \frac{3}{4} = \frac{2}{3}\cdot(x - (-4))$$

To write the equation in slope-intercept form, solve for y.

$$y - \frac{3}{4} = \frac{2}{3}\cdot x + \frac{8}{3}$$

$$y = \frac{2}{3}\cdot x + \frac{41}{12}$$

51. To write the equation, we need to determine the slope. Since the line is to be perpendicular to 4x - 2y = 8 the lines will have slopes which are negative reciprocals of one another.
Finding the slope of 4x - 2y = 8:

$$4x - 2y = 8$$

$$-2y = - 4x + 8$$

$$y = 2x - 4$$

$$m = 2$$

The slope of the equation we wish to write is -1/2. Using the point slope formula:

$$y - 4 = \frac{-1}{2}\cdot(x - (-2))$$

Writing the equation in standard form:

$$y - 4 = \frac{-1}{2}\cdot x - 1$$

$$2y - 8 = -x - 2$$

$$2y + x = 6$$

53. To write the equation of the line we need to determine the slope. Since the line we are writing the equation for is to be perpendicular to 1/2x = y - 6 the slopes will be negative reciprocals of one another. Finding the slope of 1/2x = y - 6:

$$\frac{1}{2}\cdot x = y - 6$$

$$\frac{1}{2}\cdot x + 6 = y$$

$$m = \frac{1}{2} \;\square$$

The slope of the equation we wish to write is -2. Using the point-slope formula:

$$y - (-4) = -2\left[x - \left[\frac{-2}{3}\right]\right]$$

To write the equation in slope intercept form, solve for y.

$$y + 4 = -2\left[x + \frac{2}{3}\right]$$

$$y + 4 = -2x - \frac{4}{3}$$

$$y = -2x - \frac{16}{3}$$

55. To write the equation we need to determine botha a point on the line and the slope. An x-intercept of 1/2 implies the point (1/2,0) is on the line. A y-intercept of -1/4 implies the point (0,-1/4) is also on the line. We now have two points. We can use the slope formula to find the slope.

$$m = \frac{0 - \left[\frac{-1}{4}\right]}{\frac{1}{2} - 0} = \frac{\frac{1}{4}}{\frac{1}{2}}$$

$$= \frac{1}{4}\cdot\frac{2}{1} = \frac{1}{2}$$

Using the point-slope formula and either of the two points:

$$y - 0 = \frac{1}{2}\cdot\left[x - \frac{1}{2}\right]$$

Writing the equation in standard form:

$$y = \frac{1}{2}\cdot x - \frac{1}{4}$$

$$4y = 2x - 1$$

$$4y - 2x = -1 \text{ or } 2x - 4y = 1$$

57. To write the equation we need to determine the slope by finding the slope of the parallel line with the given intercepts. An x-intercept of -3 implies the point (-3,0). A y-intercept of 2 implies the point (0,2).

$$m = \frac{0 - 2}{-3 - 0} = \frac{2}{3}$$

Since the lines are parallel, the will have the same slope. Therefore, the slope of the line we wish to write the equation for is 2/3. Using this information along with the given point (4,-2) and the point-slope formula:

$$y - (-2) = \frac{2}{3}\cdot(x - 4)$$

To write the equation in slope-intercept form, solve for y.

$$y + 2 = \frac{2}{3}\cdot x - \frac{8}{3}$$

$$y = \frac{2}{3}\cdot x - \frac{14}{3}$$

59. . To write the equation we need to determine the slope by finding the slope of the perpendicular line with the given intercepts. An x-intercept of 2 implies the point (2,0). A y-intercept of -3 implies the point (0,-3).

$$m = \frac{0 - (-3)}{2 - 0} = \frac{3}{2}$$

The lines are prependicular so the slope of the line we wish to write the equation for will be the negative reciprical of 3/2 which is -2/3. Using this information along with the given point on the line of (6,2) and the point-slope formula:

$$y - 2 = \frac{-2}{3}\cdot(x - 6)$$

To write the equation in slope-intercept form, solve for y.

$$y - 2 = \frac{-2}{3}\cdot x + 4$$

$$y = \frac{-2}{3}\cdot x + 6$$

61. To write the equation we need to first determie the slope by finding the slope of the perpendicular line through (-2,1/2) and (4,3).

$$m = \frac{\frac{1}{2} - 3}{-2 - 4} = \frac{\frac{-5}{2}}{\frac{-6}{1}}$$

$$= \frac{-5}{2}\cdot\left[\frac{-1}{6}\right] = \frac{5}{12}$$

The lines are prependicular so the slope of the line we wish to write the equation for will be the negative reciprical of 5/12 which is -12/5. Using this information along with the given point on the line of (6,-2) and the point-slope formula:

$$y - (-2) = \frac{-12}{5}\cdot(x - 6)$$

Writing the equation in standard form:

$$y + 2 = \frac{-12}{5}\cdot x + \frac{72}{5}$$

$$5y + 10 = -12x + 72$$

$$5y + 12x = 62$$

63. a) Standard form: $-x + y = 2$

b) Slope intercept form: (Slove for y.)

$y = x + 2$

c) Point-slope form: From part b, we see that $m = 1$ and the y-intercept is 2 so the point (0,2) is on the line.

$y - 2 = 1(x - 0)$

65. Write each equation in slope-intercept form by solving for y. Determine the slope of each line.

a) If the lines have the same slope, the lines are parallel.

b) If the lines have slopes which are negative recipricals, they are perpendicular.

67. The direction of the inequality is reversed.

69. To find the x-intercept, let $y = 0$ and solve for x. To find the y-intercept, let $x = 0$ and solve for y.

## EXERCISE SET 3.5

1. The ordered pairs represent a function since no two have the same first coordinate and a different second coordinate.
   Domain = {1,2,3,4,5}
   Range = {4,2,5,3,1}

3. The ordered pairs represent a function since no two have the same first coordinate and a different second coordinate.
   Domain = {3,5,1,4,2,7}
   Range = {-1,0, 4,2,5}

5. The ordered pairs represent a relation but not a function since the ordered pairs (5,0) and (5,2) have the same first coordinate and a different second.
   Domain = {1,2,3,5}
   Range = {-4,-1,0,1,2}

7. The ordered pairs represent a function since no two have the same first coordinate and a different second coordinate.
   Domain = {-2,0,1/2,2,3,5}
   Range = {-3,-1,0,2/3,2,5}

9. The ordered pairs represent a relation but not a function since the ordered pairs (1,5), (1,2) and (1,0) have the same first coordinate and a different second.
   Domain = {1,2,6}
   Range = {-3,0,2,5}

11. The ordered pairs represent a relation but not a function since the ordered pairs (2,2) and (2,-7) have the same first coordinate and a different second.
   Domain = {0,1,2}
   Range = {-7,-1,2,3}

13. The graph represents a relation since it fails the vertical line test.
   Domain: $\{x \mid -2 \le x \le 2\}$
   Range: $\{y \mid -2 \le y \le 2\}$

15. The graph represents a function since it passes the vertical line test.
   Domain: All real Numbers.
   Range: $\{y \mid y \ge 0\}$

17. The graph represents a function since it passes the vertical line test.
   Domain: {-1,0,1,2,3}
   Range: {-1,0,1,2,3}

19. The graph represents a function since it passes the vertical line test.
   Domain: All Real Numbers
   Range: All Real Numbers.

21. The graph represents a relation since it fails the vertical line test.
   Domain: {-2}
   Range: All Real Numbers

23. The graph represents a function since it passes the vertical line test.
   Domain: All Real Numbers.
   Range: $\{y \mid -5 \le y \le 5\}$

25. $f(x) = 2x + 7$
   a) $f(3) = 2(3) + 7 = 13$
   b) $f(-2) = 2(-2) + 7 = 3$

27. $f(x) = 5x - 6$
   a) $f(2) = 5(2) - 6 = 4$
   b) $f(3) = 5(3) - 6 = 9$

29. $f(x) = 3 - 2x$
   a) $f(2) = 3 - 2(2) = -1$
   b) $f(1/2) = 3 - 2(1/2)$
   $= 3 - 1$
   $= 2$

31. $$f(x) = \frac{2}{3} \cdot x - 3$$
   a) $$f(3) = \frac{2}{3} \cdot 3 - 3 = -1$$
   b) $$f(-12) = \frac{2}{3} \cdot (-12) - 3 = -8 - 3 = -11$$

33. $f(x) = 3x + 1$

| | Ordered Pairs |
|---|---|
| $f(0) = 3(0) + 1 = 1$ | (0,1) |
| $f(1) = 3(1) + 1 = 4$ | (1,4) |
| $f(-1) = 3(-1) + 1 = -2$ | (-1,-2) |

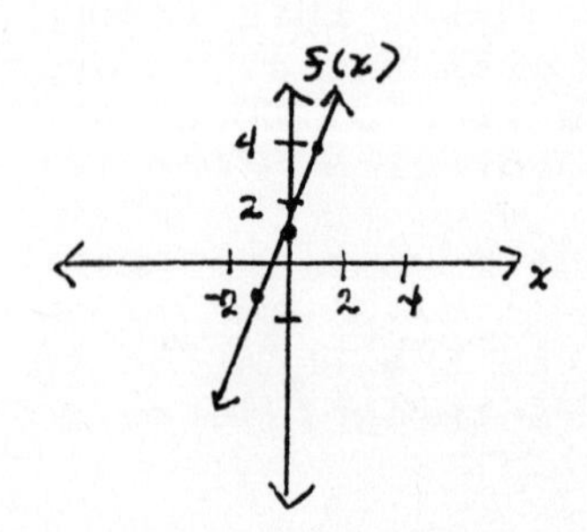

35. $f(x) = 2x - 1$

| | Ordered Pairs |
|---|---|
| $f(0) = 2(0) - 1 = -1$ | $(0,-1)$ |
| $f(1) = 2(1) - 1 = 1$ | $(1,1)$ |
| $f(2) = 2(2) - 1 = 3$ | $(2,3)$ |

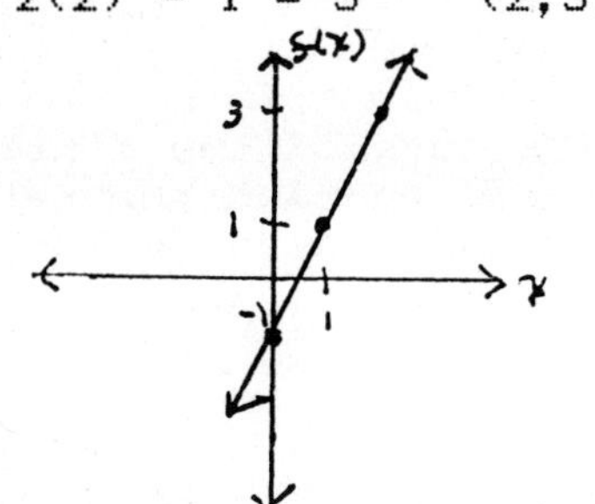

37. $f(x) = -x - 2$

| | Ordered Pairs |
|---|---|
| $f(0) = -0 - 2 = -2$ | $(0,-2)$ |
| $f(1) = -1 - 2 = -3$ | $(1,-3)$ |
| $f(-1) = -(-1)-2 = -1$ | $(-1,-1)$ |

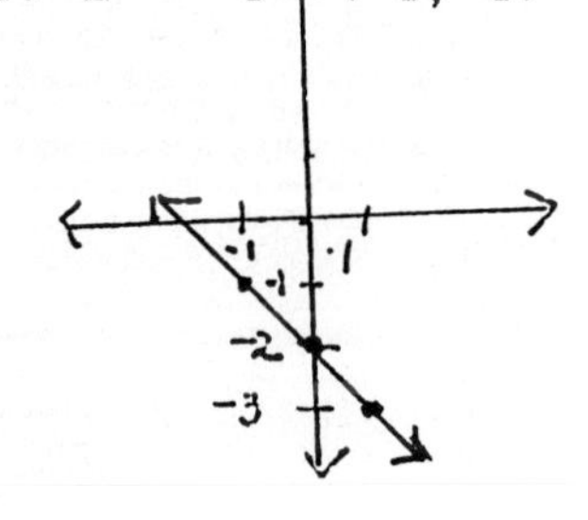

39. $f(x) = \frac{1}{2}x + 3$

| | | Ordered Pairs |
|---|---|---|
| $f(0) = \frac{1}{2}0 + 3$ | $= 3$ | $(0,3)$ |
| $f(2) = \frac{1}{2}2 + 3$ | $= 4$ | $(2,4)$ |
| $f(-2) = \frac{1}{2}(-2) + 3$ | $= 2$ | $(-2,2)$ |

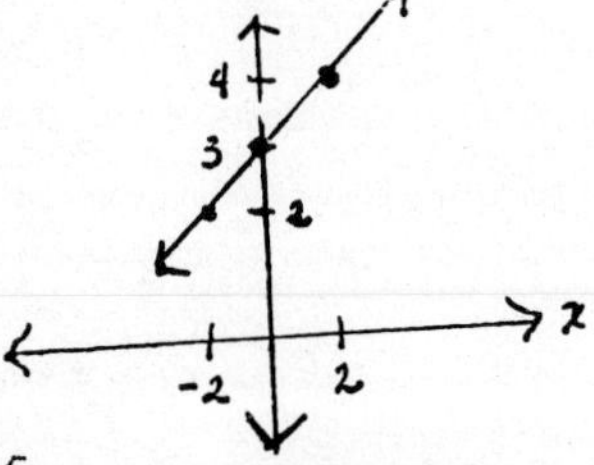

41. $g(x) = 2x - 6$

| | Ordered Pairs |
|---|---|
| $g(0) = 2(0) - 6 = -6$ | $(0,-6)$ |
| $g(1) = 2(1) - 6 = -4$ | $(1,-4)$ |
| $g(2) = 2(2) - 6 = -2$ | $(2,-2)$ |

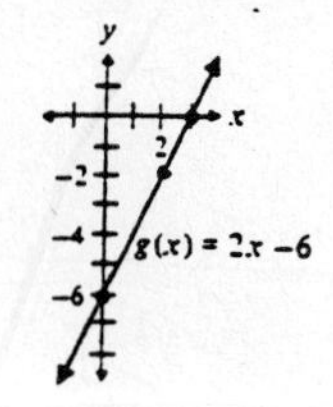


43. a) $f(Q) = -.00004Q + 4.25$

$f(10000) = -.00004(10000) + 4.25$
$= 3.85$

Ordered Pair: $(10000,3.85)$

$f(30,000) = -.00004(30,000) + 4.25$
$= 3.05$

Ordered Pair: $(30000,3.05)$

$f(60000) = -.00004(60000) + 4.25$
$= 1.85$

Ordered Pair: $(60000,1.85)$

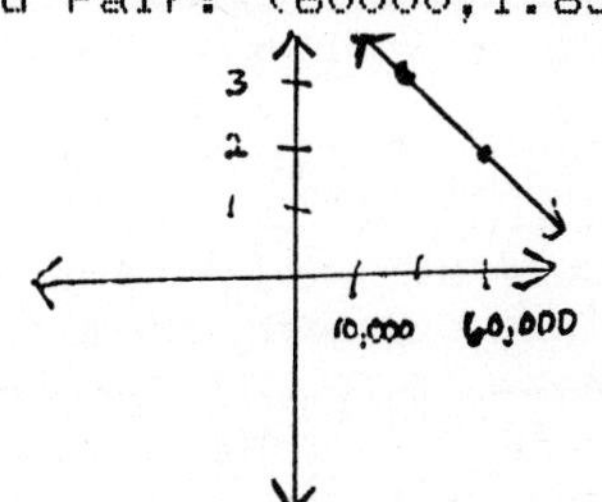


b) Approximate price per bushel would be $2.65.

45. A relation is any set of ordered pairs.

47. No. A function is a relation where no two distinct ordered pairs have the same first coordinate. An example of a relation that is not a function is (1,2), (1,3), (1,4).

49. If a vertical line can be drawn at any value of x so that the line intersects the graph at two or more places, then that value of x does not have a unique corresponding value of y. This implies the graph is not a function.

51. The range of the function is set of values that represent the dependent variable.

53. The graph of $f(x) = ax + b$, a does not equal zero, is a straight line. From this we know that both the domain and range is all real values.

55. $\frac{3}{4}x + \frac{1}{5} = \frac{2}{3}(x - 2)$

$60 \cdot \frac{3}{4}x + 60 \cdot \frac{1}{5} = 60 \cdot \frac{2}{3}(x - 2)$

$45x + 12 = 40(x - 2)$

$45x + 12 = 40x - 80$

$12 = -5x - 80$

$92 = -5x$

$-\frac{92}{5} = x$

57\. Let $x$ = the rate of the second train.
$x + 15$ = the rate of the first train.
Since the first train leaves three hours prior to the second, $t = 6$ for the first train and $t = 3$ for the second.

| Train | Distance = | Rate x | Time |
|---|---|---|---|
| #1 | 6(x + 15) | x + 15 | 6 |
| #2 | 3(x) | x | 3 |

The trains are traveling in the same direction and the distance between them is 270 miles.

57\. cont.
Therefore:

$6(x + 15) - 3x = 270$
$6x + 90 - 3x = 270$
$3x + 90 = 270$
$3x = 180$
$x = 60$
$x + 15 = 75$

The rate of the first train is 75 mph and the rate of the second is 60 mph.

EXERCISE SET 3.6

1\. $x > 3$
Graph the line $x = 3$ with a dashed line. Use (0,0) as a check point. Since $0 \not> 3$, shade the side of the graph which does not contain (0,0)

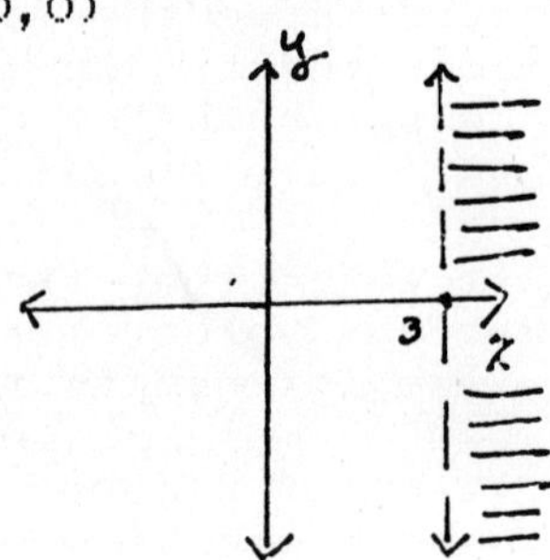


3\. $x \geq \frac{5}{2}$

Graph the line $x = 5/2$ using a solid line. Use (0,0) as a check point.

Since $0 \not\geq \frac{5}{2}$ shade the area of the graph which does not contain (0,0)

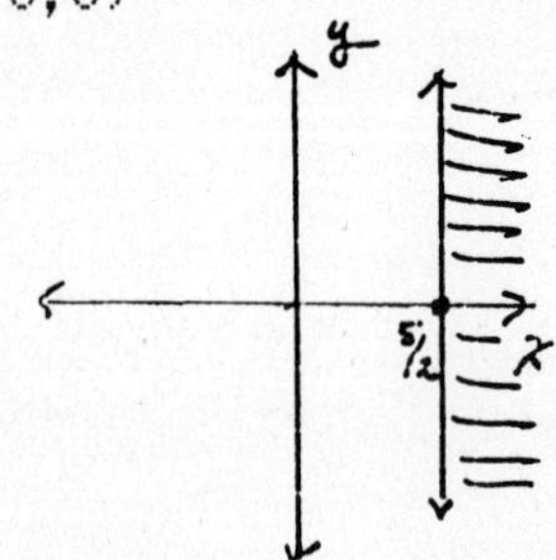


5\. $y \geq 2x$
Graph the line $y = 2x$. Use (1,0) as a check point. Since $0 \not\geq 2$ shade the area of the graph which does not contain (1,0)

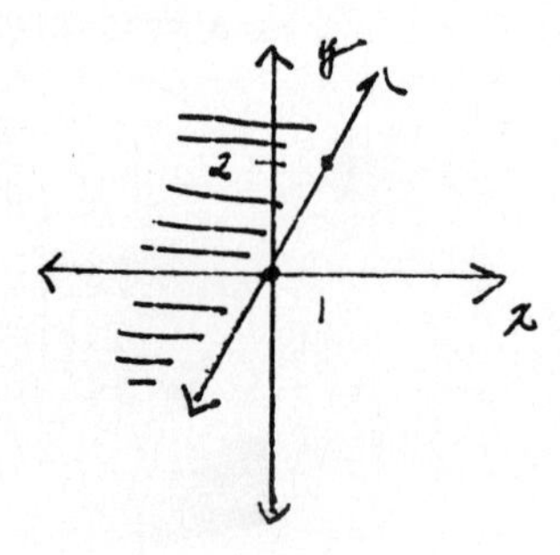


7\. $y < 2x + 1$
Graph the line $y = 2x + 1$ using a dashed line. Use (0,0) as a check point. Since $0 < 2(0) + 1$, shade the area of the graph which contains (0,0).

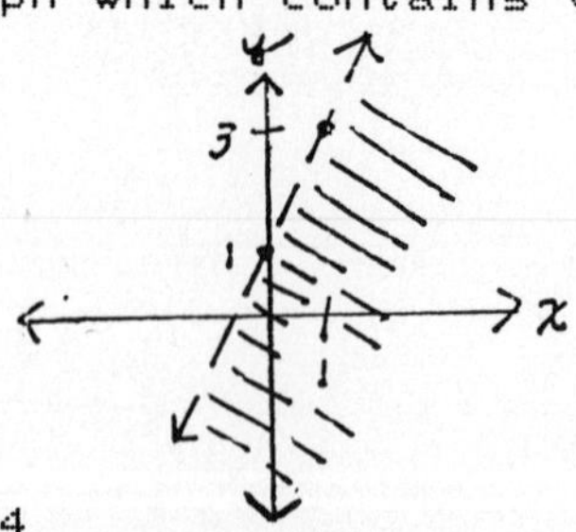


9\. $y < -3x + 4$
Graph $y = -3x + 4$ using a dashed line. Use (0,0) as a check point. Since $0 < -3(0) + 4$ shade the area of the graph which contains (0,0)

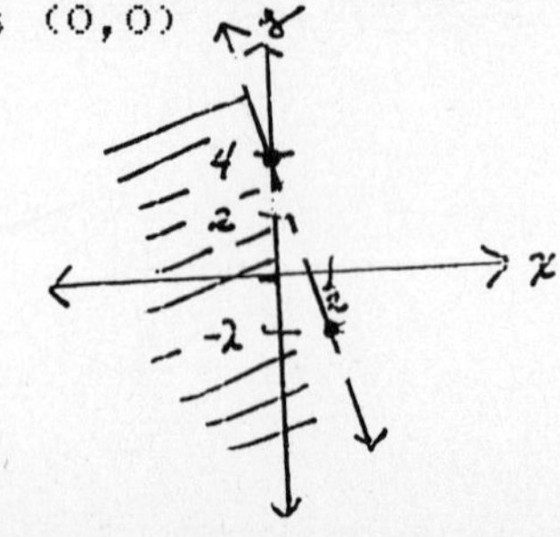

11. $y \geq \frac{1}{2}x - 4$

Graph the line y = 1/2(x) − 4. Use (0,0) as a checkpoint. Since $0 \geq \frac{1}{2}(0) - 4$ shade the area containing the point (0,0).

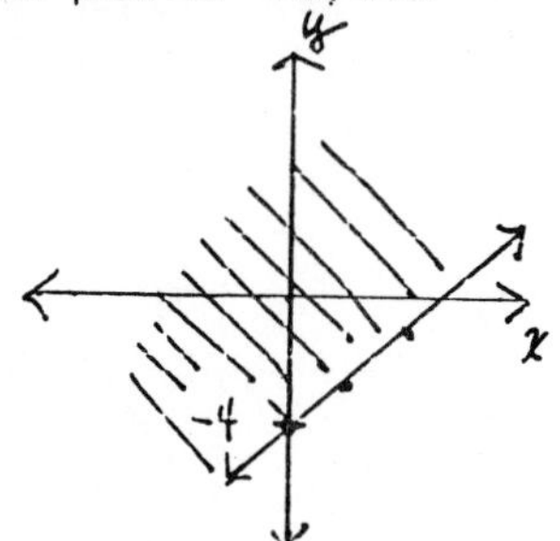


13. $y \leq \frac{1}{3}x + 6$

Graph the line y = 1/3(x) + 6. Use (0,0) as a check point. Since

$$0 \leq \frac{1}{3}(0) + 6$$

shade the area of the graph containing the point (0,0).

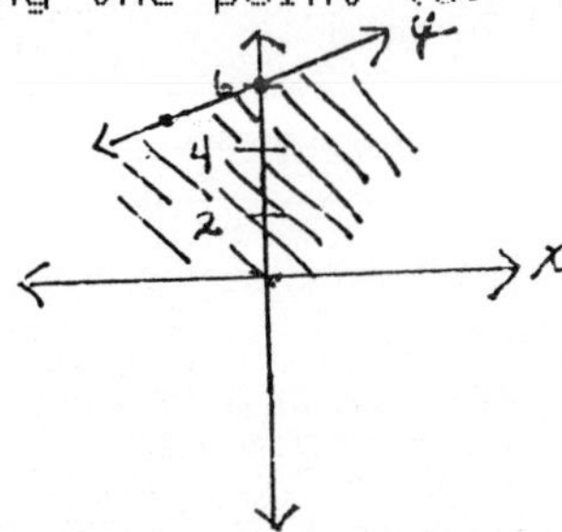


15. $y \leq -3x + 5$

Graph the line y = −3x + 5. Use (0,0) as a check point. Since $0 \leq -3(0) + 5$ shade the area containing the point (0,0)

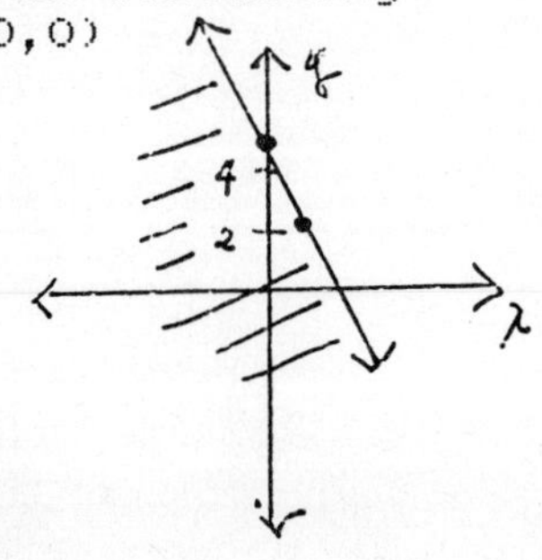


17. $y > 5x - 9$

Graph the line y = 5x − 9 using a dashed line. Use the point (0,0) as a check point. Since $0 > 5(0) - 9$, shade the area containing (0,0).

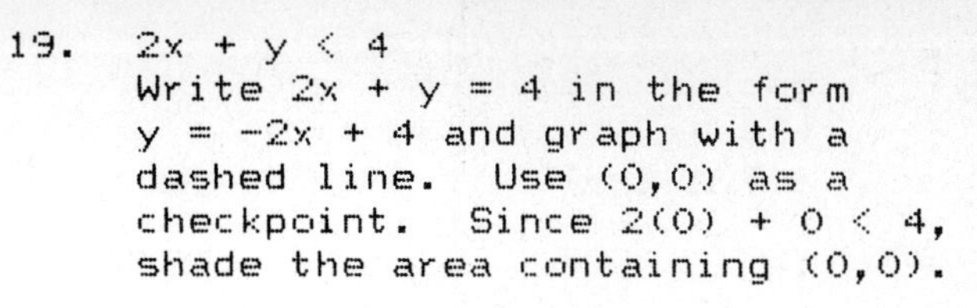

19. $2x + y < 4$

Write 2x + y = 4 in the form y = −2x + 4 and graph with a dashed line. Use (0,0) as a checkpoint. Since $2(0) + 0 < 4$, shade the area containing (0,0).

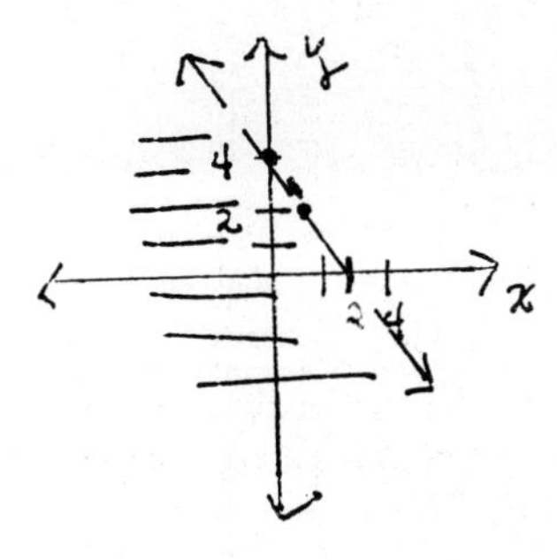


21. $2x \leq 5y + 10$

Write 2x = 5y + 10 in the form $y = \frac{2}{5}x - 2$ and graph. Use the point (0,0) as a checkpoint. Since

$$2(0) \leq 5(0) + 10$$

shade the area containing (0,0).

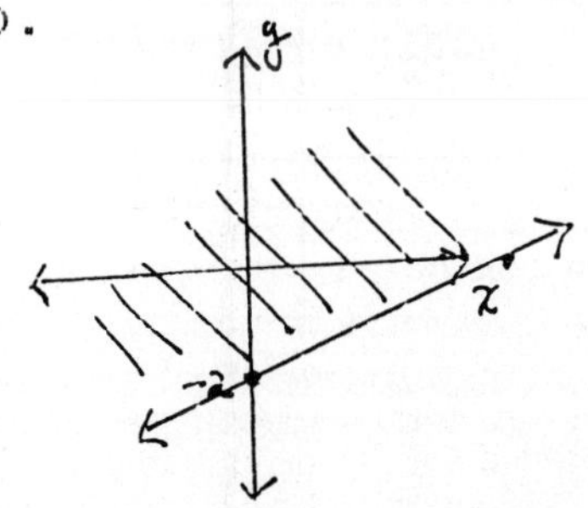


23. $\frac{-1}{6}x - \frac{1}{3}y > \frac{2}{3}$

Clear the fractions by multiplying each term by the least common multiple 6.

$$6 \cdot \frac{-1}{6}x - 6 \cdot \frac{1}{3}y > 6 \cdot \frac{2}{3}$$

$-x - 2y > 4$

Solve for y.

$-2y > x + 4$

$$y < \frac{-1}{2}x - 2$$

Graph the resulting equation with a dashed line. Use (0,0) as a check point. Since

$$\frac{-1}{6}0 - \frac{1}{3}0 \not> \frac{2}{3}$$

shade the region which does not contain the point (0,0).

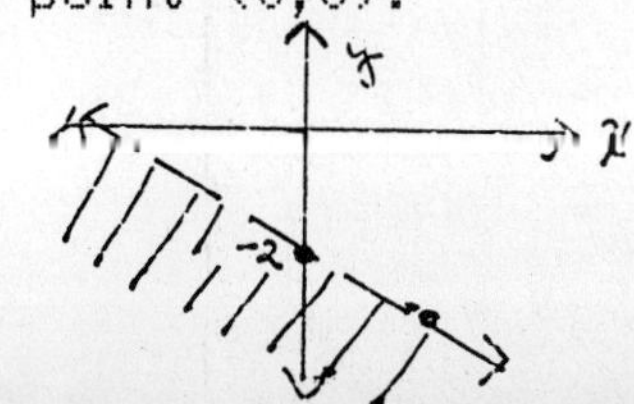

25. The two inequalities include "equal to" so the points on the line are included as solutions.

27.
$$\frac{\frac{2}{3}(x - 4)}{5} = \frac{x + 8}{6}$$

$$6 \cdot \left[\frac{2}{3}\right](x - 4) = 5(x + 8)$$

27. cont.

$$4(x - 4) = 5x + 40$$
$$4x - 16 = 5x + 40$$
$$-16 = x + 40$$
$$-56 = x$$

29. Let x equal the original cost.

$$x - (.1x) - 2 = 12.15$$
$$.9x - 2 = 12.15$$
$$.9x = 14.15$$
$$x = \$15.72$$

The original cost of the CD was $15.72

JUST FOR FUN

1. $y < |x|$
Graph $y = |x|$ using a dashed line. Use (1,0) as a check point. Since $0 < |1|$, shade the area containing the point (1,0).

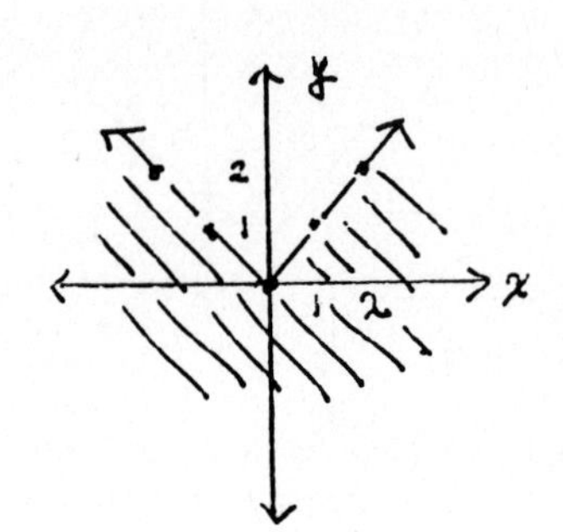


REVIEW EXERCISES

1.

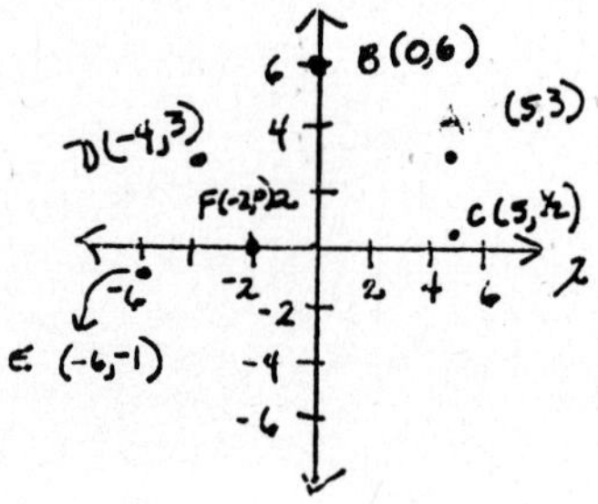

3.
$$d = \sqrt{(6 - 2)^2 + (2 - (-1))^2}$$
$$= \sqrt{4^2 + 3^2}$$
$$= \sqrt{25} = 5$$

$$\text{midpoint} = \left[\frac{6 + 2}{2}, \frac{2 + (-1)}{2}\right]$$
$$= \left[4, \frac{1}{2}\right]$$

5.
$$d = \sqrt{(-4 - (-2))^2 + (3 - 5)^2}$$
$$= \sqrt{(-2)^2 + (-2)^2}$$
$$= \sqrt{4 + 4}$$

5. cont.
$$= \sqrt{4 + 4}$$
$$= \sqrt{8} \approx 2.83$$

$$\text{midpoint} = \left[\frac{-4 + (-2)}{2}, \frac{3 + 5}{2}\right]$$
$$= (-3, 4)$$

7.
$$d = \sqrt{(-3 - (-3))^2 + (5 - (-8))^2}$$
$$= \sqrt{0^2 + 13^2}$$
$$= \sqrt{169} = 13$$

$$\text{midpoint} = \left[\frac{-3 + (-3)}{2}, \frac{5 + (-8)}{2}\right]$$
$$= \left[-3, \frac{-3}{2}\right]$$

9. $x = -2$

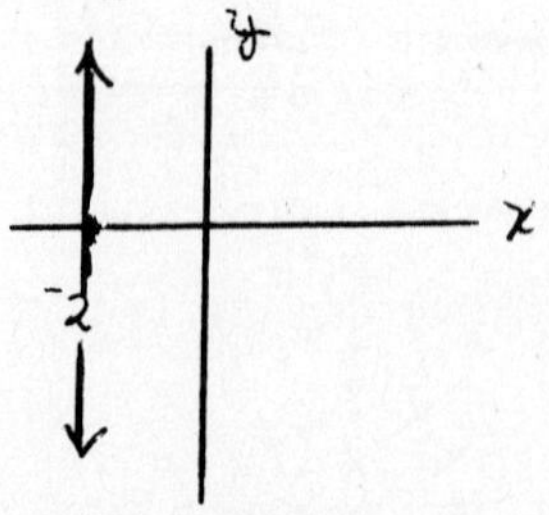

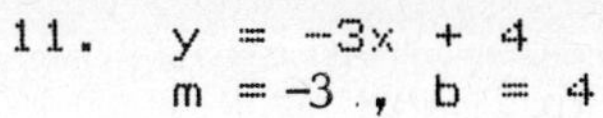

11. $y = -3x + 4$
$m = -3$, $b = 4$

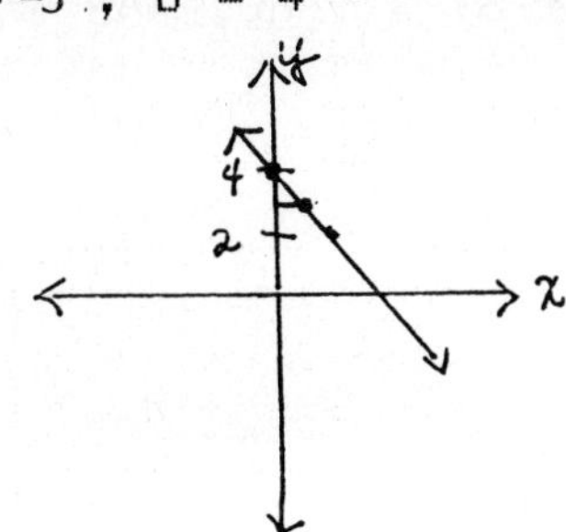


13. $2x - 3y = 12$

| x | y |
|---|---|
| 0 | -4 |
| 6 | 0 |

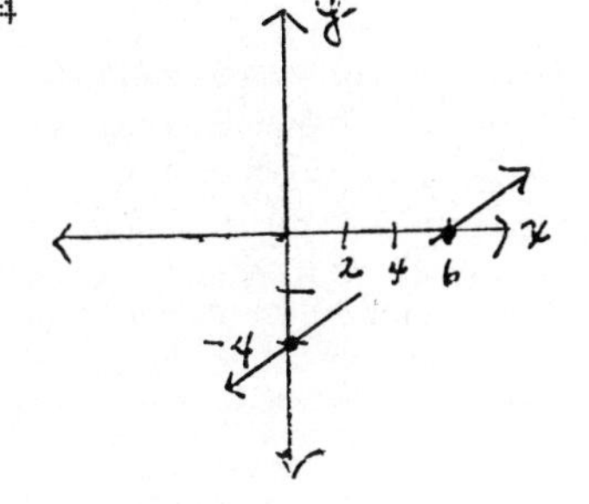


15. $5x - 2y = 10$

| x | y |
|---|---|
| 0 | -5 |
| 2 | 0 |

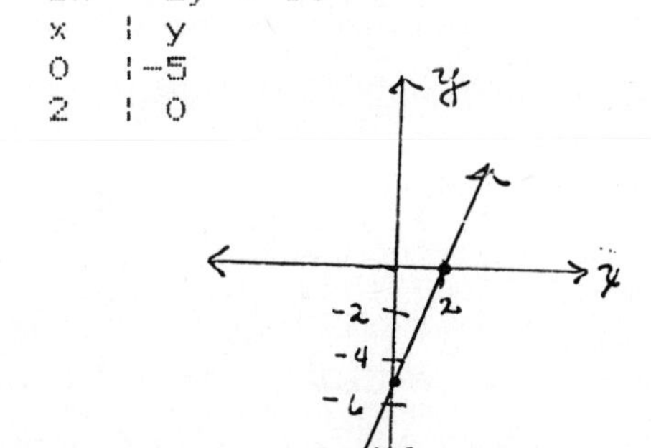


17. $25x - 50y = 200$

| x | y |
|---|---|
| 0 | -4 |
| 8 | 0 |

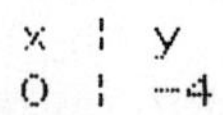

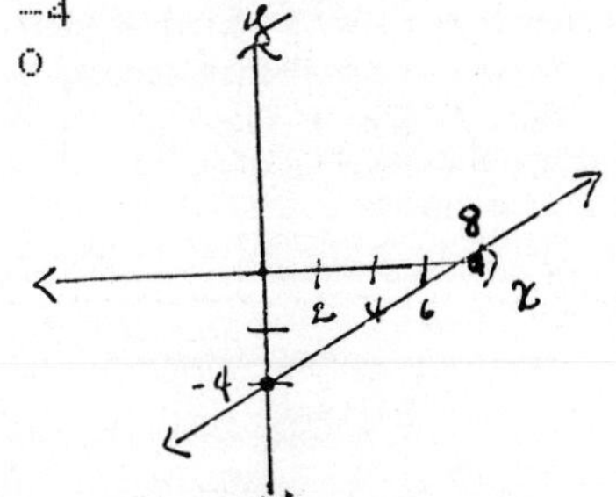


19. $\frac{2}{3}x = \frac{1}{4}y + 20$

| x | y |
|---|---|
| 0 | -80 |
| 30 | 0 |

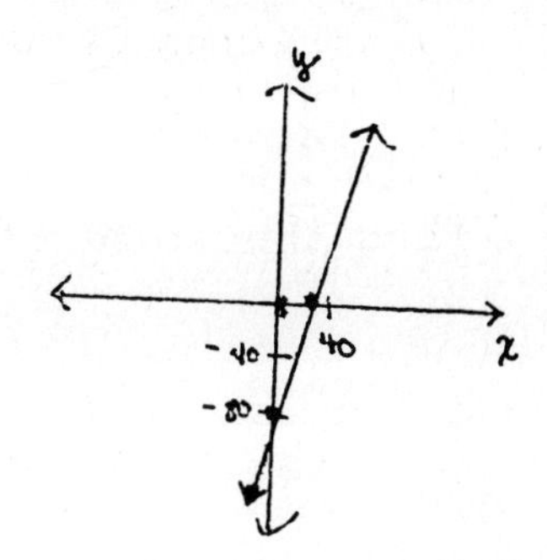


21. $y = -4x + 1/2$
$m = -4$, $b = 1/2$

23. $3x + 5y = 12$
Rewrite the equation in slope-intercept form by solving for y.

$3x + 5y = 12$

$5y = -3x + 12$

$$y = \frac{-3}{5}x + \frac{12}{5}$$

$$m = \frac{-3}{5} \qquad b = \frac{12}{5}$$

25. $36x - 72\,y = 144$
Rewrite the equation in slope-intercept form by solving for y.

$36x - 72y = 144$

$-72y = -36x + 144$

$$y = \frac{1}{2}x - 2$$

$$m = \frac{1}{2} \qquad b = -2$$

27. $y = 6$.
The graph of $y = 6$ is a horizontal line through (0,6). Since the line crosses the y-axis at 6 $b = 6$. Since the line is horizontal, $m = 0$.

29. $$m = \frac{6 - (-1)}{4 - 5}$$

$$= \frac{7}{-1} = -7$$

29. $$m = \frac{5 - 6}{-3 - 0}$$

$$= \frac{-1}{-3} = \frac{1}{3}$$

33. Find the slope of $L_1$

$$m = \frac{3 - (-3)}{4 - 0} = \frac{3}{2}$$

Find the slope of $L_2$

$$m = \frac{-1 - (-2)}{1 - 2} = \frac{1}{-1} = -1$$

Since the slopes are not equal and are not negative reciprocals, the lines are neither perpendicular nor parallel.

35. Find the slope of $L_1$

$$m = \frac{0 - 3}{4 - 1} = \frac{-3}{3} = -1$$

Find the slope of $L_2$

$$m = \frac{2 - 3}{5 - 6} = \frac{-1}{-1} = 1$$

The slopes are negative reciprocals. Therefore the lines are prependicular.

37.
$$\frac{a - 2}{5 - 4} = 1$$

$$\frac{a - 2}{1} = 1$$

$$a - 2 = 1$$
$$a = 3$$

39.
$$\frac{-1 - y}{-2 - 4} = -6$$

$$\frac{-1 - y}{-6} = -6$$

$$-1 - y = 36$$
$$-y = 37$$
$$y = -37$$

41. $y = 3$

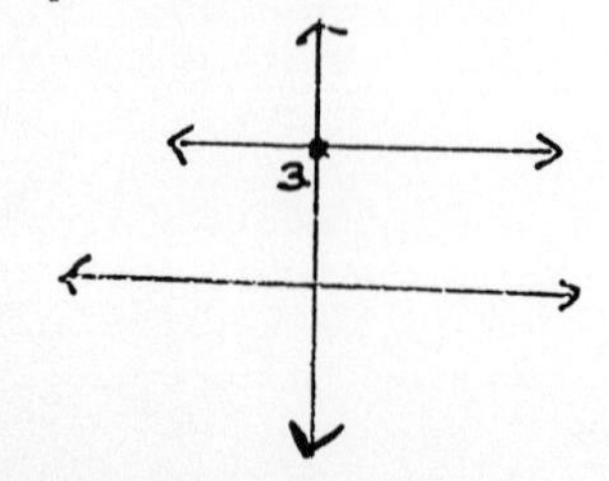


43. Using the slope-intercept form of a line, and the the information that $m = -1/2$ and $b = 2$:

$$y = -\frac{1}{2} \cdot y + 2$$

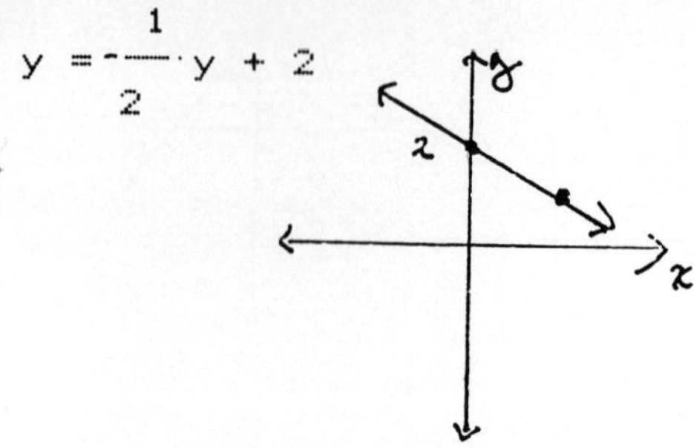


45. Determine the slope of each line by rewriting each equation in slope-intercept form.

$$2x - 3y = 9$$
$$-3y = -2x + 9$$
$$y = \frac{2}{3} \cdot x - 3$$

$$m_1 = \frac{2}{3}$$

$$-3x - 2y = 6$$
$$-2y = 3x + 6$$
$$y = \frac{-3}{2} \cdot x - 3$$

$$m_2 = \frac{-3}{2}$$

Since $-3/2$ is the negative reciprocal of $2/3$, the lines are perpendicular.

42. Determine the slope of each line by rewriting each equation in the slope-intercept form.

$$4x = 6y + 3$$
$$4x - 3 = 6y$$
$$y = \frac{2}{3} \cdot x - \frac{1}{2}$$

$$m_1 = \frac{2}{3}$$

$$-2x = -3y + 10$$
$$-2x - 10 = -3y$$
$$y = \frac{2}{3} \cdot x + \frac{10}{3}$$

$$m_2 = \frac{2}{3}$$

Since the slopes are equal, the lines are parallel.

49. Write each equation in slope-intercept form.

$$4x - 2y = 10$$
$$-2y = -4x + 10$$
$$y = 2x - 5$$
$$m_1 = 2$$

$$-2x + 4y = -8$$
$$4y = 2x - 8$$
$$y = \frac{1}{2} \cdot x - 2$$

$$m_2 = \frac{1}{2}$$

The lines are neither perpendicular nor parallel.

51. Using the point-slope formula along with the given point (3,2) and m = -2/3:

$$y - 2 = \frac{-2}{3} \cdot (x - 3)$$

Write the equation in slope-intercept form by solving for y.

$$y - 2 = \frac{-2}{3} \cdot x + 2$$

$$y = \frac{-2}{3} \cdot x + 4$$

53. To write the equation, we need to first determine the slope. Using the slope formula along with the two given points:

$$m = \frac{-4 - 3}{0 - (-2)} = \frac{-7}{2}$$

Using this information along with the point-slope formula along with one of the given points:

$$y - (-4) = \frac{-7}{2} \cdot (x - 0)$$

Write the equation in slope-intercept form by solving for y.

$$y + 4 = \frac{-7}{2} \cdot x$$

$$y = \frac{-7}{2} \cdot x - 4$$

55. To write the equation we need to determine the slope. Since the line we are writing the equation for is

55. cont.
parallel to 2x - 5y = 6, the lines will have the same slope. Determine the slope of 2x - 5y = 6 by solving for y.

$$2x - 5y = 6$$
$$-5y = -2x + 6$$
$$y = \frac{2}{5} \cdot x - \frac{6}{5}$$

$$m = \frac{2}{5}$$

Using the point-slope formula:

$$y - (-2) = \frac{2}{5} \cdot (x - 4)$$

Write the equation in slope-intercept form by solving for y.

$$y + 2 = \frac{2}{5} \cdot x - \frac{8}{5}$$

$$y = \frac{2}{5} \cdot x - \frac{18}{5}$$

57. To write the equation, we need to determine the slope. Since the line is to be perpendicular to the line 4x - 2y = 8, the slope will be the negative reciprocale of the slope of 4x - 2y = 8.

$$4x - 2y = 8$$
$$-2y = -4x + 8$$
$$y = 2x - 4$$
$$m = 2$$

Therefore, the slope of the line we wish to write the equation for will be -1/2. Using this information along with the given point and the point slope formula, we can write the equation.

$$y - 2 = \frac{-1}{2} \cdot (x - 4)$$

To write the equation in slope-intercept form, solve for y.

$$y - 2 = \frac{-1}{2} \cdot x + 2$$

$$y = \frac{-1}{2} \cdot x + 4$$

59. Let the x-axis represent the interest rates and the y-axis represent the amount of interest earned.

rate principal x rate x time= interest
0% 12000(0)(1) = 0
Ordered pair: (0,0)
10% 12000(.10)(1) = 120
Ordered pair: (.10,120)
20% 12000(.20)(1) = 240
Ordered pair: (.20,240)

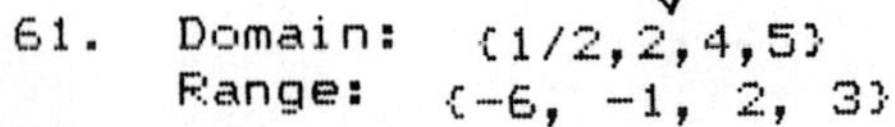

61. Domain: {1/2,2,4,5}
Range: {-6, -1, 2, 3}

63. Domain: $\{x \mid -2 \le x \le 2\}$
Range: $\{y \mid -1 \le y \le 1\}$

65. Domain: All Real Numbers
Range: All Real Numbers

67. Function

69. Not a function

71. Function

73. Function

75. Not a function

77. $f(x) = -2x + 5$
a) $f(-1) = -2(-1) + 5$
$= 7$
b) $f(1/2) = -2(1/2) + 5$
$= 4$

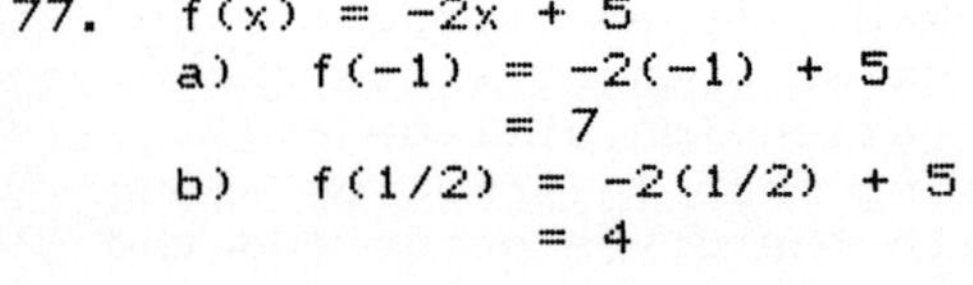

79. $f(x) = 6 - 2x$
a) $f(0) = 6 - 2(0)$
$= 6$
b) $f(-5) = 6 - 2(-5)$
$= 16$

81. $f(x) = 3x - 5$
$m = 3$ and $b = -5$

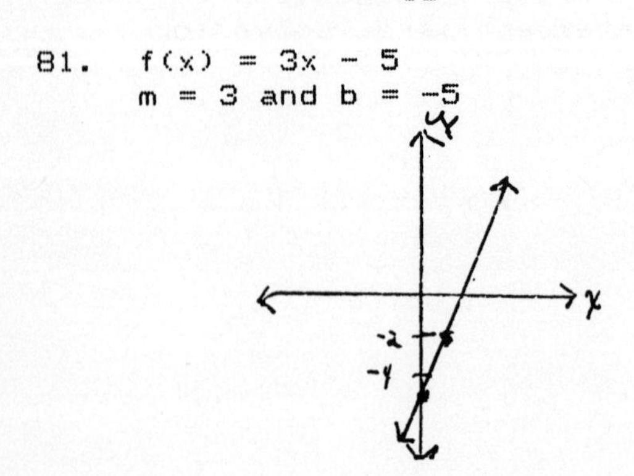


83. $g(x) = \frac{1}{2}x + 2$

$m = \frac{1}{2}$ b = 2

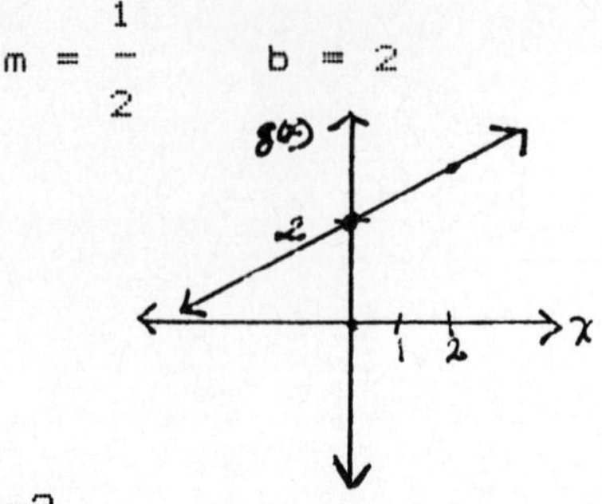


85. $y \ge -3$
Graph the line y = -3 using a solid line. Use (0,0) as a check point. Since $0 \ge -3$ shade the area containing the point (0,0).

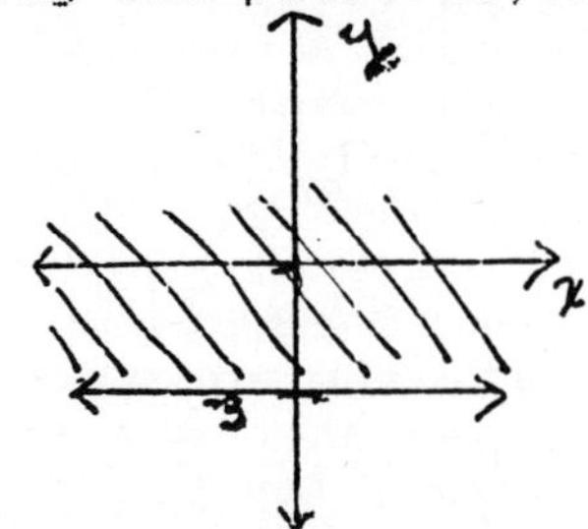


87. $y < 3x$
Graph the line y = 3x using a dashed line. Use the point (1,1) as a check point.
Since $1 < 3(1)$, shade the area containing the point (1,1).

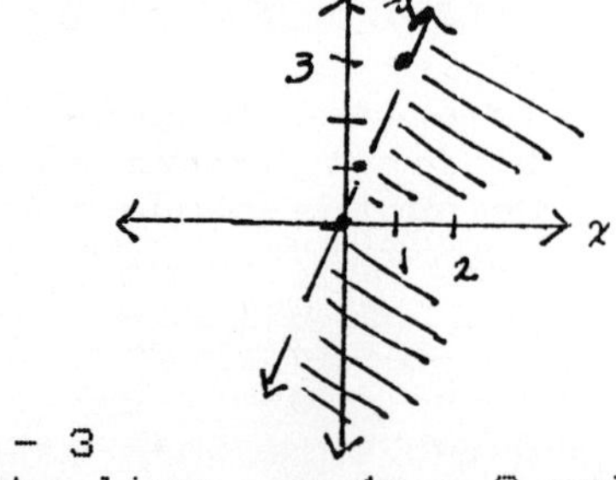


89. $y \le 4x - 3$
Graph the line y = 4x - 3 using a solid line. Use the point (0,0) as a check point. Since $0 \not\le 4\cdot(0) - 3$ shade the area of the graph that does not contain (0,0).

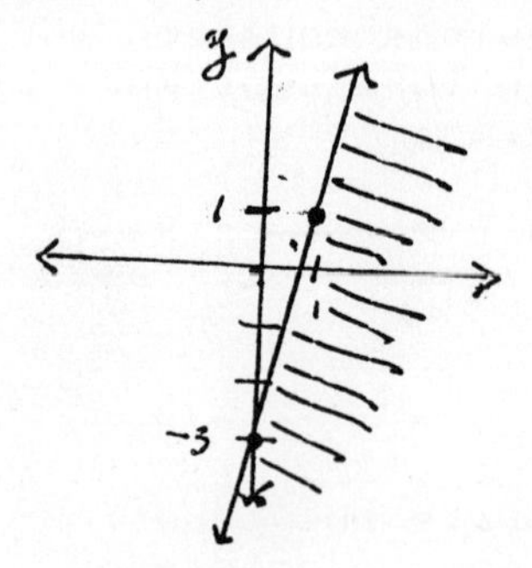

91. $y < -x + 4$
Graph the line $y = -x + 4$ using a dashed line. Use the point (0,0) as a check point. Since $0 < -0 + 4$, shade the area of the graph which contains (0,0).

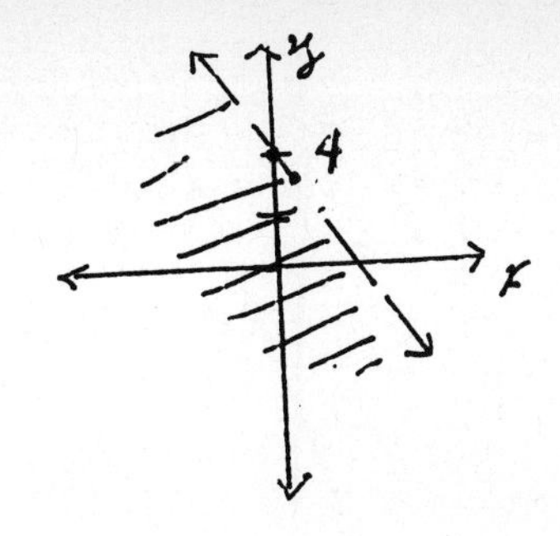


## PRACTICE TEST

1. The length of the line segment is the distance between (1,3) and (-2,-1). Using the distance formula:

$$d = \sqrt{(1 - (-2))^2 + (3 - (-1))^2}$$

$$= \sqrt{3^2 + 4^2}$$

$$= \sqrt{25} = 5$$

The length of the line segment is 5 units.
Using the midpoint formula:

$$\text{midpoint} = \left[\frac{1 + (-2)}{2}, \frac{3 + (-1)}{2}\right]$$

$$= \left[\frac{-1}{2}, 1\right]$$

2. To find the slope and the y-intercept, write the equation in slope-intercept form.

$$4x - 9y = 15$$
$$-9y = -4x + 15$$
$$y = \frac{4}{9}\cdot x - \frac{5}{3}$$

The slope is 4/9 and the y-intercept is -5/3.

3. Two points on the line are (0,-3) and (1,0). To write the equation we must determine the slope of the line. Using the slope formula:

$$m = \frac{-3 - 0}{0 - 1} = 3$$

Using the point-slope formula:

$$y - 0 = 3(x - 1)$$

To write the equation in slope-intercept form, solve for y.

$$y = 3x - 3$$

4. Using the point-slope formula, a slope of 4, and the point on the line (-1,3):

$$y - 3 = 4(x - (-1))$$
$$y - 3 = 4x + 4$$
$$y = 4x + 7$$

5. To write the equation, we must determine the slope. Using the given points on the line and the slope formula:

$$m = \frac{-1 - 2}{3 - (-4)} = \frac{-3}{7}$$

Using the point-slope formula and one of the given points:

$$y - (-1) = \frac{-3}{7}\cdot(x - 3)$$

To write the equation in slope-intercept form, solve for y.

$$y + 1 = \frac{-3}{7}\cdot x + \frac{9}{7}$$

$$y = \frac{-3}{7}\cdot x + \frac{2}{7}$$

6. To write the equation of the line we need to determine the slope. The line we wish to write the equation for is perpendicular to $2x + 3y = 6$. Therefore, the lines will have slopes which are negative reciprocals of one another. Find the slope of $2x + 3y = 6$ by solving for y.

$$2x + 3y = 6$$
$$3y = -2x + 6$$
$$y = \frac{-2}{3}\cdot x + 2$$
$$m = \frac{-2}{3}$$

The slope of the line we wish to write the equation for is 3/2. Using the point-slope formula and the point (-1,4):

$$y - 4 = \frac{3}{2}\cdot(x - (-1))$$

To write the equation in slope-intercept form, solve for y.

$$y - 4 = \frac{3}{2}\cdot x + \frac{3}{2}$$
$$y = \frac{3}{2}\cdot x + \frac{11}{2}$$

7. $y = 2x - 2$
$m = 2,\ b = -1$

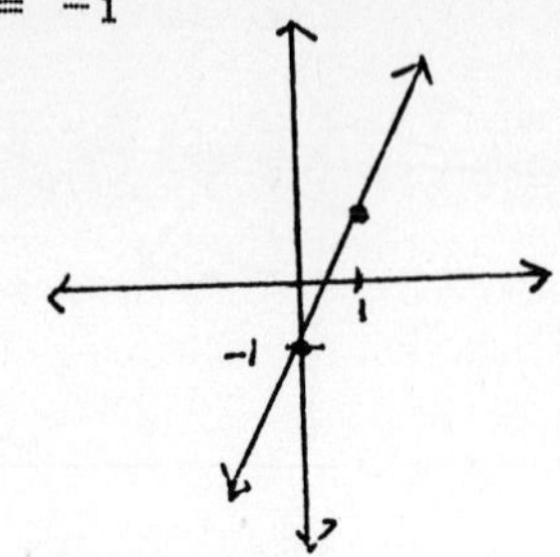

8. $2x + 3y = 10$
$3y = -2x + 10$
$$y = \frac{-2}{3}x + \frac{10}{3}$$

| x value | y value |
|---|---|
| 0 | $y = \frac{-2}{3}\cdot 0 + \frac{10}{3}$ <br> $y = \frac{10}{3}$ |
| 1 | $y = \frac{-2}{3}\cdot 1 + \frac{10}{3}$ <br> $y = \frac{8}{3}$ |
| 2 | $y = \frac{-2}{3}\cdot 2 + \frac{10}{3}$ <br> $y = \frac{6}{3} = 2$ |

Ordered Pairs:
(0,10/3)
(1,8/3)
(2,2)

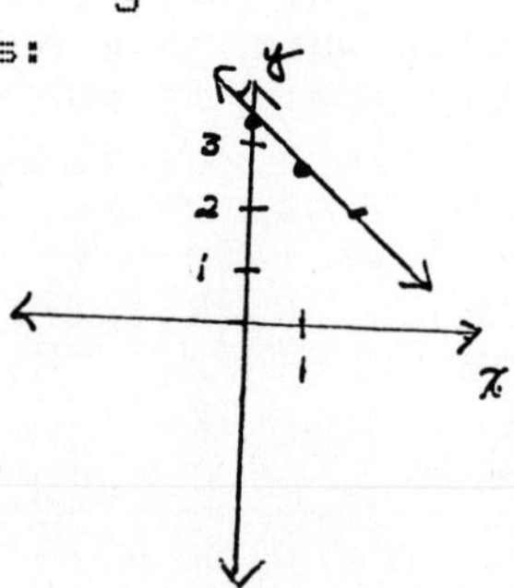

9. a) $f(c) = 2.5c - 800$

$m = 2.5 = 2\ 1/2 = 5/2$
$b = -800$

9 b) Using the graph, approximately 1900 tapes would generate a profit of $4000.

c) Using the graph, the break even point occurs at approximately 320 tapes.

10. Domain: {4,2,1/2,6}
Range: {0,-3,2,9}

11. a) Function
b) The graph does not pass the vertical line test. Therefore, it is a relation and not a function.

12. $f(x) = 3x - 5$
a) $f(2) = 3(2) - 5 = 1$
b) $f(-4) = 3(-4) - 5 = -17$

13.
$$f(x) = \frac{2}{3}x + 1$$
$$m = \frac{2}{3} \qquad b = 1$$

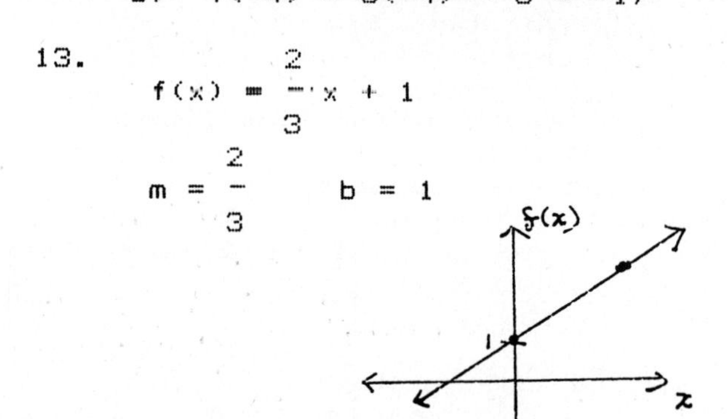

14. $y \geq -3x + 5$
Graph the line $y = -3x + 5$.
Use the point (0,0) as a check point. Since $0 \not\geq -3(0)+5$, shade the region which does not contain (0,0).

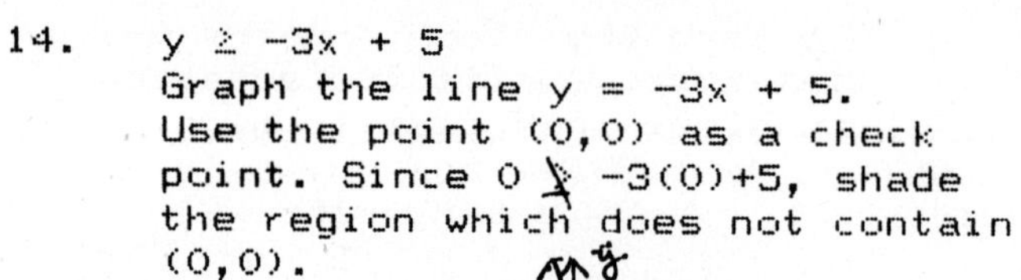

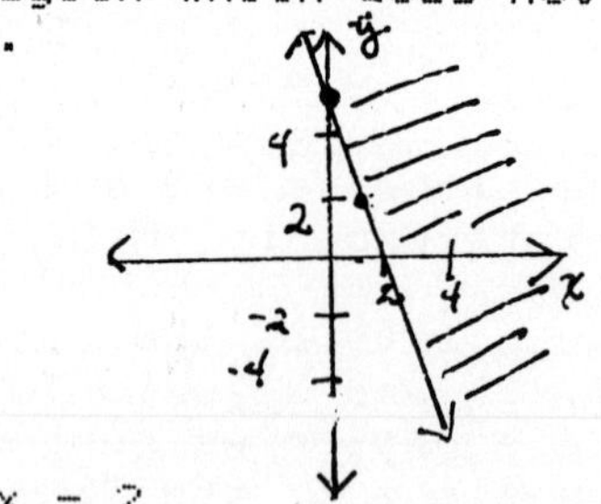

15. $y \leq 4x - 2$
Graph the line $y = 4x - 2$. Use the point (0,0) as a check point. Since $0 \not\leq 4(0) - 2$, shade the area which does not contain (0,0).

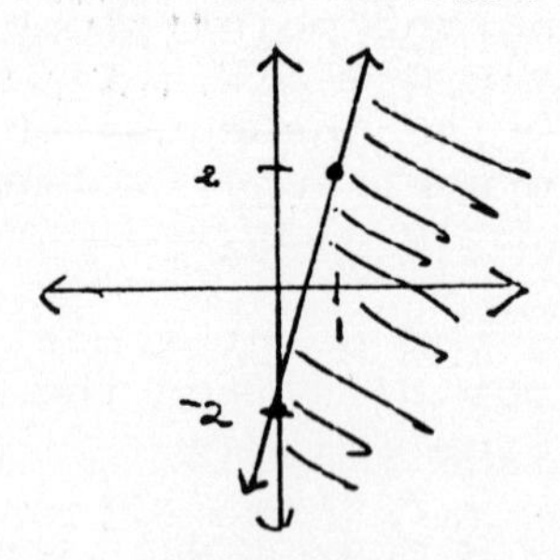

# CHAPTER 4

## EXERCISE SET 4.1

1. $y = -6x$
   $y = -2x + 8$
   a) Substitute (0,0) into each equation. $0 = -6(0)$ but $0 \neq -2(0) + 8$.
   b) Substitute (−4,16) into each equation. $16 \neq -6(-4)$
   c) Substitute (−2,12) into each equation. $12 = -6(-2)$ and $12 = -2(-2) + 8$.
   (−2,12) satisfies the system.

3. $y = 2x + 4$
   $y = 2x - 1$
   a) (0,4) $4 = 2(0) + 4$ but $4 \neq 2(0) - 1$.
   b) (3,10) $10 = 2(3) + + 4$ but $10 \neq 2(3) - 1$.
   c) (−2,0) $0 \neq 2(-2) - 1$
   None of the three points is a solution to the system.

5. $0.5y = -.5x + 2$
   $2y = -2x + 8$
   a) (2,5) $.5(5) \neq -.5(2) + 2$
   b) (1,3) $.5(3) = -.5(1) + 2$ and $2(3) = -2(1) + 8$
   c) (5,−1) $.5(-1) = -.5(5) + 2$ and $2(-1) = -2(5) + 8$
   The point (1,3) and (5,−1) are solutions to the system.

7. $2x + 3y = 6$
   $-2x + 5 = y$
   a) (1/2,5/3) $2(1/2) + 3(5/3) = 6$ but $-2(1/2) + 5 \neq 5/3$
   b) (2,1) $2(2) + 3(1) \neq 16$
   c) (9/4,1/2) $2(9/4) + 3(1/2) = 6$ and $-2(9/4) + 5 = 1/2$
   The point (9/4,1/2) satisfies the system.

9. $4x + y - 3z = 1$
   $2x - 2y + 6z = 11$
   $-6x + 3y + 12z = -4$
   a) (2,−1, −2)
   $4(2) + (-1) - 3(-2) \neq 1$
   b) (1/2,2,1)
   $2(1/2) -2(2) + 6(1) \neq 11$
   c) (1/2,−3,2/3)
   $4(1/2) + (-3) -3(2/3) \neq 1$
   None of the points is a solution the the system.

11. Write each equation in slope intercept form by solving for y
    $2y = -x + 5$ $\quad x - 2y = 1$
    $y = \frac{-1}{2}x + \frac{5}{2}$ $\quad y = \frac{2}{3}x - 2$
    $m = -1/2$ $\quad m = 2/3$
    Since the slopes are not the same. Therefore, the system is consistent and there is one solution.

13. Write each equation in slope intercept form by solving for y.
    $3y = 2x + 3$ $\quad y = \frac{2}{3}x - 2$
    $y = \frac{2}{3}x + 1$
    $m = 2/3,\ b = 1$ $\quad m = 2/3,\ b = -2$
    Since the slopes are the same but the y-intercepts differ, the system in inconsistent and has no solution.

15. Write each equation in slope intercept form by solving for y.
    $2x - 3y = 4$ $\quad 3x - 2y = -2$
    $y = \frac{2}{3}x - \frac{4}{3}$ $\quad y = \frac{3}{2}x + 1$
    $m = 2/3$ $\quad m = 3/2$
    Since the slopes differ, the system is consistent and has one solution.

17. Write each equation in slope intercept form by solving for y.
    $2x = 3y + 4$ $\quad 6x - 9y = 12$
    $y = \frac{2}{3}x - \frac{4}{3}$ $\quad y = \frac{2}{3}x - \frac{4}{3}$
    Both the slopes and y-intercepts are equal. The equations represent the same line so the system is dependent and there are an infinite number of solutions.

19. Write each equation in slope intercept form by solving for y.
    $y = \frac{3}{2}x + \frac{1}{2}$ $\quad 3x - 2y = \frac{-1}{2}$
    $y = \frac{3}{2}x + \frac{1}{4}$
    The equations have the same slopes but different y-intercepts. Therefore, the lines are parallel. The system in inconsistent and there is no solution.

21. Graph each equation.
    $y = x + 4$ $\quad y = -x + 2$

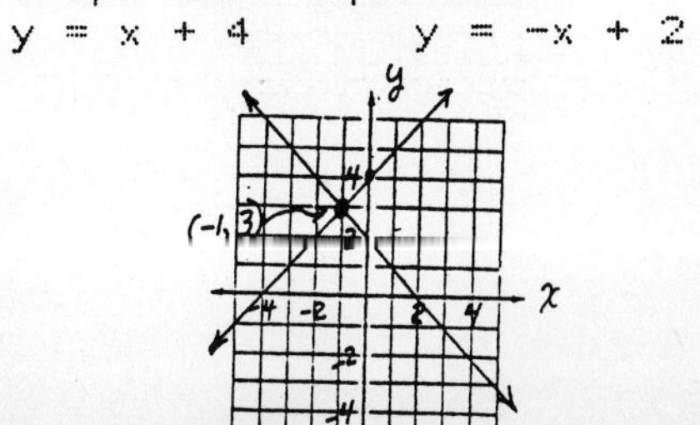

23. Graph each equation.
$y = 2x - 1 \qquad 2y = 4x + 6$

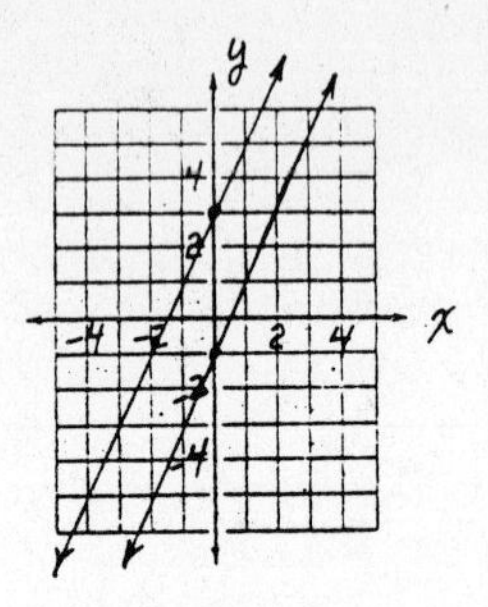


The lines are parallel. The system in inconsistent and there is no solution.

25. Graph each equation.
$2x + 3y = 6 \qquad 4x = -6y + 12$

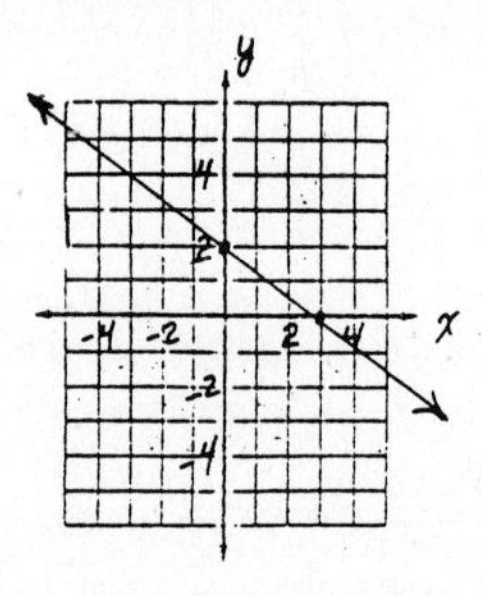


The equations are the same line. The system is dependent and there are an infinite number of solutions.

27. Graph each equation.
$x + 3y = 4 \qquad x = 1$
Solution: (1,1)

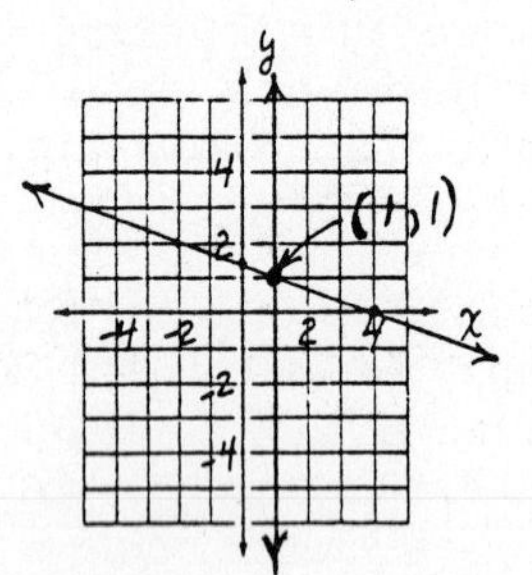


29. Graph each equation.
$y = -5x + 5 \qquad y = 2x - 2$
Solution: (1,0)

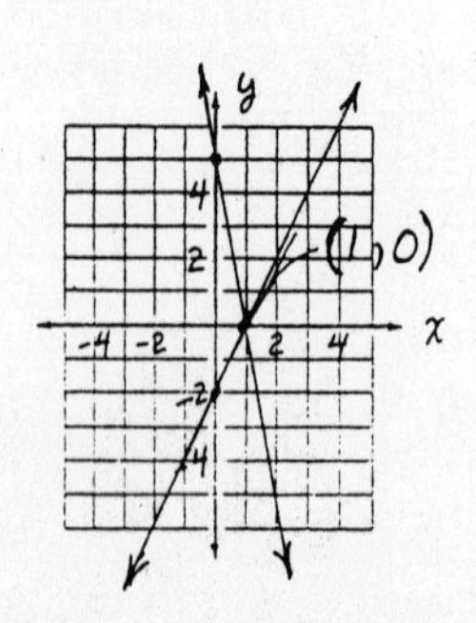


31. Substitute $x = 2y + 1$ for x in the second equation and solve for y.
$(2y + 1) + 2y = 9$
$4y + 1 = 9$
$4y = 8$
$y = 2$
Substitute $y = 2$ into $x = 2y + 1$.
$x = 2(2) + 1 = 5$
Solution: (5,2)

33. Substitute $x = y$ for x in the equation $x + y = 6$
$(y) + y = 6$
$2y = 6$
$y = 3$
Substitute $y = 3$ into $x = y$.
$x = 3$
Solution: (3,3)

35. Solve $2x + y = 3$ for y.
$y = -2x + 3$
Substitute $y = -2x + 3$ for y in $2x + y + 5 = 0$.
$2x + (-2x + 3) + 5 = 0$
$8 = 0$
Since 8 does not equal 0, the system is inconsistent and there is no solution.

37. Substitue $x = 1/2$ into the equation

$x + \frac{1}{3}\cdot y + 6 = 0$ and solve for y.

$$\frac{1}{2} + \frac{1}{3}\cdot y + 6 = 0$$

$$6\cdot\frac{1}{2} + 6\cdot\frac{1}{3}\cdot y + 6\cdot 6 = 0\cdot 6$$

$$3 + 2y + 36 = 0$$
$$2y + 39 = 0$$
$$2y = -39$$
$$y = \frac{-39}{2}$$

Solution: $\left[\frac{1}{2}, \frac{-39}{2}\right]$

39. Solve for x: $x - \frac{1}{2}\cdot y = 2$

$$x = 2 + \frac{1}{2}\cdot y$$

Substitute for x in $y = 2x - 4$.

$$y = 2\left[2 + \frac{1}{2}\cdot y\right] - 4$$

$y = 4 + y - 4$
$0 = 0$
There are an infinite number of solution.

41. Substitue $y = 3x + 5$ for $y$

$3x + (3x + 5) = -1$

$6x = -6$

$x = -1$

Substitute $x = -1$ in $y = 3x + 5$

$y = 3(1) + 5 = 2$

Solution: $(-1,2)$

43. Substitute $y = 2x - 13$ for $y$ in $-4x - 7 = 9y$.

$-4x - 7 = 9(2x - 13)$

$-4x - 7 = 18x - 117$

$-22x = 110$

$x = 5$

Substitute $x = 5$ into $y = 2x - 13$.

$y = 2(5) - 13 = -3$

Solution: $(5,-3)$

45. Solve $5 = y - 3x$ for $y$.

$y = 3x + 5$

Substitute $y = 3x + 5$ into $5x - 2y = -7$ and solve for $x$.

$5x - 2(3x + 5) = -7$

$5x - 6x - 10 = -7$

$-x = 3$

$x = -3$

Substitute $x = -3$ into $y = 3x + 5$ and solve for $y$.

$y = 3(-3) + 5 = -4$

Solution: $(-3,-4)$

47. Substitute $x = 3y + 5$ in

$$y = \frac{2}{3}\cdot x + \frac{1}{2}$$

$$y = \frac{2}{3}\cdot(3y + 5) + \frac{1}{2}$$

$$6\cdot y = 6\cdot\frac{2}{3}\cdot(3y + 5) + 6\cdot\frac{1}{2}$$

$6y = 12y + 20 + 3$

$6y = 12y + 23$

$-23 = 6y$

$$\frac{-23}{6} = y$$

Substitute $y = -23/6$ in $x = 3y + 5$.

$$x = 3\left[\frac{-23}{6}\right] + 5$$

$$= \frac{-23}{2} + \frac{10}{2} = \frac{-13}{2}$$

Solution: $\left[\frac{-13}{2}, \frac{-23}{6}\right]$

49. Solve for $x$ in $\frac{1}{2}\cdot x - \frac{1}{3}\cdot y = 2$

$$6\cdot\frac{1}{2}\cdot x - 6\cdot\frac{1}{3}\cdot y = 6\cdot 2$$

$3x - 2y = 12$

$3x = 2y + 12$

$$x = \frac{2}{3}\cdot y + 4$$

Substitute for $x$ in $\frac{1}{4}\cdot x + \frac{2}{3}\cdot y = 6$

$$\frac{1}{4}\left[\frac{2}{3}\cdot y + 4\right] + \frac{2}{3}\cdot y = 6$$

$$\frac{1}{6}\cdot y + 1 + \frac{2}{3}\cdot y = 6$$

$$6\cdot\frac{1}{6}\cdot y + 6\cdot 1 + 6\cdot\frac{2}{3}\cdot y = 6\cdot 6$$

$y + 6 + 4y = 36$

$5y = 30$

$y = 6$

Substitute $y = 6$ in $x = \frac{2}{3}\cdot y + 4$

$$x = \frac{2}{3}\cdot 6 + 4$$

$x = 4 + 4 = 8$

Solution: $(8,6)$

51.

$$\begin{array}{l} x + y = -2 \\ x - y = 4 \\ \hline 2x \quad = 2 \\ x \quad = 1 \end{array}$$

Substitute $x = 1$ into either equation and solve for $y$.

$1 + y = -2$

$y = -3$

Solution: $(1,-3)$

53.

$$\begin{array}{r} -x + y = 5 \\ x + 2y = 1 \\ \hline 3y = 6 \\ y = 2 \end{array}$$

Substitute $y = 2$ into either equation and solve for $x$.

$-x + 2 = 5$

$-x = 3$

$x = -3$

Solution: $(-3,2)$

55. $3x + 2y = 15$
$x - 2y = -7$
---
$4x = 8$
$x = 2$
Substitute $x = 2$ into either equation.
$3(2) + 2y = 15$
$6 + 2y = 15$
$2y = 9$
$y = 9/2$
Solution: $(2,9/2)$

57. 1) $3x + y = 6$
2) $-6x - 2y = 10$
Multiply equation 1) by 2. Add the result to equation 2) to eliminate y and solve for x.
1) $2(3x + y = 6)$
$6x + 2y = 12$

1) $6x + 2y = 12$
2) $-6x - 2y = 10$
---
$0 = 22$
Since 0 never equals 22, the system is inconsistent and there is no solution.

59. 1) $2x + y = 6$
2) $3x - 2y = 16$
Multiply equation 1) by 2. Add the result to equation 2) to eliminate y and solve for x.
1) $2(2x + y = 6)$
$4x + 2y = 12$

1) $4x + 2y = 12$
2) $3x - 2y = 16$
---
$7x = 28$
$x = 4$
Substitute $x = 4$ into either equation 1) or 2) and solve for y.
1) $2(4) + y = 6$
$y = -2$
Solution: $(4,-2)$

62. 1) $2x - 5y = 13$
2) $5x + 3y = 17$
Multiply equation 1) by 3 and equation 2 by 5. Add the resulting equations and solve for x.
1) $3(2x - 5y = 13)$
$6x - 15y = 39$
2) $5(5x + 3y = 17)$
$25x + 15y = 85$

1) $6x - 15y = 39$
2) $25x + 15y = 85$
---
$31x = 124$
$x = 4$

61 cont.
Substitute $x = 4$ into either equation 1) or 2) and solve for y.
1) $2(4) - 5y = 13$
$8 - 5y = 13$
$-5y = 5$
$y = -1$
Solution: $(4,-1)$

63. $3y = 2x + 4$
$3y = 2x + 4$
The equations represent the same line. Therefore, the system is dependent and there are an infinite number of solutions.

65. 1) $4x - 3y = 8$
2) $-3x + 4y = 9$
Multiply equation 1) by 3 and equation 2) by 4. Add the results to eliminate x and solve for y.
1) $3(4x - 3y = 8)$
$12x - 9y = 24$
2) $4(-3x + 4y = 9)$
$-12x + 16y = 36$

1) $12x - 9y = 24$
2) $-12x + 16y = 36$
---
$7y = 60$
$y = 60/7$
Substitute $y = 60/7$ into either equation 1) or 2) and solve for x.
1)
$$4x - 3\left[\frac{60}{7}\right] = 8$$

$$7 \cdot 4x - 7 \cdot 3\left[\frac{60}{7}\right] = 7 \cdot 8$$
$28x - 180 = 56$
$28x = 236$
$x = 236/28 = 59/7$
Solution: $(59/7,60/7)$

67. Write each equation in standard form.
1) $3x + 4y = 2$
2) $2x + 5y = -1$
Multiply equation 1) by 2 and equation 2) by $-3$. Add the results to eliminate x and solve for y.
1) $6x + 8y = 4$
2) $-6x - 15y = 3$
---
$-7y = 7$
$y = -1$
Substitute $y = -1$ into either equation 1) or 2) and solve for x.
1) $3x + 4(-1) = 2$
$3x = 6$
$x = 2$
Solution: $(2,-1)$

69\. Write each equation in standard form.

1) $5x + 2y = -3$
2) $4x - 7y = 3$

Multiply equation 1) by 7 and equation 2) by 2 to eliminate y and solve for x.

1) $35x + 14y = -21$
2) $8x - 14y = 6$

$43x = -15$

$x = -15/43$

Substitute $x = -15/43$ into either equation 1) or 2) and solve for y.

1) $5\left[\dfrac{-15}{43}\right] + 2y = -3$

$43 \cdot 5\left[\dfrac{-15}{43}\right] + 43 \cdot 2y = 43 \cdot (-3)$

$-75 + 86y = -129$

$86y = -54$

$y = -54/86 = -27/43$

Solution: $(-15/43, -27/43)$

71\. 1) $4x + 5y = 3$
2) $2x - 3y = 4$

Multiply equation 2) by $-2$. Add the result to equation 1 to eliminate x and solve for y.

2) $-2(2x - 3y = 4)$

$-4x + 6y = -8$

1) $4x + 5y = 3$
2) $-4x + 6y = -8$

$11y = -5$

$y = -5/11$

Substitititute $y = -5/11$ into either equation and solve for x.

1) $4x + 5\left[\dfrac{-5}{11}\right] = 3$

$11 \cdot 4x + 11 \cdot 5\left[\dfrac{-5}{11}\right] = 11 \cdot 3$

$44x - 25 = 33$

$44x = 58$

$x = 58/44 = 29/22$

Solution: $(29/22, -5/11)$

73\. 1) $.2x + .5y = 1.6$
2) $-.3x + .4y = -.1$

Multiply equation 1) by 3 and equation 2) by 2. Add the result to eliminate x and solve for y.

1) $.6x + 1.5y = 4.8$
2) $-.6x + .8y = -.2$

$2.3y = 4.6$

$y = 2$

73\. cont.

Subsitute $y = 2$ into either equation 1) or 2) and solve for y.

1) $.2x + .5(2) = 1.6$

$.2x + 1 = 1.6$

$.2x = .6$

$x = 3$

Solution: $(3,2)$

75\. 1) $2.1x - .6y = 8.40$
2) $-1.5x - .3y = -6.00$

Multiply equation 2) by $-2$. Add the result to equation 1) eliminate y and solve for x.

1) $2.1x - .6y = 8.4$
2) $3\ x + .6y = 12$

$5.1\ x = 20.4$

$x = 4$

Substitute $x = 4$ into either equation 1) or 2) and solve for y.

1) $2.1(4) - .6y = 8.4$

$8.4 - .6y = 8.4$

$-.6y = 0$

$y = 0$

Solution: $(4,0)$

77\. 1) $2x - \dfrac{1}{3} y = 6$

2) $5x - y = 4$

Multiply equation 1) by $-3$. Add the result to equation 2 to eliminate y and solve for x.

1) $-6x + y = -18$
2) $5x - y = 4$

$-x = -14$

$x = 14$

Substitute $x = 14$ into either equation 1) or 2) and solve for y.

2) $5(14) - y = 4$

$70 - y = 4$

$-y = -66$

$y = 66$

Solution: $(14,66)$

79\. 1) $\dfrac{x}{3} = 4 - \dfrac{y}{4}$

2) $3x = 4y$

Write equation in standard form.

1) $12\left[\dfrac{x}{3} = 4 - \dfrac{y}{4}\right]$

$4x = 48 - 3y$

1) $4x + 3y = 48$
2) $3x - 4y = 0$

Multiply equation 1) by 4 and equation 2) by 3 to eliminate y and solve for x.

79. cont.

1) $16x + 12y = 192$
2) $9x - 12y = 0$

$25x = 192$
$x = 192/25$

Substitute $x = 192/25$ into either equation 1) or equation 2) to solve for y.

2)
$$3\left[\frac{192}{25}\right] = 4y$$

$$25 \cdot 3\left[\frac{192}{25}\right] = 25 \cdot 4y$$

$576 = 100\,y$
$y = 576/100 = 144/25$
Solution: $(192/25, 144/25)$

81. 1)
$$\frac{x}{5} + \frac{y}{2} = 4$$

2)
$$\frac{2x}{3} - y = \frac{8}{3}$$

Multiply equation 1) by 10 and equation 2) by $-3$. Add the results to eliminate x and solve for y.

1) $2x + 5y = 40$
2) $-2x + 3y = -8$

$8y = 32$
$y = 4$

Substitute $y = 4$ into either equation 1) or 2) to solve for x.

1) $2x + 5(4) = 40$
$2x = 20$
$x = 10$

Solution: $(10,4)$

83. 1) $2(2x + y) = -2x + y + 8$
2) $3(x - y) = -(y + 1)$

Write equation in standard form.

1) $6x + y = 8$
2) $3x - 2y = -1$

Multiply equation 1) by 2. Add the result to equation 2) to eliminate y and solve for x.

1) $12x + 2y = 16$
2) $3x - 2y = -1$

$15x = 15$
$x = 1$

Substitute $x = 1$ into either equation 1) or 2) and solve for x.

1) $2(2(1) + y) = -2(1) + y + 8$
$4 + 2y = 6 + y$
$y = 2$

Soluiton: $(1,2)$

85. Write both equations in slope intercept form and compare their slopes and y-intercepts. If the slopes equal and the y-intercepts differ, the lines are parallel. The system is inconsistent. If the slopes equal and the y-intercepts also equal, the lines are the same. The system is dependent. If the lines have different slopes, the system is consistent.

87. At some point during the solving process, both sides of the equation will be the same.

89. a) yes
b) yes

Both the rational and irrational numbers are subgroups of the real numbers.

91.
$$A = P\left[1 + \frac{r}{n}\right]^t$$

$P = 500$, $r = .08$, $n = 2$
$t = 1$

$$A = 500\left[1 + \frac{.08}{2}\right]^1$$

$= 500(1 + .04)$
$= 500(1.04)$
$= 520$

JUST FOR FUN

1.

1)
$$\frac{x + 2}{2} - \frac{y + 4}{3} = 4$$

2)
$$\frac{x + y}{2} = \frac{1}{2} + \frac{x - y}{3}$$

Write each equation in standard form.

1)
$$6 \cdot \frac{x + 2}{2} - 6 \cdot \frac{y + 4}{3} = 6 \cdot 4$$

$3(x + 2) - 2(y + 4) = 24$
$3x + 6 - 2y - 8 = 24$

1) $3x - 2y = 26$

2)
$$6 \cdot \frac{x + y}{2} = 6 \cdot \frac{1}{2} + 6 \cdot \frac{x - y}{3}$$

JUST FOR FUN 1. cont.

$3(x + y) = 3 + 2(x - y)$
$3x + 3y = 3 + 2x - 2y$
2) $x + 5y = 3$
Multiply equation 2) by $-3$. Add the results to equation 1) to eliminate x and solve for y.

1) $3x - 2y = 26$
2) $-3x - 15y = -9$

$-17y = 17$
$y = -1$

Substitute $y = -1$ into either equation 1) or 2) to solve for x.

2) $x + 5(-1) = 3$
$x - 5 = 3$
$x = 8$
Solution: $(8,-1)$

3. 1) $\frac{3}{a} + \frac{4}{b} = -1$

2) $\frac{1}{a} + \frac{6}{b} = 2$

Multiply equation 2) by $-3$. Add the result to equation 1) to eliminate a and solve for b.

2) $(-3)\cdot\left[\frac{1}{a} + \frac{6}{b} = 2\right]$

EXERCISE SET 4.2

1. 1) $x = 1$
2) $2x + y = 4$
3) $-3x - y + 4z = 15$
Substitute $x = 1$ into equation 2) to solve for y.
2) $2(1) + y = 4$
$y = 2$
Substitute $x = 1$ and $y = 2$ into equation 3) and solve for z.
3) $-3(1) - 2 + 4z = 15$
$-5 + 4z = 15$
$4z = 20$
$z = 5$
Solution: $(1,2,5)$

3. 1) $5x - 6z = -17$
2) $3x - 4y + 5z = -1$
3) $2z = -6$
Solve equation 3) for z.
3) $2z = -6$
$z = -3$
Substitute $z = -3$ into equation 1) and solve for x.

JUST FOR FUN 2. CONT.

2) $\frac{-3}{a} - \frac{18}{b} = -6$

1) $\frac{3}{a} + \frac{4}{b} = -1$

$\frac{-14}{b} = -7$

$-14 = -7b$
$2 = b$

Substitute $b = 2$ into either equation 1) or 2) and solve for a.

1) $\frac{3}{a} + \frac{4}{2} = -1$

$\frac{3}{a} + 2 = -1$

$a\cdot\frac{3}{a} + a\cdot 2 = a\cdot(-1)$

$3 + 2a = -a$

$3 = -3a$

$a = -1$

Solution: $(-1,2)$

3. cont.
1) $5x - 6(-3) = -17$
$5x + 18 = -17$
$5x = -35$
$x = -7$
Substitute $x = -7$ and $z = -3$ into equation 2) to solve for y.
2) $3(-7) - 4y + 5(-3) = -1$
$-21 - 4y - 15 = -1$
$-4y - 36 = -1$
$-4y = 35$
$y = -35/4$
Solution: $(-7, -35/4, -3)$

5. 1) $x + 2y = 6$
2) $3y = 9$
3) $x + 2z = 12$
Solve for y in equation 2):
$3y = 9 \quad y = 3$
Substitute $y = 3$ into equation 1) and solve for x
1) $x + 2(3) = 6$
$x = 0$
Substitute $x = 0$ and into equation 3) to solve for z.
3) $0 + 2z = 12 \quad z = 6$
Solution: $(0,3,6)$

7\. 1) $x + y - z = -3$
2) $x + z = 2$
3) $2x - y + 2z = 3$
Add eqautions 1) and 3) to eliminiate y.
1) $x + y - z = -3$
3) $2x - y + 2z = 3$

4) $3x + z = 0$
Multiply equation 2) by -1. Add the result to equation 4 to eliminate z and solve for x.
2) $-x - z = -2$
4) $3x + z = 0$

$2x = -2$
$x = -1$
Substitute $x = -1$ into equation 2) to solve for z.
2) $-1 + z = 2$
$z = 3$
Substitute $x = -1$ and $z = 3$ into equation 1) to solve for y.
1) $-1 + y - 3 = -3$
$y = 1$
Solution: (-1,1,3)

9\. 1) $x - 2z = -5$
2) $-y + 3z = 3$
3) $-2x + z = 4$
Multiply equation 3) by 2. Add the result to equation 1 to eliminate z.
3) $-4x + 2z = 8$
1) $x - 2z = -5$

$-3x = 3$
$x = -1$
Substitute $x = -1$ into equation 1) and solve for z.
1) $-1 - 2z = -5$
$-2z = -4$
$z = 2$
Substitute $z = 2$ into equation 2) and solve for y.
2) $-y + 3(2) = 3$
$-y = -3$
$y = 3$
Solution: (-1,3,2)

11\. 1) $x + y + z = 4$
2) $x - 2y - z = 1$
3) $2x - y - 2z = -1$
Add equations 1) and 2) to eliminate z.
1) $x + y + z = 4$
2) $x - 2y - z = 1$

$2x - y = 5$
Multiply equation 1) by 2. Add the result to equation 3)

11\. cont.
to eliminate z.
1) $2x + 2y + 2z = 8$
3) $2x - y - 2z = -1$

5) $4x + y = 7$
Add equations 4) and 5) to eliminate y and solve for x.
4) $2x - y = 5$
5) $4x + y = 7$

$6x = 12$
$x = 2$
Substitute $x = 2$ into equation 4) or 5) and solve for y.
5) $4(2) + y = 7$
$y = -1$
Substitute $x = 2$ and $y = -1$ into equations 1),2),or 3) and solve for z.
1) $2 + (-1) + z = 4$
$1 + z = 4$
$z = 3$
Solution: (2, -1, 3)

13\. 1) $2x - 2y + 3z = 5$
2) $2x + y - 2z = -1$
3) $4x - y - 3z = 0$
Add equations 2) and 3) to eliminate y.
2) $2x + y - 2z = -1$
3) $4x - y - 3z = 0$

4) $6x - 5z = -1$

Multiply equation 2) by 2. Add to result to equation 1) to eliminate y.
1) $2x - 2y + 3z = 5$
2) $4x + 2y - 4z = -2$

5) $6x - z = 3$

Multiply equation 5) by -1. Add the result to equation 4) to eliminate x and solve for z.
4) $6x - 5z = -1$
5) $-6x + z = -3$

$-4z = -4$
$z = 1$
Substitute $z = 1$ into either equation 4) or 5) to solve for x.
5) $6x - 1 = 3$
$6x = 4$
$x = 2/3$
Substitute $x = 2/3$ and $z = 1$ into equation 1), 2) or 3) and solve for y.
1) $2\left[\frac{2}{3}\right] - 2y + 3(1) = 5$

$y = \frac{-1}{3}$
Solution: (2/3, -1/3, 1)

15. 1) $x + 2y - 3z = 5$
2) $x + y + z = 0$
3) $3x + 4y + 2z = -1$

Eliminate x by adding equations 1 and 2.

1) $x + 2y - 3z = 5$
2) $-x - y - z = 0$

4) $y - 4z = 5$

Multiply equation 1 by −3. Add the result to equation 3 to eliminate x.
1) $-3x - 6y + 9z = -15$
3) $3x + 4y + 2z = -1$

5) $-2y + 11z = -16$

Multiply equation 4 by 2. Add the result to equation 5 to eliminate y and solve for z.

4) $2y - 8z = 10$
5) $-2y + 11z = -16$

$3z = -6$
$z = -2$

Substitute z = −2 into equation 4 to solve for y.
4) $y - 4(-2) = 5$
$y + 8 = 5$
$y = -3$

Substitute y = −3 and z = −2 into equation 1, 2, or 3 to solve for x.
1) $x + 2(-3) - 3(-2) = 5$
$x - 6 + 6 = 5$
$x = 5$

Solution: (5,−3,−2)

17. 1) $2x + 2y - z = 2$
2) $3x + 4y + z = -4$
3) $5x - 2y - 3z = 5$

Add equations 1 and 2 to eliminate z.
1) $2x + 2y - z = 2$
2) $3x + 4y + z = -4$

4) $5x + 6y = -2$

Multiply equation 2 by 3. Add the result to equation 3 to eliminate z.
2) $9x + 12y + 3z = -12$
3) $5x - 2y - 3z = 5$

5) $14x + 10y = -7$

Multiply equation 4 by 5 and equation 5 by −3 to eliminate y and solve for x.

18. cont.
4) $25x + 30y = -10$
5) $-42x - 30y = 21$

$-17x = 11$
$x = -11/17$

Substitute x = −11/17 into either equation 4 or 5 to solve for y.
4) $5(-11/17) + 6y = -2$
$17(-55/17) + 17(6y) = 17(-2)$
$-55 + 102y = -34$
$102y = 21$
$y = 21/102$
$= 7/34$

Substitute x = −11/17 and y = 7/34 into equation 1,2,or 3 to solve for z.

1) $$2\left[\frac{-11}{17}\right] + 2\left[\frac{7}{34}\right] - z = 2$$

$$17\left[\frac{-22}{17}\right] + 17\left[\frac{7}{17}\right] - 17\cdot z = 17\cdot 2$$

$-22 + 7 - 17z = 34$
$-15 - 17z = 34$
$-17z = 49$

$$z = \frac{-49}{17}$$

Solution: $\left[\frac{-11}{17}, \frac{7}{34}, \frac{-49}{17}\right]$

19. 1) $$\frac{-1}{4}\cdot x + \frac{1}{2}\cdot y - \frac{1}{2}\cdot z = -2$$

2) $$\frac{1}{2}\cdot x + \frac{1}{3}\cdot y - \frac{1}{4}\cdot z = 2$$

3) $$\frac{1}{2}\cdot x - \frac{1}{2}\cdot y + \frac{1}{4}\cdot z = 1$$

Eliminate the fraction by multiplying equation 1 by 4, equation 2 by 12, and equation 3 by 4.

1) $-x + 2y - 2z = -8$
2) $6x + 4y - 3z = 24$
3) $2x - 2y + z = 4$

Multiply equation 1 by 6. Add the result to equation 2 to eliminate x.
1) $-6x + 12y - 12z = -48$
2) $6x + 4y - 3z = 24$

4) $16y - 15z = -24$

19. cont.

Multiply equation 1) by 2. Add the result to equation 3 to eliminate x.

1) $-2x + 4y - 4z = -16$
2) $2x - 2y + z = 4$

---

5) $2y - 3z = -12$

Multiply equation 5 by $-8$. Add the result to equation 4 to eliminate y and solve for z.

4) $16y - 15z = -24$
5) $-16y + 24z = 96$

---

$9z = 72$
$z = 8$

Substitute $z = 8$ into equation 4 or 5 to solve for y.

5) $2y - 3(8) = -12$
$2y - 24 = -12$
$2y = 12$
$y = 6$

Substitute $y = 6$ and $z = 8$ into equations 1,2,or 3 to solve for x.

1) $-x + 2(6) -2(8) = -8$
$-x + 12 - 16 = -8$
$-x - 4 = -8$
$-x = -4$
$x = 4$

Solution: (4,6,8)

21. 1) $x - \frac{2}{3} \cdot y - \frac{2}{3} \cdot z = -2$

2) $\frac{2}{3} \cdot x + y - \frac{2}{3} \cdot z = \frac{1}{3}$

3) $\frac{-x}{4} + y - \frac{z}{4} = \frac{3}{4}$

Eliminate the fractions by multiplying equation 1 by 3, equation2 by 3 and equation 3 by 4.

1) $3x - 2y - 2z = -6$
2) $2x + 3y - 2z = 1$
3) $-x + 4y - z = 3$

Multiply equation 1 by $-1$,add equations 1 and 2 to eliminate z.

1) $-3x + 2y + 2z = 6$
2) $2x + 3y - 2z = 1$

---

4) $-x + 5y = 7$

Multiply equation 3 by $-2$. Add the result to equation 1 to eliminate z.

1) $3x - 2y - 2z = -6$
3) $2x - 8y + 2z = -6$

---

5) $5x - 10y = -12.$

21. cont.

4) $-5x + 25y = 35$
5) $5x -10y = -12$

---

$15 y = 23$
$y = 23/15$

Substitute $y = 23/15$ into equation 4 or 5 to solve for x.

4) $-x - 5\left[\frac{23}{15}\right] = -7$

$-x - \left[\frac{23}{3}\right] = -7$

$3 \cdot (-x) - 3 \cdot \left[\frac{23}{3}\right] = 3(-7)$

$-3x - 23 = -21$
$-3x = -2$
$x = 2/3$

Substitute $x = 2/3$ and $y = 23/15$ into equation 1,2,or 3 to solve for z.

1) $\frac{2}{3} - \frac{2}{3}\left[\frac{23}{15}\right] - \frac{2}{3} \cdot z = -2$

$\frac{2}{3} - \frac{46}{45} - \frac{2}{3} \cdot z = -2$

$30 - 46 - 30z = -90$
$-16 - 30z = -90$
$-30z = -74$
$z = 74/30 = 37/15$

Solution: $\left[\frac{2}{3}, \frac{23}{15}, \frac{37}{15}\right]$

23. No point is common to all three planes. The system is inconsistent because there is no point which satisfies all three equations.

25. A straight line is common to all three planes. Therefore there are an infinite number of points common to all three planes. The system is dependent.

27. 1) $3x - 4y + z = 4$
2) $x + 2y + z = 4$
3) $-6x + 8y - 2z = -8$

Add equations 1 and 2 to eliminate z.
1) $-3x + 4y - z = -4$
2) $x + 2y + z = 4$

4) $-2x + 6y = 0$
Multiply equation 2 by 2. Add the result to equation 3 to eliminate z.
1) $2x + 4y + 2z = 8$
3) $-6x + 8y - 2z = -8$

5) $-4x + 12y = 0$
Multiply equation 4 by -2. Add the result to eqaution 5.
4) $4x - 12y = 0$
5) $-4x + 12y = 0$

$0 = 0$
The system is dependent. There are an infinite number of solutions.

29. 1) $x + 3y + 2z = 6$
2) $x - 2y - 2z = 8$
3) $-3x - 9y - 6z = -4$
Multiply equation 1 by 3. Add the result to equation 3.
1) $3x + 9y + 6z = 18$
3) $-3x - 9y - 6z = -4$

$0 = 14$
Since 0 cannot equal 14, the system is inconsistent. There is no solution.

31. Let $t - 1/6$ = time for Philippa and $t$ = time for Cameron.

| | distance | = rate | x time |
|---|---|---|---|
| Philippa | $5(t - 1/6)$ | 5 | $(t - 1/6)$ |
| Cameron | $3t$ | 3 | $t$ |

a) When Philippa catches Cameron they will have traveled the same distance. Therefore,
$5(t - 1/6) = 3t$
$5t - 5/6 = 3t$
$2t = 5/6$
$12t = 5$
$t = 5/12$
$t = 5/12$ of an hour.
$\frac{5}{12} \cdot 60\text{min} = 25\text{min}.$
Cameron will be overtaken by his mother 25 minutes after he leaves.
b) The distance traveled is $3t$. Therefore: $3(5/12) = 1.25$ miles.

33. $\left|\frac{3x - 4}{2}\right| - 1 < 5$

$\left|\frac{3x - 4}{2}\right| < 6$

$-6 < \frac{3x - 4}{2} < 6$

$-12 < 3x - 4 < 12$
$-8 < 3x < 16$
$-8/3 < x < 16/3$
$\{x \mid -8/3 < x < 16/3\}$

JUST FOR FUN

1. 1) $3a + 2b - c = 0$
2) $2a + 2c + d = 5$
3) $a + 2b - d = -2$
4) $2a - b + c + d = 2$
Multiply equation 1 by 2. Add the result to equation 2 to eliminate c.
1) $6a + 4b - 2c = 0$
2) $2a + 2c + d = 5$

5) $8a + 4b + d = 5$

Add equations 1 and 4 to eliminate c.
1) $3a + 2b - c = 0$
4) $2a - b + c + d = 2$

6) $5a + b + d = 2$

Add equations 2 and 3 to eliminate d.
2) $2a + 2c + d = 5$
3) $a + 2b - d = -2$

7) $3a + 2b + 2c = 3$
Multiply equation 1 by 2. Add the result to equation 7 to eliminate c.
1) $6a + 4b - 2c = 0$
7) $3a + 2b + 2c = 3$

8) $9a + 6b = 3$
Add equations 5 and 6 to eliminate d.
5) $-8a - 4b - d = -5$
6) $5a + b + d = 2$

9) $-3a - 3b = -3$
Multiply equation 9 by 2. Add the result to 8 to eliminate b and solve for a.
9) $-6a - 6b = -6$
8) $9a + 6b = 3$

$3a = -3$
$a = -1$

JUST FOR FUN CONT.

Substitute $a = -1$ into equation 8 and solve for b.

8) $9(-1) + 6b = 3$
$6b = 12$
$b = 2$

Substitute $a = -1$ and $b = 2$ into equation 1 and solve for c.

1) $3(-1) + 2(2) - c = 0$
$1 - c = 0$
$c = 1$

Subsitute $a = -1$ and $b = 2$ into equation 3 to solve for d.

3) $-1 + 2(2) - d = -2$
$3 - d = -2$
$-d = -5$
$d = 5$

Solution: $(-1,2,1,5)$

EXERCISE SET 4.3

1. Let x and y be the two numbers.
$x + y = 73$
$x = 3y - 15$
Substitute $x = 3y - 15$ into $x + y = 73$
$(3y - 15) + y = 73$
$4y - 15 = 73$
$4y = 88$
$y = 22$
Substitute $y = 22$ into $x = 3y - 15$.
$x = 3(22) - 15$
$x = 51$
The two numbers are 22 and 51.

3. Let x and y be the two numbers.
$x - y = 25$
$x = 3y - 1$
Substitute $x = 3y - 1$ into $x - y = 25$.
$(3y - 1) - y = 25$
$2y - 1 = 25$
$2y = 26$
$y = 13$
Substitute $y = 13$ into $x = 3y - 1$ to solve for x.
$x = 3(13) - 1$
$x = 38$
The two numbers are 13 and 38.

5. Let a and b be the measure of the two angles.
$a + b = 180$
$a = 3b - 28$
Substitute $a = 3b - 28$ into $a + b = 180$.
$(3b - 28) + b = 180$
$4b - 28 = 180$
$4b = 208$
$b = 52$
Substitute $b = 52$ into $a = 3b - 28$ to solve for a.
$a = 3(52) - 28$
$a = 128$
The angles are 128 and 52 degrees.

7. Let x be the longer and y be the shorter of two pieces of rope.
$x + y = 50$
$x = 3y + 2$
Substitute $x = 3y + 2$ into $x + y = 50$.
$(3y + 2) + y = 50$
$4y + 2 = 50$
$4y = 48$
$y = 12$
Substitute $y = 12$ into $x = 3y + 2$ to solve for x.
$x = 3(12) + 2$
$x = 38$
The two pieces of rope are 38 and 12 feet long.

9. Let x = the agency fee and y = the mileage fee.
1) $2x + 100y = 60$
2) $3x + 400y = 115$
Multiply equation 1 by $-4$. Add the result to equation 2 to eliminate y and solve for x.
1) $-8x - 400y = -240$
2) $3x + 400y = 115$
$-5x = -125$
$x = 25$
Substitute $x = 25$ into either equation 1 or 2 and solve for y.
1) $2(25) + 100y = 60$
$50 + 100y = 60$
$100y = 10$
$y = .10$
The daily fee is $25 and the milage fee is 10 cents per mile.

11. Let $x$ = the amount of 5% solution and $y$ = the amount of 25% solution.

1) $x + y = 1000$
2) $.05x + .25y = 100$

Multiply equation 1 by $-.25$. Add the result to equation 2 to eliminate $y$ and solve for $x$.

1) $-.25x - .25y = -250$
2) $.05x + .25y = 100$

$-.2x = -150$
$x = 750$

Substitute $x = 750$ into $x + y = 1000$ to solve for $y$.

$750 + y = 1000$
$y = 250$

Mr. Fiora needs to mix 250ml of 25% solution and 750ml of 5% solution to obtain 1000ml of 10% solution.

13. Let $x$ = the amount of 5% butterfat milk and $y$ = the amount of skim milk (0% butterfat).

1) $x + y = 100$
2) $.05x + 0y = .035(100)$

Solve equation 2 for $x$.

2) $.05x + 0 = 3.5$
$x = 70$

Substitute $x = 70$ into equation 1 to solve for $y$.

1) $70 + y = 100$
$y = 30$

Mario needs to mix 70 gallons of 5% milk with 30 gallons of skim to obtain 100 gallons of 3.5% milk.

15. Let $x$ = the amount of heavy cream and $y$ = the amount of half and half.

1) $x + y = 16$
2) $.36x + .105y = .2(16)$
$.35x + .105y = 3.2$

Multiply equation 1) by $-.105$. Add the result to equation 2 to eliminate $y$ and solve for $x$.

1) $-.105x - .105y = -1.68$
2) $.36x + .105y = 3.2$

$.255x = 1.52$
$x = 5.96$

Substitute $x = 5.96$ into equation 1 and solve for $y$.

1) $5.96 + y = 16$
$y = 10.04$

Steve needs to mix 5.96oz of heavy cream with 10.04oz of half and half to obtain 16oz of 20% cream.

17. Let $x$ = the rate of the slower car and $y$ = the rate of the second car.

17. cont.

$y = x + 5$
$4x + 4y = 420$

Substitute $y = x + 5$ into $4x + 4y = 420$ to solve for $x$.

$4x + 4(x + 5) = 420$
$8x + 20 = 420$
$8x = 400$
$x = 50$ mph

Substitute $x = 50$ into $y = x + 5$ and solve for $y$.

$y = 50 + 5$
$y = 55$ mph

19. Let $x$ = the number of Apple shares and $y$ = the number of Loew's shares.

$x = 3y$
$35x + 20y = 6250$

Substitute $x = 3y$ into $35x + 20y = 6250$ to solve for $y$.

$35(3y) + 20y = 6250$
$105y + 20y = 6250$
$125y = 6250$
$y = 50$

Substitute $y = 50$ into $x = 3y$ to solve for $x$.

$x = 3(50) = 150$

Geraldo should purchase 150 shares of Apple stock and 50 shares of Loew's stock.

21. Let $x$ = the pounds of almonds and $y$ = the pounds of walnuts.

1) $x + y = 30$
2) $6x + 5.40y = 5.80(30)$
$6x + 5.40y = 174$

Multiply equation 1 by $-5.4$. Add the result to equation 2 to eliminate $y$ and solve for $x$.

1) $-5.4x - 5.4y = -162$
2) $6x + 5.4y = 174$

$.6x = 12$
$x = 20$

Substitute $x = 20$ into equation 1 and solve for $y$.

1) $20 + y = 30$
$y = 10$

20 lbs of almonds should be mixed with 10 lbs of walnuts.

23. Let $x$ = the number of dimes and $y$ = the number of quarters.

1) $x + y = 25$
2) $.1x + .25y = 3.55$

Multiply equation 1 by $-.25$. Add the result to equation 2 to eliminate $y$ and solve for $x$.

1) $-.25x - .25y = -6.25$
2) $.1x + .25y = 3.55$

$-.15x = -2.7$
$x = 18$

Substitute $x = 18$ into equation 1 to solve for $y$.

1) $18 + y = 25$
$y = 7$

There are 18 dimes and 7 quarters in the collections.

25. Let $x$ = the amount of deductions Mrs. Clar claims and $y$ = the amount Mr. Clar claims.

1) $x + y = 12400$
2) $26200 - y = 22450 - x$
   or $x - y = -3750$

Add equations 1 and 2 to eliminate y and solve for x.

1) $x + y = 12400$
2) $x - y = -3750$

$$2x = 8650$$
$$x = 4325$$

Substitute $x = 4325$ into equation 1 and solve for y.

1) $4325 + y = 12400$
   $y = 8075.$

Mrs. Clar should claim $4325 and Mr. Clar should claim $8075.

27. Substitute the two points into the equation $y = ax + b$.

1) $7 = a(-1) + b$ or $7 = -a + b$
2) $4 = a(1/2) + b$ or $8 = a + 2b$

Add equations 1 and 2 to eliminate a and solve for b.

1) $7 = -a + b$
2) $8 = a + 2b$

$$15 = 3b$$
$$b = 5$$

Substitute $b = 5$ into equation 1 or 2 and solve for a.

2) $8 = a + 2(5)$
   $a = -2$

The equation of the line is $y = -2x + 5$.

29. Let x and y represent the two numbers.

1) $2x + \frac{1}{2}y = 35$ or
   $4x + y = 70$

2) $\frac{1}{2}x + \frac{1}{3}y = 15$ or
   $3x + 2y = 90$

Multiply equation 1 by -2. Add the result to equation 2 to eliminate y and solve for x.

1) $-8x - 2y = -140$
2) $3x + 2y = 90$

$$-5x = -50$$
$$x = 10$$

Substitute $x = 10$ into equation 1 or 2 to solve for y.

2) $3(10) + 2y = 90$
   $2y = 60$
   $y = 30$

The numbers are 10 and 30.

31. Let $x$ = the grams of mix A and $y$ = the grams of mix b.

1) $.1x + .2y = 20$
2) $.06x + .02y = 6$

Multiply equation 2 by -10. Add the result to equation 1 to eliminate y and solve for x.

1) $.1x + .2y = 20$
2) $-.6x - .2y = -60$

$$-.5x = -40$$
$$x = 80$$

Substitute $x = 80$ into equation 1 or 2 and solve for y.

1) $.1(80) + .2y = 20$
   $8 + .2y = 20$
   $.2y = 12$
   $y = 60$

The scientist should feed the animals 80 g of A and 60 g of B.

33. Let $x$ = the amount of one alloy and $y$ = the amount of the second.

1) $.7x + .4y = .6(300)$
   or $.7x + .4y = 180$
2) $.3x + .6y = .4(300)$
   or $.3x + .6y = 120$

Multiply equation 1 by -3 and equation 2 by 2. Add the results to eliminate y and solve for x.

1) $-2.1x - 1.2y = -540$
2) $.6x + 1.2y = 240$

$$-1.5x = -300$$
$$x = 200$$

Substitute $x = 200$ into either equation 1 or 2 and solve for y.

1) $.7(200) + .4y = 180$
   $140 + .4y = 180$
   $.4y = 40$
   $y = 100$

200 grams of the first alloy should be mixed with 100 grams of the second.

35. Let $x$ = the number of upfront floor tickets, $y$ = the farther back floor tickets, and $z$ = the number of balcony tickets.

1) $x = 2y$
2) $z = y - 6$
3) $z = x - 21$

Substitute $x = 2y$ into equation 3. Multiply equation 3 by -1. Add the results to eliminate z and solve for y.

2) $z = y - 6$
3) $-z = -2y + 21$

$$0 = -y + 15$$
$$y = 15$$

Substitute $y = 15$ into equation 1 and 2 to solve for x and z.

1) $x = 2(15) = 30$
2) $z = 15 - 6 = 9$

The prices of the tickets are $30, $15, and $9.

37. Let x, y, and z represents the three angles.

1) $x + y + z = 180$

2) $x = \frac{2}{3}y$ □ or $3x = 2y$
$3x - 2y = 0$

3) $z = 3y - 30$

Multiply equation 1) by 2. Add the result to equation 2 to eliminate y.

1) $2x + 2y + 2z = 360$
2) $3x - 2y \quad = 0$

---

4) $5x + 2z = 360$

Multiply equation 2 by 3 and equation 3 by $-2$. Add the results to eliminate y.

2) $9x = 6y$
3) $-2z = -6y + 60$

---

5) $9x - 2z = 60$

Add equations 4 and 5 to eliminate z and solve for x.

4) $5x + 2z = 360$
5) $9x - 2z = 60$

---

$14x = 420$
$x = 30$

Substitute $x = 30$ into equation 4 or 5 to solve for z.

4) $5(30) + 2z = 360$
$2z = 210$
$z = 105$

Substitute $z = 105$ into equation 3 to solve for y.

3) $105 = 3y - 30$
$135 = 3y$
$y = 45$

The angles are 30, 45, and 105 degrees.

39. Substitute the points into the equation.

1) $6 = a2^2 + b(2) + c$
$6 = 4a + 2b + c$

2) $17 = a\left[3\right]^2 + b(3) + 3$
$17 = 9a + 3b + c$

3) $-3 = a(-1)^2 + b(-1) + c$
$-3 = a - b + c$

Multiply equation 1 by $-1$ Add the result to equations 2 and 3 to eliminate c.

1) $-6 = -4a - 2b - c$
2) $17 = 9a + 3b + c$

---

4) $11 = 5a + b$

1) $-6 = -4a - 2b - c$
3) $-3 = a - b + c$

---

5) $-9 = -3a - 3b$

39. cont.

Multiply equation 4 by 3. Add the result to equation 5 to eliminate b and solve for a.

4) $33 = 15a + 3b$
5) $-9 = -3a - 3b$

---

$24 = 12a$
$a = 2$

Substitute $a = 2$ into equation 4 or 5 to solve for b.

4) $11 = 5(2) + b$
$1 = b$

Substitute $a = 2$ and $b = 1$ into equation 1, 2, or 3 to solve for c.

1) $6 = 4(2) + 2(1) + c$
$6 = 10 + c$
$-4 = c$

The equation is

$y = 2x^2 + x - 4$

41. Let x = the amount of 10% solution, y = the amount of 12% solution and z = the amount of 20% solution.

1) $x + y + z = 8$
2) $.1x + .12y + .2z = .13(8)$
or $.1x + .12y + .2z = 1.04$
3) $z = x - 2$

Substitute $z = x - 2$ into equation 1 and 2 to eliminate z.

1) $x + y + (x - 2) = 8$
4) $2x + y = 10$

2) $.1x + .12y + .2(x - 2) = 1.04$
5) $.3x + .12y = 1.44$

Multiply equation 4 by $-.12$. Add the result to equation 5 to eliminate y and solve for x.

4) $-.24x - .12y = -1.2$
5) $.3x + .12y = 1.44$

---

$.06x = .24$
$x = 4$

Substitute $x = 4$ into equation 3 to solve for z.

3) $z = 4 - 2$
$z = 2$

Substitute $x = 4$ and $z = 2$ into equation 1 and solve for y.

1) $4 + y + 2 = 8$
$y = 2$

4 liters of 10% solution, 2 liters of 12% solution, and 2 liters of 20% solution should be mixed.

43. Let $x$ = the number of children's chairs, $y$ = the number of standard chairs and $z$ = the number of executive chairs.

1) $5x + 4y + 7z = 154$
2) $3x + 2y + 5z = 94$
3) $2x + 2y + 4z = 76$

Multiply equation 2 by −2. Add the result to equation 1 to eliminate y.

1) $5x + 4y + 7z = 154$
2) $-6x - 4y - 10z = -188$

4) $-x - 3z = -34$

Multiply equation 3 by −1. Add the result to equation 2 to eliminate y.

2) $3x + 2y + 5z = 94$
3) $-2x - 2y - 4z = -76$

5) $x + z = 18$

Add equations 4 and 5 to eliminate x and solve for x.

4) $-x - 3z = -34$
5) $x + z = 18$

$-2z = -16$
$z = 8$

Substitute $z = 8$ into equation 5 and solve for x.

5) $x + 8 = 18$
$x = 10$

Substitute $x = 10$ and $z = 8$ into equation 1,2,or 3 and solve for y.

1) $5(10) + 4y + 7(8) = 154$
$50 + 4y + 56 = 154$
$4y = 48$
$y = 12$

To opperate at full capacity Donaldson's should make 10 children's chairs, 12 standard chairs, and 8 executive chairs.

45.

1) $I_A + I_B + I_C = 0$

2) $-8I_B + 10I_C = 0$

3) $4I_A - 8I_B = 6$

Multiply equation 1 by −4. Add the result to equation 3.

1) $-4I_A - 4I_B - 4I_C = 0$

3) $4I_A - 8I_B = 6$

4) $-12I_b - 4I_c = 6$

45. cont.

Multiply equation 2 by 2 and equation 4 by 5. Add the results.

2) $-16I_B + 20I_C = 0$

4) $-60I_B - 20I_C = 30$

$-76I_B = 30$

$$I_B = \frac{-30}{76} = \frac{-15}{38}$$

Substitute $I_B = \frac{-15}{38}$ into equation 2.

2) $$-8\left[\frac{-15}{38}\right] + 10I_C = 0$$

$$10I_C = \frac{-60}{19}$$

$$I_C = \frac{-6}{19}$$

Substitute $I_B = \frac{-15}{38}$ into equation 3.

3) $$4I_A - 8\left[\frac{-15}{38}\right] = 6$$

$$4I_A + \frac{60}{19} = \frac{144}{19}$$

$$4I_A = \frac{54}{19}$$

$$I_A = \frac{27}{38}$$

The current in brances A, B, and C are 27/38, −15/38, and −6/19 respectively.

47. $$\frac{1}{2}x + \frac{2}{5}xy + \frac{1}{8}y \qquad x = -2,\ y = 5$$

$$\frac{1}{2}(-2) + \frac{2}{5}(-2)(5) + \frac{1}{8}(5)$$

$$= -1 - 4 + \frac{5}{8}$$

$$= -5 + \frac{5}{8}$$

$$= \frac{-40}{8} + \frac{5}{8} = \frac{-35}{8}$$

49. Find the slope of the line through the two points.

$$m = \frac{-8 - (-4)}{2 - 6}$$

$$= \frac{-4}{-4} = 1$$

49. cont.

Using the point slope formula and one of the given points:

$y-(-4) = 1(x - 6)$

$y + 4 = x - 6$

$y = x - 10$

JUST FOR FUN

1. a) Since we wish the pull towards the food to equal the pull away,

$$\frac{-1}{5} \cdot d + 70 = \frac{-4}{3} \cdot d + 230$$

$-3d + 1050 = - 20d + 3450$

$17d + 1050 = 3450$

$17d = 2400$

$$d = \frac{2400}{17} \approx 141.2$$

b) Substitute d = 100 into both equations.

Pull towards food =

$$-\left[\frac{1}{5}\right] \cdot (100) + 70 = 50$$

Pull away from shock =

$$\frac{-4}{3} \cdot 100 + 230 = 96.67$$

Since the pull away is greater, we conclude the rat will pull away.

EXERCISE SET 4.4

1. $x + 2y = 5$

$x - 2y = 1$

Calculate $D, D_x, D_y$

$$D = \begin{vmatrix} 1 & 2 \\ 1 & -2 \end{vmatrix} = 1(-2) - (1)(2) = -4$$

$$D_x = \begin{vmatrix} 5 & 2 \\ 1 & -2 \end{vmatrix} = 5(-2)-(1)(2) = -12$$

$$D_y = \begin{vmatrix} 1 & 5 \\ 1 & 1 \end{vmatrix} = 1(1)-5(1) = -4$$

$$x = \frac{D_x}{D} = \frac{-12}{-4} = 3$$

$$y = \frac{D_y}{D} = \frac{-4}{-4} = 1$$

Solution: (3,1)

3. $x - 2y = -1$

$x + 3y = 9$

Calculate: $D, D_x, D_y$

$$D = \begin{vmatrix} 1 & -2 \\ 1 & 3 \end{vmatrix} = 1(3)-1(-2) = 5$$

$$D_x = \begin{vmatrix} -1 & -2 \\ 9 & 3 \end{vmatrix} = -1(3)-9(-2) = 15$$

3. cont.

$$D_y = \begin{vmatrix} 1 & -1 \\ 1 & 9 \end{vmatrix} = 1(9)-1(-1)= 10$$

$$x = \frac{D_x}{D} = \frac{15}{5} = 3$$

$$y = \frac{D_y}{D} = \frac{10}{5} = 2$$

Solution: (3,2)

5. $3x + 4y = 8$

$2x - 3y = 9$

$$D = \begin{vmatrix} 3 & 4 \\ 2 & -3 \end{vmatrix} = 3(-3)-2(4) = -17$$

$$D_x = \begin{vmatrix} 8 & 4 \\ 9 & -3 \end{vmatrix} = 8(-3)-9(4) = -60$$

$$D_y = \begin{vmatrix} 3 & 8 \\ 2 & 9 \end{vmatrix} = 3(9) - 2(8) = 11$$

$$x = \frac{D_x}{D} = \frac{60}{17}$$

$$y = \frac{D_y}{D} = \frac{-11}{17}$$

Solution: (60/17,-11/17)

7. Write equat equation in standard form:

$2x - y = 5$

$6x + 2y = -5$

$$D = \begin{vmatrix} 2 & -1 \\ 6 & 2 \end{vmatrix} = 2(2)-6(-1) = 10$$

$$D_x = \begin{vmatrix} 5 & -1 \\ -5 & 2 \end{vmatrix} = 5(2)-(-5)(-1) = 5$$

$$D_y = \begin{vmatrix} 2 & 5 \\ 6 & -5 \end{vmatrix} = 2(-5)-6(5) = -40$$

$$x = \frac{D_x}{D} = \frac{5}{10} = \frac{1}{2}$$

$$y = \frac{D_y}{D} = \frac{-40}{10} = -4$$

Solution: $(1/2, -4)$

9. Write both equations in standard form.

$3x + 4y = -6$

$5x + 3y = 1$

$$D = \begin{vmatrix} 3 & 4 \\ 5 & 3 \end{vmatrix} = 3(3)-5(4) = -11$$

$$D_x = \begin{vmatrix} -6 & 4 \\ 1 & 3 \end{vmatrix} = -6(3)-4(1) = -22$$

$$D_y = \begin{vmatrix} 3 & -6 \\ 5 & 1 \end{vmatrix} = 3(1)-5(6) = 33$$

$$x = \frac{D_x}{D} = \frac{-22}{-11} = 2$$

$$y = \frac{D_y}{D} = \frac{33}{-11} = -3$$

Solution: $(2, -3)$

11. $6.3x - 4.5y = -9.9$

$-9.1x + 3.2y = -2.2$

$$D = \begin{vmatrix} 6.3 & -4.5 \\ -9.1 & 3.2 \end{vmatrix} = 6.3(3.2)-(-4.5)(-9.1) = -20.79$$

$$D_x = \begin{vmatrix} -9.9 & -4.5 \\ -2.2 & 3.2 \end{vmatrix} = -9.9(3.2)-(-4.5)(-2.2) = -41.58$$

$$D_y = \begin{vmatrix} 6.3 & -9.9 \\ -9.1 & -2.2 \end{vmatrix} = 6.3(-2.2)-(-9.9)(-9.1) = -33.88$$

$$x = \frac{D_x}{D} = \frac{-41.58}{-20.78} \approx 2$$

11. cont.

$$y = \frac{D_y}{D} = \frac{-103.95}{-20.78} \approx 5$$

Solution: $(2,5)$

13. $x + y - z = -3$

$x + 0y + z = 2$

$2x - y + 2z = 3$

$$D = \begin{vmatrix} 1 & 1 & -1 \\ 1 & 0 & 1 \\ 2 & -1 & 2 \end{vmatrix}$$ Expanding by the second row:

$$-1\cdot\begin{vmatrix} 1 & -1 \\ -1 & 2 \end{vmatrix} + 0\cdot\begin{vmatrix} 1 & -1 \\ 2 & 2 \end{vmatrix} + -1\cdot\begin{vmatrix} 1 & 1 \\ 2 & -1 \end{vmatrix}$$

$$= -1(2-1)+0-1(-1-2) = -1+3 = 2$$

$$D_x = \begin{vmatrix} -3 & 1 & -1 \\ 2 & 0 & 1 \\ 3 & -1 & 2 \end{vmatrix}$$ Expanding by the second column:

$$-1\cdot\begin{vmatrix} 2 & 1 \\ 3 & 2 \end{vmatrix} + 0\cdot\begin{vmatrix} -3 & -1 \\ 3 & 2 \end{vmatrix} - (-1)\cdot\begin{vmatrix} -3 & -1 \\ 2 & 1 \end{vmatrix}$$

$$= -1(4-3) + 0 + 1(-3-(-2)) = -1-1 = -2$$

$$D_y = \begin{vmatrix} 1 & -3 & -1 \\ 1 & 2 & 1 \\ 2 & 3 & 2 \end{vmatrix}$$ Expansion by the first row:

$$1\cdot\begin{vmatrix} 2 & 1 \\ 3 & 2 \end{vmatrix} - (-3)\cdot\begin{vmatrix} 1 & 1 \\ 2 & 2 \end{vmatrix} + (-1)\cdot\begin{vmatrix} 1 & 2 \\ 2 & 3 \end{vmatrix}$$

$$= 1(4-3) + 3(2-2) - 1(3-4) = 1 + 0 + 1 = 2$$

$$D_z = \begin{vmatrix} 1 & 1 & -3 \\ 1 & 0 & 2 \\ 2 & -1 & 3 \end{vmatrix}$$ Expansion by the second row:

$$-1\begin{vmatrix} 1 & -3 \\ -1 & 3 \end{vmatrix} + 0 - 2\cdot\begin{vmatrix} 1 & 1 \\ 2 & -1 \end{vmatrix}$$

$$= -1(3-3) + 0 - 2(-1-2) = 0 - 2(-3) = 6$$

$$x = \frac{D_x}{D} = \frac{-2}{2} = -1$$

$$y = \frac{D_y}{D} = \frac{2}{2} = 1$$

$$z = \frac{D_z}{D} = \frac{6}{2} = 3$$

Solution: $(-1,1,3)$

15. Write each equation in standard form.

$-x + y + 0z = 1$
$0x + y - z = 2$
$x + 0y + z = -2$

$$D = \begin{vmatrix} -1 & 1 & 0 \\ 0 & 1 & -1 \\ 1 & 0 & 1 \end{vmatrix}$$ Expansion by row one:

$$-1\begin{vmatrix} 1 & -1 \\ 0 & 1 \end{vmatrix} - 1\begin{vmatrix} 0 & -1 \\ 1 & 1 \end{vmatrix} + 0$$

$= -1(1-0) -1(0+1) = -2$

$$D_x = \begin{vmatrix} 1 & 1 & 0 \\ 2 & 1 & -1 \\ -2 & 0 & 1 \end{vmatrix}$$ Expansion by row one:

$$1 \cdot \begin{vmatrix} 1 & -1 \\ 0 & 1 \end{vmatrix} - 1 \cdot \begin{vmatrix} 2 & -1 \\ -2 & 1 \end{vmatrix} + 0$$

$= 1(1-0)-1(2-2) = 1$

$$D_y = \begin{vmatrix} -1 & 1 & 0 \\ 0 & 2 & -1 \\ 1 & -2 & 1 \end{vmatrix}$$ Expansion by row one:

$$-1\begin{vmatrix} 2 & -1 \\ -2 & 1 \end{vmatrix} - 1 \cdot \begin{vmatrix} 0 & -1 \\ 1 & 1 \end{vmatrix} + 0$$

$= -1(2-2) -1(0+1) = -1$

$$D_z = \begin{vmatrix} -1 & 1 & 1 \\ 0 & 1 & 2 \\ 1 & 0 & -2 \end{vmatrix}$$ Expansion by column one

$$-1 \cdot \begin{vmatrix} 1 & 2 \\ 0 & -2 \end{vmatrix} - 0 + 1 \cdot \begin{vmatrix} 1 & 1 \\ 1 & 2 \end{vmatrix}$$

$= -1(-2 - 0) + 1(2 - 1)$
$= 2 + 1 = 3$

$$x = \frac{D_x}{D} = \frac{1}{-2}$$

$$y = \frac{D_y}{D} = \frac{-1}{-2} = \frac{1}{2}$$

$$z = \frac{D_z}{D} = \frac{3}{-2}$$

Solution: $\left[\frac{-1}{2}, \frac{1}{2}, \frac{-3}{2}\right]$

17. $2x + 2y + 2z = 0$
$-x - 3y + 7z = 15$
$3x + y + 4z = 21$

$$D = \begin{vmatrix} 2 & 2 & 2 \\ -1 & -3 & 7 \\ 3 & 1 & 4 \end{vmatrix}$$ Expansion by row one:

$$2 \cdot \begin{vmatrix} -3 & 7 \\ 1 & 4 \end{vmatrix} - 2 \cdot \begin{vmatrix} -1 & 7 \\ 3 & 4 \end{vmatrix} + 2 \cdot \begin{vmatrix} -1 & -3 \\ 3 & 1 \end{vmatrix}$$

$= 2(-12-7) - 2(-4-21) + 2(-1+9)$
$= 2(-19) - 2(-25) + 2(8)$
$= 28$

$$D_x = \begin{vmatrix} 0 & 2 & 2 \\ 15 & -3 & 7 \\ 21 & 1 & 4 \end{vmatrix}$$ Expansion by row one:

$$0 \cdot \begin{vmatrix} -3 & 7 \\ 1 & 4 \end{vmatrix} - 2 \cdot \begin{vmatrix} 15 & 7 \\ 21 & 4 \end{vmatrix} + 2 \cdot \begin{vmatrix} 15 & -3 \\ 21 & 1 \end{vmatrix}$$

$= 0-2(-87) + 2(78)$
$= 330$

$$D_y = \begin{vmatrix} 2 & 0 & 2 \\ -1 & 15 & 7 \\ 3 & 21 & 4 \end{vmatrix}$$ Expansion by row one:

$$2 \cdot \begin{vmatrix} 15 & 7 \\ 21 & 4 \end{vmatrix} - 0 + 2\begin{vmatrix} -1 & 15 \\ 3 & 21 \end{vmatrix}$$
$= 2(60-147) + 2(-21-45)$
$= 2(-87) + 2(-66) = -306$

$$D_z = \begin{vmatrix} 2 & 2 & 0 \\ -1 & -3 & 15 \\ 3 & 1 & 21 \end{vmatrix}$$ Expansion by row one:

$$2 \cdot \begin{vmatrix} -3 & 15 \\ 1 & 21 \end{vmatrix} - 2 \cdot \begin{vmatrix} -1 & 15 \\ 3 & 21 \end{vmatrix} + 0$$

$= 2(-63-15) - 2(-21-45)$
$= 2(-78) - 2(-66) = -24$

$$x = \frac{D_x}{D} = \frac{330}{28} = \frac{165}{14}$$

$$y = \frac{D_y}{D} = \frac{-306}{28} = \frac{-153}{14}$$

$$z = \frac{D_z}{D} = \frac{-24}{28} = \frac{-6}{7}$$

Solution: $\left[\frac{165}{14}, \frac{-153}{14}, \frac{-6}{7}\right]$

19. $x - y + 2z = 3$
$x - y + z = 1$
$2x + y + 2z = 2$

$$D = \begin{vmatrix} 1 & -1 & 2 \\ 1 & -1 & 1 \\ 2 & 1 & 2 \end{vmatrix}$$ Expansion by row one:

$$1\begin{vmatrix} -1 & 1 \\ 1 & 2 \end{vmatrix} - (-1)\begin{vmatrix} 1 & 1 \\ 2 & 2 \end{vmatrix} + 2\begin{vmatrix} 1 & -1 \\ 2 & 1 \end{vmatrix}$$

$= 1(-2-1) + 1(0) + 2(1+2)$
$= -3 + 6 = 3$

$$D_x = \begin{vmatrix} 3 & -1 & 2 \\ 1 & -1 & 1 \\ 2 & 1 & 2 \end{vmatrix}$$ Expansion by row two:

$$-1\begin{vmatrix} -1 & 2 \\ 1 & 2 \end{vmatrix} + (-1)\begin{vmatrix} 3 & 2 \\ 2 & 2 \end{vmatrix} - 1\begin{vmatrix} 3 & -1 \\ 2 & 1 \end{vmatrix}$$

$= -2(-2-2) -1(6-4) - 1(3+2)$
$= 4 - 2 - 5 = -3$

$$D_y = \begin{vmatrix} 1 & 3 & 2 \\ 1 & 1 & 1 \\ 2 & 2 & 2 \end{vmatrix}$$ Expansion by row one:

$$1\begin{vmatrix} 1 & 1 \\ 2 & 2 \end{vmatrix} - 3\begin{vmatrix} 1 & 1 \\ 2 & 2 \end{vmatrix} + 2\begin{vmatrix} 1 & 1 \\ 2 & 2 \end{vmatrix}$$

$1(0) -3(0) + 2(0) = 0$

$$D_z = \begin{vmatrix} 1 & -1 & 3 \\ 1 & -1 & 1 \\ 2 & 1 & 2 \end{vmatrix}$$ Expansion by row one

$$1\begin{vmatrix} -1 & 1 \\ 1 & 2 \end{vmatrix} - (-1)\begin{vmatrix} 1 & 1 \\ 2 & 2 \end{vmatrix} + 3\begin{vmatrix} 1 & -1 \\ 2 & 1 \end{vmatrix}$$

$= (-2-1) + 1(0) + 3(1+2)$
$= -3 + 9 = 6$

$$x = \frac{D_x}{D} = \frac{-3}{3} = -1$$

$$y = \frac{D_y}{D} = \frac{0}{3} = 0$$

$$z = \frac{D_z}{D} = \frac{6}{3} = 2$$

Solution: $(-1,0,2)$

21. $x + 2y + z = 1$
$x - y + z = 1$
$2x + y + 2z = 2$

$$D = \begin{vmatrix} 1 & 2 & 1 \\ 1 & -1 & 1 \\ 2 & 1 & 2 \end{vmatrix}$$ Expansion by row one

$$1\begin{vmatrix} -1 & 1 \\ 1 & 2 \end{vmatrix} - 2\begin{vmatrix} 1 & 1 \\ 2 & 2 \end{vmatrix} + 1\begin{vmatrix} 1 & -1 \\ 2 & 1 \end{vmatrix}$$

$= 1(-2-1) - 2(2-2) + 1(1+2)$
$= 0$

The system is either dependent (if $D_x$, $D_y$, and $D_z$ all equal zero) or inconsistent (if at least one of the three is not zero.

$$D_x = \begin{vmatrix} 1 & 2 & 1 \\ 1 & -1 & 1 \\ 2 & 1 & 2 \end{vmatrix}$$ Expansion by row one:

$$1\begin{vmatrix} -1 & 1 \\ 1 & 2 \end{vmatrix} - 2\begin{vmatrix} 1 & 1 \\ 2 & 2 \end{vmatrix} + 1\begin{vmatrix} 1 & -1 \\ 2 & 1 \end{vmatrix}$$

$= 1(-2-1) - 2(2-2) +1(1+2) = 0$

$$D_y = \begin{vmatrix} 1 & 1 & 1 \\ 1 & 1 & 1 \\ 2 & 2 & 2 \end{vmatrix}$$ Expansion by row one:

$$1\begin{vmatrix} 1 & 1 \\ 2 & 2 \end{vmatrix} - 1\begin{vmatrix} 1 & 1 \\ 2 & 2 \end{vmatrix} + 1\begin{vmatrix} 1 & 1 \\ 2 & 2 \end{vmatrix}$$

$= 0$

$$D_z = \begin{vmatrix} 1 & 2 & 1 \\ 1 & -1 & 1 \\ 2 & 1 & 2 \end{vmatrix}$$ = 0, since the determinant is identical to that of D.

Since the determinant for all four is zero, the system is dependent and there are infinitely many solutions.

23. $1.1x + 2.3y - 4.0z = -9.2$
$- 2.3x + 0y + 4.6z = 6.9$
$0x - 8.2y - 7.5z = -6.8$

$$D = \begin{vmatrix} 1.1 & 2.3 & -4 \\ -2.3 & 0 & 4.6 \\ 0 & -8.2 & -7.5 \end{vmatrix}$$

Expansion by column one:

$$1.1\begin{vmatrix} 0 & 4.6 \\ -8.2 & -7.5 \end{vmatrix} - (-2.3)\begin{vmatrix} 2.3 & -4 \\ -8.2 & -7.5 \end{vmatrix}$$

23. cont.

$D = 1.1(37.72) + 2.3(-50.05)$
$= -73.623$

$$D_x = \begin{vmatrix} -9.2 & 2.3 & -4 \\ 6.9 & 0 & 4.6 \\ -6.8 & -8.2 & -7.5 \end{vmatrix}$$

Expansion by row two:

$$-6.9 \cdot \begin{vmatrix} 2.3 & -4 \\ -8.2 & -7.5 \end{vmatrix} + 0$$

$$-4.6 \cdot \begin{vmatrix} -9.2 & 2.3 \\ -6.8 & -8.2 \end{vmatrix}$$

$= -6.9(-50.05) - 4.6(91.08)$
$= -73.623$

$$D_y = \begin{vmatrix} 1.1 & -9.2 & -4 \\ -2.3 & 6.9 & 4.6 \\ 0 & -6.8 & -7.5 \end{vmatrix}$$

Expansion by column one:

$$1.1 \cdot \begin{vmatrix} 6.9 & 4.6 \\ -6.8 & -7.5 \end{vmatrix} - (-2.3) \cdot \begin{vmatrix} -9.2 & -4 \\ -6.8 & -7.5 \end{vmatrix}$$

$= 1.1(-20.47) + 2.3(41.80$
$= 73.623$

$$D_z = \begin{vmatrix} 1.1 & 2.3 & -9.2 \\ -2.3 & 0 & 6.9 \\ 0 & -8.2 & -6.8 \end{vmatrix}$$

Expansion by column one:

$$1.1 \cdot \begin{vmatrix} 0 & 6.9 \\ -8.2 & -6.8 \end{vmatrix} - (-2.3) \cdot \begin{vmatrix} 2.3 & -9.2 \\ -8.2 & -6.8 \end{vmatrix}$$

$= 1.1(56.58) + 2.3(-91.08)$
$= -147.246$

$$x = \frac{D_x}{D} = \frac{-73.623}{-73.623} = 1$$

$$y = \frac{D_y}{D} = \frac{73.623}{-73.623} = -1$$

$$z = \frac{D_y}{D} = \frac{-147.246}{-73.623} = 2$$

Solution: (1,-1,2)

25. Determinate: A square array of numbers enclosed between two vertical bars.

Second order determinate: A determinante that has two rows and two columns of elements.

Third order determinate: A determinante that has three rows and three columns of elements.

27.

$$\begin{vmatrix} a_1 & b_1 \\ a_2 & b_2 \end{vmatrix} = a_1 \cdot b_2 - a_2 \cdot b_1$$

$$\begin{vmatrix} a_2 & b_2 \\ a_1 & b_1 \end{vmatrix} = a_2 \cdot b_1 - a_1 \cdot b_2$$

Interchanging the rows will have the effect of changing the sign on the determinante

29. $3x + 4y = 8$

| x | y |
|---|---|
| 0 | 2 |
| 1 | 5/4 |
| 2 | 1/2 |

31. Write the expression in slope intercept form by solving for y.

$3x + 4y = 8$
$4y = -3x + 8$

$$y = \frac{-3}{4} \cdot x + 2 \qquad m = \frac{-3}{4} \qquad b = 2$$

EXERCISE SET 4.5

1. Graph each inequality and find the point of intersection.

$x - y > 2$

| x | y |
|---|---|
| 0 | −2 |
| 2 | 0 |

Use (0,0) as a checkpoint. Since $0 - 0 \ngtr 2$, shade the area of the graph which does not contain (0,0)

$y < -2x + 3$

| x | y |
|---|---|
| 0 | 3 |
| 1 | 1 |

Use (0,0) as a checkpoint. Since $0 < -2(0) + 3$, shade the area of the graph which contains (0,0)

The solution lies in the area where the shaded regions overlap.

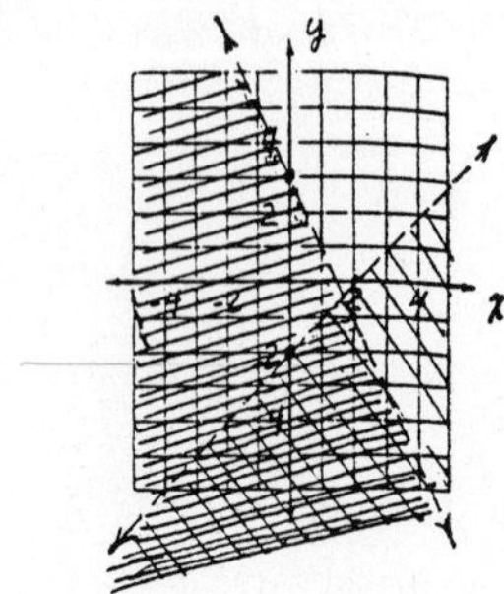

3. Graph each inequaltiy.

$y \le x - 4$

| x | y |
|---|---|
| 0 | −4 |
| 4 | 0 |

Use (0,0) as a checkpoint. Since $0 \nleq 0 - 4$ shade the region which does not include (0,0)

$y < -2x + 4$

| x | y |
|---|---|
| 0 | 4 |
| 2 | 0 |

Use (0,0) as a checkpoint. Since $0 < -2(0) + 4$, shade the region which contains (0,0).

The solution lies in the area where the shaded regions overlap.

5. Graph each inequality.

$y < x$

| x | y |
|---|---|
| 0 | 0 |
| 1 | 1 |

Use (3,1) as a checkpoint. Since $1 < 3$, shade the region which contains (3,1).

$y \ge 3x + 2$

| x | y |
|---|---|
| 0 | 2 |
| −1 | −1 |

Use (0,0) as a checkpoint. Since $0 \ngeq 3\cdot(0) + 2$ shade the region which does not contain (0,0).

The solution lies in the area where the shaded regions overlap.

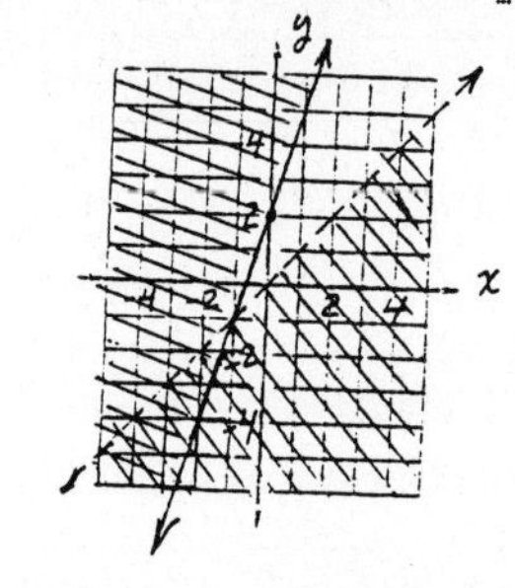

7. Graph each inequality.

$4x - 2y < 6$

| x | y |
|---|---|
| 0 | −3 |
| 3/2 | 0 |

Use (0,0) as a checkpoint. Since $4(0) - 2(0) < 6$, shade the area of the graph which contains (0,0)

$y \le -x + 4$

Use (0,0) as a checkpoint. Since $0 \le 0 + 4$ shade the area of the graph which contains (0,0).

The solution lies in the area where the shaded regions overlap.

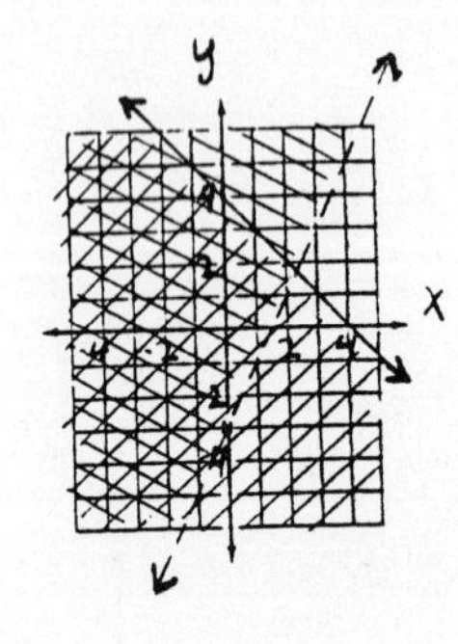

9. Graph each inequality.

$-4x + 5y < 20$

| x | y |
|---|---|
| 0 | 4 |
| -5 | 0 |

Use (0,0) as a checkpoint.
Since $-4(0) + 5(0) < 20$,
shade the region of the graph which contains (0,0).

$x \ge -3$

Use (0,0) as a checkpoint.
Since $0 \not\le -3$ shade the area of the graph which does not contain (0,0).
The solution lies in the area where the shaded regions overlap.

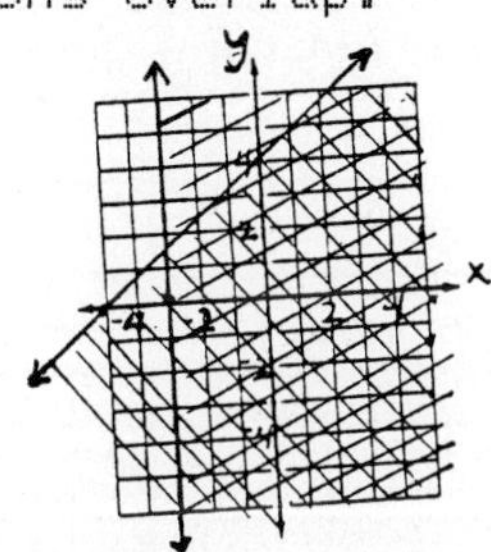


11. Graph each inequality.

$x \le 4$

Use (0,0) as a checkpoint.
Since $0 \le 4$ shade the area which contains (0,0).

$y \ge -2$

Use (0,0) as a checkpoint.
Since $0 \ge -2$ shade the area which contains (0,0).
The solution lies in the area where the shaded regions overlap.

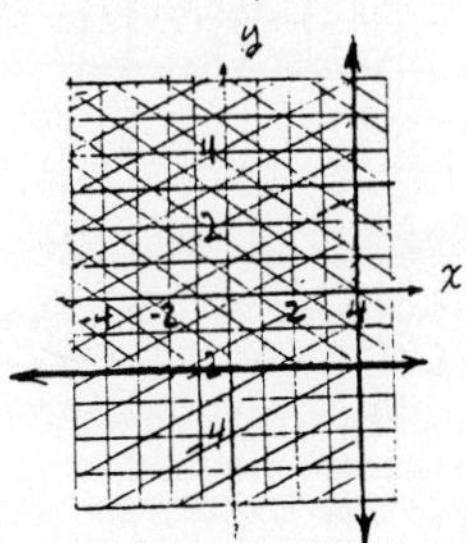


13. Graph each inequality.

$5x + 2y > 10$

| x | y |
|---|---|
| 0 | 5 |
| 2 | 0 |

Use (0,0) as a checkpoint.
Since $5(0) + 2(0) \not> 10$
shade the area of the graph which does not contain (0,0).

$3x - y > 3$

| x | y |
|---|---|
| 0 | -3 |
| 1 | 0 |

Use (0,0) as a checkpoint.
Since $3(0) - 0 \not> 3$, shade the area of the graph which does not contains (0,0).

13. cont.
The solution lies in the area where the shaded regions overlap.

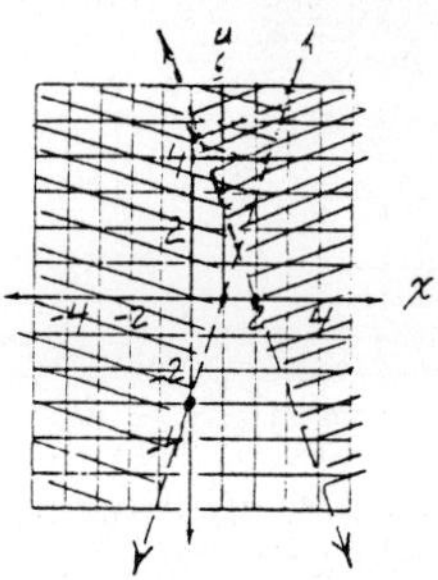


15. Graph each inequality.

$-2x < y + 4$

| x | y |
|---|---|
| 0 | -4 |
| -2 | 0 |

Use (0,0) as a checkpoint.
Since $-2(0) < 0 + 4$ shade the area which contains (0,0).

$3x \ge y$

| x | y |
|---|---|
| 0 | 0 |
| 1 | 3 |

Use (1,0) as a checkpoint. Since $3 \cdot 1 \ge 0$ shade the area which contains (1,0). The solution lies in the area where the shaded regions overlap.

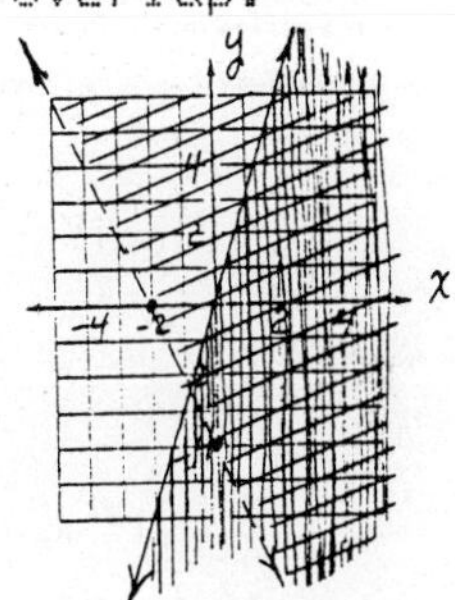


17. Graph each inequality.

$\frac{1}{2}x + 3y > 6$

| x | y |
|---|---|
| 0 | 2 |
| 2 | 5/3 |

Use (0,0) as a checkpoint. Since $\frac{1}{2} \cdot 0 + 3 \cdot (0) \not> 6$ shade the area of the graph which does not contain (0,0).

$y < 3x - 4$

| x | y |
|---|---|
| 0 | -4 |
| 1 | -1 |

Check (0,0). Since $0 \not< 3(0) - 4$ shade the area of the graph which does not contain (0,0).
The soltuion lies in the area where the shaded regions overlap.

17. cont.

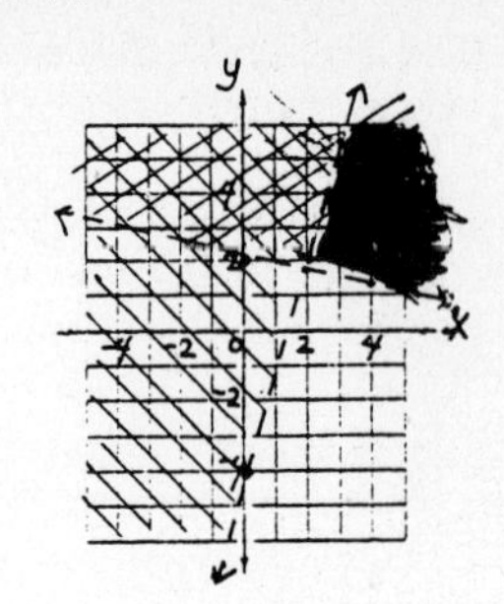

19. Graph each inequality.

$x \geq 0$

Check (0,0). Since $0 \geq 0$ shade the area of the graph which contains (0,0).

$y \geq 0$

Check (0,0). Since $0 \geq 0$ shade the area of the graph which contains (0,0).

$5x + 4y \leq 20$

| x | y |
|---|---|
| 0 | 5 |
| 4 | 0 |

Use (0,0) as a checkpoint. Since $5 \cdot (0) + 4 \cdot (0) \leq 20$ shade the area of the graph which contains (0,0).

$x + 2y \leq 6$

| x | y |
|---|---|
| 0 | 3 |
| 6 | 0 |

Use (0,0) as a checkpoint. Since $0 + 2 \cdot 0 \leq 6$ shade the area of the graph which contains (0,0).

The solution lies in the area where the shaded regions overlap.

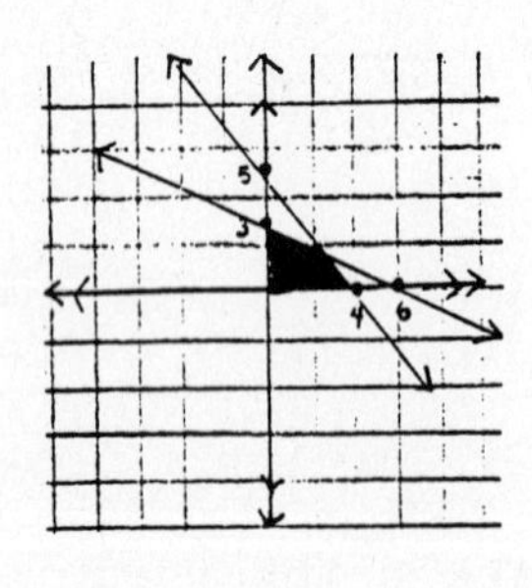

21. Graph each inequality. The solution lies in the area where the shaded regions overlap.

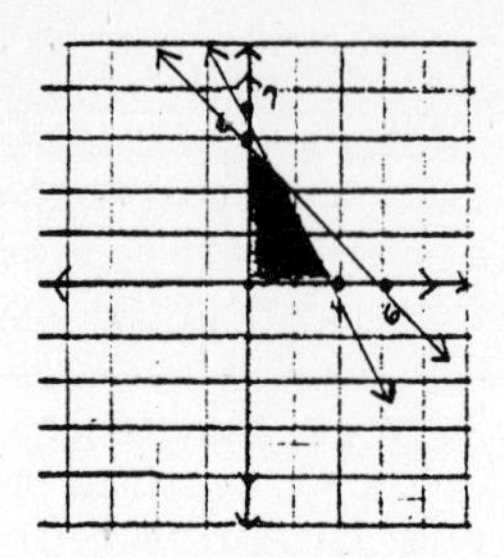

23. Graph each inequality. The solution lies in the area where the shaded regions overlap.

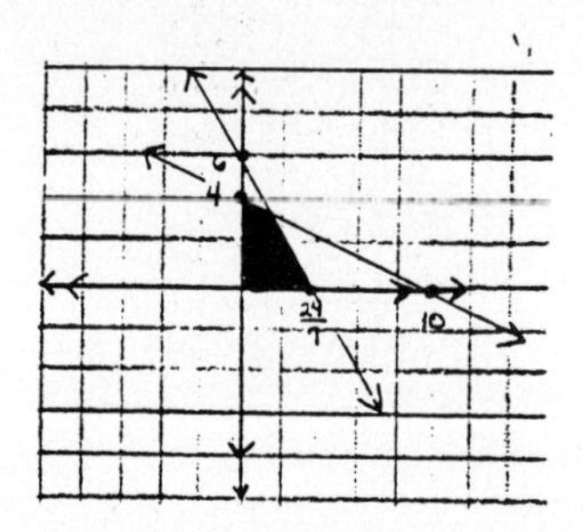

25. Graph each inequality. The solution lies in the area where the shaded regions overlap.

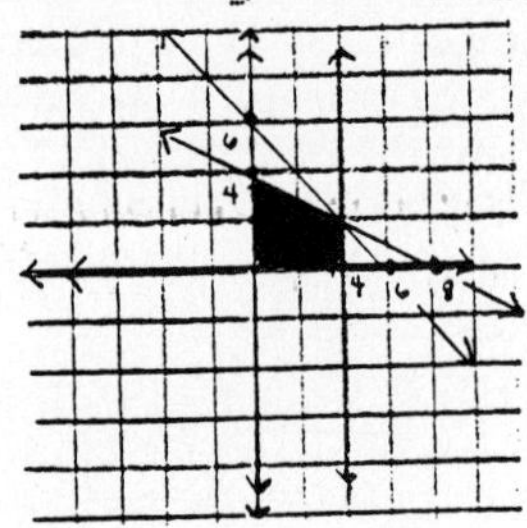

27. Graph each inequality. The solution lies in the area where the shaded regions overlap.

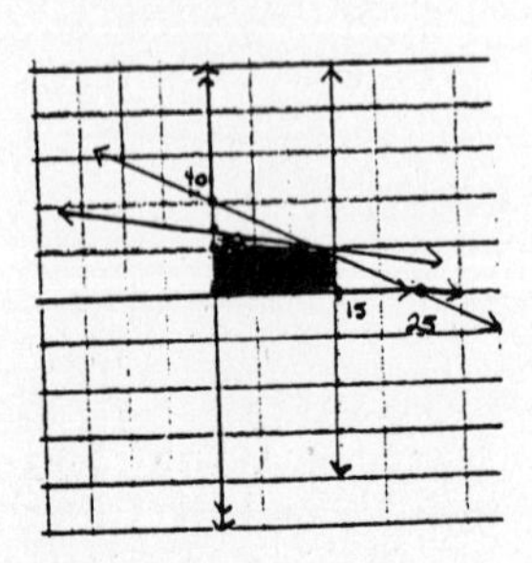

29. Graph each inequality.
$|y| > 2$
To graph this inequality, graph the inequalities $y > 2$ and $y < -2$.
Use (0,0) as a checkpoint.
Since $|0| \not> 2$, shade the two regions of the graph which do not contain the point (0,0).
$y \le x + 3$
Check (0,0). Since $0 \le 0 + 3$ shade the area of the graph which contains (0,0). The solution lies in the area of the graph where the shaded regions overlap.

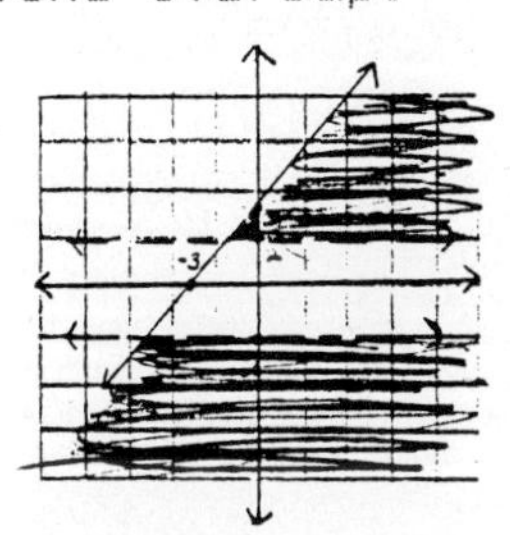

31. Graph each inequality. To graph $|y| < 4$, graph $y > -4$ and $y < 4$. Use (0,0) as a checkpoint.
Since $|0| < 4$, shade the area of the graph which contains (0,0).
$y \ge -2x + 2$
Use (0,0) as a checkpoint.
Since $0 \not\ge -2 \cdot 0 + 2$ shade the area of the graph which does not contain (0,0). The solution lies in the area where the shaded regions overlap.

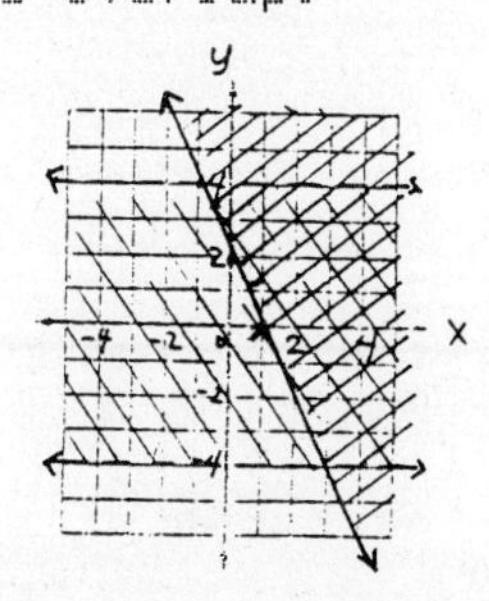

33. Graph each inequality.
To graph $|x| \ge 1$ graph $x \ge 1$ and $x \le -1$.
Use (0,0) as a checkpoint.
Since $|0| \not\ge 1$ shade The two areas which do not contain the point (0,0).
To graph $|y| \ge 2$ graph $y \ge 2$ and $y \le -2$.
Use (0,0) as a checkpoint.

33. cont.
Since $|0| \not\ge 2$ shade the two areas of the graph which do not contain (0,0). The solution lies in the area of the graph were the shaded regions overlap.

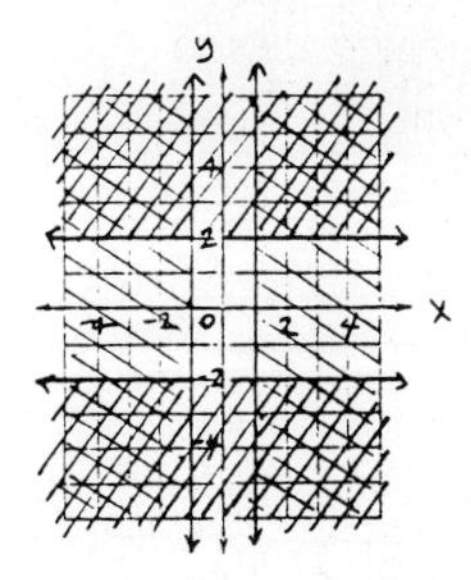

35. To graph $|x + 2| < 3$, rewrite the inequality as $-3 < x + 2 < 3$. Solve for x: $-5 < x < 1$. Graph $x > -5$ and $x < 1$. Use (0,0) as a checkpoint. Since $|0 + 2| < 3$, shade the area of the graph which contains (0,0).
To graph $|y| > 4$, graph the inequalities $y < -4$ and $y > 4$. Use (0,0) as a checkpoint. Since $|0| \not> 4$, shade the region above the line $y = 4$ and below the the line $y = -4$. The solution lies in the area of the graph where the shaded regions overlap.

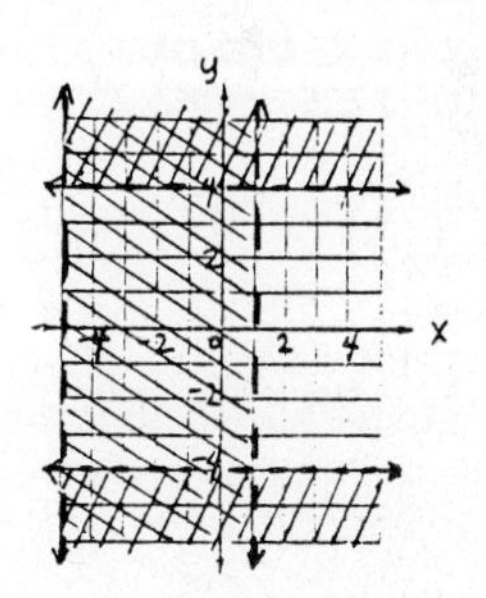

37. To graph $|x - 2| > 1$, rewrite the inequality as $x - 2 < -1$ or $x - 2 > 1$.
Solve each for x:
$x < 1$ or $x > 3$. Graph these two inequalities. Use (0,0) as a checkpoint. Since $|0 - 2| > 1$, shade the area of the graph which contains (0,0).
To graph $y > -2$, graph the line $y = -2$ using a dashed line.
Use (0,0) as a checkpoint.
Since $0 \not> -2$, shade the area of the graph which does not contain the point (0,0).

37. cont.
The solution lies in the area where the shaded regions overlap.

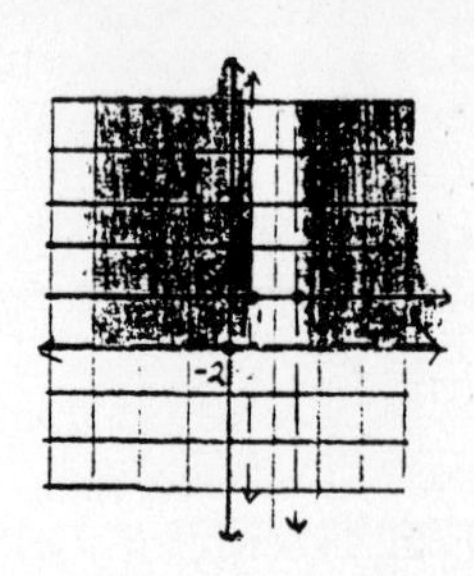

39. To graph $|x - 3| > 4$, rewrite the inequality as $x - 3 < -4$ or $x - 3 > 4$. Solve for x: $x < -1$ or $x > 7$ Graph these two inequalities. Use (0,0) as a checkpoint. Since $|0 - 3| \not> 4$, shade the areas of the graph to the left of the line $x = -1$ and to the right of the line $x = 7$.
To graph $|y + 1| \le 3$ rewrite the inequality and solve for y:

$$-3 \le y + 1 \le 3$$
$$-4 \le y \le 2$$

Graph $y \ge -4$ and $y \le 2$ □
Use (0,0) as a checkpoint. Since $|0 + 1| \le 3$ shade the region of the graph which contains (0,0). The solution lies in the intersection of the two shaded areas.

39. cont.

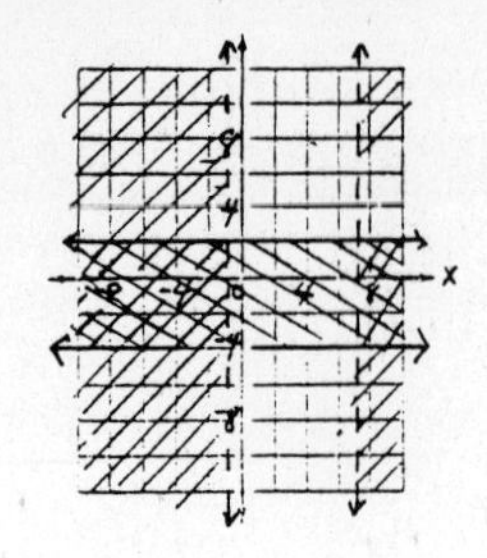

41. $f_1 d_1 + f_2 d_2 = f_3 d_3$

Solve for $f_2$ □

$$f_1 d_1 - f_1 d_1 + f_2 d_2 = f_3 d_3 - f_1 d_1$$

$$\frac{f_2 d_2}{d_2} = \frac{f_3 d_3 - f_1 d_1}{d_2}$$

$$f_2 = \frac{f_3 d_3 - f_1 d_1}{d_2}$$

43. $f(x) = \frac{2}{3}x - 4$

Domain: all reals
Range: all reals

JUST FOR FUN

1. To graph $|2x - 3| - 1 > 3$ begin by solving for x.

$|2x - 3| > 4$

$2x - 3 < -4$ or $2x - 3 > 4$
$2x < -1$ $\quad$ $2x > 7$
$x < -1/2$ $\quad$ $x > 7/2$

Graph the inequalities $x < -1/2$ and $x > 7/2$.
Use (0,0) as a checkpoint. Since $|2(0) - 3| - 1 \not> 3$ shade the area to the left of the line $x = -1/2$ and to the right of the lin $x = 7/2$.
To graph $|y - 2| < 3$, solve for y.

$$-3 < y - 2 < 3$$
$$-1 < y < 5$$

Graph $y > -1$ and $y < 5$. Use (0,0) as a checkpoint. Since $|0 - 2| < 3$ shade the area of the graph containing (0,0). The solution lies in the intersection of the two shaded areas.

JUST FOR FUN, 1. cont.

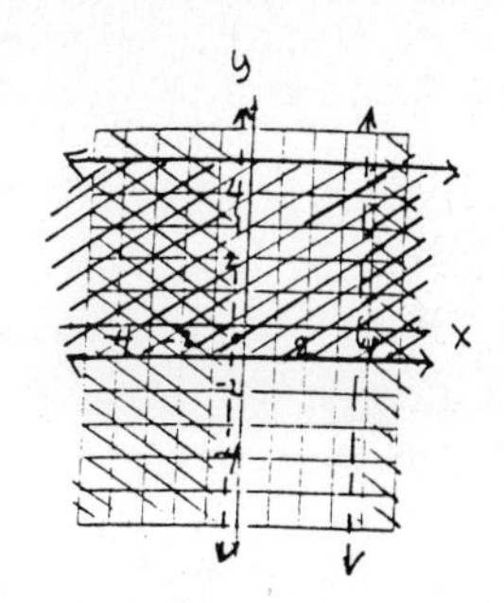


3. To graph $y < |x|$, graph $y = |x|$ using a dashed line. Use (1,0) as a checkpoing. Since $0 < |1|$ shade the area of the graph which contains the point (1,0).
Graph $y < 4$. Use (0,0) as a checkpoint. Since $0 < 4$, shade the region below the line $y = 4$.
The solution lies in the area where the shaded regions overlap.

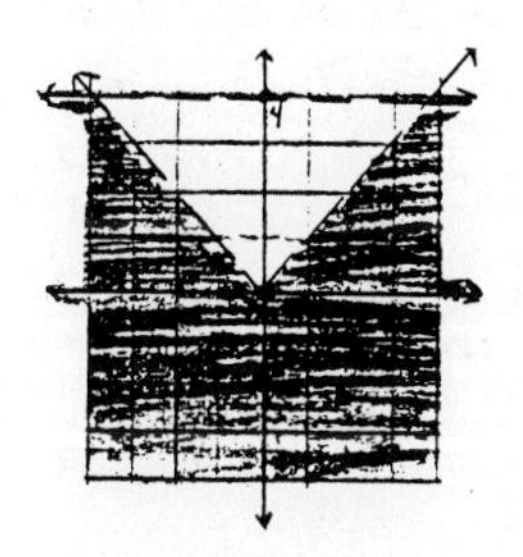

REVIEW EXERCISES

1. $x + 2y = 8$    $3x + 6y = 12$

$2y = -x + 8$    $6y = -3x + 12$

$y = \frac{-1}{2}x + 4$    $y = \frac{-1}{2}x + 2$

$m = \frac{-1}{2}$    $m = \frac{-1}{2}$

$b = 4$    $b = 2$

The slopes equal but the y-intercepts differ. The lines are parallel. Therefore, the system is inconsistent and there is no solution.

3. $y = \frac{1}{2}x + 4$    $x + 2y = 8$

$2y = -x + 8$

$y = \frac{-1}{2}x + 4$

$m = \frac{1}{2}$    $m = \frac{-1}{2}$

The slopes differ. Therefore, the system is consistent and has one solution.

5. $y = x + 3$

$y = 2x + 5$

Graph each equation.
The point where the lines intersect is the solution.

5. cont.

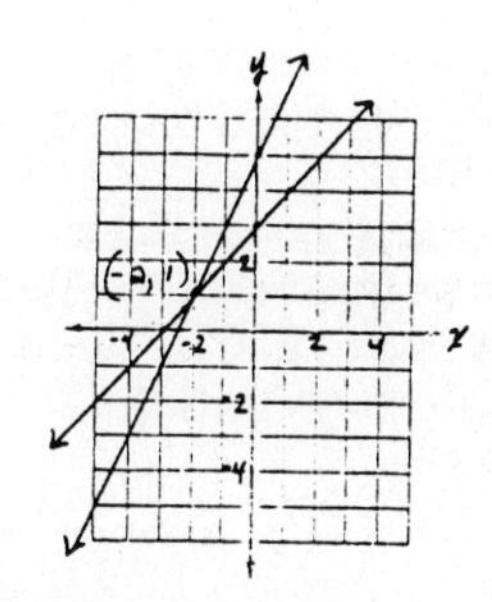


7. $2x + 2y = 8$

$2x - y = -4$

Graph each equation. The point where the lines intersect is the solution.

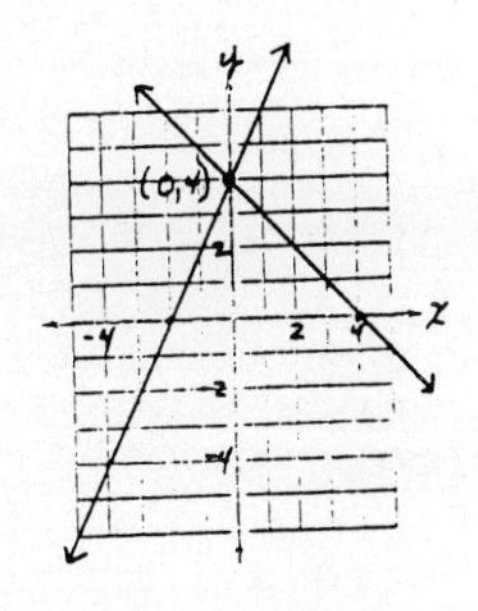

9. $y = 2x + 1$
$y = 3x - 2$
Substitute $y = 2x + 1$ into $y = 3x - 2$ to eliminate y and solve for x.
$2x + 1 = 3x - 2$
$1 = x - 2$
$3 = x$
Substitute $x = 3$ into $y = 3x-2$.
$y = 3(3) - 2$
$y = 7$
Solution: (3,7)

11. $y = 2x - 8$
$2x - 5y = 0$
Substitute $y = 2x - 8$ into $2x - 5y = 0$ and solve for x.
$2x -5(2x - 8) = 0$
$2x - 10x + 40 = 0$
$-8x + 40 = 0$
$-8x = -40$
$x = 5$
Substitute $x = 5$ into $y = 2x - 8$.
$y = 2(5) - 8$
$y = 2$
Solution: (5,2)

13. $2x + y = 5$
$3x + 2y = 8$
Solve $2x + y = 5$ for y:
$y = -2x + 5$. Substitute $y = -2x + 5$ into $3x + 2y = 8$.
$3x + 2(-2x + 5) = 8$
$3x - 4x + 10 = 8$
$-x = -2$
$x = 2$
Substitute $x = 2$ into $2x + y = 5$ and solve for y.
$2(2) + y = 5$
$y = 1$
Solution: (2,1)

15. $3x + y = 17$
$2x - 3y = 4$
Solve $3x + y = 17$ for y:
$y = -3x + 17$
Substitute $y = -3x + 17$ into $2x - 3y = 4$ and solve for x.
$2x -3(-3x + 17) = 4$
$2x + 9x - 51 = 4$
$11x = 55$
$x = 5$
Substitute $x = 5$ into $3x + y = 17$ and solve for y.
$3(5) + y = 17$
$y = 2$
Solution: (5,2)

17. Add the equations to eliminate y and solve for x:
$x + y = 6$
$x - y = 10$
----------
$2x = 16$
$x = 8$

17. cont.
Substitute $x = 8$ into either equation and solve for y:
$8 + y = 6$
$y = -2$
Solution: (8,-2)

19. 1) $2x + 3y = 4$
2) $x + 2y = -6$
Multiply equation 2 by -2. Add the result to equation 1 to eliminate x and solve for y:
1) $2x + 3y = 4$
2) $-2x - 4y = 12$
----------------
$-y = 16$
$y = -16$
Substitute $y = -16$ into equation 1 or 2 and solve for x.
2) $x + 2(-16) = -6$
$x - 32 = -6$
$x = 26$
Solution: (26,-16)

21. 1) $4x - 3y = 8$
2) $2x + 5y = 8$
Multiply equation 2 by -2. Add the result to equation 1 to eliminate y and solve for x.
1) $4x - 3y = 8$
2) $-4x -10y = -16$
----------------
$-13y = -8$
$y = 8/13$
Substitute $y = 8/13$ into either equation and solve for y.
1)
$$4x - 3\left[\frac{8}{13}\right] = 8$$
$$4x - \frac{24}{13} = \frac{104}{13}$$
$$\frac{1}{4}\cdot 4x = \frac{128}{13}\cdot\frac{1}{4}$$
$$x = \frac{32}{13}$$
Solution: $\left[\frac{32}{13}, \frac{8}{13}\right]$

23\. 1) $x + \frac{2}{5}y = \frac{9}{5}$

2) $x - \frac{3}{2}y = -2$

Multiply equation 1 by 5 and equation 2 by 2 to eliminate the fractions.

1) $5x + 2y = 9$
2) $2x - 3y = -4$

Multiply equation 1 by 3 and equation 2 by 2 to eliminate y and solve for x.

1) $15x + 6y = 27$
2) $4x - 6y = -8$

---

$19x = 19$
$x = 1$

Substitute x = 1 into equation 1 or 2 and solve for y.

1) $15(1) + 6y = 27$
$15 + 6y = 27$
$6y = 12$
$y = 2$

Solution: (1,2)

25\. 1) $y = \frac{-3}{4}x + \frac{10}{4}$

2) $x + \frac{5}{4}y = \frac{7}{2}$

Multiply equations 1) and 2) by 4.

1) $4y = -3x + 10$
2) $4x + 5y = 14$

Rewrite equation 1) in standard form.

1) $3x + 4y = 10$

Multiply equation 1) by −4 and equation 2) by 3. Add the results to eliminate x.

1) $-12x - 16y = -40$
2) $12x + 15y = 42$

---

$-y = 2$
$y = -2$

Substitute y = −2 into equation 2) and solve for x:

$x + \frac{5}{4}(-2) = \frac{7}{2}$

$x - \frac{5}{2} = \frac{7}{2}$

$x = \frac{12}{2} = 6$

Solution: (6, −2)

27\. 1) $x + 2y = 12$
2) $4x = 8$
3) $3x - 4y + 5z = 20$

Solve equation 2 for x:

$4x = 8$
$x = 2$

Substitute x = 2 into equation 1 and solve for y.

1) $2 + 2y = 12$
$2y = 10$
$y = 5$

Substitute x = 2 and y = 5 into equation 3 and solve for z.

3) $3(2) - 4(5) + 5z = 20$
$5z = 34$
$z = 34/5$

Solution: $\left[2, 5, \frac{34}{5}\right]$

29\. 1) $x + 5y + 5z = 6$
2) $3x + 3y - z = 10$
3) $x + 3y + 2z = 5$

Multiply equation 1 by −1. Add the result to equation 3 to eliminate x.

1) $-x - 5y - 5z = -6$
3) $x + 3y + 2z = 5$

---

4) $-2y - 3z = -1$

Multiply equation 1 by −3. Add the result to equation 2 to eliminate x.

1) $-3x - 15y - 15z = -18$
2) $3x + 3y - z = 10$

---

5) $-12y - 16z = -8$

Multiply equation 4 by −6. Add the result to equation 5 to eliminate y and solve for z.

4) $12y + 18z = 6$
5) $-12y - 16z = -8$

---

$2z = -2$
$z = -1$

Substitute z = −1 into equation 4 or 5 and solve for y.

4) $-2y - 3(-1) = -1$
$-2y = -4$
$y = 2$

Substitute y = 2 and z = −1 into equation 1,2,or 3 and solve for x.

1) $x + 5(2) + 5(-1) = 6$
$x = 1$

Solution: (1,2,−1)

31. 1) $3y - 2z = -4$
2) $3x - 5z = -7$
3) $2x + y = 6$
Multiply equation 2) by 2 and equation 3 by -3. Add the resulting equations to eliminate x:
2) $6x - 10z = -14$
3) $-6x - 3y = -18$
---
4) $-3y - 10z = -32$
Add equations 1 and 4 to eliminate y and solve for z.
1) $3y - 2z = -4$
4) $-3y - 10z = -32$
---
$-12z = -36$
$z = 3$
Substitute $z = 3$ into equation 1 and solve for y.
1) $3y - 2(3) = -4$
$3y = 2$
$y = 2/3$
Substitute $z = 3$ into equation 2 and solve for x.
2) $3x - 5(3) = -7$
$3x = 8$
$x = 8/3$

Solution: (8/3, 2/3, 3)

33. Let x and y = the two numbers.
$x + y = 48$
$y = 2x - 3$
Substitute $y = 2x - 3$ into $x + y = 48$ and solve for x.
$x + (2x - 3) = 48$
$3x = 51$
$x = 17$
Substitute $x = 17$ into
$y = 2x - 3$
$y = 2(17) - 3$
$y = 31$
The two numbers are 17 and 31.

35. Let x = the speed of the plane and y = speed of the wind.
Statement: The plane travels 600 mph with the wind.
Equation: $x + y = 600$
Statement: The plane travels 530 mph against the wind.
Equation: $x - y = 530$
Add the two equations to eliminate y and solve for x.
$x + y = 600$
$x - y = 530$
---
$2x = 1130$
$x = 565$
Substitute $x = 565$ into $x + y = 600$ and solve for y.
$565 + y = 600$ and $y = 35$.
The speed of the plane is 565mph and the speed of the wind is 35mph.

37. Let x = the number of adult tickets and y = the number of children's tickets.
1) $x + y = 650$
2) $7.5x + 5.5y = 4395$
Solve equation 1 for x.
$x = 650 - y$
Substitute $x = 650 - y$ into equation 2 and solve for y.
2) $7.5(650 - y) + 5.5y = 4395$
$4875 - 7.5y + 5.5y = 4395$
$-2y = -480$
$y = 240$
Substitute $y = 240$ into equation 1 and solve for x.
1) $x + 240 = 650$
$x = 410$
410 adult and 240 children's tickets were sold.

39. Let x = the amount in the 10% account y = the amount in the 8% account and z = the amount in the 6% account.
1) $x + y + z = 40000$
2) $y = x - 5000$
3) $.1x + .08y + .06z = 3500$
Substitute $y = x - 5000$ into equation 1 and 3.
1) $x + x - 5000 + z = 40000$
$2x + z = 45000$
3) $.1x + .08(x - 5000) + .06z = 3500$
$.18x + .06z = 3900$
Multiply equation 1 by -.06. Add the result to equation 3 to eliminate z and solve for x.
1) $-.12x - .06z = -2700$
3) $.18x + .06z = 3900$
---
$.06x = 1200$
$x = 20000$
Substitute $x = 20000$ into equation 2 and solve for y.
2) $y = 20000 - 5000 = 15000$
Substitute $x = 20000$ and $y = 15000$ into equation 1 and solve for z.
1) $20000 + 15000 + z = 40000$
$z = 5000$
$20,000 was invested at 10%, $15,000 was invested at 8% and $5000 was invested at 6%.

41. $3x + 5y = -2$
$5x + 3y = 2$
Compute $D, D_x, D_y$

$$D = \begin{vmatrix} 3 & 5 \\ 5 & 3 \end{vmatrix} = 3(3) - 5(5) = -16$$

$$D_x = \begin{vmatrix} -2 & 5 \\ 2 & 3 \end{vmatrix} = -2(3) - 5(2) = -16$$

$$D_y = \begin{vmatrix} 3 & -2 \\ 5 & 2 \end{vmatrix} = 3(2) - 5(-2) = 16$$

41. cont.

$$x = \frac{D_x}{D} = \frac{-16}{-16} = 1$$

$$y = \frac{D_y}{D} = \frac{16}{-16} = -1$$

Solution: $(1,-1)$

43. $x + y + z = 8$
$x - y - z = 0$
$x + 2y + z = 9$

$$D = \begin{vmatrix} 1 & 1 & 1 \\ 1 & -1 & -1 \\ 1 & 2 & 1 \end{vmatrix}$$ Expansion by row one:

$= 1(-1 + 2) - 1(-1 - 1)$
$\quad + 1(2 - (-1))$
$= 1 - 2 + 3 = 2$

$$D_x = \begin{vmatrix} 8 & 1 & 1 \\ 0 & -1 & -1 \\ 9 & 2 & 1 \end{vmatrix}$$ Expansion by column one:

$= 8(-1-(-2)) + 9(-1-(-1))$
$= 8(1) + 9(0) = 8$

$$D_y = \begin{vmatrix} 1 & 8 & 1 \\ 1 & 0 & -1 \\ 1 & 9 & 1 \end{vmatrix}$$ Expansion by row two:

$= -1(8 - 9) -(-1)(9 - 8)$
$= 1 + 1 = 2$

$$D_z = \begin{vmatrix} 1 & 1 & 8 \\ 1 & -1 & 0 \\ 1 & 2 & 9 \end{vmatrix}$$ Expansion by row two:

$= -1(9 - 16) -1(9 - 8)$
$= -1(-7) -1(1) = 6$

$$x = \frac{D_x}{D} = \frac{8}{2} = 4$$

$$y = \frac{D_y}{D} = \frac{2}{2} = 1$$

$$z = \frac{D_z}{D} = \frac{6}{2} = 3$$

Solution: $(4,1,3)$

45. $y + 3z = 4$
$-x - y + 2z = 0$
$x + 2y + z = 1$

$$D = \begin{vmatrix} 0 & 1 & 3 \\ -1 & -1 & 2 \\ 1 & 2 & 1 \end{vmatrix}$$ Expansion by row one:

$= -1(-1 - 2) + 3(-2 + 1)$
$= 0$

Since $D = 0$ the system either has an infinite number of solutions (if $D_x$, $D_y$, and, $D_z$ all equal zero) or no solution (if at least one of these three determinants does not equal zero).

$$D_x = \begin{bmatrix} 4 & 1 & 3 \\ 0 & -1 & 2 \\ 1 & 2 & 1 \end{bmatrix}$$ Expansion by column one:

$= 4(-1 - 4) + 1(2 + 3) = -15$

Since $D = 0$ but $D_x$ does not equal zero, there is no solution.

47. Graph each inequality by graphing $5x - 2y = 10$ with a solid line and $3x + 2y = 6$ with a dashed line.
$5x - 2y \leq 10$ □
Use $(0,0)$ as a checkpoint.
Since $5 \cdot (0) - 2 \cdot (0) \leq 10$ shade the area of the graph which contains $(0,0)$.
$3x + 2y > 6$
Use $(0,0)$ as a checkpoint.
Since $3(0) + 2(0) \not> 6$ shade the area of the graph which does not contain $(0,0)$. The solution lies in the area of the graph where the shaded areas overlap.

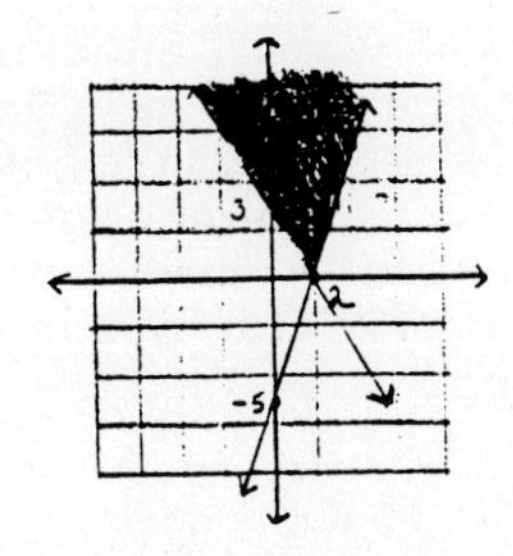

49. Graph the inequalities by graphing $y = -x + 4$ with a solid line and $2x + 2y = 6$ with a dashed line.
$y \le -x + 4$
Use (0,0) as a checkpoint.
Since $0 \le -0 + 4$ shade the area of the graph containing the point (0,0).
$2x + 4y > 6$
Use (0,0) as a checkpoint.
Since $2(0) + 4(0) \not> 6$ shade the area of the graph which does not contain (0,0). The solution lies in the area where the shaded regions overlap.

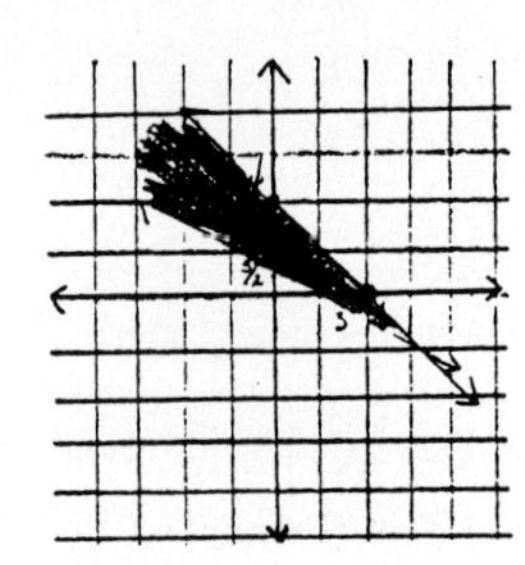

51. Graph the inequalities by graphing $x = 0$, $y = 0$, $2x + y = 6$ and $4x + 5y = 20$ using a solid line.
$x \ge 0$
Use (1,0) as a checkpoint.
Since $1 \ge 0$ shade the area of the graph to the right of the line $x = 0$.
$y \ge 0$
Use (0,1) as a checkpoint.
Since $1 \ge 0$ shade the area of the graph above the line $y = 0$.
$2x + y \le 6$
Use (0,0) as a checkpoint.
Since $2 \cdot (0) + 0 \le 6$ shade the region of the graph which contains (0,0).
$4x + 5y \le 20$
Use (0,0) as a checkpoint.
Since $4 \cdot (0) + 5 \cdot (0) \le 20$ shade the region of the graph which contains (0,0).
The solution lies in the area where the shaded regions overlap.

51. cont.

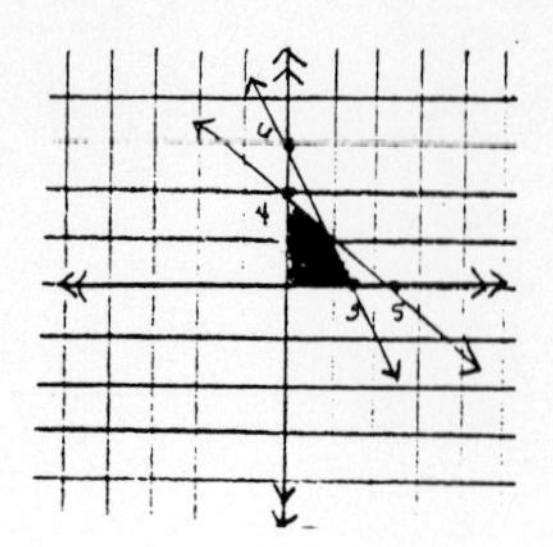

53. To graph $|x| > 4$, rewrite the inequality as $x < -4$ and $x > 4$. Graph the inequalities $x = -4$ and $x = 4$.
Use (0,0) as a checkpoint.
Since $|0| \not> 4$ shade the area of the graph to the left of the line $x = -4$ and to the right of $x = 4$.
To graph $|y - 2| \le 3$ by rewriting the inequality as $-3 \le y - 2 \le 3$ and solving for y:
$-3 \le y - 2 \le 3$
$-1 \le y \le 5$
Graph $y \ge -1$ and $y \le 5$.
Use (0,0) as a checkpoint.
Since $|0 - 2| \le 3$ shade the area of the graph between the lines $y = -1$ and $y = 5$.
The solution lies in the area where the shaded regions overlap.

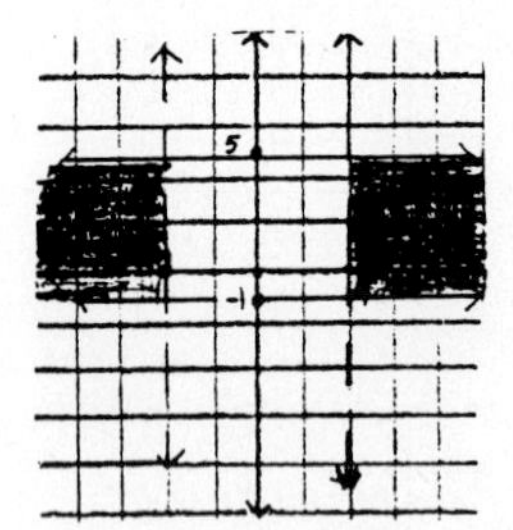

PRACTICE TEST

1. Find the slope of each line by solving the equations for y:

$4x + 3y = -6$ $\qquad$ $6y = 8x + 4$

$3y = -4x - 6$

$y = \frac{-4}{3}\cdot x - 2$ $\qquad$ $y = \frac{4}{3}\cdot x + \frac{2}{3}$

$m = \frac{-4}{3}$ $\qquad$ $m = \frac{4}{3}$

Since the slopes are different, the system is consistent and there will be one solution.

2. Find the slope of each line by solving the equations for y:

$5x + 3y = 9$ $\qquad$ $5x - 3y = 9$

$3y = -5x + 9$ $\qquad$ $-3y = -5x + 9$

$y = \frac{-5}{3}\cdot x + 3$ $\qquad$ $y = \frac{5}{3}\cdot x - 3$

Since the slopes are different, the system is consistent and there will be one solution.

3. Graph each equation. The solution will be the point of intersection.

$y = 3x - 2$

| x | y |
|---|---|
| 0 | -2 |
| 1 | 1 |

$y = -2x + 8$

| x | y |
|---|---|
| 0 | 8 |
| 4 | 0 |

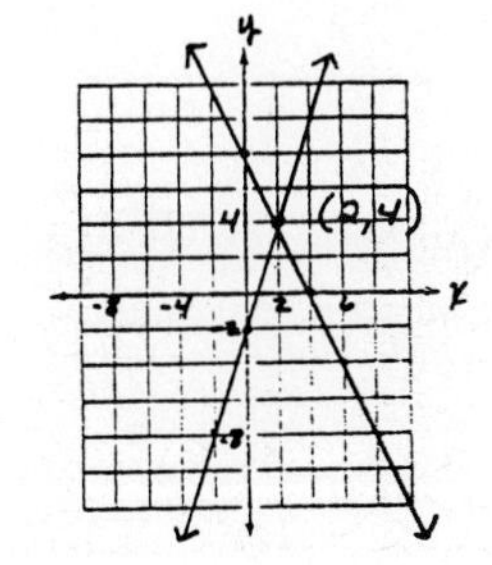


4. 1) $y = 4x - 5$

2) $y = 2x + 7$

Substitute equation 1 for y in equation 2.

2) $4x - 5 = 2x + 7$

$2x - 5 = 7$

$2x = 12$

$x = 6$

Substitute $x = 6$ into equation 1 or 2 and solve for y:

1) $y = 4(6) - 5 = 19$

Solution: (6,19)

5. 1) $3x + y = 8$

2) $x - y = 6$

Solve equation 1) for y.

1) $y = -3x + 8$

Substitute $y = -3x + 8$ for y in equation 2.

2) $x - (-3x + 8) = 6$

$4x - 8 = 6$

$4x = 14$

$x = 7/2$

Substitute $x = 7/2$ into equation 1 to solve for y.

$3\cdot\frac{7}{2} + y = 8$

$\frac{21}{2} + y = \frac{16}{2}$

$y = \frac{-5}{2}$

Solution: $\left(\frac{7}{2}, \frac{-5}{2}\right)$

6. 1) $2x + y = 5$

2) $x + 3y = -10$

Multiply equation 2 by -2. Add the result to equation 2 and solve for y.

1) $2x + y = 5$

2) $-2x - 6y = 20$

$-5y = 25$

$y = -5$

Substitute $y = -5$ into equation 1 or 2 and solve for x.

2) $x + 3(-5) = -10$

$x = 5$

Solution: (5,-5)

7. 1) $\frac{3}{2}\cdot x + y = 6$

2) $x - \frac{5}{2}\cdot y = -4$

Multiply each equation by 2 to eliminate the fractions.

1) $3x + 2y = 12$

2) $2x - 5y = -8$

Multiply equation 1) by 5 and equation 2) by 2 to eliminate y and solve for x.

1) $15x + 10y = 60$

2) $4x - 10y = -16$

$19x = 44$

$x = 44/19$

Substitute $x = 44/19$ into equation 1 or 2 and solve for y.

7. cont.

1) $\frac{3}{2}\cdot\frac{44}{19} + y = 6$

$\frac{66}{19} + y = \frac{114}{19}$

$y = \frac{48}{19}$

Solution: $\left[\frac{44}{19}, \frac{48}{19}\right]$

8. Let x = the pounds of cashews, y = the pounds of peanuts.
Statement: mix to get 20 lbs of mixture
Equation: $x + y = 20$
Statement: Cost for the peanuts plus the cost for the cashews = cost of the mix.
$4x + 2.5y = 20(3.00)$
$4x + 2.5y = 60$
Solve the system:
1) $x + y = 20$
2) $4x + 2.5y = 60$
"Multiply equation 1 by −4. Add the result to equation 2 to solve for y:
1) $-4x - 4y = -80$
2) $4x + 2.5y = 60$

$-1.5y = -20$
$y = 13\ 1/3$ or $40/3$
Substitute y = 40/3 into equation 1 to solve for x.
1) $x + \frac{40}{3} = 20$

$x = \frac{60}{3} - \frac{40}{3}$

$x = \frac{20}{3} = 6\ \frac{2}{3}$

Max should mix 13 1/3 lbs of peanuts with 6 2/3 lbs of cashews.

9. 1) $x = 2$
2) $2x + 3y = 10$
3) $-x + 3y - 2z = 10$
Substitute x = 2 into equation 2 to solve for y.
2) $2(2) + 3y = 10$
$3y = 6$
$y = 2$
Substitute x = 2 and y = 2 into equation 3 to solve for z.

9 cont.
3) $-2 + 3(2) - 2z = 10$
$4 - 2z = 10$
$-2z = 6$
$z = -3$
Solution: (2,2,−3)

10. 1) $x + y + z = 2$
2) $-2x - y + z = 1$
3) $x - 2y - z = 1$
Add equation 1 and 3 to eliminate z:
1) $x + y + z = 2$
3) $x - 2y - z = 1$

4) $2x - y = 3$
Multiply equation 1 by −1. Add the result to equation 2 to eliminate z.
1) $-x - y - z = -2$
2) $-2x - y + z = 1$

5) $-3x - 2y = -1$
Multiply equation 4 by −2. Add the result to equation 5 to eliminate y and solve for x.
4) $-4x + 2y = -6$
5) $-3x - 2y = -1$

$-7x = -7$
$x = 1$
Substitute x = 1 into equation 4 or 5 and solve for y.
4) $2(1) - y = 3$
$2 - y = 3$
$-y = 1$
$y = -1$
Substitute x = 1 and y = −1 into equation 1,2,or 3 and solve for z.
1) $1 - 1 + z = 2$
$z = 2$
Solution: (1,−1,2)

11. Graph each inequality:
$y + 3x \le 6$

| x | y |
|---|---|
| 0 | 6 |
| 2 | 0 |

Use (0,0) as a checkpoint. Since $0 + 3\cdot 0 \le 6$ shade the region which contains (0,0).
$2x + y > 4$

| x | y |
|---|---|
| 0 | 4 |
| 2 | 0 |

Use (0,0) as a checkpoint. Since $2(0) + 0 \not> 4$, shade the region which does not contain (0,0)
The solution lies in the region of the graph where the shaded areas overlap.

11. cont.

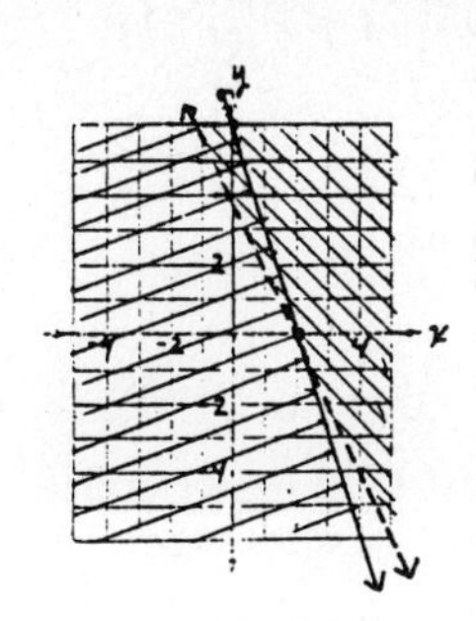

12. Graph each inequality.

$3x + 2y < 9$

| x | y |
|---|---|
| 3 | 0 |
| 0 | 9/2 |

Use (0,0) as a checkpoint.
Since $3(0) + 2(0) < 9$
shade the region which
contains (0,0)

$-2x + 5y \leq 10$

| x | y |
|---|---|
| 0 | 2 |
| -5 | 0 |

Use (0,0) as a checkpoint.
Since $-2 \cdot 0 + 5 \cdot 0 \leq 10$
shade the region which
contains (0,0)
The solution lies in
the region where the
shaded regions overlap.

12. cont.

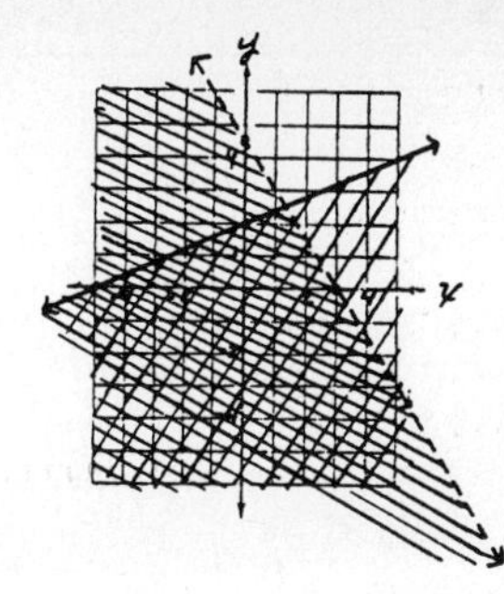

13. To graph $|x| > 3$, rewrite the inequality as $x < -3$ or $x > 3$. Graph the lines $x = 3$ and $x = -3$. Use (0,0) as a checkpoint. Since $|0| \not> 3$, shade the region to the left of $x = -3$ and to the right of $x = 3$. To graph $|y| \leq 1$ rewrite the inequality as $-1 \leq y \leq 1$. Graph the lines $y = -1$ and $y = 1$. Use (0,0) as a checkpoint. Since $|0| \leq 1$ shade the region between the lines $y = -1$ and $y = 1$. The solution lies in the area where the shaded regions overlap.

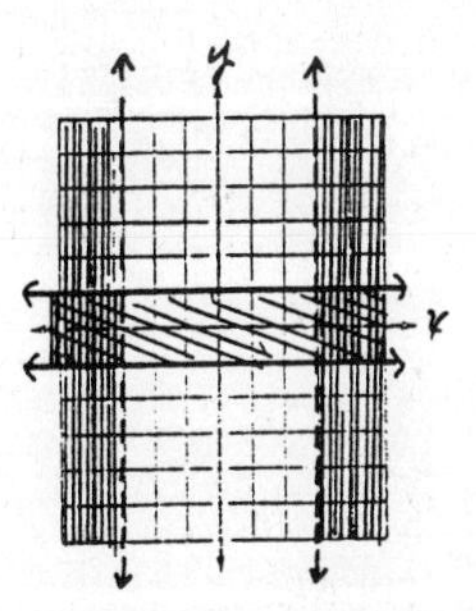

CUMMULATIVE REVIEW TEST

1.
$24 - 4(2 - (5 - 2))^2 - 6$
$24 - 4(2 - 3)^2 - 6$
$24 - 4(-1)^2 - 6$
$24 - 4(1) - 6$
$24 - 4 - 6$
$6 - 6 = 0$

2. a) 9, 1
b) 1/2, −4, 9, 0, −4.63, 1
c) $\frac{1}{2}, -4, 9, 0, \sqrt{3}, -4.63, 1$

3. $-|-8|, -1, 5/8, 3/4, |-4|, |-10|$

4.
$-(3-2(x - 4)) = 3(x - 6)$
$-(3 - 2x + 8) = 3x - 18$
$-(-2x + 11) = 3x - 18$
$2x - 11 = 3x - 18$
$-11 = x - 18$
$7 = x$

6.
$|4x - 3| + 2 = 10$
$|4x - 3| = 8$
$4x - 3 = -8$ or $4x - 3 = 8$
$4x = -5$ $\quad$ $4x = 11$
$x = -5/4$ $\quad$ $x = 11/4$
Solution set: $\{-5/4, 11/4\}$

7. $R = 3(a + b)$ Solve for b.
$R = 3a + 3b$
$R - 3a = 3b$
$\frac{R - 3a}{3} = \frac{3b}{3}$
$\frac{R - 3a}{3} = b$

8.
$0 < \frac{3x - 2}{4} \leq 8$
$0 < 3x - 2 \leq 32$
$2 < 3x \leq 34$
$\frac{2}{3} < x \leq \frac{34}{3}$
$\{x \mid \frac{2}{3} < x \leq \frac{34}{3}\}$

9.

$$d = \sqrt{(1 - (-3))^2 + (5 - 2)^2}$$
$$= \sqrt{4^2 + 3^2}$$
$$= \sqrt{16 + 9}$$
$$= \sqrt{25} = 5$$

$$\text{midpoint} = \left[\frac{1 + (-3)}{2}, \frac{5 + 2}{2}\right]$$
$$= \left[-1, \frac{7}{2}\right]$$

10. $-2x + 4y = 12$

| x | y |
|---|---|
| 0 | 3 |
| -6 | 0 |

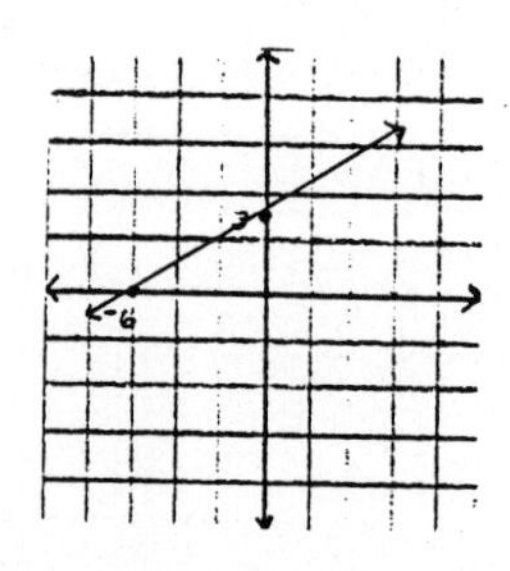


11. $2y = 3x - 8$

| x | y |
|---|---|
| 0 | -4 |
| 8/3 | 0 |

12. To write the equation of a line, we must first determine the slope. Since the line we wish to write an equation for is parallel to $2x - 3y = 8$, the lines will have the same slope.
Determine the slope of $2x - 3y = 8$ by solving for y.

$2x - 3y = 8$

$-3y = -2x + 8$

$$y = \frac{2}{3}x - \frac{8}{3}$$

12. cont.

$$m = \frac{2}{3}$$

Using $m = 2/3$ ,the point-slope formula, and the given point (2,3):

$$y - 3 = \frac{2}{3}(x - 2)$$

$$y = \frac{2}{3}x - \frac{4}{3} + \frac{9}{3}$$

$$y = \frac{2}{3}x + \frac{5}{3}$$

13. a) The graph passes the vertical line test. Therefore, this is the graph of a function.
b) The graph passes the vertical line test. Therefore, this is the graph of a function.
c) The graph fails the vertical line test. Therefore, this is not the graph of a function.

14. $6x - 3y < 12$
Graph the line $6x - 3y = 12$ using a dashed line.

| x | y |
|---|---|
| 0 | -4 |
| 2 | 0 |

Use (0,0) as a checkpoint. Since $6(0) - 3(0) < 12$ shade the area which contains (0,0)

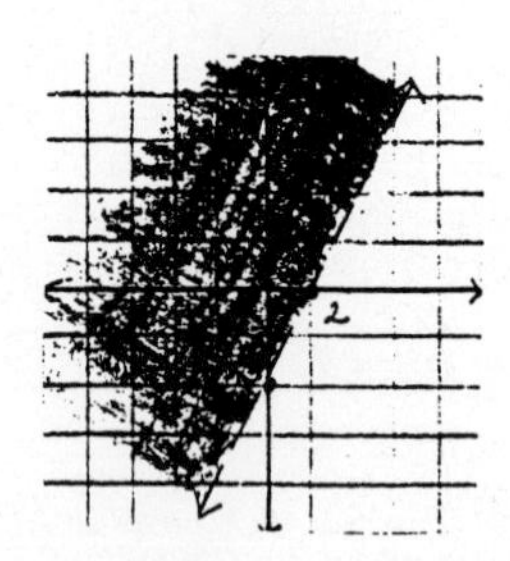


15. $3x + y = 6$
$y = 2x + 1$
Substutite $y = 2x + 1$ into
$3x + y = 6$
$3x + (2x + 1) = 6$
$5x + 1 = 6$
$5x = 5$
$x = 1$
Substitute $x = 1$ into $y = 2x + 1$.
$y = 2(1) + 1 = 3$
Solution: (1,3)

16. 1) $5x + 4y = 10$
2) $3x + 5y = -7$
Multiply equation 1 by $-3$ and equation 2 by 5. Add the result to eliminate x and solve for y.
1) $-15x - 12y = -30$
2) $15x + 25y = -35$

$$13y = -65$$
$$y = -5$$

Substitute $y = -5$ into either equation 1 or 2 and solve for x.
1) $5x + 4(-5) = 10$
$5x = 30$
$x = 6$
Solution: $(6,-5)$

17. 1) $x - 2y = 0$
2) $2x + z = 7$
3) $y - 2z = -5$
Multiply equation 1 by $-2$. Add the result to equation 2 to eliminate x.
1) $-2x + 4y = 0$
2) $2x + z = 7$

4) $4y + z = 7$
Multiply equation 4 by 2. Add the result to equation 3 to eliminate z and solve for y.
4) $8y + 2z = 14$
3) $y - 2z = -5$

$$9y = 9$$
$$y = 1$$

Substitute $y = 1$ into equation 1 to solve for x.
1) $x - 2(1) = 0$
$x = 2$
Substitute $y = 1$ into equation 3 and solve for z.
3) $1 - 2z = -5$
$-2z = -6$
$z = 3$
Solution: $(2,1,3)$

18. Let x, y, and z represent the angles in a triangle. Since the sum of the angles in any triangle is 180 degrees:
1) $x + y + z = 180$
2) $z = 9x$
3) $y = x + 70$
Substitute $z = 9x$ and $y = x + 70$ into equation 1 to solve for x.
1) $x + x + 70 + 9x = 180$
$11x + 70 = 180$
$11x = 110$
$x = 10$

18. cont.
Substitute $x = 10$ into equations 2 and 3 to solve for the remaining angles.
2) $z = 9(10) = 90$
3) $y = 10 + 70 = 80$
The angles are 10, 80, and 90 degrees.

19. Let x = Judy's time.
$x + \frac{1}{2}$ □ = Dawn's time.

| | Distance | = rate | x time |
|---|---|---|---|
| Judy | $6x$ | 6 | $x$ |
| Dawn | $4\left[x + \frac{1}{2}\right]$ | 4 | $\left[x + \frac{1}{2}\right]$ |

Judy will catch Dawn when they have traveled the same distance.

$$6x = 4\left[x + \frac{1}{2}\right]$$

$$6x = 4x + 2$$
$$2x = 2$$
$$x = 1$$

Judy will catch Dawn in 1 hour.

20. Let x = the high prices seats and y = the cheaper seats.
1) $x + y = 1000$
2) $20x + 16y = 18400$
Solve equation 1 for y and substitute that value into equation 2.
1) $y = 1000 - x$
2) $20x + 16(1000 - x) = 18400$
$20x + 16000 - 16x = 18400$
$4x = 2400$
$x = 600$
Substitute $x = 600$ into equation 1.
$600 + y = 1000$
$y = 400$
600 seats were sold at \$20 each and 400 seats were sold at \$16 each.

CHAPTER 5

EXERCISE SET 5.1

1. $3^{-2} = \frac{1}{3^2} = \frac{1}{9}$

3. $1^{-2} = \frac{1}{1^2} = 1$

5. $5y^{-3} = \frac{5}{y^3}$

7. $\frac{1}{x^{-4}} = \frac{1}{\left[\frac{1}{x^4}\right]} = 1\cdot\frac{x^4}{1} = x^4$

9. $\frac{2x}{y^{-3}} = \frac{2x}{\left[\frac{1}{y^3}\right]} = \frac{2x}{1}\cdot\frac{y^3}{1} = 2xy^3$

11. $\frac{5x^{-2}\cdot y^{-3}}{2z^{-1}} = \frac{5\cdot\frac{1}{x^2}\cdot\frac{1}{y^3}}{\frac{2}{1}\cdot\frac{1}{z}}$

$= \frac{\left[\frac{5}{x^2\cdot y^3}\right]}{\left[\frac{2}{z^1}\right]} = \frac{5}{x^2\cdot y^3}\cdot\frac{z}{2}$

$= \frac{5z}{2x^2\cdot y^3}$

13. $\frac{\left[5x^{-2}\cdot y^{-3}\right]}{z^{-4}} = \frac{\left[\frac{5}{1}\cdot\frac{1}{x^2}\cdot\frac{1}{y^3}\right]}{\left[\frac{1}{z^4}\right]}$

$= \left[\left[\frac{5}{x^2\cdot y^3}\right]\right]\cdot\left[\frac{z^4}{1}\right] = \frac{5z^4}{x^2\cdot y^3}$

15. $\frac{4^{-1}\cdot x^{-1}}{y} = \frac{\left[\frac{1}{4}\cdot\frac{1}{x}\right]}{\frac{y}{1}}$

$= \left[\left[\frac{1}{4x}\right]\cdot\frac{1}{y} = \frac{1}{4xy}\right.$

17. $x^0 = 1$

19. $4x^0 = 4(1) = 4$

21. $-(7x)^0 = -(7)^0\cdot x^0$

$= -1(1) = -1$

23. $-3x^0 = -3(1) = -3$

25. $-(a + b)^0 = -1$

27. $3x^0 + 4y^0 = 3(1) + 4(1)$

$= 3 + 4 = 7$

29. $6^3\cdot 6^{-4} = 6^{3+(-4)}$

$= 6^{-1} = \frac{1}{6}$

31. $x^2\cdot x = x^{2+1}$

$= x^3$

33. $x^{6} \cdot x^{-2} = x^{6+(-2)}$

$= x^{4}$

35. $\dfrac{3^{4}}{3^{2}} = 3^{4-2} = 3^{2} = 9$

37. $\dfrac{5^{2}}{5^{-2}} = 5^{2-(-2)} = 5^{4} = 625$

39. $\dfrac{x^{-9}}{x^{2}} = x^{-9-2} = x^{-11} = \dfrac{1}{x^{11}}$ □

41. $\dfrac{x^{0}}{x^{-3}} = x^{0-(-3)} = x^{3}$

43. $\dfrac{3y^{-2}}{y^{-7}} = 3y^{-2-(-7)}$

$= 3y^{5}$

45. $\dfrac{4x^{-3}}{x^{-1}} = 4x^{-3-(-1)}$

$= 4x^{-2} = \dfrac{4}{x^{2}}$

47. $2x^{-4} \cdot 6x^{-3} = 2(6)x^{-4+(-3)}$

$= 12x^{-7} = \dfrac{12}{x^{7}}$

49. $\left[-3y^{-2}\right] \cdot \left[-y^{3}\right] = 3y^{-2+3}$

$= 3y$

51. $\left[2x^{-3} \cdot y^{-4}\right]\left[6x^{-4} \cdot y^{7}\right]$

$= 2(6)x^{-3+(-4)} \cdot y^{-4+7}$

$= 12x^{-7} \cdot y^{3}$

51. cont.

$= \dfrac{12y^{3}}{x^{7}}$

53. $\left[5x^{2} \cdot y^{-2} \cdot z^{4}\right]\left[-2x^{5} \cdot y^{2} \cdot z\right]$

$= 5(-2)x^{2+5} \cdot y^{-2+2} \cdot z^{4+1}$

$= -10x^{7} \cdot z^{5}$ □

55. $\left[2x^{4} \cdot y^{7} \cdot z^{9}\right]\left[4x^{3} \cdot y^{-5} \cdot z^{-12}\right]$

$= 2(4)x^{4+3} \cdot y^{7+(-5)} \cdot z^{9+(-12)}$

$= 8x^{7} \cdot y^{2} \cdot z^{-3}$

$= \dfrac{8x^{7} \cdot y^{2}}{z^{3}}$

57.

$\dfrac{27x^{5} \cdot y^{-4}}{9x^{3} \cdot y^{2}} = \left[\dfrac{27}{9}\right] \cdot x^{5-3} \cdot y^{-4-(-2)}$

$= 3x^{2} \cdot y^{-6} = \dfrac{3x^{2}}{y^{6}}$

59. $\dfrac{9xy^{-4}}{3x^{-2}y} = \left[\dfrac{9}{3}\right] \cdot x^{1-(-2)} \cdot y^{-4-1}$

$= 3x^{3} \cdot y^{-5} = \dfrac{3x^{3}}{y^{5}}$

61. $\dfrac{\left[2x^{4}\right]\left[6xy^{3}\right]}{4y^{3}} = \dfrac{12x^{4+1} \cdot y^{3}}{4y^{3}}$

$= \left[\dfrac{12}{4}\right] \cdot x^{5} \cdot y^{3-3}$

$= 3x^{5}$

63. $$\frac{\left[-3x^{-1}y^{-2}\right]\left[2x^{4}y^{-3}\right]}{6xy^{4}}$$

$$= \frac{-6x^{-1+4}y^{-2+(-3)}}{6xy^{4}}$$

$$= \frac{-6x^{3}y^{-5}}{6xy^{4}} = \left[\frac{-6}{6}\right]x^{3-1}y^{-5-4}$$

$$= -x^{2}y^{-9} = \frac{-x^{2}}{y^{9}}$$

65. $$\frac{\left[4x^{-5}y^{-2}\right]\left[3x^{2}y^{-5}\right]}{24x^{3}y^{-4}}$$

$$= \frac{12x^{-5+2}y^{-2+(-5)}}{24x^{3}y^{-4}}$$

$$= \frac{12x^{-3}y^{-7}}{24x^{3}y^{-4}}$$

$$= \left[\frac{12}{24}\right]x^{-3-3}y^{-7-(-4)}$$

$$= \frac{x^{-6}y^{-3}}{2} = \frac{1}{2x^{6}y^{3}}$$

67. $$\left[\frac{2x^{5}}{y^{-3}}\right]\cdot\left[\frac{3xy^{-2}}{z^{-3}}\right]$$

67. cont.

$$= \frac{6x^{5+1}y^{-2}}{y^{-3}z^{-3}} = \frac{6x^{6}y^{-2}}{y^{-3}z^{-3}}$$

$$= 6x^{6}y^{-2-(-3)}z^{3}$$

$$= 6x^{6}yz^{3}$$

69. $$\left[\frac{3x^{-2}y^{-2}}{x^{4}y^{-5}}\right]\cdot\left[\frac{2x^{3}y^{5}}{9x^{-2}y^{3}}\right]$$

$$= \frac{6x^{-2+3}y^{-2+5}}{9x^{4+(-2)}y^{-5+3}}$$

$$= \frac{6xy^{3}}{9x^{2}y^{-2}}$$

$$= \left[\frac{6}{9}\right]x^{1-2}y^{3-(-2)}$$

$$= \frac{2x^{-1}y^{5}}{3} = \frac{2y^{5}}{3x}$$

71. $$\left[\frac{x^{4}y^{-3}}{x^{-4}y^{-3}z}\right]\cdot\left[\frac{x^{3}y^{4}z}{x^{5}y^{-2}z^{-3}}\right]$$

$$= \frac{x^{4+3}y^{-3+4}z}{x^{-4+5}y^{-3+(-2)}z^{1+(-3)}}$$

$$= \frac{x^{7}yz}{xy^{-5}z^{-2}} = x^{7-1}y^{1-(-5)}z^{1-(-}$$

$$= x^{6}y^{6}z^{3}$$

73. $x^{4a} x^{3a+4} = x^{4a+3a+4}$

$= x^{7a+4}$

75. $w^{5b-2} w^{2b+3}$

$= w^{5b-2+2b+3}$

$= w^{7b+1}$

77. $\frac{x^{2w+3}}{x^{w-4}} = x^{(2w+3)-(w-4)}$

$= x^{w+7}$

79. $\frac{\left[x^{3p+5}\right]\left[x^{2p-3}\right]}{x^{4p-1}}$

$= x^{(3p+5)+(2p-3)-(4p-1)}$

$= x^{3p+2p-4p+5-3+1} = x^{p+3}$

81. $3700 = 3700.$

$= 3.700x10^{3}$

83. $900 = 900.$

$= 9.x10^{2}$

85. $.047 = 4.7x\ 10^{-2}$

87. $19000 = 19000.$

$= 1.9x10^{4}$

89. $.00000186 = 1.86x10^{-6}$

91. $.00000914 = 9.14x10^{-6}$

93. $5.2x10^{3} = 5.2(1000)$
$= 5200$

95. $4x10^{7} = 4(10,000,000)$
$= 40,000,000$

97. $2.13x10^{-5} = 2.13(.00001)$
$= .0000213$

99. $3.12x10^{-1} = 3.12(.1)$
$= .312$

101. $9x10^{6} = 9(1000000)$
$= 9,000,000$

103 $5.35x10^{2} = 5.35(100)$
$= 535$

105. $\left[5x10^{3}\right]\left[3x10^{4}\right] = 15x10^{3+4}$

$= 15x10^{7} = 150,000,000$

107. $\left[1.6x10^{-2}\right]\left[4x10^{-3}\right] = 6.4x10^{-5}$

$= .000064$

109. $\frac{8.4x10^{-6}}{4x10^{-4}} = \left[\frac{8.4}{4}\right] 10^{-6-(-4)}$

$= 2.1x10^{-2} = .021$

111. $\frac{4x10^{5}}{2x10^{4}} = \left[\frac{4}{2}\right] x10^{5-4}$

$= 2x10^{1} = 20$

113. $(700,000)(6,000,000)$
$= 4,200,000,000,000$

115. $(.003)(.00015)$

$= \left[3x10^{-3}\right]\left[1.5x10^{-4}\right]$

$= 4.5x10^{-3+(-4)}$

$= 4.5x10^{-7}$ □

117. $\frac{1400000}{700} = \frac{1.4x10^{6}}{7x10^{2}}$

$= .2x10^{6-2} = .2x10^{4}$

117. cont.

$$= \left[2\times10^{-1}\right]\times10^{4}$$

$$= 2\times10^{-1+4} = 2.0\times10^{3}$$

119.

$$\frac{.0000426}{200} = \frac{4.26\times10^{-5}}{2\times10^{2}}$$

$$= 2.13\times10^{-5-2}$$

$$= 2.13\times10^{-7}$$

121. Let t equal the time necessary to travel the distance.

$$93000000 = 3100t$$

$$\frac{93000000}{3100} = t$$

$$30{,}000 \text{ hrs.} = t$$

123. If $x^{-1} = 5$ then

$$\frac{1}{x} = 5$$

Solving for x:

$$1 = 5x$$

$$\frac{1}{5} = x$$

125. $\sqrt[3]{-125} = -5$

127. Let x = the unknown number.

$$\frac{2x}{5} = 8$$

$$5 \cdot \frac{2x}{5} = 5 \cdot 8$$

$$2x = 40$$

$$x = 20$$

The number is 20.

JUST FOR FUN

1. a)

0 gives $10^{0} = 1$

1 gives $10^{1} = 10$

2 gives $10^{2} = 100$

3 gives $10^{3} = 1000$

4 gives $10^{4} = 10000$

5 gives $10^{5} = 100{,}000$

6 gives $10^{6} = 1{,}000{,}000$

7 gives $10^{7} = 10{,}000{,}000$

8 gives $10^{8} = 100{,}000{,}000$

9 gives $10^{9} = 1{,}000{,}000{,}000$

10 gives $10^{10} = 10{,}000{,}000{,}000$

b) An earthquake that measures 6 gives $10^{6} = 1{,}000{,}000$. An earthquake that measures 2 gives $10^{2} = 100$ Since $10^{4} \cdot 10^{2} = 10^{6}$, then an earthquake that measures 6 is 10,000 times more intense than one that measures 2 on the Richter scale.

c) The Coalinga earthquake measured 6.9 on the Richter scale. The San Francisco earthquake measured 8.3 on the Richter scale. The difference between 8.3 and 6.9 is 1.4. Therefore, the San Francisco earthquake was $10^{1.4} \approx 25.1$ times more intense than the Coalinga quake.

EXERCISE SET 5.2

1. $$x^0 \cdot x^4 = x^{0+4} = x^4$$

3. $$\left[3^2\right]^2 = 3^{2\cdot 2} = 3^4 = 81$$

5. $$\left[2^3\right]^{-2} = 2^{3\cdot(-2)} = 2^{-6} = \frac{1}{2^6} = \frac{1}{64}$$

7. $$\left[y^0\right]^3 = y^{0\cdot 3} = y^0 = 1$$

9. $$(-x)^2 = (-x)(-x) = x^2$$

11. $$(-x)^{-3} = \frac{1}{(-x)(-x)(-x)} = \frac{1}{-x^3} = -\left[\frac{1}{x^3}\right]$$

13. $$3\left[x^4\right]^{-2} = 3\left[x^{-8}\right] = \frac{3}{x^8}$$

15. $$\left[3x^2\right]^3 = 3^3 \cdot x^{2\cdot 3} = 27x^6$$

17. $$\left[\frac{1}{2}\right]^{-3} = \frac{1}{\left[\frac{1}{2}\right]^3} = \frac{1}{\left[\frac{1}{8}\right]} = 1\cdot\frac{8}{1} = 8$$

19. $$\left[-3x^2 \cdot y\right]^4 = (-3)^4 \cdot x^{2\cdot 4} \cdot y^4 = 81x^8 \cdot y^4$$

21. $$\left[4x^2 \cdot y^{-2}\right]^2 = 4^2 \cdot x^{2\cdot 2} \cdot y^{-2\cdot 2} = 16x^4 \cdot y^{-4} = \frac{16x^4}{y^4}$$

23. $$\left[2x^3 \cdot y\right]^{-3} = \frac{1}{\left[2x^3 \cdot y\right]^3} = \frac{1}{2^3 \cdot x^{3\cdot 3} \cdot y^3} = \frac{1}{8x^9 \cdot y^3}$$

25. $$\left[-4x^{-4} \cdot y^5\right]^{-3} = \frac{1}{\left[-4x^{-4} \cdot y^5\right]^3} = \frac{1}{(-4)^3 \cdot x^{-4\cdot 3} \cdot y^{5\cdot 3}}$$

25. cont.

$$= \frac{1}{-64 \cdot x^{-12} \cdot y^{15}}$$

$$= \frac{x^{12}}{-64y^{15}}$$

27.

$$\left[\frac{6x}{y^2}\right]^2 = \frac{6^2 \cdot x^2}{y^{2 \cdot 2}}$$

$$= \frac{36x^2}{y^4}$$

29.

$$\left[\frac{2x^4 \cdot y^5}{x^2}\right]^3 = \frac{2^3 \cdot x^{4 \cdot 3} \cdot y^{5 \cdot 3}}{x^{2 \cdot 3}}$$

$$= \frac{8x^{12} \cdot y^{15}}{x^6}$$

$$= 8x^{12-6} \cdot y^{15}$$

$$= 8x^6 \cdot y^{15}$$

31.

$$\left[\frac{20x^3 \cdot y^4}{4x \cdot y^4}\right]^3 = \left[5x^{3-4} \cdot y^{4-1}\right]^3$$

$$= \left[5x^{-1} \cdot y^3\right]^3$$

$$= 5^3 \cdot x^{-1 \cdot 3} \cdot y^{3 \cdot 3}$$

$$= 125x^{-3} \cdot y^9$$

$$= \frac{125y^9}{x^3}$$

33.

$$\left[\frac{3xy^5}{6xy^5}\right]^{-2} = \left[\frac{x^{1-1} \cdot y^{5-5}}{2}\right]^{-2}$$

$$= \left[\frac{x^0 \cdot y^0}{2}\right]^{-2}$$

$$= \left[\frac{1}{2}\right]^{-2}$$

$$= \frac{1^{-2}}{2^{-2}}$$

$$= \frac{2^2}{1} = 4$$

35.

$$\left[\frac{4x^{-2} \cdot y}{x^{-5}}\right]^3 = \left[4x^{-2-(-5)} \cdot y\right]^3$$

$$= \left[4x^3 \cdot y\right]^3$$

$$= 4^3 \cdot x^{3 \cdot 3} \cdot y^3$$

$$= 64x^9 \cdot y^3$$

37.

$$\left[\frac{6x^2 \cdot y}{3xz}\right]^{-3} = \left[2x^{2-1} \cdot y \cdot z^{-1}\right]^{-3}$$

$$= \left[2xyz^{-1}\right]^{-3}$$

$$= 2^{-3} \cdot x^{-3} \cdot y^{-3} \cdot z^3$$

$$= \frac{z^3}{8x^3 \cdot y^3}$$

39.

$$[2x^3 y][6x^4 y^3]^2$$

$$= [2x^3 y]\cdot[36x^8 y^6]$$

$$= 72x^{3+8} y^{1+6}$$

$$= 72x^{11} y^7$$

41.

$$[4x^2 y^5]^2 \cdot [6x^4 y^3]$$

$$= 4^2 \cdot x^{2\cdot 2} y^{5\cdot 2} \cdot 6 \cdot x^4 \cdot y^3$$

$$= 16x^4 y^{10} \cdot 6 \cdot x^4 y^3$$

$$= 96x^{4+4} y^{10+3}$$

$$= 96x^8 y^{13}$$

43.

$$\left[\frac{3x^2 y^{-2}}{y^3}\right]^2 \cdot \left[\frac{xy^2}{3}\right]^{-2}$$

$$= [3x^2 y^{-2-3}]^2 \cdot \frac{x^{-2} y^{2\cdot(-2)}}{3^{-2}}$$

$$= 3^2 \cdot x^{2\cdot 2} y^{-5\cdot 2} \cdot x^{-2} y^{-4} \cdot 3^2$$

$$= 9 \cdot x^4 y^{-10} \cdot x^{-2} y^{-4} \cdot 9$$

$$= 81x^{4+(-2)} y^{-10+(-4)}$$

$$= 81x^2 y^{-14} = \frac{81x^2}{y^{14}}$$

45.

$$\left[\frac{3x^2 y^5}{2z^{-1}}\right] \cdot \left[\frac{2z^3}{3xy^4}\right]^4$$

45. cont.

$$\frac{3x^2 y^5 z^1}{2} \cdot \frac{2^4 \cdot z^{3\cdot 4}}{3^4 \cdot x^4 \cdot y^{4\cdot 4}}$$

$$= \frac{8x^2 y^5 z^1 z^{12}}{27x^4 y^{16}}$$

$$= \frac{8x^2 y^5 z^{13}}{27x^4 y^{16}}$$

$$= \frac{8x^{2-4} y^{5-16} z^{13}}{27}$$

$$= \frac{8x^{-2} y^{-11} z^{13}}{27}$$

$$= \frac{8z^{13}}{27x^2 y^{11}}$$

47.

$$\frac{[4x^2 y^{-3}]^2 \cdot [xy^5]^{-3}}{[6x^4 y^5]^3}$$

$$= \frac{16x^4 y^{-6} \cdot x^{-3} y^{-15}}{216x^{12} y^{15}}$$

$$= \frac{16x^{4-3} y^{-6-15}}{216x^{12} y^{15}}$$

$$= \frac{16x^{1-12} y^{-21-15}}{216}$$

$$= \frac{2}{27x^{11} y^{36}}$$

49.

$$\left[\frac{x^{2} \cdot y^{-3} \cdot z^{4}}{x^{-1} \cdot y^{2} \cdot z^{3}}\right]^{-1} \cdot \left[\frac{x \cdot y^{2} \cdot z}{x^{-3} \cdot y^{-7} \cdot z^{3}}\right]^{3}$$

$$= \left[x^{2-(-1)} \cdot y^{-3-2} \cdot z^{4-3}\right]^{-1} \cdot \left[x^{-1-(-3)} \cdot y^{2-(-7)} \cdot z^{1-3}\right]^{3}$$

$$= \left[x^{3} \cdot y^{-5} \cdot z^{1}\right]^{-1} \cdot \left[x^{4} \cdot y^{9} \cdot z^{-2}\right]^{3}$$

$$= x^{-3} \cdot y^{5} \cdot z^{-1} \cdot x^{12} \cdot y^{27} \cdot z^{-6}$$

$$= x^{-3+12} \cdot y^{5+27} \cdot z^{-1-6}$$

$$= x^{9} \cdot y^{32} \cdot z^{-7} = \frac{x^{9} \cdot y^{32}}{z^{7}}$$

51.

$$\left[\frac{6x^{4} \cdot y^{-6} \cdot z^{4}}{2xy^{-6} \cdot z^{-2}}\right]^{-2} \left[\frac{-x^{4} \cdot y^{3}}{2z^{4}}\right]^{-1}$$

$$= \left[3x^{4-1} \cdot y^{-6-(-6)} \cdot z^{4-(-2)}\right]^{-1} \cdot \left[\frac{-x^{-4} \cdot y^{-3}}{2^{-1} \cdot z^{-4}}\right]$$

$$= \left[3x^{3} \cdot z^{6}\right]^{-2} \cdot \frac{-2z^{4}}{x^{4} \cdot y^{3}}$$

$$= \frac{\left[3^{-2} \cdot x^{-6} \cdot z^{-12}\right]\left[-2z^{4}\right]}{x^{4} \cdot y^{3}}$$

$$= \frac{-2 \cdot x^{-6-4} \cdot z^{-12+4}}{3^{2} \cdot y^{3}}$$

51. cont.

$$= \frac{-2}{9x^{10} \cdot y^{3} \cdot z^{8}}$$

53.

$$\frac{\left[4x^{-1} \cdot y^{-2}\right]^{-3} \cdot \left[xy^{3}\right]^{2}}{\left[3x^{-1} \cdot y^{3}\right]^{2}}$$

$$= \frac{4^{-3} \cdot x^{3} \cdot y^{6} \cdot x^{2} \cdot y^{6}}{3^{2} \cdot x^{-2} \cdot y^{6}}$$

$$= \frac{4^{-3} \cdot x^{3+2} \cdot y^{6+6}}{9x^{-2} \cdot y^{6}}$$

$$= \frac{4^{-3} \cdot x^{5} \cdot y^{12}}{9x^{-2} \cdot y^{6}}$$

$$= \frac{x^{5-(-2)} \cdot y^{12-6}}{9 \cdot 4^{3}}$$

$$= \frac{x^{7} \cdot y^{6}}{576}$$

55.

$$\frac{\left[5y^{3} \cdot z^{-4}\right]^{-1} \cdot \left[2y^{-3} \cdot z^{2}\right]^{-1}}{\left[3y^{3} \cdot z^{-2}\right]^{-1}}$$

$$= \frac{5^{-1} \cdot y^{-3} \cdot z^{4} \cdot 2^{-1} \cdot y^{3} \cdot z^{-2}}{3^{-1} \cdot y^{-3} \cdot z^{2}}$$

$$= \left[\frac{3y^{-3+3-(-3)} \cdot z^{4+(-2)-2}}{5 \cdot 2}\right]$$

$$= \frac{3y^{3} \cdot z^{0}}{10} = \frac{3y^{3}}{10}$$

57.

$$x^{-m}\cdot\left[x^{3m+2}\right]^{2}$$

$$= x^{-m}\cdot x^{2\cdot(3m+2)}$$

$$= x^{-m}\cdot x^{6m+4}$$

$$= x^{-m+6m+4} = x^{5m+4}$$

59.

$$\left[b^{5y-2}\right]^{y}\cdot b^{5y} = b^{y(5y-2)}\cdot b^{5y}$$

$$= b^{5y^2-2y+5y} = b^{5y^2+3y}$$

61.

$$\frac{\left[m^{-5y+2}\right]\cdot\left[m^{2y+3}\right]}{m\cdot\left[m^{4y+1}\right]}$$

$$= \frac{m^{-5y+2+2y+3}}{m^{1+4y+1}}$$

$$= \frac{m^{-3y+5}}{m^{4y+2}}$$

$$= m^{(-3y+5)-(4y+2)}$$

$$= m^{-7y+3}$$

63. Represent the unknown exponents as a and b:

$$\left[\frac{x^{2}\cdot y^{-2}}{x^{-3}\cdot y^{-1}}\right]^{2}\left[\frac{x^{a}\cdot y^{3}}{x^{2}\cdot y^{b}}\right]$$

Simplify:

$$= \left[x^{2-(-3)}\cdot y^{-2-(-1)}\right]^{2}$$

$$\left[x^{a-2}\cdot y^{3-b}\right]$$

63. cont.

$$= \left[x^{5}\cdot y^{-1}\right]^{2}\cdot x^{a-7}\cdot y^{3-b}$$

$$= x^{10}\cdot y^{-2}\cdot x^{a-7}\cdot y^{3-b}$$

$$= x^{10+a-7}\cdot y^{-2+3-b}$$

$$= x^{3+a}\cdot y^{1-b}$$

Since we want the total expression to euqal 1, the exponent on x must equal 0 and the exponent on y must equal 0. Therefore:

$3 + a = 0$ and $1 - b = 0$

$a = -3$ $\quad$ $b = 1$

x should be raised to -3 and y should be raised to the first power.

65. Let a,b and c represent the missing exponents:

$$\left[\frac{x^{-4}\cdot y^{-2}\cdot z^{7}}{x^{-2}\cdot y^{5}\cdot z^{2}}\right]^{-4}\cdot\left[\frac{x^{a}\cdot y^{5}\cdot z^{-2}}{x^{4}\cdot y^{b}\cdot z^{c}}\right]$$

Simplify:

$$= \left[x^{-4-(-2)}\cdot y^{-2-5}\cdot z^{7-2}\right]^{-4}$$

$$\left[x^{a-4}\cdot y^{5-b}\cdot z^{-2-c}\right]$$

$$= \left[x^{-2}\cdot y^{-7}\cdot z^{5}\right]^{-4}\cdot\left[x^{a-4}\cdot y^{5-b}\cdot z^{-2-c}\right]$$

$$= \left[x^{8}\cdot y^{28}\cdot z^{-20}\cdot x^{a-4}\cdot y^{5-b}\cdot z^{-2-c}\right]$$

$$= x^{8+a-4}\cdot y^{28+5-b}\cdot z^{-20-2-c}$$

$$= x^{4+a}\cdot y^{33-b}\cdot z^{-22-c}$$

Since we want the total expression to equal one, the exponent on x, y, and z must equal 0. Therefore:

$4 + a = 0$ $\quad$ $33 - b = 0$ $\quad$ $-22 - c = 0$

$a = -4$ $\quad$ $b = 33$ $\quad$ $c = -22$

67. 1) $x + 3y = 10$
2) $2x - 4y = -10$
Solve equation 1 for x:
1) $x = -3y + 10$
Substitute $x = -3y + 10$ into equation 2 to solve for y.

2) $2(-3y + 10) - 4y = -10$
$-6y + 20 - 4y = -10$
$-10y = -30$
$y = 3$

Substitute $y = 3$ into equation 1 and solve for x:
1) $x + 3(3) = 10$
$x = 1$
Solution: (1,3)

"69. Graph each equation. The point of intersection is the solution.

| $x + y = 1$ | | $-3x + 2y = 12$ | |
|---|---|---|---|
| x | y | x | y |
| 0 | 1 | 0 | 6 |
| 1 | 0 | -4 | 0 |

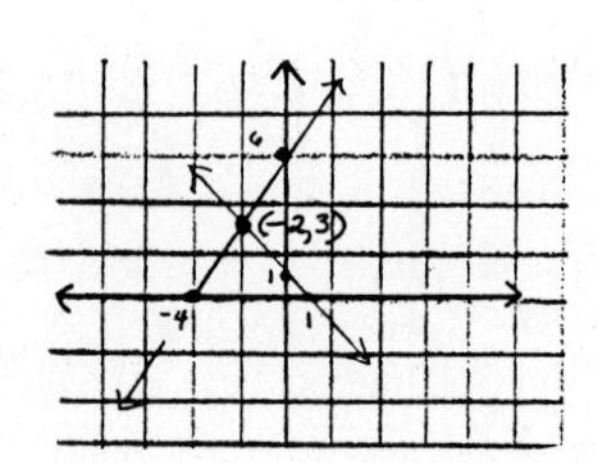


JUST FOR FUN

1.
$$\left[\frac{x^{\frac{1}{2}}}{x^{-1}}\right]^{\frac{3}{2}} = \frac{x^{\frac{1}{2}\cdot\frac{3}{2}}}{x^{-1\cdot\frac{3}{2}}}$$

$$= \frac{x^{\frac{3}{4}}}{x^{\frac{-3}{2}}}$$

JUST FOR FUN 1. cont.

$$= x^{\left[\frac{3}{4}\right]-\left[\frac{-3}{2}\right]}$$

$$= x^{\left[\frac{3}{4}\right]+\frac{6}{4}}$$

$$= x^{\left[\frac{9}{4}\right]}$$

3.
$$\left[\frac{x^{4}}{x^{\frac{-1}{2}}}\right]^{-1} = \frac{x^{-4}}{x^{\frac{1}{2}}}$$

$$= x^{-4-\frac{1}{2}}$$

$$= x^{\left[\frac{-8}{2}\right]-\frac{1}{2}}$$

$$= x^{\frac{-9}{2}} = \frac{1}{x^{\left[\frac{9}{2}\right]}}$$

5.
$$\left[\frac{x^{\frac{1}{2}}\cdot y^{4}}{x^{-3}\cdot y^{\frac{5}{2}}}\right]^{2} = \frac{x^{1}\cdot y^{8}}{x^{-6}\,y^{5}}$$

$$= \left[x^{1-(-6)}\cdot y^{8-5}\right]$$

$$= x^{7}\cdot y^{3}$$

EXERCISE SET 5.3

1. Since $5y$ has only one term, it is a monomial.

3. Since $-10$ has only one term, it is a monomial.

5. The polynomial $4 - 5z$ has two terms so it is a binomial.

7. The polynomial $8x^2 - 2x + 8y^2$ has three terms and is a trinomial.

9. $3x^{1/2} + 2xy$ is not a polynomial because it contains a fractional exponent.

11. $2xy + 5y^2$ has two terms and is a binomial.

13. $-8 - 4x - x^2 = -x^2 - 4x - 8$
Second degree

15. $6y^2 + 3xy + 10x^2$
$= 10x^2 + 3xy + 6y^2$
Second degree

17. $5 + 2x^3 + x^2 - 3x$
$= 2x^3 + x^2 - 3x + 5$
Third degree

19. $-2x^4 + 5x^2 - 4$ is in descending order.
Fourth degree

21. $5xy^2 + 3x^2y - 6 - 2x^3$
$= -2x^3 + 3x^2y + 5xy^2 - 6$
Third degree

23. $(6x + 3) + (x - 5)$
$= 6x + 3 + x - 5$
$= 7x - 2$

25. $(3x - 4) - (2x + 2)$
$= 3x - 4 - 2x - 2$
$= x - 6$

27. $(-12x - 3) - (-5x - 7)$
$= -12x - 3 + 5x + 7$
$= -7x + 4$

29. $\left[x^2 - 6x + 3\right] - (2x + 5)$
$= x^2 - 6x + 3 - 2x - 5$
$= x^2 - 8x - 2$

31. $(x + y - z) + (2x - y + 3)$
$= x + y - z + 2x - y + 3$
$= 3x - z + 3$

33. $(5x - 7) + \left[2x^2 + 3x + 12\right]$
$= 5x - 7 + 2x^2 + 3x + 12$
$= 2x^2 + 8x + 5$

35. $\left[6y^2 - 6y + 4\right] - \left[-2y^2 - y + 7\right]$
$= 6y^2 - 6y + 4 + 2y^2 + y - 7$
$= 8y^2 - 5y - 3$

37. $\left[-2x^2 + 4x - 5\right] - \left[5x^2 + 3x + 7\right]$
$= -2x^2 + 4x - 5 - 5x^2 - 3x - 7$
$= -7x^2 + x - 12$

39. $\left[5x^2 - x + 12\right] - (x + 5)$
$= 5x^2 - x + 12 - x - 5$
$= 5x^2 - 2x + 7$

41. $\left[-3x^3 + 4x^2y + 3xy^2\right] + \left[2x^3 - x^2y + xy^2\right]$
$= -3x^3 + 4x^2y + 3xy^2 + 2x^3 - x^2y + xy^2$
$= -3x^3 + 3x^2y + 4xy^2$

43. $\left[-2xy^2 + 4\right] - \left[-7xy^2 + 12\right]$
$= -2xy^2 + 4 + 7xy^2 - 12$
$= 5xy^2 - 8$

45. $\left[x^2 + xy - y^2\right] + \left[2x^2 - 3xy + y^2\right]$
$= x^2 + xy - y^2 + 2x^2 - 3xy + y^2$
$= 3x^2 - 2xy$

47. $\left[4x^3 - 6x^2 + 5x - 7\right] - \left[2x^2 + 6x - 3\right]$

$= 4x^3 - 6x^2 + 5x - 7 - 2x^2 - 6x + 3$

$= 4x^3 - 8x^2 - x - 4$

49. $\left[4x^2y + 2x - 3\right] + \left[3x^2y - 5x + 5\right]$

$= 4x^2y + 2x - 3 + 3x^2y - 5x + 5$

$= 7x^2y - 3x + 2$

51. $\left[9x^3 - 4\right] - \left[x^2 + 5\right]$

$= 9x^3 - 4 - x^2 - 5$

$= 9x^3 - x^2 - 9$

53. $\left[x^2 - 2x + 4\right] + (3x + 12)$

$= x^2 - 2x + 4 + 3x + 12$

$= x^2 + x + 16$

55. $(3x + 5) - (4x - 6)$

$= 3x + 5 - 4x + 6$

$= -x + 11$

57. $\left[-2x^2 + 4x - 12\right] + \left[-x^2 - 2x\right]$

$= -2x^2 + 4x - 12 - x^2 - 2x$

$= -3x^2 + 2x - 12$

59. $\left[2x^2 - 4x + 8\right] - \left[5x^2 - 6\right]$

$= 2x^2 - 4x + 8 - 5x^2 + 6$

$= -3x^2 - 4x + 14$

61. $\left[3x^2 + 4x - 5\right] + \left[4x^2 + 3x - 8\right]$

$= 3x^2 + 4x - 5 + 4x^2 + 3x - 8$

$= 7x^2 + 7x - 13$

63. $\left[9y^2 - 3y\right] - \left[-6y^2 + 3y - 4\right]$

$= 9y^2 - 3y + 6y^2 - 3y + 4$

$= 15y^2 - 6y + 4$

65. $\left[6x^2 + 3xy\right] + \left[-2x^2 + 4xy + 3y\right]$

$= 6x^2 + 3xy - 2x^2 + 4xy + 3y$

$= 4x^2 + 7xy + 3y$

67. $\left[x^3 - 6\right] - \left[-4x^2 + 6x\right]$

$= x^3 - 6 + 4x^2 - 6x$

$= x^3 + 4x^2 - 6x - 6$

69. $\left[4x^2 + 3x + y^2\right] + \left[4x - 3y - 5y^2\right]$

$= 4x^2 + 3x + y^2 + 4x - 3y - 5y^2$

$= 4x^2 + 7x - 3y - 4y^2$

71. $\left[-2x^2y + 6xy^2 + 8\right] - \left[5x^2y + 8\right]$

$= -2x^2y + 6xy^2 + 8 - 5x^2y - 8$

$= -7x^2y + 6xy^2$

73. The signs of all the terms in the polynomial being subtracted change to the opposite sign.

75.

$-4 < \frac{6 - 3x}{2} \le 5$

$-8 < 6 - 3x \le 10$

$-14 < -3x \le 4$

$\frac{14}{3} > x \ge \frac{-4}{3}$ or

$\frac{-4}{3} \le x < \frac{14}{3}$

$\{x \mid \frac{-4}{3} \le x < \frac{14}{3}\}$

77. 1) $-2x + 3y + 4z = 17$
2) $-5x - 3y + z = -1$
3) $-x - 2y + 3z = 18$

Multiply equation 3 by -2. Add the result to equation 1 to eliminate x.

1) $-2x + 3y + 4z = 17$
3) $2x + 4y - 6z = -36$

---

4) $7y - 2z = -19$

Multiply equation 3 by -5. Add the result to equation 2 to eliminate x.

77. cont.

2) $-5x - 3y + z = -1$
3) $5x + 10y - 15z = -90$

---

5) $7y - 14z = -91$

Multiply equation 4 by -1. Add the result to equation 5 to eliminate y and solve for z.

4) $-7y + 2z = 19$
5) $7y - 14z = -91$

---

$-12z = -72$
$z = 6$

Substitute z = 6 into equation 4 or 5 and solve for y.

4) $7y - 2(6) = -19$
$7y = -7$
$y = -1$

Substitute y = -1 and z = 6 into equation 1,2,or 3 and solve for x.

1) $-2x + 3(-1) + 4(6) = 17$
$-2x - 3 + 24 = 17$
$-2x + 21 = 17$
$-2x = -4$
$x = 2$

Solution: (2, -1, 6)

EXERCISE SET 5.4

1.
$$(4xy)(6xy^4) = 4(6)x \cdot x \cdot y \cdot y^4 = 24x^{1+1}y^{1+4} = 24x^2y^5$$

3.
$$\left[\frac{5}{9}x^2y^5\right]\left[\frac{1}{5}x^5y^3z^2\right] = \frac{5}{9}\cdot\frac{1}{5}x^{2+5}y^{5+3}z^2 = \frac{1}{9}x^7y^8z^2$$

5.
$$-2x\left[x^2 - 2x + 5\right] = -2x \cdot x^2 - (-2x)(2x) - 2x(5) = -2x^3 + 4x^2 - 10x$$

7.
$$5x^2\left[-4x^2 + 6x - 4\right]$$

7. cont.
$$= 5x^2\left[-4x^2\right] + 5x^2(6x) + 5x^2(-4) = -20x^4 + 30x^3 - 20x^2$$

9.
$$-3x^2y\left[-2x^4y^2 + 3xy^3 + 4\right]$$
$$= \left[-3x^2y\right]\left[-2x^4y^2\right] + \left[-3x^2y\right]\left[3xy^3\right] + \left[-3x^2y\right](4)$$
$$= 6x^6y^3 - 9x^3y^4 - 12x^2y$$

11.
$$\frac{2}{3}yz\left[3x + 4y - 9y^2\right]$$
$$= \frac{2}{3}yz(3x) + \frac{2}{3}yz(4y) + \frac{2}{3}yz\left[-9y^2\right]$$
$$= 2xyz + \frac{8}{3}y^2z - 6y^3z$$

13.
$$(x - 4)(x + 5) = \overset{F}{x(x)} + \overset{O}{5x} \overset{I}{-4x} \overset{L}{-4(5)} = x^2 + x - 20$$

15.
$$(3x+1)(x-3) = \overset{F}{3xx} + \overset{O}{3x(-3)} + \overset{I}{x} + \overset{L}{1(-3)} = 3x^2 - 9x + x - 3 = 3x^2 - 8x - 3$$

17.
$$(x - y)(x + y) = \overset{F}{xx} + \overset{O}{xy} - \overset{I}{yx} \overset{L}{-yy} = x^2 - y^2$$

19.
$$\left[2x^2 - 3\right](2x - 3)$$
$$= \overset{F}{2x^2(2x)} + \overset{O}{2x^2(-3)} + \overset{I}{(-3)(2x)} - \overset{L}{3(-3)}$$
$$= 4x^3 - 6x^2 - 6x + 9$$

21.
$$(4 - x)\left[3 + 2x^2\right] = 4(3) + 4\left[2x^2\right] - x(3) - x\left[2x^2\right]$$

21. cont.

$$= 12 + 8x^2 - 3x - 2x^3$$

$$= -2x^3 + 8x^2 - 3x + 12$$

23. $\left[\frac{1}{2}x + 2y\right]\left[2x - \frac{1}{3}y\right]$

$$= \frac{1}{2}x(2x) + \frac{1}{2}x\left[\frac{-1}{3}y\right] + 2y(2x)$$

$$+ \quad 2y\left[\frac{-1}{3}y\right]$$

$$= x^2 - \frac{1}{6}xy + 4xy - \frac{2}{3}y^2$$

$$= x^2 - \frac{1}{6}xy + \frac{24}{6}xy - \frac{2}{3}y^2$$

$$= x^2 + \frac{23}{6}xy - \frac{2}{3}y^2$$

25. $\left[4x^2 - 3y\right]\left[2y^2 - 3x\right]$

$$= 4x^2\left[2y^2\right] + 4x^2(-3x) - 3y\left[2y^2\right]$$

$$- \; 3y(-3x)$$

$$= 8x^2y^2 - 12x^3 - 6y^3 + 9xy$$

$$= -12x^3 + 8x^2y^2 + 9xy - 6y^3$$

27.

$$\begin{array}{r} x^2 - 3x + 2 \\ x - 4 \\ \hline -4x^2 + 12x - 8 \\ x^3 - 3x^2 + 2x \qquad \\ \hline x^3 - 7x^2 + 14x - 8 \end{array}$$

Multiply the top expression by -4.
Multiply the top expression by x.
Add like terms.

29.

$$\begin{array}{r} y^2 - 3y + 4 \\ 3 - 2y \\ \hline -2y^3 + 6y^2 - 8y \qquad \\ 3y^2 - 9y + 12 \\ \hline -2y^3 + 9y^2 - 17y + 12 \end{array}$$

Multiply the top expression by -2y.
Multiply the top expression by 3
Add like terms.

31.

$$\begin{array}{r} -2x^2 - 4x + 1 \\ 7x - 3 \\ \hline 6x^2 + 12x - 3 \\ -14x^3 - 28x^2 + 7x \qquad \\ \hline -14x^3 - 22x^2 + 19x - 3 \end{array}$$

33.

$$\begin{array}{r} x^2y - 3xy^2 \\ x + 2y \\ \hline 2x^2y^2 - 6xy^3 \\ x^3y - 3x^2y^2 \qquad \\ \hline x^3y - x^2y^2 - 6xy^3 \end{array}$$

35.

$$\begin{array}{r} 2a^2 - ab + 2b^2 \\ a - 3b \\ \hline -6a^2b + 3ab^2 - 6b^3 \\ 2a^3 - a^2b + 2ab^2 \qquad \\ \hline 2a^3 - 7a^2b + 5ab^2 - 6b^3 \end{array}$$

37.

$$\begin{array}{r} x^3 - 2x^2 + 5x - 6 \\ 2x^2 - 3x + 4 \\ \hline 4x^3 - 8x^2 + 20x - 24 \\ -3x^4 + 6x^3 - 15x^2 + 18x \qquad \\ 2x^5 - 4x^4 + 10x^3 - 12x^2 \qquad\qquad \\ \hline 2x^5 - 7x^4 + 20x^3 - 35x^2 + 28x - 24 \end{array}$$

39.

$$(3x - 1)^3 = (3x - 1)(3x - 1)(3x - 1)$$

F O I L

$$= \left[9x^2 - 3x - 3x + 1\right](3x - 1)$$

$$= \left[9x^2 - 6x + 1\right](3x - 1)$$

Now multiply the expressions vertically:

$$\begin{array}{r} 9x^2 - 6x + 1 \\ 3x - 1 \\ \hline -9x^2 + 6x - 1 \\ 27x^3 - 18x^2 + 3x \quad\quad \\ \hline 27x^3 - 27x^2 + 9x - 1 \end{array}$$

Thus

$$(3x - 1)^3 = 27x^3 - 27x^2 + 9x - 1$$

41.

$$(x + 4)(x - 4) = x^2 - (4)^2$$
$$= x^2 - 16$$

43.

$$(2x - 1)(2x + 1) = (2x)^2 - (1)^2$$
$$= 4x^2 - 1$$

45.

$$(2x - 3y)^2 = (2x)^2 - 2(2x)(3y) + (3y)^2$$
$$= 4x^2 - 12xy + 9y^2$$

47.

$$(2x + 5y)^2 = (2x)^2 + 2(2x)(5y) + (5y)^2$$
$$= 4x^2 + 20xy + 25y^2$$

49.

$$\left[2y^2 - 5w\right]^2 = \left[2y^2\right]^2 - 2\left[2y^2\right](5w) + (5w)^2$$
$$= 4y^4 - 20y^2 w + 25w^2$$

51.

$$\left[5m^2 + 2n\right]\left[5m^2 - 2n\right] = \left[5m^2\right]^2 - (2n)^2$$
$$= 25m^4 - 4n^2$$

53.

$$\left[y + (4 - 2x)\right]^2$$
$$= y^2 + 2(y)(4 - 2x) + (4 - 2x)^2$$
$$= y^2 + 8y - 4xy$$
$$+ \left[4^2 - 2(4)(2x) + (2x)^2\right]$$
$$= y^2 + 8y - 4xy + 16 - 16x + 4x^2$$

55.

$$\left[4 - (x - 3y)\right]^2$$
$$= 4^2 - 2(4)(x - 3y) + (x - 3y)^2$$
$$= 16 - 8x + 24y$$
$$+ \left[x^2 - 2(x)(3y) + (3y)^2\right]$$
$$= 16 - 8x + 24y + x^2 - 6xy + 9y^2$$

57.

$$\left[(x + y) + 4\right]^2$$
$$= (x + y)^2 + 2(x + y)(4) + 4^2$$
$$= x^2 + 2xy + y^2 + 8x + 8y + 16$$

59.

$$\left[(x - 3y) - 5\right]^2$$
$$= (x - 3y)^2 - 2(x - 3y)(5) + 5^2$$
$$= x^2 - 6xy + 9y^2 - 10x + 30y + 25$$

61. $(8r^5s^4)(-3rs^9)$

$= (8)(-3)r^{5+1}s^{4+9}$

$= -24r^6s^{13}$

63. $3x(x^2 + 3x - 1)$

$= 3x(x^2) + 3x(3x) + 3x(-1)$

$= 3x^3 + 9x^2 - 3x$

65. $\left(\frac{1}{3}x - \frac{2}{5}y\right)\left(\frac{1}{2}x - y\right)$

$= \frac{1}{3}x\left(\frac{1}{2}x\right) + \frac{1}{3}x(-y) - \frac{2}{5}y\left(\frac{1}{2}x\right) - \frac{2}{5}y(-y)$

$= \frac{1}{6}x^2 - \frac{1}{3}xy - \frac{1}{5}xy + \frac{2}{5}y^2$

$= \frac{1}{6}x^2 - \frac{5}{15}xy - \frac{3}{15}xy + \frac{2}{5}y^2$

$= \frac{1}{6}x^2 - \frac{8}{15}xy + \frac{2}{5}y^2$

69. $\frac{-3}{5}x^2y\left(\frac{-2}{3}xy^4 + \frac{1}{9}xy + 3\right)$

$= \frac{-3}{5}x^2y\left(\frac{-2}{3}xy^4\right)$

$\frac{-3}{5}x^2y\left(\frac{1}{9}xy\right) - \frac{3}{5}x^2y(3)$

$= \frac{2}{5}x^3y^5 - \frac{1}{15}x^3y^2 - \frac{9}{5}x^2y$

71. $\left(2x - \frac{3}{4}\right)\left(2x + \frac{3}{4}\right)$

$= (2x)^2 - \left(\frac{3}{4}\right)^2$

71. cont.

$= 4x^2 - \frac{9}{16}$

73. $(4x - 5y)^2$

$= (4x)^2 - 2(4x)(5y) + (5y)^2$

$= 16x^2 - 40xy + 25y^2$

75.

$$\begin{array}{r} 2x^2 + 4x - 3 \\ x + 3 \\ \hline 6x^2 + 12x - 9 \\ 2x^3 + 4x^2 - 3x \quad\quad \\ \hline 2x^3 + 10x^2 + 9x - 9 \end{array}$$

77.

$$\begin{array}{r} x^2 - x + 4 \\ 5x + 4 \\ \hline 4x^2 - 4x + 16 \\ 5x^3 - 5x^2 + 20x \quad\quad \\ \hline 5x^3 - x^2 + 16x + 16 \end{array}$$

79.

$$\begin{array}{r} 3x^2 + 4xy - 2y^2 \\ 2x - 3y \\ \hline 6x^3 + 8x^2y - 4xy^2 \quad\quad \\ -9x^2y - 12xy^2 + 6y^3 \\ \hline 6x^3 - x^2y - 16xy^2 + 16y^3 \end{array}$$

81. $\left(\frac{1}{2}r + \frac{1}{4}s\right)^2$

$= \left(\frac{1}{2}r\right)^2 + 2\left(\frac{1}{2}r\right)\left(\frac{1}{4}s\right) + \left(\frac{1}{4}s\right)^2$

$= \frac{1}{4}r^2 + \frac{1}{4}rs + \frac{1}{16}s^2$

83. $\frac{2}{3}x^2y^4\left[\frac{3}{5}xy^3 - \frac{1}{4}x^4y + 2xy^3z^5\right]$

$= \frac{2}{3}x^2y^4\left[\frac{3}{5}xy^3\right] + \frac{2}{3}x^2y^4\left[\frac{-1}{4}x^4y\right] + \frac{2}{3}x^2y^4\left[2xy^3z^5\right]$

$= \frac{2}{5}x^3y^7 - \frac{1}{6}x^6y^5 + \frac{4}{3}x^3y^7z^5$

85. $\left[w + (3x + 4)\right]\left[w - (3x + 4)\right] = w^2 - (3x + 4)^2$

$= w^2 - \left[(3x)^2 + 2(3x)(4) + 4^2\right]$

$= w^2 - \left[9x^2 + 24x + 16\right]$

$= w^2 - 9x^2 - 24x - 16$

87.

$$\begin{array}{r} a^2 - ab + b^2 \\ a + b \\ \hline a^2b - ab^2 + b^3 \\ a^3 - a^2b + ab^2 \phantom{+ b^3} \\ \hline a^3 \phantom{- a^2b + ab^2} - b^3 \end{array}$$

89.

$$\begin{array}{r} a^2 - 2ab + 4b^2 \\ a + 2b \\ \hline 2a^2b - 4ab^2 + 8b^3 \\ a^3 - 2a^2b + 4ab^2 \phantom{+ 8b^3} \\ \hline a^3 \phantom{- 2a^2b + 4ab^2} + 8b^3 \end{array}$$

91. $(x + 3)^3 = (x + 3)^2 \cdot (x + 3)$

$= \left[x^2 + 2x(3) + 3^2\right](x + 3)$

$= \left[x^2 + 6x + 9\right](x + 3)$

$= x^3 + 6x^2 + 9x + 3x^2 + 18x + 27$

$= x^3 + 9x^2 + 27x + 27$

93. $\left[(3m + 2) + n\right]\left[(3m + 2) - n\right]$

$= (3m + 2)^2 - n^2$

$= (3m)^2 + 2(3m)(2) + 2^2 - n^2$

$= 9m^2 + 12m + 4 - n^2$

95. $\left[(5x + 1) + 6y\right]\left[(5x + 1) - 6y\right]$

$= (5x + 1)^2 - (6y)^2$

$= \left[(5x)^2 + 2(5x)(1) + 1^2\right] - 36y^2$

$= 25x^2 + 10x + 1 - 36y^2$

97. $\dfrac{3^3 \sqrt[3]{-27} + 3^0}{3 - 3(3) + 3 - 3}$

$= \dfrac{27 - (-3) + 1}{3 - 9 + 1}$

$= \dfrac{-31}{5}$

99. Let x = the amount in the 5% account and y = the amount in the 6% account.

1) $x + y = 10000$
2) $.05x + .06y = 560$

Multiply equation 1 by -.05. Add the result to equation 2 to eliminate x and solve for y.

$$\begin{aligned} 1)\quad -.05x - .05y &= -500 \\ 2)\quad .05x + .06y &= 560 \\ \hline .01y &= 60 \\ y &= 6000 \end{aligned}$$

Substitute y = 6000 into equation 1 and solve for x.

$$\begin{aligned} 1)\quad x + 6000 &= 10000 \\ x &= 4000 \end{aligned}$$

$4000 was invested at 5% and $6000 was invested at 6%.

JUST FOR FUN

1. $$\left[(y + 1) - (x + 2)\right]^2$$

$$= (y + 1)^2 - 2(y + 1)(x + 2) + (x + 2)^2$$

$$= y^2 + 2y + 1 - 2(xy + 2y + x + 2) + x^2 + 4x + 4$$

$$= y^2 + 2y + 1 - 2xy - 4y - 2x - 4 + x^2 + 4x + 4$$

$$= y^2 - 2y - 2xy + x^2 + 2x + 1$$

EXERCISE SET 5.5

1. $$\frac{6x + 8}{2} = \frac{6x}{2} + \frac{8}{2}$$

$$= 3x + 4$$

3. $$\frac{4x^2 + 2x}{2x} = \frac{4x^2}{2x} + \left[\frac{2x}{2x}\right]$$

$$= 2x + 1$$

5. $$\frac{12x^2 - 4x - 8}{4} = \frac{12x^2}{4} - \frac{4x}{4} - \frac{8}{4}$$

$$= 3x^2 - x - 2$$

7. $$\frac{4x^5 - 6x^4 + 12x^3}{4x^2}$$

$$= \frac{4x^5}{4x^2} - \frac{6x^4}{4x^2} + \frac{12x^3}{4x^2}$$

$$= x^3 - \frac{3}{2}x^2 + 3x$$

9. $$\frac{4x^2y^2 - 8xy^3 + 3y^4}{2y^2}$$

$$= \frac{4x^2y^2}{2y^2} - \frac{8xy^3}{2y^2} + \frac{3y^4}{2y^2}$$

$$= 2x^2 - 4xy + \frac{3}{2}y^2$$

11. $$\frac{6x^2y - 12x^3y^2 + 9y^3}{2xy^2}$$

$$= \frac{6x^2y}{2xy^2} - \frac{12x^3y^2}{2xy^2} + \frac{9y^3}{2xy^2}$$

$$= \frac{3x}{y} - 6x^2 + \frac{9y}{2x}$$

13.

$$\begin{array}{r|l} & x + 3 \\ \hline x + 1 & x^2 + 4x + 3 \\ & x^2 + x \\ \hline & \quad 3x + 15 \\ & \quad 3x + 15 \\ \hline & \qquad 0 \end{array}$$

Thus $$\frac{x^2 + 4x + 3}{x + 1} = x + 3$$

15.

$$\begin{array}{r} 2x+3 \\ x+5\,\overline{)\,2x^2+13x+15} \\ \underline{2x^2+10x\phantom{+15}} \\ 3x+15 \\ \underline{3x+15} \\ 0 \end{array}$$

Thus $\dfrac{2x^2+13x+15}{x+5} = 2x+3$

17.

$$\begin{array}{r} 3x+2 \\ 2x-1\,\overline{)\,6x^2+x-2} \\ \underline{6x^2-3x\phantom{-2}} \\ 4x-2 \\ \underline{4x-2} \\ 0 \end{array}$$

Thus $\dfrac{6x^2+x-2}{2x-1} = 3x+2$

19.

$$\begin{array}{r} 2x+3 \\ 2x-3\,\overline{)\,4x^2+0x-9} \\ \underline{4x^2-6x\phantom{-9}} \\ 6x-9 \\ \underline{6x-9} \\ 0 \end{array}$$

Thus $\dfrac{4x^2-9}{2x-3} = 2x+3$

21.

$$\begin{array}{r} x^2+2x+3 \\ x+1\,\overline{)\,x^3+3x^2+5x+4} \\ \underline{x^3+x^2\phantom{+5x+4}} \\ 2x^2+5x\phantom{+4} \\ \underline{2x^2+2x\phantom{+4}} \\ 3x+4 \\ \underline{3x+3} \\ 1 \end{array}$$

21. cont.

Thus $\dfrac{x^3+3x^2+5x+4}{x+1}$

$= x^2+2x+3+\dfrac{1}{x+1}$

23.

$$\begin{array}{r} 3x^2-3x+1 \\ 3x+2\,\overline{)\,9x^3-3x^2-3x+4} \\ \underline{9x^3+6x^2\phantom{-3x+4}} \\ -9x^2-3x\phantom{+4} \\ \underline{-9x^2-6x\phantom{+4}} \\ 3x+4 \\ \underline{3x+2} \\ 2 \end{array}$$

Thus $\dfrac{9x^3-3x^2-3x+4}{3x+2}$

$= 3x^2-3x+1+\dfrac{2}{3x+2}$

25.

$$\begin{array}{r} 2x^2+x-2 \\ 2x-1\,\overline{)\,4x^3+0x^2-5x+0} \\ \underline{4x^3-2x^2\phantom{-5x+0}} \\ 2x^2-5x\phantom{+0} \\ \underline{2x^2-x\phantom{+0}} \\ -4x+0 \\ \underline{-4x+2} \\ -2 \end{array}$$

Thus $\dfrac{4x^3-5x}{2x-1}$

$= 2x^2+x-2-\dfrac{2}{2x-1}$

27.

$$\begin{array}{r}
2x^3 - 6x + 4 \\
2x^2 + 0x - 3 \,\overline{\big)\, 4x^5 + 0x^4 - 18x^3 + 8x^2 + 18x - 12} \\
4x^5 + 0x^4 - 6x^3 \qquad\qquad\qquad\qquad \\
\hline
-12x^3 + 8x^2 + 18x \qquad\quad \\
-12x^3 + 0x^2 + 18x \qquad\quad \\
\hline
8x^2 + 0x - 12 \\
8x^2 + 0x - 12 \\
\hline
0
\end{array}$$

Thus $\dfrac{4x^5 - 18x^3 + 8x^2 + 18x - 12}{2x^2 - 3} = 2x^3 - 6x + 4$

29. $\dfrac{6x^2 + 3x + 12}{2x}$

$= \dfrac{6x^2}{2x} + \dfrac{3x}{2x} + \dfrac{12}{2x}$

$= 3x + \dfrac{3}{2} + \dfrac{6}{x}$

31.

$$\begin{array}{r}
2x + 5 \\
x - 2 \,\overline{\big)\, 2x^2 + x - 10} \\
2x^2 - 4x \qquad \\
\hline
5x - 10 \\
5x - 10 \\
\hline
0
\end{array}$$

Thus $\dfrac{2x^2 + x - 10}{x - 2}$

$= 2x + 5$

33. $\dfrac{12x^3 + 6x^2 + 3x + 9}{6x^2}$

$= \dfrac{12x^3}{6x^2} + \dfrac{6x^2}{6x^2} + \dfrac{3x}{6x^2} + \dfrac{9}{6x^2}$

33. cont.

$= 2x + 1 + \dfrac{1}{2x} + \dfrac{3}{2x^2}$

35. $\dfrac{-5x^3y^2 + 10xy - 6}{10x}$

$= \dfrac{-5x^3y^2}{10x} + \dfrac{10xy}{10x} - \dfrac{6}{10x}$

$= \dfrac{-x^2y}{2} + y - \dfrac{3}{5x}$

37.

$$\begin{array}{r}
3x^2 + 2x + 1 \\
3x - 2 \,\overline{\big)\, 9x^3 + 0x^2 - x + 3} \\
9x^3 - 6x^2 \qquad\qquad \\
\hline
6x^2 - x \qquad \\
6x^2 - 4x \qquad \\
\hline
3x + 3 \\
3x - 2 \\
\hline
5
\end{array}$$

Thus $\dfrac{9x^3 - x + 3}{3x - 2}$

$= 3x^2 + 2x + 1 + \dfrac{5}{3x - 2}$

39. $$\frac{3xyz + 6xyz^2 - 9x^3y^5z^7}{6xy}$$

$$= \frac{3xyz}{6xyz} + \frac{6xyz^2}{6xy} - \frac{9x^3y^5z^7}{6xy}$$

$$= \frac{z}{2} + z^2 - \frac{3x^2y^4z^7}{2}$$

41.

$$\begin{array}{r|l} & 2x^2 - 6x + 3 \\ \hline x^2 - x + 5 & 2x^4 - 8x^3 + 19x^2 - 33x + 15 \\ & 2x^4 - 2x^3 + 10x^2 \\ \hline & \quad -6x^3 + 9x^2 - 33x \\ & \quad -6x^3 + 6x^2 - 30x \\ \hline & \qquad 3x^2 - 3x + 15 \\ & \qquad 3x^2 - 3x + 15 \\ \hline & \qquad\qquad 0 \end{array}$$

Thus $$\frac{2x^4 - 8x^3 + 19x^2 - 33x + 15}{x^2 - x + 5} = 2x^2 - 6x + 3$$

43.

$$\begin{array}{r|l} & x^3 + x^2 - 6 \\ \hline 2x^2 + 0x - 3 & 2x^5 + 2x^4 - 3x^3 - 15x^2 + 18 \\ & 2x^5 + 0x^4 - 3x^3 \\ \hline & \quad 2x^4 + 0x^3 - 15x^2 \\ & \quad 2x^4 + 0x^3 - 3x^2 \\ \hline & \qquad -12x^2 + 18 \\ & \qquad -12x^2 + 18 \\ \hline & \qquad\qquad 0 \end{array}$$

Thus $$\frac{2x^5 + 2x^4 - 3x^3 - 15x^2 + 18}{2x^2 - 3} = x^3 + x^2 - 6$$

45. a) $ax + by = c$
b) $y = mx + b$
c) $y - y_1 = m(x - x_1)$

47.

$$f\left(\frac{-2}{3}\right) = \frac{1}{2}\left(\frac{-2}{3}\right) + \frac{3}{7}$$

$$= \frac{-1}{3} + \frac{3}{7}$$

$$= \frac{-7}{21} + \frac{9}{21} = \frac{2}{21}$$

JUST FOR FUN

1.

$$
\begin{array}{r|l}
 & 2x^2 + 3xy - y^2 \\
\hline
x - 2y & 2x^3 - x^2y - 7xy^2 + 2y^3 \\
 & 2x^3 - 4x^2y \\
 & \overline{\quad 3x^2y - 7xy^2} \\
 & \quad 3x^2y - 6xy^2 \\
 & \overline{\qquad -xy^2 + 2y^3} \\
 & \qquad -xy^2 + 2y^3 \\
 & \overline{\qquad\qquad 0}
\end{array}
$$

Thus $\dfrac{2x^3 - x^2y - 7xy^2 + 2y^3}{x - 2y} = 2x^2 + 3xy - y^2$

3.

$$
\begin{array}{r|l}
 & x^2 + \overset{*}{\tfrac{2}{3}}x + \overset{**}{\tfrac{4}{9}} \\
\hline
3x - 2 & 3x^3 + 0x^2 + 0x - 5 \\
 & 3x^3 - 2x^2 \\
 & \overline{\quad 2x^2 + 0x} \\
 & \quad 2x^2 - \tfrac{4}{3}x \\
 & \overline{\qquad \tfrac{4}{3}x - 5} \\
 & \qquad \tfrac{4}{3}x - \tfrac{8}{9} \\
 & \overline{\qquad\quad -\tfrac{37}{9}}
\end{array}
$$

*
To find this number, solve the following:
What number times 3 = 2? If we let a = the unknown number,
$3a = 2$
$a = 2/3$
Therefore, multiply $3x - 2$ by $2x/3$ to perform the next line of division.

**
To find this number, solve the following:
What number times 3 = 4/3?
If we let b = the unknown number,
$3b = \dfrac{4}{3}$ then $9b = 4$ and $b = 4/9$
Therefore, multiply $3x - 2$ by $4/9$ to perform the next line of division.

Thus

$$\frac{3x^3 - 5}{3x - 2} = x^2 + \frac{2}{3}x + \frac{4}{9} - \frac{37}{9(3x - 2)}$$

EXERCISE SET 5.6

1. Synthetic division by 2:

| 2 \| | 1 | 1 | -6 |
|---|---|---|---|
| | | 2 | 6 |
| | 1 | 3 | 0 |

Thus $\frac{x^2 + x - 6}{x - 2} = x + 3$

3. Synthetic division by -6:

| -6 \| | 1 | 5 | -6 |
|---|---|---|---|
| | | -6 | 6 |
| | 1 | -1 | 0 |

Thus $\frac{x^2 + 5x - 6}{x + 6} = x - 1$

5. Synthetic division by 3:

| 3 \| | 1 | 5 | -12 |
|---|---|---|---|
| | | 3 | 24 |
| | 1 | 8 | 12 |

Thus $\frac{x^2 + 5x - 12}{x - 3} = x + 8 + \frac{12}{x - 3}$

7. Synthetic division by 4:

| 4 \| | 3 | -7 | -10 |
|---|---|---|---|
| | | 12 | 20 |
| | 3 | 5 | 10 |

Thus $\frac{3x^2 - 7x - 10}{x - 4} = 3x + 5 + \frac{10}{x - 4}$

9. Synthetic division by 1:

| 1 \| | 4 | -3 | 2 | 0 |
|---|---|---|---|---|
| | | 4 | 1 | 3 |
| | 4 | 1 | 3 | 3 |

Thus

$\frac{4x^3 - 3x^2 + 2x}{x - 1} = 4x^2 + x + 3 + \frac{3}{x - 1}$

11. Synthetic division by -3:

| -3 \| | 3 | 7 | -4 | 12 |
|---|---|---|---|---|
| | | -9 | 6 | -6 |
| | 3 | -2 | 2 | 6 |

Thus $\frac{3x^3 + 7x^2 - 4x + 12}{x + 3}$

$= 3x^2 - 2x + 2 + \frac{6}{x + 3}$

13. Synthetic division by -1:

| -1 \| | 5 | -6 | 3 | -6 |
|---|---|---|---|---|
| | | -5 | 11 | -14 |
| | 5 | -11 | 14 | -20 |

Thus $\frac{5x^3 - 6x^2 + 3x - 6}{x + 1}$

$= 5x^2 - 11x + 14 - \frac{20}{x + 1}$

15. Synthetic division by -4:

| -4 \| | 1 | 0 | 0 | 0 | 16 |
|---|---|---|---|---|---|
| | | -4 | 16 | -64 | 256 |
| | 1 | -4 | 16 | -64 | 272 |

Thus $\frac{x^4 + 16}{x + 4}$

$= x^3 - 4x^2 + 16x - 64 + \frac{272}{x + 4}$

17. Synthetic division by -1:

| -1 \| | 1 | 1 | 0 | 0 | 0 | -10 |
|---|---|---|---|---|---|---|
| | | -1 | 0 | 0 | 0 | 0 |
| | 1 | 0 | 0 | 0 | 0 | -10 |

Thus $\frac{y^5 + y^4 - 10}{y + 1} = y^4 - \frac{10}{y + 1}$

19. Synthetic division by 1/3:

| 1/3 \| | 3 | 2 | -4 | 1 |
|---|---|---|---|---|
| | | 1 | 1 | -1 |
| | 3 | 3 | -3 | 0 |

Thus $\frac{3x^3 + 2x^2 - 4x + 1}{x - \frac{1}{3}}$

$= 3x^2 + 3x - 3$

21. Synthetic division by 1/2:

| 1/2 \| | 2 | -1 | 2 | -3 | 1 |
|---|---|---|---|---|---|
| | | 1 | 0 | 1 | -1 |
| | 2 | 0 | 2 | -2 | 0 |

Thus $\frac{2x^4 - x^3 + 2x^2 - 3x + 1}{x - \frac{1}{2}}$

$= 2x^3 + 2x - 2$

23.
$$-1 < \frac{4(3x - 2)}{3} \le 5$$

$$-3 < 12x - 8 \ge 5$$

$$5 < 12x \le 23$$

$$\frac{5}{12} < x \le \frac{23}{12}$$

25. $20x - 60y = 120$

| x | y |
|---|---|
| 0 | -2 |
| 6 | 0 |

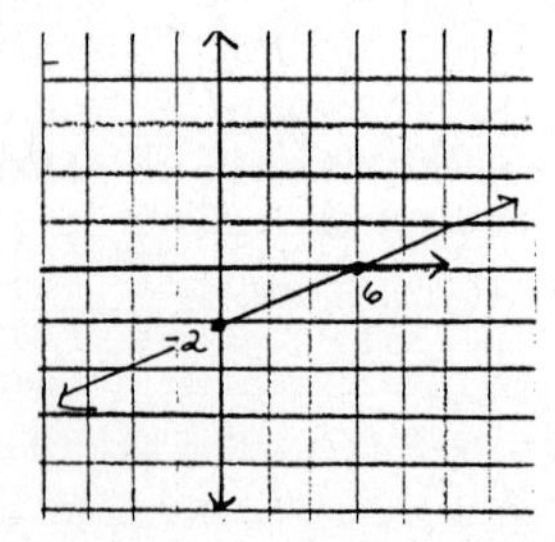


JUST FOR FUN

1.
```
.4|  .2   -.4   .32   -.64
           .08  -.128  .0768
   --------------------------
     .2   -.32  .192  -.5632
```

Thus
$$\frac{.2x^3 - .4x^2 + .32x - .64}{x - .4}$$

$$= .2x^2 - .32x + .192 - \frac{.5632}{x - .4}$$

EXERCISE SET 5.7

1. a) $f(1) = 3(1) - 1 = 2$
   b) $f(-3) = 3(-3) - 1 = -10$
   c) $f(a) = 3a - 1$

3. a) $f(4) = 2(4) + 3 = 11$
   b) $f(-2) = 2(-2) + 3 = -1$
   c) $f(a + b) = 2(a + b) + 3$
   $= 2a + 2b + 3$

5. a)
$$f(2) = \frac{1}{2}(2)^2 - 2 + 4$$
$$= 2 - 2 + 4 = 4$$

5. cont.

b)
$$f(-3) = \frac{1}{2}(3)^2 - 3 + 4$$
$$= \frac{9}{2} + 1 = \frac{11}{2}$$

c)
$$f(-1) = \frac{1}{2}(-1)^2 - (-1) + 4$$
$$= \frac{1}{2} + 1 + 4 = \frac{11}{2}$$

7. a)
$$f(-1) = 2(-1)^3 - (-1) + 3$$
$$= -2 + 1 + 3 = 2$$

b)
$$f(2) = 2(2)^3 - 2 + 3$$
$$= 16 - 2 + 3 = 17$$

c)
$$f\left[\frac{2}{3}\right] = 2\left[\frac{2}{3}\right]^3 - \frac{2}{3} + 3$$
$$= \frac{16}{27} - \frac{2}{3} + 3$$
$$= \frac{16}{27} - \frac{18}{27} + \frac{81}{27} = \frac{79}{27}$$

9. a)
$$f(-4) = -(-4)^2 + 4(-4) - 3$$
$$-16 - 16 - 3 = -35$$

b)
$$f(.3) = -.3^2 + 4(.3) - 3$$
$$= -.09 + 1.2 - 3 = -1.89$$

c)
$$f(c) = -c^2 + 4c - 3$$

13. a)
$$f(-2) = -(-2)^2 - 2(-2) + 5$$
$$= -4 + 4 + 5 = 5$$

b)
$$f(c) = -c^2 - 2c + 5$$

c)
$$f(c + 2) = -(c + 2)^2 - 2(c + 2) + 5$$
$$= -\left[c^2 + 4c + 4\right] - 2c - 4 + 5$$
$$= -c^2 - 4c - 4 - 2c - 4 + 5$$
$$= -c^2 - 6c - 3$$

15. a)
$$f(5) = 2(5)^2 - 3(5) + 5$$
$$= 50 - 15 + 5 = 40$$

15. cont.

b)

$$f(-x) = 2(-x)^2 - 3(-x) + 5$$
$$= 2x^2 + 3x + 5$$

c)

$$f(2x) = 2(2x)^2 - 3(2x) + 5$$
$$= 2\left[4x^2\right] - 6x + 5$$
$$= 8x^2 - 6x + 5$$

17.

a) $f(h) = h^2 + 3h - 4$

b)

$$f(h + 4) = (h + 4)^2 + 3(h + 4) - 4$$
$$= h^2 + 8h + 16 + 3h + 12 - 4$$
$$= h^2 + 11h + 24$$

c)

$$f(a + h) = (a + h)^2 + 3(a + h) - 4$$
$$= a^2 + 2ah + h^2 + 3a + 3h - 4$$

19.

a) $f(h) = 2h^2 - 3h + 1$

b)

$$f(x + 3) = 2(x + 3)^2 - 3(x + 3) + 1$$
$$= 2x^2 + 12x + 18 - 3x - 9 + 1$$
$$= 2x^2 + 9x + 10$$ □

c)

$$f(x + h) = 2(x + h)^2 - 3(x + h) + 1$$
$$= 2\left[x^2 + 2xh + h^2\right] - 3x - 3h + 1$$
$$= 2x^2 + 4xh + 2h^2 - 3x - 3h + 1$$

21.

a) $s = f(10) = 10^2 + 10$
$= 110$

b) $s = f(15) = 15^2 + 15$
$= 240$

23.

a) $v = f(6) = 3.2(6) + .45$
$= 19.65$ m/s

$h = g(6) = 1.6(6)^2 + .45(6)$
$= 60.3$ m

b) $v = f(2.5) = 3.2(2.5) + .45(6)$
$= 8.45$ m/s

$h = g(2.5) = 1.6(2.5)^2 + .45(2.5)$
$= 11.125$

25.

a) $d = f(50) = .18(50) + .01(50)^2$
$= 34$ m

25. cont.

b) $d = f(25) = .18(25) + .01(25)^2$
$= 10.75$m

27.

a) $s = f(2) = 3.4(2) - .3(2)^2$
$= 6.8 - 1.2 = 5.6$ mc

b) $s = f(4.2) = 3.4(4.2) - .3(4.2)^2$
$= 14.28 - 5.292 = 8.988$cm

29.

a) $N = f(6) = \frac{1}{3}\cdot(6)^3 + \frac{1}{2}\cdot(6)^2 + \frac{1}{6}\cdot(6)$

$= 72 + 18 + 1 = 91$

b) $N = f(8) = \frac{1}{3}\cdot(8)^3 + \frac{1}{2}\cdot(8)^2 + \frac{1}{6}\cdot(8)$
$= 170.667 + 32 + 1.333 = 204$

31. a) $V = f(2) = 7.2(2) + 4.6$
$= 19$ ft/sec

$h = g(2) = -.8(2)^2 - 10(2) + 70$
$= 46.8$ ft

b) $V = f(5) = 7.2(5) + 4.6$
$= 40.6$ ft/sec

$h = g(5) = -.8(5)^2 - 10(5) + 70$
$= 0$ ft

33. No. The expression contains fractional exponents and a polynomial must have whole number exponents.

35. Let x = the price of the home.

$$600 \leq 9.14\frac{x}{1000}$$

$$600 \leq .00915\,x$$

$$65573.77 \leq x$$

The house must not cost more than $65,573.77.

37. The slope of a line is the ratio of the vertical change to the horizontal change between any two points on the line.

## JUST FOR FUN

1.

$$f(x + 3) = (x + 3)^3 - 2(x + 3)^2 + 6(x + 3) + 3$$

$$= (x + 3)^2 (x + 3) - 2(x + 3)^2 + 6(x + 3) + 3$$

$$= \left[x^2 + 6x + 9\right](x + 3) - 2\left[x^2 + 6x + 9\right] + 6x + 18 + 3$$

$$= x^3 + 9x^2 + 27x + 27 - 2x^2 - 12x - 18 + 6x + 18 + 3$$

$$= x^3 + 7x^2 + 21x + 30$$

3. a) The area of a square is $s^2$ where s is the length of one side of the square. Since the length of one side is equal to d in the figure, the area of the square can be rewritten as $d^2$

The area of the circle is $\pi r^2$ where r is the radius of the circle. Recall that the radius is 1/2 the diameter. Therefore, the area of the circle can be rewritten as

$$\pi \cdot \left[\frac{1}{2} d\right]^2 = \frac{1}{4} \pi d^2$$

where d is the diameter. The difference between the area of the square and the area of the circle is the area of the shaded region.

Area of the shaded region =

$$A(d) = d^2 - \frac{1}{4} \pi d^2$$

b) $$A(4) = 4^2 - \frac{1}{4} \pi 4^2$$

$$= 16 - 4(3.14) \approx 3.44$$

c) $$A(6) = 6^2 - \frac{1}{4} \pi 6^2$$

$$= 36 - 9(3.14) \approx 7.74$$

## EXERCISE SET 5.8

1. $y = x^2 + 2x - 7$

$a = 1$, $b = 2$, $c = -7$
The parabola opens up since $a > 0$.

Axis of symmetry:

$$x = \frac{-b}{2a} = \frac{-2}{2 \cdot (1)} = -1$$

$x = -1$

Vertex: Since -1 is the x coordinate of the vertex, f(-1) is the y-coordinate.

$$f(-1) = (-1)^2 + 2(-1) - 7$$
$$= 1 - 2 - 7 = -8$$

Thus (-1,-8) is the vertex.

3. $y = -x^2 + 4x - 6$
$a = -1$, $b = 4$, $c = -6$
The parabola opens downward since $a < 0$.
Axis of symmetry:

$$x = \frac{-b}{2a} = \frac{-4}{2 \cdot (-1)} = 2$$

$x = 2$

Vertex: Since 2 is the x coordinate of the vertex, f(2) is the y coordinate.

$$f(2) = -2^2 + 4(2) - 6$$
$$= -4 + 8 - 6 = -2$$

Thus (2,-2) is the vertex.

5. $y = -3x^2 + 6x + 8$
$a = -3$, $b = 6$, $c = 8$
The parabola opens downward since $a < 0$.
Axis of symmetry:

$$x = \frac{-b}{2a} = \frac{-6}{2 \cdot (-3)} = 1$$

$x = 1$

Vertex: Since 1 is the x coordinate of the vertex, f(1) is the y coordinate.

$$f(1) = -3(1)^2 + 6(1) + 8$$
$$= -3 + 6 + 8 = 11$$

Thus (1,11) is the vertex.

7. $y = -4x^2 - 8x - 12$

$a = -4 \quad b = -8 \quad c = -12$

The parabola opens downward since $a < 0$.

Axis of Symmetry:

$$x = \frac{-(-8)}{2(-4)} = -1$$

$x = -1$ is the equation of the axis of symmetry

Vertex: Since $-1$ is the x coordinate of the vertex, $f(-1)$ must be the y coordinate.

$$f(-1) = -4(-1)^2 - 8(-1) - 12$$
$$= -4 + 8 - 12 = -8$$

The vertex is $(-1,-8)$.

9. $y = x^2 - x + 2$

$a = 1,\ b = -1,\ c = 2$

Since $a > 0$, the parabola opens up.

Axis of Symmetry:

$$x = \frac{-(-1)}{2(1)} = \frac{1}{2}$$

$x = \frac{1}{2}$ is the equation of the axis of symmetry.

Vertex: Since $1/2$ is the x coordinate of the vertex, then $f(1/2)$ must be the y coordinate.

$$f\left[\frac{1}{2}\right] = \left[\frac{1}{2}\right]^2 - \frac{1}{2} + 2$$

$$= \frac{1}{4} - \frac{2}{4} + \frac{8}{4} = \frac{7}{4}$$

The vertex is $\left[\frac{1}{2}, \frac{7}{4}\right]$

11. $y = 4x^2 + 12x - 5$

$a = 4,\ b = 12,\ c = -5$

Since $a > 0$, the parabola opens upward.

Axis of Symmetry:

$$x = \frac{-12}{2(4)} = \frac{-3}{2}$$

11. cont.

$x = \frac{-3}{2}$ is the equation of the axis of symmetry.

Vertex: Since $-3/2$ is the x coordinate of the vertex, $f(-3/2)$ must be the y coordinate.

$$f\left[\frac{-3}{2}\right] = 4\left[\frac{-3}{2}\right]^2 + 12\left[\frac{-3}{2}\right] - 5$$

$$= 4\left[\frac{9}{4}\right] - 18 - 5$$

$$= 9 - 18 - 5 = -14$$

The vertex is $\left[\frac{-3}{2}, -14\right]$

13. $y = x^2 - 1$

$a = 1,\ b = 0,\ c = -1$

Since $a > 0$, the parabola opens up.

Axis of Symmetry:

$$x = \frac{0}{2(1)} = 0$$

$x = 0$ is the axis of symmetry.

$f(0) = 0^2 - 1 = -1$

Vertex: $(0,-1)$

Domain: All reals

Range: $\{y \mid y \geq -1\}$

| x | y |
|---|---|
| -2 | 3 |
| -1 | 0 |
| 0 | -1 |
| 1 | 0 |
| 2 | 3 |

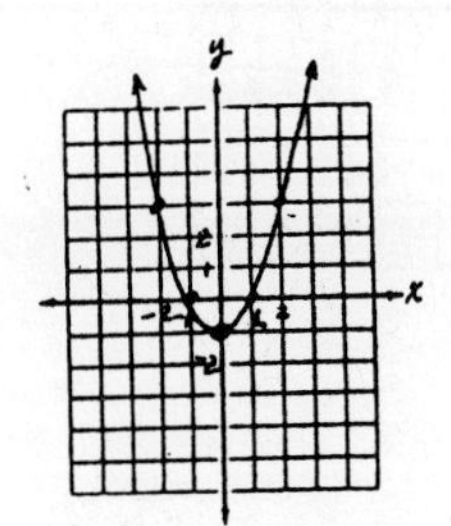

15.

$y = -x^2 + 3$

$a = -1 \quad b = 0 \quad c = 3$

Since $a < 0$ the parabola opens downward.

Axis of Symmetry:

$$x = \frac{0}{2(-1)} = 0$$

$f(0) = -0^3 + 3 = 3$

Vertex: (0, 3)

Domain: All reals

Range: $\{y \mid y \leq 3\}$

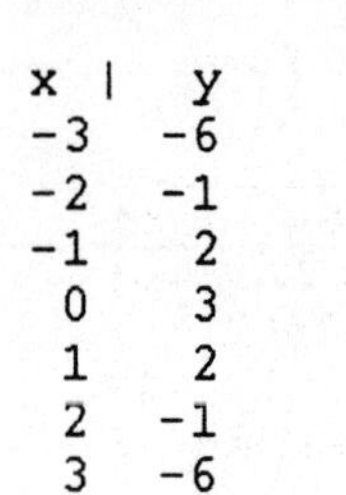

| x | y |
|---|---|
| -3 | -6 |
| -2 | -1 |
| -1 | 2 |
| 0 | 3 |
| 1 | 2 |
| 2 | -1 |
| 3 | -6 |

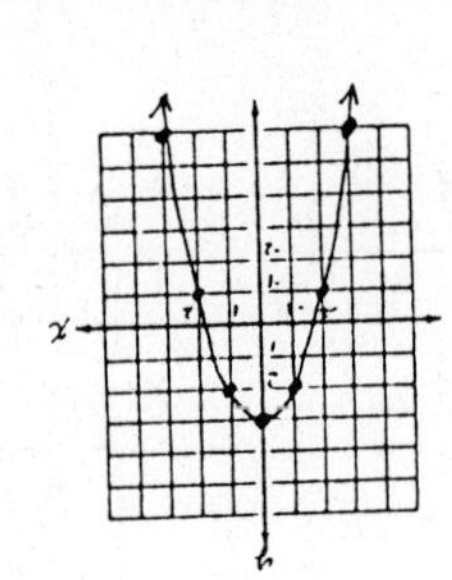


17.

$y = x^2 + 2x - 15$

$a = 1,\ b = 2,\ c = -15$

Since $a > 0$, the parabola opens upward.

Axis of symmetry:

$$x = \frac{-2}{2(1)} = -1$$

$f(-1) = (-1)^2 + 2(-1) - 15$

$= 1 - 2 - 15 = -16$

Vertex: (-1,-16)

Domain: All reals

Range: $\{y \mid y \geq -16\}$

| x | y |
|---|---|
| -6 | 9 |
| -5 | 0 |
| -4 | -7 |
| -2 | -15 |
| -1 | -16 |
| 0 | -15 |
| 3 | 0 |
| 4 | 9 |

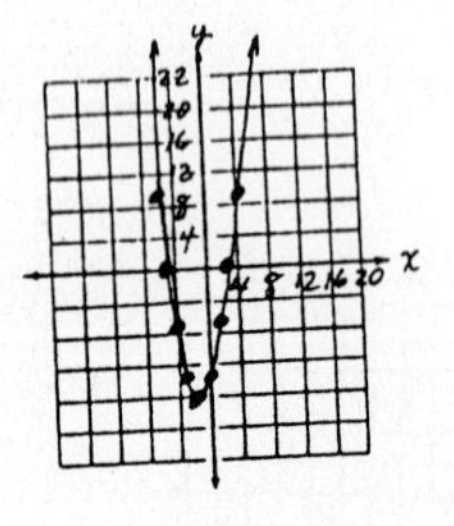


19.

$y = -x^2 + 4x - 5$

$a = -1 \quad b = 4$

Since $a < 0$, the parabola opens downward.

Axis of symmetry:

19. cont.

$$x = \frac{-4}{2(-1)} = 2$$

$f(2) = -(2)^2 + 4(2) - 5$

$= -4 + 8 - 5 = -1$

Vertex: (2, -1)

Domain: All reals

Range: $\{y \mid y \leq -1\}$

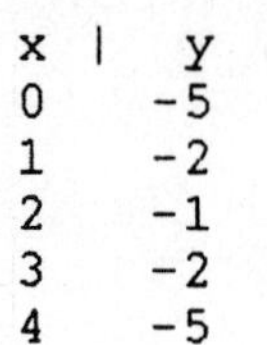

| x | y |
|---|---|
| 0 | -5 |
| 1 | -2 |
| 2 | -1 |
| 3 | -2 |
| 4 | -5 |

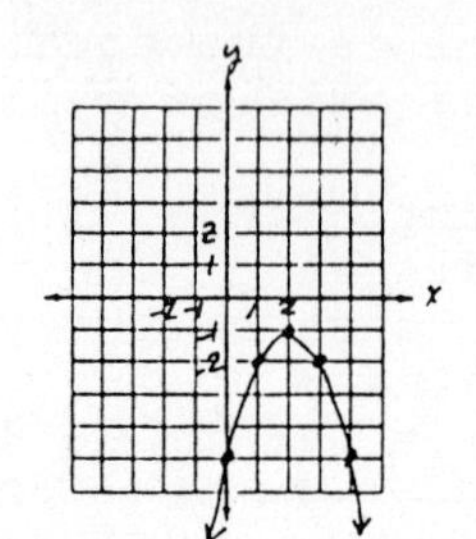


21.

$y = x^2 - 6x + 4$

Parabola opens upward

$$x = \frac{-(-6)}{2(1)} = 3$$

$f(3) = 3^2 - 6(3) + 4$

$= 9 - 18 + 4 = -5$

Vertex: (3,-5)

Domain: All Reals

Range: $\{y \mid y \geq -5\}$

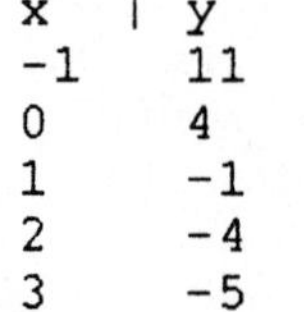
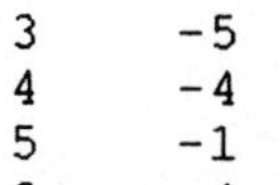

| x | y |
|---|---|
| -1 | 11 |
| 0 | 4 |
| 1 | -1 |
| 2 | -4 |
| 3 | -5 |
| 4 | -4 |
| 5 | -1 |
| 6 | 4 |
| 7 | 11 |

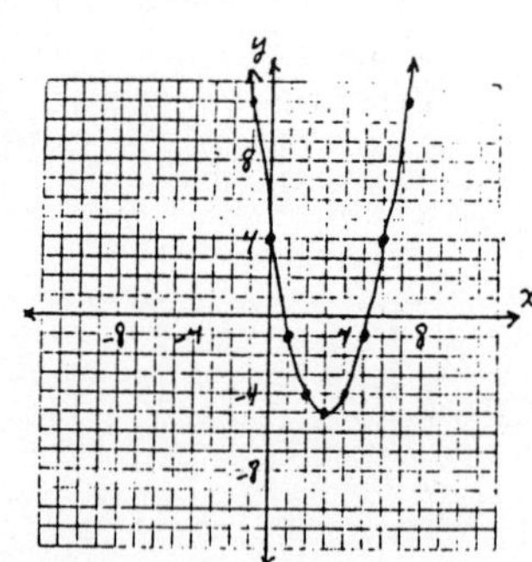


23.

$y = x^2 - 6x$

Parabola opens upward.

$$x = \frac{-(-6)}{2(1)} = 3$$

$f(3) = 3^2 - 6(3)$

$= 9 - 18 = -9$

Vertex: (3, -9)

Domain: All reals

Range: $\{y \mid y \geq -9\}$

23. cont.

| x | y |
|---|---|
| 0 | 0 |
| 1 | -5 |
| 2 | -8 |
| 3 | -9 |
| 4 | -8 |
| 5 | -5 |
| 6 | 0 |

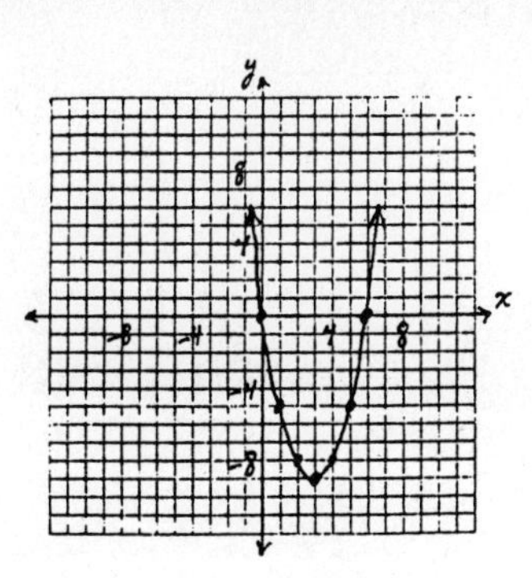

25. $f(x) = x^2 - 4x + 4$

The parabola opens upward.

$$x = \frac{-(-4)}{2(1)} = 2$$

$$f(2) = (2)^2 - 4(2) + 4$$
$$= 4 - 8 + 4 = 0$$

Vertex: (2,0)

Domain: All Reals

Range: $\{y \mid y \geq 0\}$

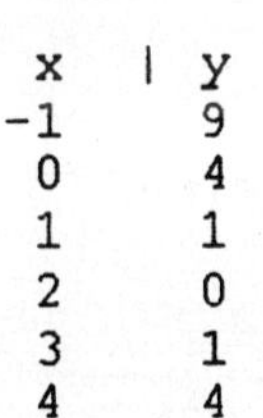

| x | y |
|---|---|
| -1 | 9 |
| 0 | 4 |
| 1 | 1 |
| 2 | 0 |
| 3 | 1 |
| 4 | 4 |
| 5 | 9 |

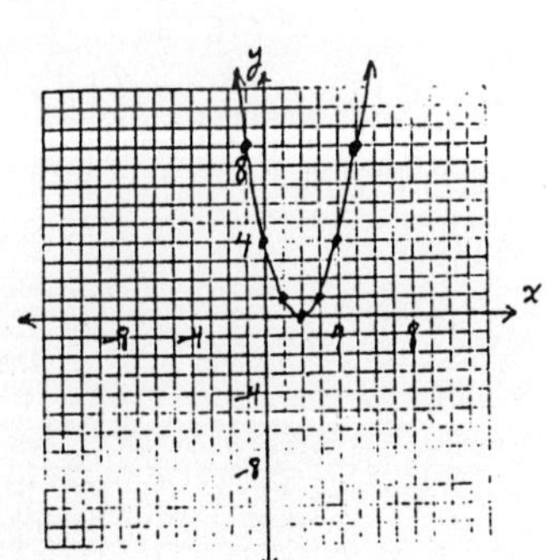

27. $f(x) - x^2 + 4x - 8$

The parabola opens downward.

Axis of Symmetry:

$$x = \frac{-4}{2(-1)} = 2$$

$$f(2) = -(2)^2 + 4(2) - 8$$
$$= -4 + 8 - 8 = -4$$

Vertex: (2,-4)

Domain: All reals

Range: $\{y \mid y \leq -4\}$

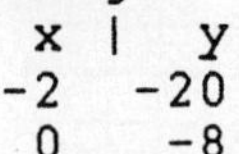
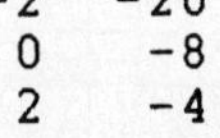
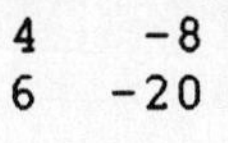

| x | y |
|---|---|
| -2 | -20 |
| 0 | -8 |
| 2 | -4 |
| 4 | -8 |
| 6 | -20 |

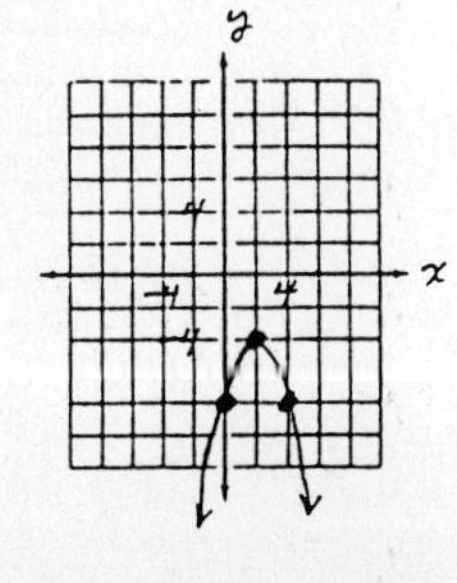

29. $f(x) = x^2 - 2x - 15$

The parabola opens up.

Axis of Symmetry:

$$x = \frac{-(-2)}{2(1)} = 1$$

$$f(1) = (1)^2 - 2(1) - 15$$
$$= 1 - 2 - 15 = -16$$

Vertex: (1,-16)

Domain: All Reals

Range: $\{y \mid y \geq -16\}$

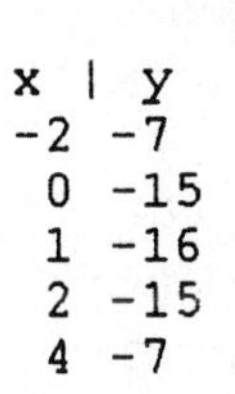

| x | y |
|---|---|
| -2 | -7 |
| 0 | -15 |
| 1 | -16 |
| 2 | -15 |
| 4 | -7 |

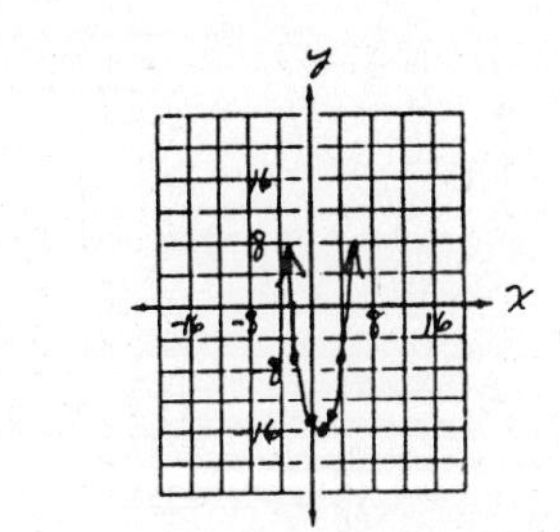

31. $f(x) = 3x^2 - 6x + 1$

The parabola opens upward.

Axis of Symmetry:

$$x = \frac{-(-6)}{2(3)} = 1$$

$$f(1) = 3(1)^2 - 6(1) + 1$$
$$= 3 - 6 + 1 = -2$$

Vertex: (1,-2)

Domain: All Reals

Range: $\{y \mid y \geq -2\}$

| x | y |
|---|---|
| -1 | 10 |
| 0 | 1 |
| 1 | -2 |
| 2 | 1 |
| 3 | 10 |

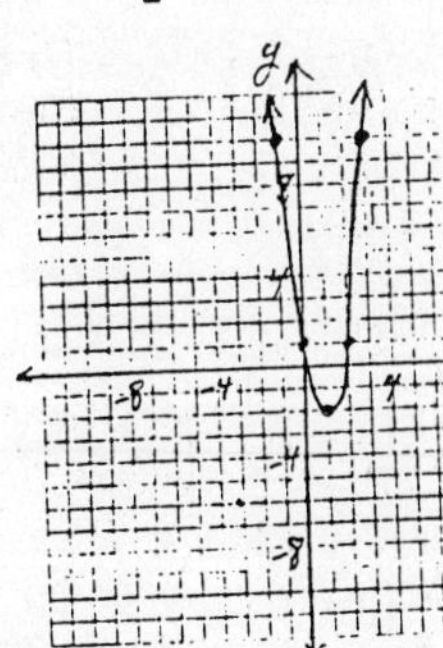

33. $f(x) = \frac{1}{2}x^2 - 2$

The parabola opens up.

Axis of Symmetry:

$$x = \frac{-0}{2\left[\frac{1}{2}\right]} = 0$$

33. cont.

$$f(0) = \frac{1}{2}0^2 - 2 = -2$$

Vertex: (0,-2)
Domain: All Reals
Range: $\{y| \quad y \geq -2\}$

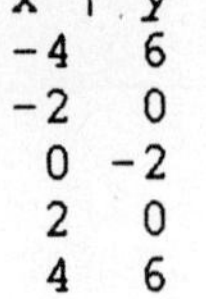

| x | y |
|---|---|
| -4 | 6 |
| -2 | 0 |
| 0 | -2 |
| 2 | 0 |
| 4 | 6 |

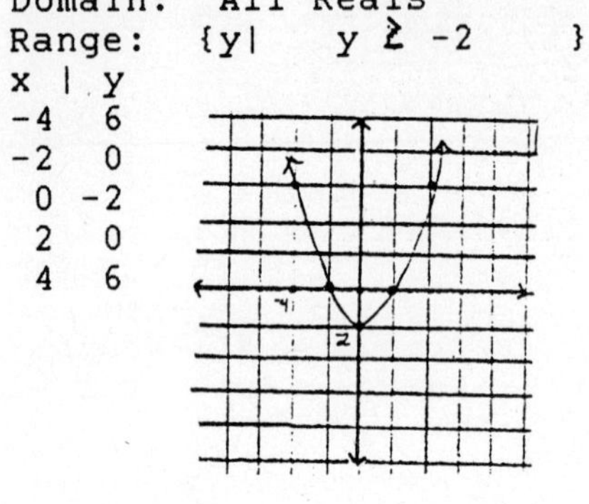

35.

$$f(x) = x^3$$

$$f(-2) = (-2)^3 = -8$$

$$f(-1) = (-1)^3 = -1$$

$$f(0) = (0)^3 = 0$$

$$f(1) = (1)^3 = 1$$

$$f(2) = (2)^3 = 8$$

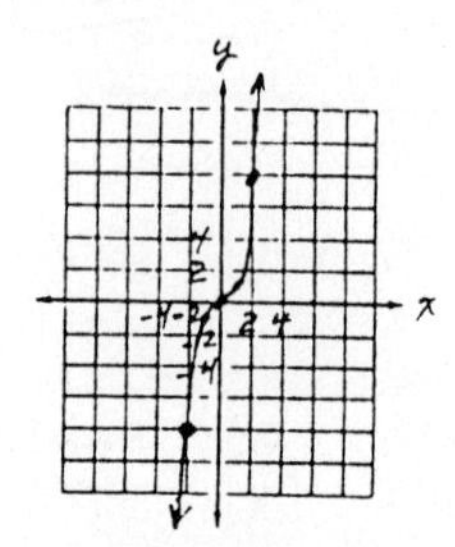


39

$$y = x^3 + x^2 - 3x - 1$$

$$f(-4) = (-4)^3 + (-4)^2 - 3(-4) - 1$$
$$= -64 + 16 + 12 - 1 = -37$$

$$f(-3) = (-3)^3 + (-3)^2 - 3(-3) - 1$$
$$= -27 + 9 + 9 - 1 = -37$$

$$f(-2) = (-2)^3 + (-2)^2 - 3(-2) - 1$$
$$= -8 + 4 + 6 - 1 = 1$$

$$f(-1) = (-1)^3 + (-1)^2 - 3(-1) - 1$$
$$= -1 + 1 + 3 - 1 = 2$$

$$f(0) = (0)^3 + (0)^2 - 3(0) - 1$$
$$= -1$$

$$f(1) = 1^3 + 1^2 - 3(1) - 1$$
$$= 1 + 1 - 3 - 1 = -2$$

$$f(2) = 2^3 + 2^2 - 3(2) - 1$$
$$= 8 + 4 - 6 - 1 = 5$$

39. cont.

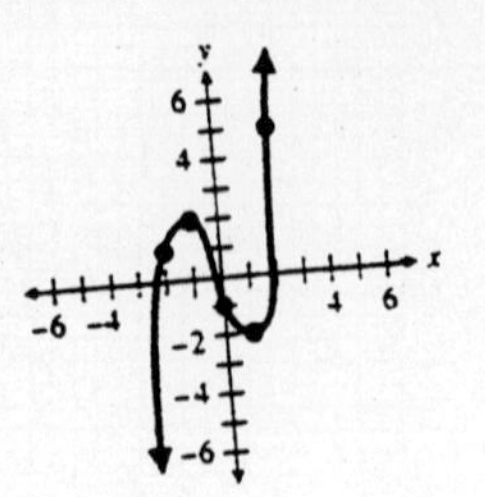


41.

$$f(x) = -x^3 + 3x$$

$$f(-2) = -(-2)^3 + 3(-2)$$
$$= 8 - 6 = 2$$

$$f(-1) = -(-1)^3 + 3(-1)$$
$$= 1 - 3 = -2$$

$$f(0) = -0^3 + 3(0) = 0$$

$$f(1) = -1^3 + 3(1) = 2$$

$$f(2) = -(2)^3 + 3(2)$$
$$= -8 + 6 = -2$$

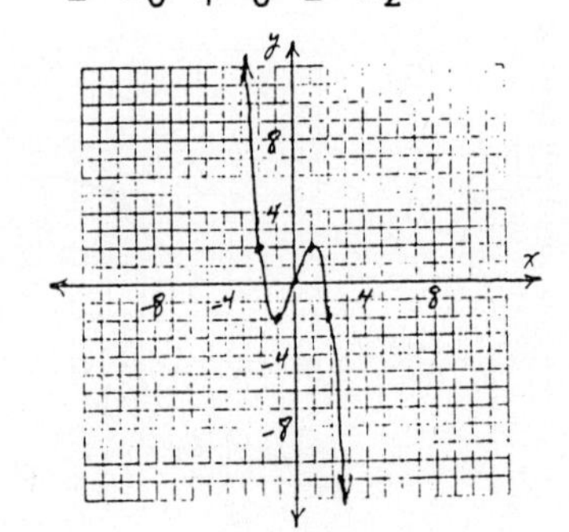


43.

$$y = -x^2 - 2$$

The parabola opens downward.
Axis of Symmetry:

$$x = \frac{-0}{2(-1)} = 0$$

$$f(0) = -(0)^2 - 2 = -2$$

Vertex: (0,-2)
Domain: All Reals
Range: $\{y| \quad y \leq -2\}$

| x | y |
|---|---|
| -2 | -6 |
| -1 | -3 |
| 0 | -2 |
| 1 | -3 |
| 2 | -6 |

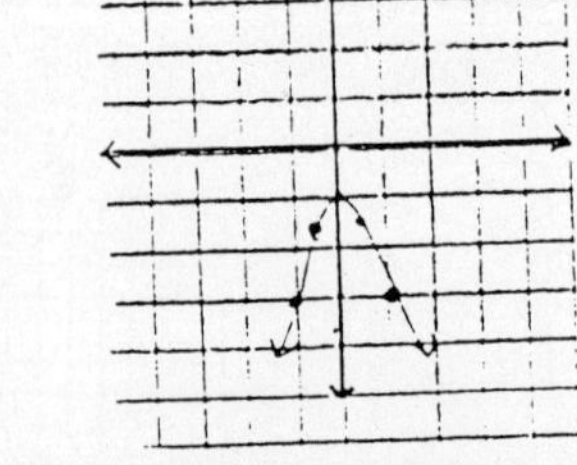

45.

$y = x^2 + 6x - 2$

a = 1, b = 6

Since a = 1, the parabola opens up.

Axis of symmatry:

$x = \frac{-6}{(2)(1)} = -3$

Vertex:

$f(3) = 3^2 + 6(3) - 2$
$= -11$

Vertex: (-3,-11)

Domain: All reals

Range: $\{y \mid y \geq -11\}$

| x | y |
|---|---|
| -5 | -7 |
| -4 | -10 |
| -3 | -11 |
| -2 | -10 |
| -1 | -7 |
| 0 | -2 |

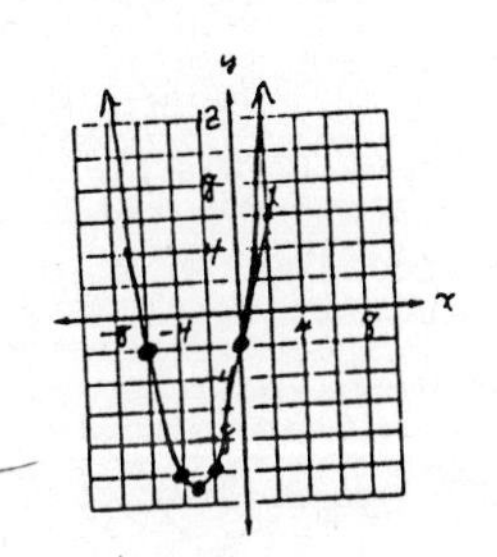

47.

$f(x) = x^3 - x^2 + 2x$

f(-2) = -16

f(-1) = -4

f(0) = 0

f(1) = 2

f(2) = 8

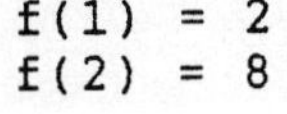

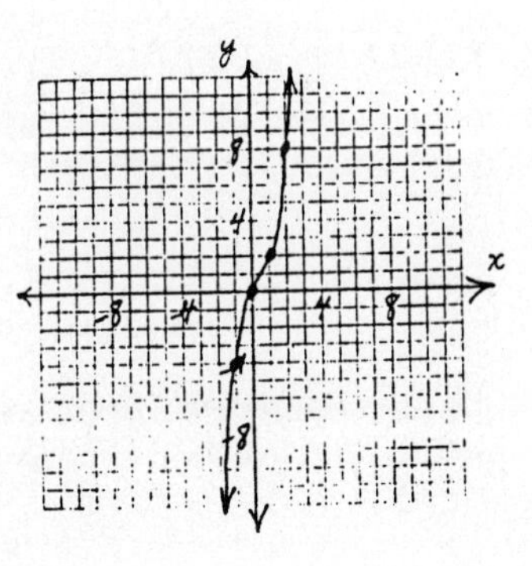

49.

$f(x) = -2x^3 + 6x^2 + 2x - 6$

f(-2) = 30

f(-1) = 0

f(0) = -6

f(1) = 0

f(2) = 6

f(3) = 0

f(4) = -30

49. cont.

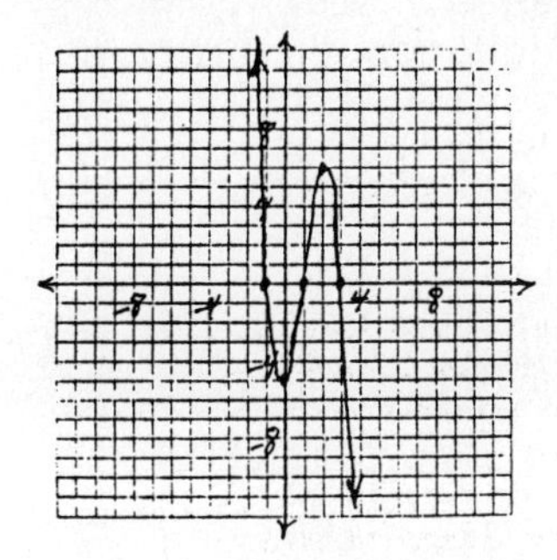

51.

$y = \frac{1}{2} \cdot x^2 - 2x - 4$

a = 1/2, b = -2

Since a = 1/2, the parabola opens up.

Axis of symmatry:

$x = \frac{-(-2)}{2\left[\frac{1}{2}\right]} = 2$

Vertex: f(2) = -6

Vertex: (2,-6)

| x | y |
|---|---|
| -2 | 2 |
| 0 | -4 |
| 2 | -6 |
| 4 | -4 |
| 6 | 2 |

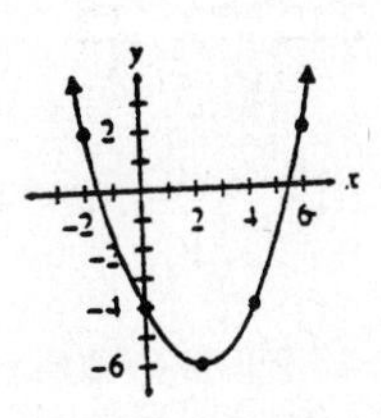

53. When the coefficient of the quadratic (squared) term is positive, the graph opens upward, if negative, the graph opens downward.

55. As x increases, y also increses. To see this, choose increasing values of x and calculate y:

| x | y = f(x) |
|---|---|
| -1 | -1 |
| 0 | 0 |
| 1 | 1 |
| 2 | 8 |

57. As x increases from -3 to 0, y decreases and as x increases from 0 to 3, y increases.

| x | y = f(x) |
|---|---|
| -3 | 81 |
| -2 | 16 |
| -1 | 1 |
| 0 | 0 |
| 1 | 1 |
| 2 | 16 |
| 3 | 81 |

59. Answers vary.

61. 1) $x - 4y = -16$
2) $2x + 3y = -10$
Multiply equation 1 by - 2 and add the results to equation 2 to solve for y.

1) $-2x + 8y = 32$
2) $2x + 3y = -10$

$11y = 22$
$y = 2$

Substitute y = 2 into equation 1 or 2 to solve for x.
1) $x - 4(2) = -16$
$x = -8$
Solution: $(-8,2)$

63. Let x = the smaller number, y = the next larger and z = the largest number.

1) $x + y + z = 12$
2) $x + y = z$ or
$x + y - z = 0$
3) $z = 2y - 4$ or
$-2y + z = -4$

Add equations 1 and 2 to eliminate z.
1) $x + y + z = 12$
2) $x + y - z = 0$

4) $2x + 2y = 12$

Add equations 2 and 3 to eliminate z.

63. cont.
2) $x + y - z = 0$
3) $-2y + z = -4$

5) $x - y = -4$

Multiply equation 5 by 2. Add the result to equation 4 to eliminate y and solve for x.

4) $2x + 2y = 12$
5) $2x - 2y = -8$

$4x = 4$
$x = 1$

Substitute x = 1 into equation 5 and solve for y.

5) $1 - y = -4$
$y = 5$

Substitute x = 1 and y = 5 into equation 1, 2, or 3 to solve for z.

2) $1 + 5 = z$
$6 = z$

The three numbers are 1,5,and 6.

JUST FOR FUN

1.
$$f(x) = x^4 - 3x^2 + 6$$
$$f(-2) = (-2)^4 - 3(-2)^2 + 6 = 16 - 3(4) + 6 = 16 - 12 + 6 = 10$$
$$f(-1) = (-1)^4 - 3(-1)^2 + 6 = 1 - 3(1) + 6 = 4$$
$$f(0) = 0^4 - 3(0) + 6 = 6$$
$$f(1) = 1^4 - 3(1)^2 + 6 = 1 - 3 + 6 = 4$$
$$f(2) = 2^4 - 3(2)^2 + 6 = 16 - 12 + 6 = 10$$

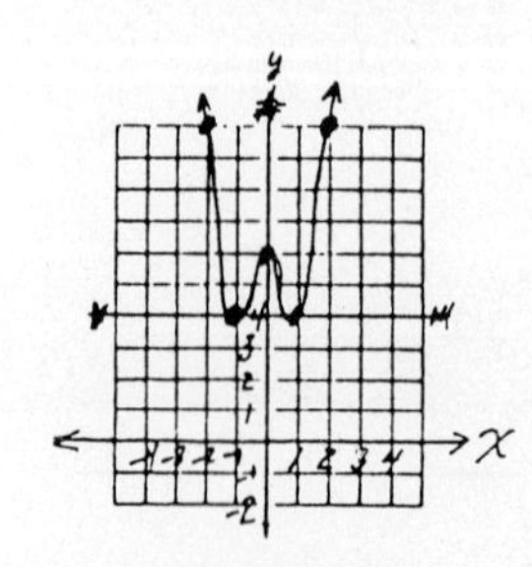

REVIEW EXERCIES

1. $4^2 \cdot 4^1 = 4^{2+1}$

$= 4^3$

$= 64$

3. $y^5 \cdot y^2 = y^{5+2}$

$= y^7$

5. $\frac{x^6}{x^2} = x^{6-2}$

$= x^4$

7. $\frac{y^5}{y^6} = y^{5-6}$

$= y^{-1}$

$= \frac{1}{y}$

9. $x^4 \cdot x^{-7} = x^{4+(-7)}$

$= x^{-3}$

$= \frac{1}{x^3}$

11. $2^{-3} \cdot 2^{-2} = 2^{-3+(-2)}$

$= 2^{-5}$

$= \frac{1}{2^5}$

$= \frac{1}{32}$

13. $\left[3x^2\right]^2 = 3^2 \cdot x^{2 \cdot 2}$

$= 9x^4$

15. $\left[\frac{3}{4}\right]^{-2} = \frac{1}{\left[\left[\frac{3}{4}\right]\right]^2}$

$= \frac{1}{\left[\frac{9}{16}\right]}$

$= \frac{16}{9}$

17. $\left[6xy^2\right]\left[-2xy^4\right] = -12x^{1+1} \cdot y^{2+4}$

$= -12x^2 \cdot y^6$

19. $\left[4x^2 \cdot y^{-3}\right]\left[2x^{-4} \cdot y^2\right]$

$= 8x^{2+(-4)} \cdot y^{-3+2}$

$= 8x^{-2} \cdot y^{-1}$

$= \frac{8}{x^2 \cdot y}$

21. $\frac{6x^{-3} \cdot y^5}{2x^2 \cdot y^{-2}} = 3x^{-3-2} \cdot y^{5-(-2)}$

$= 3x^{-5} \cdot y^7$

$= \frac{3y^7}{x^5}$

23. $\frac{\left[5x^3 \cdot y^2\right]\left[2xy^4 \cdot z\right]}{20x^4 \cdot y^{-2} \cdot z} = \frac{10x^4 \cdot y^6 \cdot z}{20x^4 \cdot y^{-2} \cdot z}$

$= \frac{x^{4-4} \cdot y^{6-(-2)} \cdot z^{1-1}}{2}$

$= \frac{y^8}{2}$

25.

$$\left[\frac{5x^2 y}{x}\right]^3 = \left[5x^{2-1} y\right]^3$$

$$= (5xy)^3$$

$$= 125x^3 \cdot y^3$$

27.

$$\left[\frac{2xy^3}{y^4}\right]^2 = \frac{2^2 \cdot x^2 \cdot y^6}{y^8}$$

$$= \left[4x^2 \cdot y^{6-8}\right]$$

$$= \frac{4x}{y^2}$$

29.

$$\left[\frac{-5x^{-2} y}{z^3}\right]^3 = \frac{(-5)^3 \cdot x^{-6} \cdot y^3}{z^9}$$

$$= \left[\frac{-125y^3}{x^6 \cdot z^9}\right]$$

31.

$$\left[\frac{9x^{-2} y}{3xy}\right]^{-3} = \left[3x^{-2-1} \cdot y^{1-1}\right]^{-3}$$

$$= \left[3 \cdot x^{-3}\right]^{-3}$$

$$= 3^{-3} \cdot x^9$$

$$= \frac{x^9}{27}$$

33.

$$\left[\frac{5x^{-2} y^3}{xy^4}\right]^3 = \left[5x^{-2-1} \cdot y^{3-4}\right]^3$$

33. cont.

$$= \left[5x^{-3} \cdot y^{-1}\right]^3$$

$$= 5^3 \cdot x^{-9} \cdot y^{-3}$$

$$= \frac{125}{x^9 \cdot y^3}$$

35.

$$\left[\frac{16x^4 \cdot y^3 \cdot z^{-2}}{4x^5 \cdot y^2 \cdot z^3}\right]^3$$

$$= \left[4x^{4-5} \cdot y^{3-2} \cdot z^{-2-3}\right]^3$$

$$= \left[4x^{-1} \cdot y \cdot z^{-5}\right]^3$$

$$= 4^3 \cdot x^{-3} \cdot y^3 \cdot z^{-15}$$

$$= \frac{64y^3}{x^3 \cdot z^{15}}$$

37.

$$\left[\frac{3x^4 \cdot y^{-2}}{6xy^{-3}}\right]^2 \left[\frac{2x^{-1} \cdot y^5}{3x^4 \cdot y^{-2}}\right]^{-3}$$

$$= \left[\frac{x^{4-1} \cdot y^{-2-(-3)}}{2}\right]^2$$

$$\left[\frac{2x^{-1-4} \cdot y^{5-(-2)}}{3}\right]^{-3}$$

$$= \left[\frac{x^3 \cdot y^1}{2}\right]^2 \left[\frac{2x^{-5} \cdot y^7}{3}\right]^{-3}$$

$$= \frac{x^6 \cdot y^2}{2^2} \cdot \frac{2^{-3} \cdot x^{15} \cdot y^{-21}}{3^{-3}}$$

37. cont.

$$= \frac{27x^{6+15}y^{2}}{32y^{21}}$$

$$= \frac{27x^{21}y^{2}}{32y^{21}}$$

$$= \frac{27x^{21}y^{2-21}}{32}$$

$$= \frac{27x^{21}}{32y^{19}}$$

39.

$$\frac{\left[4x^{-2}y^{3}\right]^{-2}\left[x^{4}y^{3}\right]^{4}}{\left[2x^{-3}y^{6}\right]^{2}}$$

$$= \frac{4^{-2}x^{4}y^{-6}x^{16}y^{12}}{2^{2}x^{-6}y^{12}}$$

$$= \frac{x^{4+16-(-6)}y^{-6+12-12}}{2^{2}4^{2}}$$

$$= \frac{x^{26}y^{-6}}{64}$$

$$= \frac{x^{26}}{64y^{6}}$$

41. $7.42 \times 10^{-5}$

43. $1.83 \times 10^{5}$

45. $\left[25 \times 10^{-3}\right]\left[1.2 \times 10^{6}\right]$

45. cont.

$= 25 \times 1.2 \times 10^{-3+6}$

$= 30 \times 10^{3}$

$= 30{,}000$

47.

$$\frac{4{,}000{,}000}{.02} = \frac{4 \times 10^{6}}{2 \times 10^{-2}}$$

$= 2 \times 10^{6-(-2)}$

$= 2 \times 10^{8} = 200{,}000{,}000$

49. $5 - x$
Binomial
$-x + 5$
First

51. $x^{2} - y^{2} + xy$
Trinomial
$x^{2} + xy - y^{2}$
Second

53. $-3 - 9x^{2}y + 6xy^{3} + 2x^{4}$
Polynomial
$2x^{4} - 9x^{2}y + 6xy^{3} - 3$
Fourth

55. $(4x + 3) + (6x - 8)$
$= 4x + 6x + 3 - 8$
$= 10x - 5$

57. $4x\left[x^{2} + 2x + 3\right]$
$= 4x^{3} + 8x^{2} + 12x$

59.

$$\frac{15y^{3} + 6y}{3y} = \frac{15y^{3}}{3y} + \frac{6y}{3y}$$

$= 5y^{2} + 2$

61. $(2x + 3)(2x - 3)$
$= (2x)^{2} - 3^{2}$
$= 4x^{2} - 9$

63.

$$\begin{array}{r} 2x - 3 \\ 3x - 1 \overline{)\,6x^2 - 11x + 3} \\ 6x^2 - 2x \\ \hline -9x + 3 \\ -9x + 3 \\ \hline 0 \end{array}$$

Thus $[6x^2 - 11x + 3] \div (2x - 3)$

$= 2x - 3$

65. $[2x^3 - 4x^2 - 3x] - [4x^2 - 3x + 9]$

$= 2x^3 - 4x^2 - 3x - 4x^2 + 3x - 9$

$= 2x^2 - 8x^2 - 9$

67.

$(3x - 2y)^2 = (3x)^2 - 2(3x)(2y) + (2y)^2$

$= 9x^2 - 12xy + 4y^2$

69.

$(5xy - 6)(5xy + 6) = (5xy)^2 - 6^2$

$= 25x^2 y^2 - 36$

71.

$[2x - 5y^2][2x + 5y^2] = (2x)^2 - [5y^2]^2$

$= 4x^2 - 25y^4$ □

73. $\dfrac{4x^3 y^2 + 8x^2 y^3 + 12xy^4}{8xy^3}$

$= \dfrac{4x^3 y^2}{8xy^3} + \dfrac{8x^2 y^3}{8xy^3} + \dfrac{12xy^4}{8xy^3}$

$= \dfrac{x^{3-1} y^{2-3}}{2} + x^{2-1} y^{3-3}$

$+ \dfrac{3x^{1-1} y^{4-3}}{2}$

73. cont.

$= \dfrac{x^2 y^{-1}}{2} + x + \dfrac{3y}{2}$

$= \dfrac{x^2}{2y} + x + \dfrac{3y}{2}$

75. $[(x + 3y) + 2][(x + 3y) - 2]$

$= (x + 3y)^2 - 2^2$

$= x^2 + 6xy + 9y^2 - 4$

77.

$$\begin{array}{r} 3x^2 + 4x - 6 \\ 2x - 3 \\ \hline -9x^2 - 12x + 18 \\ 6x^3 + 8x^2 - 12x \quad \\ \hline 6x^3 - x^2 - 24x + 18 \end{array}$$

Thus $[3x^2 + 4x - 6](2x - 3)$

$= 6x^3 - x^2 - 24x + 18$

81.

$$\begin{array}{r} x^2 y + 6xy + y^2 \\ x + y \\ \hline x^2 y^2 + 6xy^2 + y^3 \\ x^3 y + 6x^2 y + xy^2 \quad \\ \hline x^3 y + 6x^2 y + x^2 y^2 + 7xy^2 + y^3 \end{array}$$

83.

$$\begin{array}{r|rrrrrr} -1 & 2 & 0 & -10 & 0 & 1 & -1 \\ & & -2 & 2 & 8 & -8 & 7 \\ \hline & 2 & -2 & -8 & 8 & -7 & 6 \end{array}$$

Thus: $[2y^5 - 10y^3 + y - 1] \div (y + 1)$

$= 2y^4 - 2y^3 - 8y^2 + 8y - 7 + \dfrac{6}{y + 1}$

85. 

| 1/2 | 2 | 1 | 5 | -3 |
|---|---|---|---|---|
| | | 1 | 1 | 3 |
| | 2 | 2 | 6 | 0 |

Thus $\dfrac{2x^3 + x^2 + 5x - 3}{x - \frac{1}{2}}$

$= 2x^2 + 2x + 6$

87. a)

$f\left[\frac{1}{2}\right] = \left[\frac{1}{2}\right]^3 - 2\left[\frac{1}{2}\right] + 3$

$= \frac{1}{8} - 1 + 3$

$= \frac{17}{8}$

b)

$f(2) = 2^3 - 2(2) + 3$

$= 8 - 4 + 3 = 7$

89. a)

$f(a) = a^2 + 2a - 1$

$f(a + 2) = (a + 2)^2 + 2(a + 2) - 1$

$= a^2 + 4a + 4 + 2a + 4 - 1$

$= a^2 + 6a + 7$

91.

$f(x) = x^2 - 4x + 4$

$a = 1 \quad b = -4$

Parabola opens up.

Axis of symmetry:

$x = -(-4)/2(1) = 2$

Vertex:

$f(2) = 2^2 - 4(2) + 4$

$= 0$

Vertex: (2,0)

| x | y |
|---|---|
| 0 | 4 |
| 1 | 1 |
| 2 | 0 |
| 3 | 1 |
| 4 | 4 |

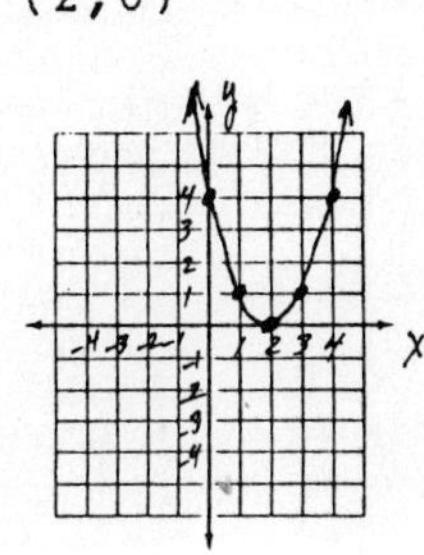


91. cont.

Domain: All Reals

Range: $\{y| \quad y \geq 0 \quad\}$

93.

$f(x) = 2x^2 - 4x + 3$

$a = 2, b = -4$

Parabola opens up.

Axis of symmetry:

$x = -(-4)/2(2) = 1$

Vertex:

$f(1) = 2(1)^2 - 4(1) + 3$

$= 1$

Vertex: (1,1)

Domain: All Reals

Range: $\{y| \quad y \geq 1 \quad\}$

| x | y |
|---|---|
| -1 | 9 |
| 0 | 3 |
| 1 | 1 |
| 2 | 3 |
| 3 | 9 |

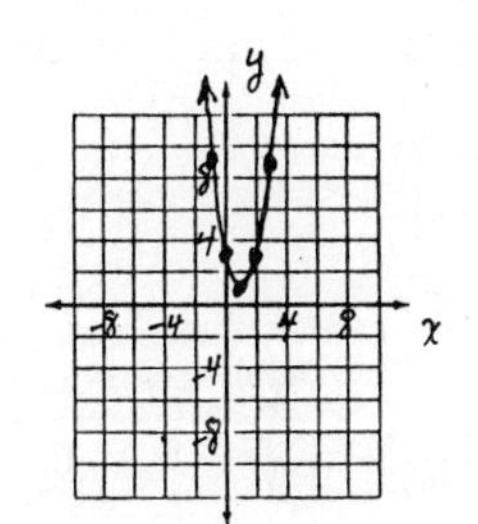


95. $y = x^3 + 1$

| x | y |
|---|---|
| -3 | -26 |
| -2 | -7 |
| -1 | 0 |
| 0 | 1 |
| 1 | 2 |
| 2 | 9 |
| 3 | 28 |

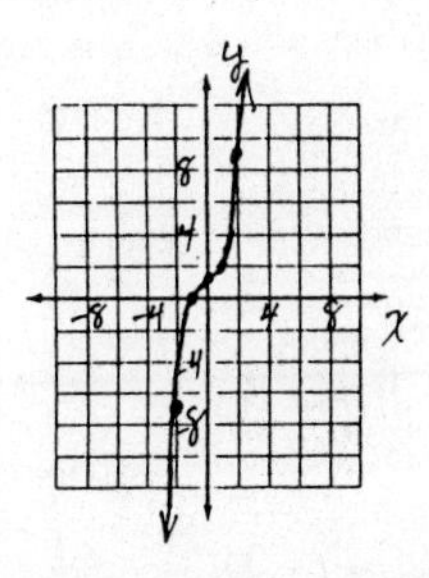


97. a) $N = f(2) = 40(20) - .2(20)^2$

$= 720$

b) $N = f(50) = 40(50) - .2(50)^2$

$= 1500$

PRACTICE TEST

1.

$$\left[\frac{3x^2y^3}{9x^5y^{-2}}\right]^2 = \left[\frac{x^{2-5}y^{3-(-2)}}{3}\right]^2$$

$$= \left[\frac{x^{-3}y^5}{3}\right]^2 = \left[\frac{x^{-6}y^{10}}{3^2}\right]$$

$$= \frac{y^{10}}{9x^6}$$

2.

$$\left[\frac{-3x^3y^{-2}}{y^5}\right]^2 \cdot \left[\frac{x^3y^4}{x^{-2}y^5}\right]^{-3}$$

$$= \left[-3x^3y^{-2-5}\right]^2 \cdot \left[x^{3-(-2)}y^{4-5}\right]^{-3}$$

$$= \left[-3x^3y^{-7}\right]^2 \cdot \left[x^5y^{-1}\right]^{-3}$$

$$= (-3)^2x^6y^{-14}x^{-15}y^3$$

$$= 9x^{-9}y^{-11} = \frac{9}{x^9y^{11}}$$

3. a) Trinomial
b) $-6x^4 - 4x^2y^3 + 2x$
c) Fifth

4. $[4x^3 - 3x - 4] - [2x^2 - 5x - 12]$

$$= 4x^3 - 3x - 4 - 2x^2 + 5x + 12$$

$$= 4x^3 - 2x^2 + 2x + 8$$ □

5. $[12x^6 - 6xy^2 + 15] - 3x$

$$= \frac{12x^6}{3x} - \frac{6xy^2}{3x} + \frac{15}{3x}$$

5. cont.

$$= 4x^{6-1} - 2x^{1-1}y^2 + \frac{5}{x}$$

$$= 4x^5 + \frac{5}{x} - 2y^2$$

6. $(3x + y)(y - 2x)$

$$= 3xy - 6x^2 + y^2 - 2xy$$

$$= -6x^2 + xy + y^2$$

7.

$$\begin{array}{r}
2x^2 + 3xy - 6y^2 \\
2x + y \\
\hline
2x^2y + 3xy^2 - 6y^3 \\
4x^3 + 6x^2y - 12xy^2 \qquad\quad \\
\hline
4x^3 + 8x^2y - 9xy^2 - 6y^3
\end{array}$$

8.

$$\begin{array}{r}
x - 5 \qquad\quad \\
2x + 3 \;\big)\; 2x^2 - 7x + 10 \\
2x^2 + 3x \qquad \\
\hline
-10x + 10 \\
-10x - 15 \\
\hline
25
\end{array}$$

Thus:

$$\left[2x^2 - 7x + 10\right] - (2x + 3)$$

$$= x - 5 + \frac{25}{2x + 3}$$

9.

$$\left[6x^2y + 3y^2 + 5x\right] - \left[4x^2y + 2x - 4y^2\right]$$

$$= 6x^2y + 3y^2 + 5x - 4x^2y - 2x + 4y^2$$

$$= 2x^2y + 3x + 7y^2$$

10.

$$\begin{array}{r|rrrr}
3 & 2 & -1 & 5 & -7 \\
 & & 6 & 15 & 60 \\
\hline
 & 2 & 5 & 20 & 53
\end{array}$$

Thus

$$\left[2x^3 - x^2 + 5x - 7\right] - (x - 3) =$$

$$2x^2 + 5x + 20 + \frac{53}{x - 3}$$

11. $3x^2 \cdot y^4 \left[-2x^5 \cdot y^2 + 6x^2 \cdot y^3 - 3x\right]$

$= -6x^{5+2} \cdot y^{4+2} + 18x^{2+2} \cdot y^{4+3}$

$-9x^{2+1} \cdot y^4$

$= -6x^7 \cdot y^6 + 18x^4 \cdot y^7 - 9x^3 \cdot y^4$

12.

$(2x + 3y)^2 = (2x)^2 + 2(2x)(3y) + (3y)^2$

$= 4x^2 + 12xy + 9y^2$

13.

| 5 \| | 3 | -12 | 0 | -60 | 4 |
|---|---|---|---|---|---|
| | | 15 | 15 | 75 | 75 |
| | 3 | 3 | 15 | 15 | 79 |

Thus

$\left[3x^4 - 12x^3 - 60x + 4\right] \div (x - 5)$

$= 3x^3 + 3x^2 + 15x + 15 + \dfrac{79}{x - 5}$

14.

$f(-3) = (-3)^3 - 2(-3) + 3$
$= -27 + 6 + 3 = -18$

15.

$f(x) = x^2 - 4x + 2$
$a = 1,\ b = -4$
Parabola opens up.
Axis of Symmetry:

$x = \left[\dfrac{-(-4)}{2(1)}\right] = 2$

Vertex:

$f(2) = 2^2 - 4(2) + 2 = -2$

Vertex: $(2,-2)$

| x | y |
|---|---|
| 0 | 2 |
| 1 | -1 |
| 2 | -2 |
| 3 | -1 |
| 4 | 2 |

Domain: All Reals
Range:
$\{y \mid y \geq -2\}$

15. cont.

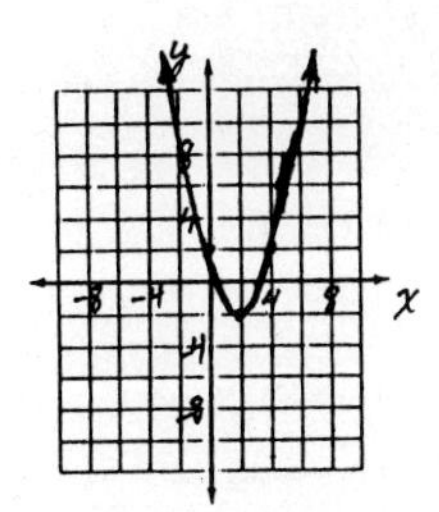

16. $P = f(8) = 12(8)^2 + 10(8) + 6000$
$= 768 + 80 + 6000$
$= 6848$

# CHAPTER 6

## EXERCISE SET 6.1

1. $8n + 8 = 8(n + 1)$

3. $13x + 5$ cannot be factored

5. $16x^2 - 12x - 6$

$= 2\left[8x^2 - 6x - 3\right]$

7. $7x^5 - 9x^4 + 3x^3$

$= x^3 \cdot \left[7x^2 - 9x + 3\right]$

9. $24y^{15} - 9y^3 + 3y$

$= 3y\left[8y^{14} - 3y^2 + 1\right]$

11. $x + 3xy^2 = x\left[1 + 3y^2\right]$

13. $6x + 5y + 5xy$ cannot be factored.

15. $40x^2 \cdot y^2 + 16xy^4 + 64xy^3$

$= 8xy^2 \cdot \left[5x + 2y^2 + 8y\right]$

17. $36xy^2 \cdot z^3 + 36x^3 \cdot y^2 \cdot z + 9x^2 \cdot yz$

$= 9xyz\left[4yz^2 + 4x^2 \cdot y + x\right]$

19. $24x^6 + 8x^4 - 4x^3 \cdot y$

$= 4x^3 \cdot \left[6x^3 + 2x - y\right]$

21. $52x^2 \cdot y^2 + 16xy^3 + 26z$

$= 2\left[26x^2 \cdot y^2 + 8xy^3 + 13z\right]$

23. $5x(2x - 5) + 3(2x - 5)$
$= (5x + 3)(2x - 5)$

25. $3x(4x - 5)^3 + 1(4x - 5)^2$

$= (4x - 5)^2 \cdot \left[3x(4x - 5) + 1\right]$

25. cont.

$= (4x - 5)^2 \cdot \left[12x^2 - 15x + 1\right]$

27. $3x(2x + 5) - 6(2x + 5)^2$

$= (2x + 5)\left[3x - 6(2x + 5)\right]$

$= (2x + 5)(3x - 12 - 30)$

$= (2x + 5)(-9x - 30)$

$= 3(2x + 5)(-3x - 10)$

29. $(3p - q)(2p - q) + (3p - q)(p - 2q)$

$= (3p - q)[2p - q + p - 2q]$

$= (3p - q)(3p - 3q)$

$= 3(3p - q)(p - q)$

31. $(x - 2)(3x + 5) - (x - 2)(5x - 4)$

$= (x - 2)[3x + 5 - (5x - 4)]$

$= (x - 2)(3x + 5 - 5x + 4)$

$= (x - 2)(-2x + 9)$

33.

$6x^3 \cdot (2x + 5) - 2x^2 \cdot (2x + 5) - (2x + 5)$

$= (2x + 5)\left[6x^3 - 2x^2 - 1\right]$

35. $4p(2r - 3)^7 - 3(2r - 3)^6$

$= (2r - 3)^6 \left[4p(2r - 3) - 3\right]$

$= (2r - 3)^6 (8pr - 12p - 3)$

37. $x^2 + 3x - 5x - 15$

$= x(x + 3) - 5(x + 3)$

$= (x - 5)(x + 3)$

39. $3x^2 + 9x + x + 3$

$= 3x(x + 3) + 1(x + 3)$

$= (3x + 1)(x + 3)$

41. $4x^2 - 2x - 2x + 1$

$= 2x(2x - 1) - 1(2x - 1)$

$= (2x - 1)(2x - 1)$

$= (2x - 1)^2$

43. $8x^2 - 4x - 20x + 10$

$= 4x(2x - 1) - 10(2x - 1)$

$= (4x - 10)(2x - 1)$

$= 2(2x - 5)(2x - 1)$

45. $2b + 2c + ab + ac$

$= 2(b + c) + a(b + c)$

$= (2 + a)(b + c)$

47. $35x^2 - 40xy + 21xy - 24y^2$

$= 5x(7x - 8y) + 3y(7x - 8y)$

$= (5x + 3y)(7x - 8y)$

49. $x^3 - 3x^2 + 4x - 12$

$= x^2(x - 3) + 4(x - 3)$

$= \left[x^2 + 4\right](x - 3)$

51. $10x^2 - 12xy - 25xy + 30y^2$

$= 2x(5x - 6y) - 5y(5x - 6y)$

$= (2x - 5y)(5x - 6y)$

53. $x^3 + 3x^2 - 2x - 6$

$= x^2(x + 3) - 2(x + 3)$

$= \left[x^2 - 2\right](x + 3)$

55. $2a^4b - 2ac^2 - 3a^3bc + 3c^3$

$= 2a\left[a^3b - c^2\right] - 3c\left[a^3b - c^2\right]$

$= (2a - 3c)\left[a^3b - c^2\right]$

57. $3p^3 + 3pq^2 + 2p^2q + 2q^3$

$= 3p\left[p^2 + q^2\right] + 2q\left[p^2 + q^2\right]$

$= (3p + 2q)\left[p^2 + q^2\right]$

59. $20p^3 - 18p^2 + 12p$

$= 2p\left[10p^2 - 9p + 6\right]$

61. $16x^2y^2z + 4x^3y - 8$

$= 4\left[4x^2y^2z + x^3y - 2\right]$

63. $5x^2 - 10x + 3x - 6$

$= 5x(x - 2) + 3(x - 2)$

$= (5x + 3)(x - 2)$

65. $14y^3z^5 - 28y^3z^6 - 9xy^2z^2$

$= y^2z^2\left[14yz^3 - 28yz^4 - 9x\right]$

67. $7x^4y^9 - 21x^3y^7z^5 - 35y^8z^9$

$= 7y^7\left[x^4y^2 - 3x^3z^5 - 5yz^9\right]$

69. $15a^2 - 18ab - 20ab + 24b^2$

$= 3a(5a - 6b) - 4b(5a - 6b)$

$= (3a - 4b)(5a - 6b)$

71. $3x(7x + 1) - 2(7x + 1)$

$= (3x - 2)(7x + 1)$

73. $6x^2 - 9xy + 2xy - 3y^2$

$= 3x(2x - 3y) + y(2x - 3y)$

$= (3x + y)(2x - 3y)$

75. $5x(x + 3)^2 - 3(x + 3)$

$= (x + 3)[5x(x + 3) - 3]$

$= (x + 3)\left[5x^2 + 15x - 3\right]$

77. $(3c - d)(c + d) - (3c - d)(c - d)$

$= (3c - d)[(c + d) - (c - d)]$

$= (3c - d)(c + d - c + d)$

$= (3c - d)(2d)$

79. $3x^5 - 15x^3 + 2x^3 - 10x$

$= 3x^3 [x^2 - 5] + 2x[x^2 - 5]$

$= [3x^3 + 2x][x^2 - 5]$

$= x[3x^2 + 2][x^2 - 5]$

81. Determine if all the terms contain a GCF. If so, factor the GCF out.

83. Let x = the price of the shirt.

$(x + .1x) - 10 = 17.50$
$1.1x - 10 = 17.5$
$1.1x = 27.5$
$x = 25$

The original price of the shirt was $25.

85. $$\frac{3x^2 - 6xy^2 + 12xy^3}{4x^2y^2}$$

$$= \frac{3x^2}{4x^2y^2} - \frac{6xy^2}{4x^2y^2} + \frac{12xy^3}{4x^2y^2}$$

$$= \frac{3}{4y^2} - \frac{3}{2x} + \frac{3y}{x}$$

JUST FOR FUN

1. $4x^2(x - 3)^3 - 6x(x - 3)^2 + 4(x - 3)$

$= (x - 3)[4x^2(x - 3)^2 - 6x(x - 3) + 4]$

1. cont.

$= (x - 3)[4x^2[x^2 - 6x + 9]$

$+ [[-6x^2 + 18x + 4]]$

$= (x - 3)(4x^4 - 24x^3 + 30x^2$

$+ 18x + 4)$

$= 2(x - 3)[2x^4 - 12x^3 + 15x^2 + 9x + 2]$

3. $4x(x + 5)^{-2} + 2x(x + 5)^{-1}$

$= (x + 5)^{-2}(4x + 2x(x + 5))$

$= (x + 5)^{-2}[4x + 2x^2 + 10x]$

$= (x + 5)^{-2}[2x^2 + 14x]$

$= (2x)(x + 5)^{-2}(x + 7)$

EXERCISE SET 6.2

1. $x^2 + 7x + 6$

$= x^2 + 6x + x + 1$

$= x(x + 6) + 1(x + 6)$

$= (x + 1)(x + 6)$

3. $p^2 - 3p - 10$

$= p^2 - 5p + 2p - 10$

$= p(p - 5) + 2(p - 5)$

$= (p + 2)(p - 5)$

5. $w^2 - 7w + 9$ Cannot be factored.

7.

$$x^2 - 34x + 64$$
$$= x^2 - 32x - 2x + 64$$
$$= x(x - 32) - 2(x - 32)$$
$$= (x - 2)(x - 32)$$

9.

$$a^2 - 18a + 45$$
$$= a^2 - 15a - 3a + 45$$
$$= a(a - 15) - 3(a - 15)$$
$$= (a - 3)(a - 15)$$

11.

$$y^2 - 9y + 15$$

Cannot be factored.

13.

$$x^2 - 4xy + 3y^2$$
$$= x^2 - 3xy - xy + 3y^2$$
$$= x(x - 3y) - y(x - 3y)$$
$$= (x - y)(x - 3y)$$

15.

$$z^2 - 7yz + 10y^2$$
$$= z^2 - 2yz - 5yz + 10y^2$$
$$= z(z - 2y) - 5y(z - 2y)$$
$$= (z - 5y)(z - 2y)$$

17.

$$5x^2 + 20x + 15$$
$$= 5\left[x^2 + 4x + 3\right]$$
$$= 5\left[x^2 + 3x + x + 3\right]$$
$$= 5(x(x + 3) + 1(x + 3))$$
$$= 5(x + 1)(x + 3)$$

19.

$$x^3 - 3x^2 - 18x$$
$$= x\left[x^2 - 3x - 18\right]$$

19. cont.

$$= x\left[x^2 - 6x + 3x - 18\right]$$
$$= x[x(x - 6) + 3(x - 6)]$$
$$= x(x + 3)(x - 6)$$

21.

$$x^3 - 5x^2 - 24x$$
$$= x\left[x^2 - 5x - 24\right]$$
$$= x\left[x^2 - 8x + 3x - 24\right]$$
$$= x[x(x - 8) + 3(x - 8)]$$
$$= x(x + 3)(x - 8)$$

23.

$$4w^2 + 13w + 3$$
$$= 4w^2 + 12w + w + 3$$
$$= 4w(w + 3) + 1(w + 3)$$
$$= (4w + 1)(w + 3)$$

25.

$$3x^2 - 11x - 6$$

Cannot be factored.

27.

$$3w^2 - 2w - 8$$

$-6 \times 4 = -24$
and
$-6 + 4 = -2$
Therefore:

$$3w^2 - 6w + 4w - 8$$
$$= 3w(w - 2) + 4(w - 2)$$
$$= (3w + 4)(w - 2)$$

29.

$$3y^2 - 2y - 5$$

$-5 \times 3 = -15$
and
$-5 + 3 = -2$
Therefore:

$$3y^2 - 5y + 3y - 5$$
$$= y(3y - 5) + 1(3y - 5)$$
$$= (y + 1)(3y - 5)$$

31. $4x^2 + 4xy - 3y^2$

$6 \times -2 = -12$
and
$6 + -2 = 4$
Therefore:

$4x^2 - 2xy + 6xy - 3y^2$

$= 2x(2x - y) + 3y(2x - y)$

$= (2x + 3y)(2x - y)$

33. $8x^2 + 2x - 20$

$= 2\left[4x^2 + x - 10\right]$

Cannot be factored further.

35. $8x^2 - 8xy - 6y^2$

$= 2\left[4x^2 - 4xy - 3y^2\right]$

$-6 \times 2 = -12$
and
$-6 + 2 = -4$
Therefore

$2\left[4x^2 - 6xy + 2xy - 3y^2\right]$

$= 2\left[2x(2x - 3y) + y(2x - 3y)\right]$

$= 2(2x + y)(2x - 3y)$

37. $x^3 y - 3x^2 y - 18xy$

$= xy\left[x^2 - 3x - 18\right]$

$= xy\left[x^2 - 6x + 3x - 18\right]$

$= xy\left[x(x - 6) + 3(x - 6)\right]$

$= xy(x + 3)(x - 6)$

39. $a^3 b + 2a^2 b - 35ab$

$= ab\left[a^2 + 2a - 35\right]$

39. cont.

$= ab\left[a^2 - 5a + 7a - 35\right]$

$= ab[a(a - 5) + 7(a - 5)]$

$= ab(a + 7)(a - 5)$

41. $6p^3 q^2 - 24p^2 q^3 - 30pq^4$

$= 6pq^2 \left[p^2 - 4pq - 5q^2\right]$

$= 6pq^2 \left[p^2 - 5pq + pq - 5q^2\right]$

$= 6pq^2 \left[p(p - 5q) + q(p - 5q)\right]$

$= 6pq^2 (p + q)(p - 5q)$

43. $35x^2 + 13x - 12$

$28 \times -15 = -420$
and
$28 + (-15) = 13$
Therefore:

$35x^2 - 15x + 28x - 12$

$= 5x(7x - 3) + 4(7x - 3)$

$= (5x + 4)(7x - 3)$

45. $8x^2 - 34x + 30$

$= 2\left[4x^2 - 17x + 15\right]$

$-12 \times (-5) = 60$
and
$-12 + (-5) = -17$
Therefore:

$2\left[4x^2 - 12x - 5x + 15\right]$

$= 2[4x(x - 3) - 5(x - 3)]$

$= 2(4x - 5)(x - 3)$

47. $x^4 + x^2 - 6$

$= x^4 - 2x^2 + 3x^2 - 6$

$= x^2 \cdot \left[x^2 - 2\right] + 3\left[x^2 - 2\right]$

$= \left[x^2 + 3\right]\left[x^2 - 2\right]$

49. $x^4 + 5x^2 + 6$

$= x^4 + 3x^2 + 2x^2 + 6$

$= x^2 \cdot \left[x^2 + 3\right] + 2\left[x^2 + 3\right]$

$= \left[x^2 + 2\right]\left[x^2 + 3\right]$

51. $6a^4 + 5a^2 - 25$

$-10 \times 15 = -150$
and
$-10 + 15 = 5$
Therefore:

$6a^4 - 10a^2 + 15a^2 - 25$

$= 2a^2 \cdot \left[3a^2 - 5\right] + 5\left[3a^2 - 5\right]$

$= \left[2a^2 + 5\right]\left[3a^2 - 5\right]$

53. $4(x + 1)^2 + 8(x + 1) + 3$

$6 \times 2 = (4)(3)$
and
$6 + 2 = 8$
Therefore:

$4(x + 1)^2 + 6(x + 1) + 2(x + 1) + 3$

$= 2(x + 1)\left[2(x + 1) + 3\right] + 1\left[2(x + 1) + 3\right]$

$= (2(x + 1) + 1)(2(x + 1) + 3)$

$= (2x + 2 + 1)\ (2x + 2 + 3)$

$= (2x + 3)(2x + 5)$

55. $6(a + 2)^2 - 7(a + 2) - 5$

$3 \times (-10) = 6(-5)$
and
$3 + (-10) = -7$

55. cont.
Therefore:

$6(a + 2)^2 - 10(a + 2) + 3(a + 2) - 5$

$= 2(a + 2)\left[3(a + 2) - 5\right]$

$+ 1\left[3(a + 2) - 5\right]$

$= \left[2(a + 2) + 1\right]\left[3(a + 2) - 5\right]$

$= (2a + 4 + 1)(3a + 6 - 5)$

$= (2a + 5)(3a + 1)$

57. $a^2 \cdot b^2 - 8ab + 15$

$= a^2 \cdot b^2 - 5ab - 3ab + 15$

$= ab(ab - 5) - 3(ab - 5)$

$= (ab - 3)(ab - 5)$

59. $3x^2 \cdot y^2 - 2xy - 5$

$-5 \times 3 = 3(-5)$
and
$-5 + 3 = -2$
Therefore:

$3x^2 \cdot y^2 - 5xy + 3xy - 5$

$= xy(3xy - 5) + 1(3xy - 5)$

$= (xy + 1)\ (3xy - 5)$

61. $2a^2 \cdot (5 - a) - 7a(5 - a) + 5(5 - a)$

$= (5 - a)\left[2a^2 - 7a + 5\right]$

$-5 \times -2 = 2(5)$
and
$-5 + (-2) = -7$
Therefore:

$(5 - a)\left[2a^2 - 5a - 2a + 5\right]$

$= (5 - a)[a(2a - 5) -1(2a - 5)]$

$= (5 - a)(a - 1)(2a - 5)$

63.

$$2x^2(x - 3) + 7x(x - 3) + 6(x - 3)$$

$$= (x - 3)\left[2x^2 + 7x + 6\right]$$

$3 \times 4 = 12$
and
$3 + 4 = 7$
Therefore:

$$(x - 3)\left[2x^2 + 3x + 4x + 6\right]$$

$$= (x - 3)[x(2x + 3) + 2(2x + 3)]$$
$$= (x - 3)(x + 2)(2x + 3)$$

65.

$$x^2 + 16x + 16$$

$$= (x + 8)^2$$

67.

$$3y^2 - 33y + 54$$

$$= 3\left[y^2 - 11y + 18\right]$$

$$= 3\left[y^2 - 9y - 2y + 18\right]$$

$$= 3\left[y(y - 9) - 2(y - 9)\right]$$

$$= 3(y - 2)(y - 9)$$

69.

$$y^4 - 7y^2 - 30$$

$$= y^4 - 10y^2 + 3y^2 - 30$$

$$= y^2\left[y^2 - 10\right] + 3\left[y^2 - 10\right]$$

$$= \left[y^2 + 3\right]\left[y^2 - 10\right]$$

71.

$$3z^4 - 14z^2 - 5$$

$-15 \times 1 = -15$
and
$-15 + 1 = -14$
Therefore:

$$3z^4 - 15z^2 + z^2 - 5$$

$$= 3z^2\left[z^2 - 5\right] + 1\left[z^2 - 5\right]$$

$$= \left[3z^2 + 1\right]\left[z^2 - 5\right]$$

73.

$$12x^2 + 16x - 3$$

$18 \times -2 = -36$
and
$18 + (-2) = 16$
Therefore:

$$12x^2 - 2x + 18x - 3$$

$$= 2x(6x - 1) + 3(6x - 1)$$

$$= (2x + 3)(6x - 1)$$

75.

$$2x^2y^2 + 3xy - 9$$

$6 \times (-3) = -18$
and
$6 + (-3) = 3$
Therefore:

$$2x^2y^2 - 3xy + 6xy - 9$$

$$= xy(2xy - 3) + 3(2xy - 3)$$

$$= (xy + 3)(2xy - 3)$$

77.

$$9x^2 - 15x - 36$$

$$= 3\left[3x^2 - 5x - 12\right]$$

$-9 \times 4 = 3(-12) = -36$
and
$-9 + 4 = -5$
Therefore:

$$3\left[3x^2 - 9x + 4x - 12\right]$$

$$= 3(3x(x - 3) + 4(x - 3))$$

$$= 3(3x + 4)(x - 3)$$

79.

$$x^2(x + 3) + 3x(x + 3) + 2(x + 3)$$

$$= (x + 3)\left[x^2 + 3x + 2\right]$$

$$= (x + 3)\left[x^2 + 2x + x + 2\right]$$

$$= (x + 3)\left[x(x + 2) + 1(x + 2)\right]$$

$$= (x + 3)(x + 1)(x + 2)$$

81.

$$5x^2 + 25xy + 20y^2$$

$$= 5\left[x^2 + 5xy + 4y^2\right]$$

$$= 5\left[x^2 + 4xy + xy + 4y^2\right]$$

$$= 5[x(x + 4y) + y(x + 4y)]$$

$$= 5(x + y)(x + 4y)$$

83.

$$15a^2 + 16a - 15$$

25 x -9 = -225
and
25 + (-9) = 16
Therefore:

$$15a^2 - 9a + 25a - 15$$

$$= 3a(5a - 3) + 5(5a - 3)$$

$$= (3a + 5)(5a - 3)$$

85.

$$20y^2 + 13y - 15$$

25 x (-12) = -300
and
25 + (-12) = 13
Therefore:

$$20y^2 - 12y + 25y - 15$$

$$= 4y(5y - 3) + 5(5y - 3)$$

$$= (4y + 5)(5y - 3)$$

87. Factor out the GCF if there is one.

89. Since 3, (4x - 5), and (2x -3) are factors of the polynomial, multiply the factors to find the polynomial.

$$3(4x - 5)(2x - 3)$$

$$= 3\left[8x^2 - 12x - 10x + 15\right]$$

$$= 3\left[8x^2 - 22x + 15\right]$$

$$= 24x^2 - 66x + 45$$

91. To find the second factor, divide the polynomial by (x - 3y).

91. cont.

$$\begin{array}{r} x + 2y \\ x - 3y \overline{\smash{)}\, x^2 - xy - 6y^2} \\ x^2 - 3xy \\ \hline 2xy - 6y^2 \\ 2xy - 6y^2 \\ \hline 0 \end{array}$$

The other factor is x + 2y

93. 1) Factor out the GCF 2.

$$2\left[4x^2 - 13x + 3\right]$$

2) Find two numbers such that their product is 4(3) =12 and their sum is -13.
-12 x (-1) = 12
and
-12 + (-1) = -13

3) Rewrite the polynomial:

$$2\left[4x^2 - 12x - x + 3\right]$$

4) Factor the polynomial by grouping.

$$2[4x(x - 3) - 1(x - 3)]$$
$$= 2(4x - 1)(x - 3)$$

95. The slope is undefined since the change in x is 0 and you cannot divide by zero.

97.

$$2x^2y - 6xy^2 - \left[3x^2y + 2xy^2 - 6\right]$$

$$= 2x^2y - 6xy^2 - 3x^2y - 2xy^2 + 6$$

$$= -x^2y - 8xy^2 + 6$$

JUST FOR FUN

1. Recall that the first line of the proof states that a = b. Subtracting b from both sides yields:

a = b
a - b = b - b
a - b = 0

In line 7 of the proof, both sides of the equation are divided by (a - b) but since a - b euqals 0, we are dividing by 0. Hence, the error occurs at line 7.

3. $12x^{2n} \cdot y^{2n} + 2x^{n} \cdot y^{n} - 2$

$= 2\left[6x^{2n} \cdot y^{2n} + x^{n} \cdot y^{n} - 1\right]$

3 x -2 = -6
and
3 + (-2) = 1
Therefore:

$= 2\left[6x^{2n} \cdot y^{2n} - 2x^{n} \cdot y^{n} + 3x^{n} \cdot y^{n} - 1\right]$

$= 2\left[2x^{n} \cdot y^{n} \cdot \left[3x^{n} \cdot y^{n} - 1\right]\right.$

$+ 1\left[3x^{n} \cdot y^{n} - 1\right]$

$= 2\left[2x^{n} \cdot y^{n} + 1\right]\left[3x^{n} \cdot y^{n} - 1\right]$

EXERCISE SET 6.3

1. $x^2 - 81 = (x + 9)(x - 9)$

3. $x^2 + 9$ □ Cannot be factored.

5. $1 - 4x^2 = (1 - 2x)(1 + 2x)$

7. $x^2 - 36y^2 = (x - 6y)(x + 6y)$

9. $x^6 - 144y^4 = \left[x^3 - 12y^2\right]\left[x^3 + 12y^2\right]$

11. $x^6 - 4 = \left[x^3 - 2\right]\left[x^3 + 2\right]$

13. $a^2 \cdot b^2 - 49c^2 = (ab - 7c)(ab + 7c)$

15. $9x^2 \cdot y^2 - 4x^2 = x^2 \cdot \left[9y^2 - 4\right]$

$= x^2 \cdot (3y + 2)(3y - 2)$

17. $3y - (x - 6)^2$

$= \left[6 - (x - 6)\right]\left[6 + (x - 6)\right]$

$= x(12-x)$

21. $x^2 + 10x + 25 = (x + 5)^2$

23. $4 + 4x + x^2 = (2 + x)^2$

25. $4x^2 - 20xy + 25y^2 = (2x - 5y)^2$

27. $9a^2 + 12a + 4 = (3a + 2)^2$

29. $w^4 + 16w^2 + 64 = \left[w^2 + 8\right]^2$

31. $(x + y)^2 + 2(x + y) + 1$

$= \left[(x + y) + 1\right]^2$

33. $a^4 - 2a^2 \cdot b^2 + b^4$

$= \left[a^2 - b^2\right]^2$

$= \left[(a + b)(a - b)\right]^2$

35. $x^2 + 6x + 9 - y^2$

$= \left[x^2 + 6x + 9\right] - y^2$

$= (x + 3)^2 - y^2$

$= \left[(x + 3) - y\right]\left[(x + 3) + y\right]$

$= (x + 3 - y)(x + 3 + y)$

37. $25 - \left[x^2 + 4x + 4\right]$

$= 5^2 - (x + 2)^2$

$= \left[5 - (x + 2)\right]\left[5 + (x + 2)\right]$

$= (3 - x)(7 + x)$ or $(-x + 3)(x + 7)$

39.

$$9a^2 - 12ab + 4b^2 - 9$$

$$= \left[9a^2 - 12ab + 4b^2\right] - 9$$

$$= (3a - 2b)^2 - 9$$

$$= \left[(3a - 2b) - 3\right]\left[(3a - 2b) + 3\right]$$

41.

$$x^3 - 27 = (x - 3)\left[x^2 + 3x + 9\right]$$

43.

$$x^3 + y^3 = (x + y)\left[x^2 - xy + y^2\right]$$

45.

$$x^3 - 8a^3 = (x - 2a)\left[x^2 + 2ax + 4a^2\right]$$

47.

$$y^2 + 1 = (y + 1)\left[y^2 - y + 1\right]$$

49.

$$27y^3 - 8x^3$$

$$= (3y - 2x)\left[9y^2 + 6xy + 4x^2\right]$$

51.

$$24x^3 - 81y^3 = 3\left[8x^3 - 27y^3\right]$$

$$= 3(2x - 3)\left[4x^2 + 6xy + 9y^2\right]$$

53.

$$5x^3 - 625y^3 = 5\left[x^3 - 125y^3\right]$$

$$= 5(x - 5y)\left[x^2 + 5xy + 25y^2\right]$$

55.

$$(x + 1)^3 + 1$$

$$= ((x + 1) + 1)\left[(x + 1)^2 - (x + 1) + 1\right]$$

$$= (x + 2)\left[x^2 + x + 1\right]$$

57.

$$(x - y)^3 - 27$$

$$= ((x - y) - 3)\left[(x - y)^2 + 3(x - y) + 9\right]$$

57. cont.

$$= (x - y - 3)\left[x^2 - 2xy + y^2 + 3x - 3y + 9\right]$$

59.

$$y^4 - 49x^2 = \left[y^2 - 7x\right]\left[y^2 + 7x\right]$$

61.

$$16y^2 - 81x^2 = (4y - 9x)(4y + 9x)$$

63.

$$25x^4 - 81y^6 = \left[5x^2 - 9y^3\right]\left[5x^2 + 9y^3\right]$$

65.

$$a^3 - 8 = (a - 2)\left[a^2 + 2a + 4\right]$$

67.

$$x^3 - 64 = (x - 4)\left[x^2 + 4x + 16\right]$$

69.

$$a^4 + 12a^2 + 36 = \left[a^2 + 6\right]^2$$

71.

$$a^4 + 2a^2b^2 + b^4 = \left[a^2 + b^2\right]^2$$

73.

$$x^2 - 2x + 1 - y^2$$

$$= \left[x^2 - 2x + 1\right] - y^2$$

$$= (x - 1)^2 - y^2$$

$$= [(x - 1) - y][(x - 1) + y]$$

75.

$$(x + y)^3 + 1$$

$$= [(x + y) + 1]\left[x^2 + 2xy + y^2 - x - y + 1\right]$$

77. The first and last terms are perfect squares while the center term is the product of 2 times the square root of the first term times the square root of the last term.

79. $cx = 2(4x)(2)$
$cx = 16x$

Therefore: $c = 16$
or

$cx = -2(4x)(2)$
$cx = -16x$
$c = -16$

81. Let a = the unknown number.

$2(a)(7) = -42$
$14a = -42$
$a = =3$

Since $d = a^2$
$d = (-3)^2 = 9$

83. a) 3, 6
b) -2, 5/9, -1.67, 0, 3, 6
c) $\sqrt{3}, -\sqrt{6}$
d) -2,5/9, -1.67,0,3,6
$\sqrt{3}, -\sqrt{6}$

85. $\{b\} \subseteq \{a,b,c,d\}$

87. Let w = the width and 2w + 2 = the length. Substitute 2w + 2 for L in the perimieter formula P = 2L + 2W

$22 = 2(2w + 2) + 2w$
$22 = 4w + 4 + 2w$
$22 = 6w + 4$
$18 = 6w$
$3ft = w$
and
length = 2(3) + 2 = 8ft

JUST FOR FUN

1.
$$x^2 - 7 = \left[x - \sqrt{7}\right]\left[x + \sqrt{7}\right]$$

3.
$$(x - 8)^2 - (x - 5)^2$$
$$= [(x - 8) - (x - 5)][(x - 8) + (x - 5)]$$
$$= (-3)(2x - 13)$$

EXERCISE SET 6.4

1.
$$3x^2 + 3x - 36 = 3\left[x^2 + x - 12\right]$$
$$= 3(x + 4)(x - 3)$$

3.
$$10s^2 - 19s - 15 = (5s - 3)(2s + 5)$$

5.
$$8r^2 - 26r + 15 = (4r - 3)(2r - 5)$$

7.
$$2x^2 - 72 = 2\left[x^2 - 36\right]$$
$$= 2(x + 6)(x - 6)$$

9.
$$5x^5 - 45x = 5x\left[x^4 - 9\right]$$
$$= 5x\left[x^2 - 3\right]\left[x^2 + 3\right]$$

11.
$$3x^3 - 3x^2 - 12x^2 + 12x$$
$$= 3x\left[x^2 - x - 4x + 4\right]$$
$$= 3x[\ x(x - 1) - 4(x - 1)\ ]$$
$$= 3x(x - 4)(x - 1)$$

13.
$$5x^4 \cdot y^2 + 20x^3 \cdot y^2 - 15x^3 \cdot y^2$$
$$-60x^2 \cdot y^2$$
$$= 5x^2 \cdot y^2 \cdot \left[x^2 + 4x - 3x - 12\right]$$
$$= 5x^2 \cdot y^2 \cdot \left[x(x + 4) - 3(x + 4)\right]$$
$$= 5x^2 \cdot y^2 \cdot (x - 3)(x + 4)$$

15.
$$x^4 - x^2 \cdot y^2 = x^2 \cdot \left[x^2 - y^2\right]$$
$$= x^2 \cdot (x + y)(x - y)$$

17.
$$x^7 \cdot y^2 - x^4 \cdot y^2 = x^4 \cdot y^2 \cdot \left[x^3 - 1\right]$$
$$= x^4 \cdot y^2 \cdot (x - 1)\left[x^2 + x + 1\right]$$

19.
$$x^5 - 16x = x\left[x^4 - 16\right]$$
$$= x\left[x^2 - 4\right]\left[x^2 + 4\right]$$
$$= x(x + 2)(x - 2)\left[x^2 + 4\right]$$

21. $4x^6 + 32y^3 = 4\left[x^6 + 8y^3\right]$

$= 4\left[x^2 + 2y\right]\left[x^4 - 2x^2 y + 4y^2\right]$

23. $2(a+b)^2 - 18 = 2\left[(a+b)^2 - 9\right]$

$= 2[(a+b) - 3][(a+b) + 3]$

25. $x^2 + 6xy + 9y^2 = (x + 3y)^2$

27. $(x+2)^2 - 4$

$= [(x+2) - 2][(x+2) + 2]$

$= x(x+4)$

29. $(2a+b)(2a-3b) - (2a+b)(a-b)$

$= (2a+b)[(2a-3b) - (a-b)]$

$= (2a+b)(a-2b)$

31. $(y+3)^2 + 4(y+3) + 4$

$= ((y+3)+2)^2$

$= (y+5)^2$

33. $45a^4 - 30a^3 + 5a^2$

$= 5a^2\left[9a^2 - 6a + 1\right]$

$= 5a^2(3a-1)^2$

35. $x^3 + \frac{1}{27} = \left[x + \frac{1}{3}\right]\left[x^2 - \frac{x}{3} + \frac{1}{9}\right]$

37. $3x^3 + 2x^2 - 27x - 18$

$= x^2(3x+2) - 9(3x+2)$

$= \left[x^2 - 9\right](3x+2)$

$= (x+3)(x-3)(3x+2)$

39. $a^3 b - 16ab^3 = ab\left[a^2 - 16b^2\right]$

$= ab(a+4b)(a-4b)$

41. $9 - \left[x^2 + 2xy + y^2\right]$

$= 9 - (x+y)^2$

$= [3 - (x+y)][3 + (x+y)]$

$= [3 - x - y)][3 + x + y]$

43. $24x^2 - 34x + 12$

$= 2\left[12x^2 - 17x + 6\right]$

$-8(-9) = 72$
and
$-8 + -9 = -17$
Therefore:

$= 2\left[12x^2 - 8x - 9x + 6\right]$

$= 2[4x(3x-2) - 3(3x-2)]$

$= 2(4x-3)(3x-2)$

45. $7x^2 - 13x + 6$

$-6(-7) = 42$
and
$-6 + (-7) = -13$
Therefore:

$= 7x^2 - 6x - 7x + 6$

$= (x-1)(7x-6)$

47. $x^4 - 81 = \left[x^2 - 9\right]\left[x^2 + 9\right]$

$= (x-3)(x+3)\left[x^2 + 9\right]$

49. $5bc - 10cx - 6by + 12xy$

$= 5c(b-2x) - 6y(b-2x)$

$= (5c-6y)(b-2x)$

51. $3x^4 - x^2 - 4$

$-4(3) = -12$
and
$-4 + 3 = -1$
Therefore:

51. cont.

$3x^4 - x^2 - 4 = 3x^4 - 4x^2 + 3x^2 - 4$

$= x^2 \left[3x^2 - 4\right] + 1\left[3x^2 - 4\right]$

$= \left[x^2 + 1\right]\left[3x^2 - 4\right]$

53.

$y^2 - \left[x^2 - 8x + 16\right]$

$= y^2 - (x - y)^2$

$= [y - (x - 4)][y + (x - 4)]$

$= (y - x + 4)(y + x - 4)$

55. $24ax + 19x + 36ay + 27y$

$= 6x(4a + 3) + 9y(4a + 3)$

$= (4a + 3)(6x + 9y)$

$= 3(4a + 3)(2x + 3y)$

57. $x^6 - 11x^3 + 30$

$-5(-6) = 30$
and
$-5 + (-6) = -11$
Therefore:

$x^6 - 11x^3 + 30$

$= x^6 - 5x^3 - 6x^3 + 30$

$= x^3 \left[x^3 - 5\right] - 6\left[x^3 - 5\right]$

$= \left[x^3 - 6\right]\left[x^3 - 5\right]$

59. $y - y^3 = y\left[1 - y^2\right]$

$= y(1 - y)(1 + y)$

61. $4x^2 y^2 + 12xy + 9 = (2xy + 3)^2$

63. $6r^2 s^2 + 2rs + 3rs + 1$

$3(-2) = -6$
and
$3 + (-2) = 1$
Therefore:

63. cont.

$6r^2 s^2 - 2rs + 3rs - 1$

$= 6r^2 s^2 - 2rs + 3rs - 1$

$= 2rs(3rs - 1) + 1(3rs - 1)$

$= (2rs + 1)(3rs - 1)$

65. a) Factor out the GCF if necessary. If the remaining polynomial is the difference of squares, sum of cubes, or difference of cubes, use the following special factoring formulas to factor:

$a^2 - b^2 = (a + b)(a - b)$

$a^3 - b^3 = (a - b)\left[a^2 + ab + b^2\right]$

$a^3 + b^3 = (a + b)\left[a^2 - ab + b^2\right]$

b) Factor out the GCF if necessary. If the remaining polynomial is a perfect square polynomial, use the special factoring formula to factor:

$a^2 + 2ab + b^2 = (a + b)^2$

$a^2 - 2ab + b^2 = (a - b)^2$

If the polynomial is not a perfect square, factor using the method described in section 6.2.

c) Factor out the GCF if necessary. Factor the remaining polynomial using factor by grouping methods.

67. $-5(x - 2) + 3 = -5x - 6$
$-5x + 10 + 3 = -5x - 6$
$-5x + 13 = -5x - 6$
$13 = -6$
Since 13 does not equal 6, there is no solution.

69. $|2x - 3| > -4$
Since the absolute value is always greater than zero, the solution is all real numbers.

EXERCISE SET 6.5

1. $x(x + 5) = 0$
$x = 0$ or $x + 5 = 0$
$x = -5$
Solution: $\{0, -5\}$

3. $5x(x + 9) = 0$
$5x = 0$ or $x + 9 = 0$
$x = 0$ $\quad x = -9$
Solution: $\{0, -9\}$

5. $(2x + 5)(x - 3)(3x + 6) = 0$
$2x + 5 = 0$ or $x - 3 = 0$ or $3x + 6 = 0$
$x = -5/2$ $\quad x = 3$ $\quad x = -2$
Solution: $\{-5/2, -2, 3\}$

7. $4x - 12 = 0$
$4(x - 3) = 0$
$x - 3 = 0$
$x = 3$
Solution: $\{3\}$

9. $-x^2 + 12x = 0$
$x(-x + 12) = 0$
$x = 0$ or $-x + 12 = 0$
$x = 12$
Solution: $\{0, 12\}$

11. $9x^2 = -18x$
$9x^2 + 18x = 0$
$9x(x + 2) = 0$
$9x = 0$ or $x + 2 = 0$
$x = 0$ $\quad x = -2$
Solution: $\{-2, 0\}$

13. $x^2 + x - 12 = 0$
$(x + 4)(x - 3) = 0$
$x + 4 = 0$ $\quad x - 3 = 0$
$x = -4$ $\quad x = 3$
Solution: $\{-4, 3\}$

15. $x(x - 12) = -20$
$x^2 - 12x = -20$
$x^2 - 12x + 20 = 0$
$(x - 10)(x - 2) = 0$
$x - 10 = 0,$ $\quad x - 2 = 0$
$x = 10$ $\quad x = 2$
Solution: $\{10, 2\}$

17. $-z^2 - 3z = -18$
$-z^2 - 3z + 18 = 0$
$z^2 + 3z - 18 = 0$
$(z + 6)(z - 3) = 0$
$z + 6 = 0$ or $z - 3 = 0$
$z = -6$ $\quad z = 3$
Solution: $\{-6, 3\}$

19. $3x^2 - 6x - 72 = 0$

$3\left[x^2 - 2x - 24\right] = 0$

19. cont.

$3(x - 6)(x + 4) = 0$

$x - 6 = 0$ or $x + 4 = 0$
$x = 6$ $\quad x = -4$
Solution: $\{-4, 6\}$

21. $x^3 + 19x^2 = 42x$

$x^3 + 19x^2 - 42x = 0$

$x\left[x^2 + 19x - 42\right] = 0$

$x(x + 21)(x - 2) = 0$

$x = 0$ or $x + 21 = 0$ or $x - 2 = 0$
$x = -21$ $\quad x = 2$
Solution: $\{-21, 0, 2\}$

23. $2y^2 + 22y + 60 = 0$

$2\left[y^2 + 11y + 30\right] = 0$

$2(y + 6)(y + 5) = 0$

$y + 6 = 0$ or $y + 5 = 0$
$y = -6$ $\quad y = -5$
Solution: $\{-6, -5\}$

25. $-16x - 3 = -12x^2$

$12x^2 - 16x - 3 = 0$

$(6x + 1)(2x - 3) = 0$
$6x + 1 = 0$ or $2x - 3 = 0$
$x = -1/6$ $\quad x = 3/2$
Solution: $\{-1/6, 3/2\}$

27. $-28x^2 + 15x - 2 = 0$

$28x^2 - 15x + 2 = 0$
$(4x - 1)(7x - 2) = 0$

$4x - 1 = 0$ or $7x - 2 = 0$
$x = 1/4$ $\quad x = 2/7$
Solution: $\{1/4, 2/7\}$

29. $3x^3 - 8x^2 - 3x = 0$
$x\left[3x^2 - 8x - 3\right] = 0$
$x(3x + 1)(x - 3) = 0$
$x = 0$ or $3x + 1 = 0$ $\quad x - 3 = 0$
$x = -1/3$ $\quad x = 3$
Solution: $\{-1/3, 0, 3\}$

31. $3p^2 = 22p - 7$

$3p^2 - 22p + 7 = 0$

$(3p - 1)(p - 7) = 0$
$3p - 1 = 0 \quad p - 7 = 0$
$p = 1/3 \quad p = 7$
Solution: {1/3,7}

33. $3r^2 + r = 2$
$(3r - 2)(r + 1) = 0$
$3r - 2 = 0$ or $r + 1 = 0$
$r = 2/3 \quad r = -1$
Solution: {-1,2/3}

35. $4x^3 + 4x^2 - 48x = 0$

$4x\left[x^2 + x - 12\right] = 0$

$4x(x + 4)(x - 3) = 0$

$4x = 0$ or $x+4=0$ or $x-3 = 0$
$x = 0 \quad x = -4 \quad x = 3$
Solution: {-4,0,3}

37. $6x^2 = 16x$

$6x^2 - 16x = 0$

$2x(3x - 8) = 0$
$2x = 0 \quad 3x - 8 = 0$
$x = 0 \quad x = 8/3$
Solution: {0, 8/3}

39. $25x^3 - 16x = 0$

$x\left[25x^2 - 16\right] = 0$

$x(5x - 4)(5x + 4) = 0$

$x = 0$ or $5x-4 = 0$ or $5x+4 = 0$
$x = 4/5 \quad x = -4/5$
Solution: {-4/5,0,4/5}

41. $(x + 4)^2 - 16 = 0$

$[ (x + 4) - 4][(x + 4) + 4] = 0$
$x(x + 8) = 0$
$x = 0 \quad x + 8 = 0$
$x = -8$
Solution: {-8, 0}

43. $(x - 7)(x + 5) = -20$

$x^2 - 2x - 35 + 20 = 0$

$x^2 - 2x - 15 = 0$

$(x - 5)(x + 3) = 0$
$x - 5 = 0$ or $x + 3 = 0$
$x = 5 \quad x = -3$
Solution: {-3,5}

45. $6a^2 - 12a^2 - 4a = 19a - 32$

$6a^2 - 23a^2 + 20 = 0$

$(3a - 4)(2a - 5) = 0$

$3a - 4 = 0$ or $2a - 5 = 0$
$a = 4/3 \quad a = 5/2$
Solution: {4/3, 5/2}

47. $(b - 1)(3b + 2) = 4b$

$3b^2 + 2b - 3b - 4b - 2 = 0$

$3b^2 - 5b - 2 = 0$

$(3b + 1)(b - 2) = 0$
$3b + 1 = 0$ or $b - 2 = 0$
$b = -1/3 \quad b = 2$

49. $2(x+2)(x-2) = (x-2)(x+3) - 2$

$2\left[x^2 - 4\right] = x^2 + x - 6 - 2$

$2x^2 - 8 = x^2 + x - 8$

$x^2 - x = 0$

$x(x - 1) = 0$
$x = 0$ or $x - 1 = 0$
$x = 1$

Solution: {0,1}

51. $2x^3 + 16x^2 + 30x = 0$

$2x[x^2 + 8x + 15] = 0$

$2x(x + 5)(x + 3) = 0$
$2x = 0$ or $x + 5 = 0$ or $x + 3 = 0$
$x = 0$ $\quad x = -5$ $\quad x = -3$
Solution: $\{-5,-3,0\}$

53. Let x and x + 1 be the unknown integers.
$x(x + 1) = 72$

$x^2 + x - 72 = 0$

$(x + 9)(x - 8) = 0$

$x + 9 = 0$ or $x - 8 = 0$
$x = -9$ $\quad x = 8$
x cannot be -9 because the numbers we are solving for are positive. Therefore x must equal 8 and x + 1 = 9. The two integers are 8 and 9.

55. Let x and x + 2 be the two integers.
$x(x + 2) = 99$

$x^2 + 2x - 99 = 0$

$(x + 11)(x - 9) = 0$
$x + 11 = 0$ or $x - 9 = 0$
$x = -11$ or $x = 9$

x cannot equal - 11 since the unknown numbers must be positive. Therefore x = 9 and x + 2 = 11.

57. Let x equal the first number and 2x - 3 equal the other.
$x(2x - 3) = 35$

$2x^2 - 3x - 35 = 0$

$(2x + 7)(x - 5) = 0$
$2x + 7 = 0$ or $x - 5 = 0$
$x = -7/2$ $\quad x = 5$

x cannot equal -7/2 since the uninown numbers must be positive. Therefore x = 5 and 2x - 3 = 7. The two numbers are 5 and 7.

59. Let w equal the width and 4w equal the length.
$36 = w(4w)$

$0 = 4w^2 - 36$

59. cont.

$0 = 4(w - 3)(w + 3)$
$w - 3 = 0$ or $w + 3 = 0$
$w = 3$ $\quad w = -3$

Since the width cannot equal a negative number, w must be 3 and the length equals 4w = 12. The rectangle is 3 ft by 12 ft.

61. Let h equal the height and h + 6 equal the base.

$$A = \frac{1}{2} \cdot bh$$

is the formula for the area of the triangle. Substitute h + 6 for b in the formula.

$$80 = \frac{1}{2} \cdot h(h + 6)$$

$$160 = h^2 + 6h$$

$$0 = h^2 + 6h - 160$$

$$0 = (h - 10)(h + 16)$$

$h - 10 = 0$ or $h + 16 = 0$
$h = 10$ or $h = -16$

Since the height cannot be a negative number, h = 10 and the base is h + 6 = 16. The height is 10cm and the base is 16cm.

63. Let x = the length of the original square.

$$(x + 4)^2 = 121$$

$$x^2 + 8x + 16 = 121$$

$$x^2 + 8x - 105 = 0$$

$(x + 15)(x - 7) = 0$
$x + 15 = 0$ or $x - 7 = 0$
$x = -15$ $\quad x = 7$

Since the length cannot be negative, the oriignal square was 7m by 7m.

65. $d = -16t^2 + 96$

a) $d = -16(1)^2 + 96$
$= 80$ ft.

b) $d = -16(1.5)^2 + 96$
$= 60$ ft.

c) $0 = -16(t)^2 + 96$

$0 = -16\left[t^2 - 6\right]$

$0 = -16\left[t - \sqrt{6}\right]\left[t + \sqrt{6}\right]$

$t - \sqrt{6} = 0$ or $t + \sqrt{6} = 0$

$t = \sqrt{6}$ $\quad$ $t = -\sqrt{6}$

Since t cannot be negative,

$t = \sqrt{6} \approx 2.449$

The ball will fall in the water in about 2.45 sec.

67. $s = -16t^2 - 48t + 640$

$0 = -16t^2 - 48t + 640$

$0 = -16\left[t^2 + 3t - 40\right]$

$0 = -16(t + 8)(t - 5)$
$t + 8 = 0$ or $t - 5 = 0$
$t = -8$ $\quad$ $t = 5$
Since t cannot be negative $t = 5$ and the TV will hit the ground in 5 seconds.

69. a) The zero factor property is used to solve this type of equation and only holds when the equation is set equal to zero.

b) $(x + 3)(x + 4) = 2$

$x^2 + 7x + 12 = 2$

$x^2 + 7x - 10 = 0$

$(x + 5)(x + 2) = 0$
$x + 5 = 0$ $\quad$ $x + 2 = 0$
$x = -5$ $\quad$ $x = -2$

71. 1) $3x + 5y = 9$
2) $2x - y = 6$
Multiply equation 2 by 5 and add the result to equation 1.

1) $3x + 5y = 9$
2) $10x - 5y = 30$
------------------
$13x = 39$
$x = 3$
Substitute $x = 3$ into equation 1 or 2 and solve for y.

1) $3(3) + 5y = 9$
$5y = 0$
$y = 0$

Solution: (3,0)

73.

$$\begin{array}{r} 3x + 4 \\ 2x - 3 \overline{)\, 6x^2 - x - 12} \\ 6x^2 - 9x \\ \hline 8x - 12 \\ 8x - 12 \\ \hline 0 \end{array}$$

Therefore:

$$\frac{6x^2 - x - 12}{2x - 3} = 3x + 4$$

JUST FOR FUN

1. Answers may vary

$$(x - 2)(x + 5) = 0$$

$$x^2 + 3x - 10 = 0$$

3. $x(x + 1)(x - 3) = 0$

$$x\left[x^2 - 2x - 3\right] = 0$$

$$x^3 - 2x^2 - 3x = 0$$

5. $(x + 3)^2 + 2(x + 3) = 24$

$$(x + 3)^2 + 2(x + 3) - 24 = 0$$

$$x^2 + 6x + 9 + 2x + 6 - 24 = 0$$

$$x^2 + 8x - 9 = 0$$

$(x + 9)(x - 1) = 0$
$x + 9 = 0$ or $x - 1 = 0$
$x = -9$ $\quad x = 1$
Solution: $\{-9, 1\}$

EXERCISE SET 6.6

1. $3x + 2y = xy + 4$ Solve for y
$3x - 4 = xy - 2y$
$3x - 4 = y(x - 2)$

$$\frac{3x - 4}{x - 2} = \frac{y(x - 2)}{x - 2}$$

$$\frac{3x - 4}{x - 2} = y$$

3. $2(x + y) = x(3 + y)$ Solve for y
$2x + 2y = 3x + xy$
$-x = xy - 2y$
$-x = y(x - 2)$

$$\frac{-x}{x - 2} = \frac{y(x - 2)}{x - 2}$$

$$\frac{-x}{x - 2} = y$$

5. $y = \dfrac{x}{x - 1}$ Solve for x

$$y(x - 1) = x \cdot \frac{x - 1}{x - 1}$$

$xy - y = x$
$xy - x = y$
$x(y - 1) = y$

$$\frac{x(y - 1)}{y - 1} = \frac{y}{y - 1}$$

$$x = \frac{y}{y - 1}$$

7. $y = \dfrac{4 - x}{x - 2}$ Solve for x.

$$y(x - 2) = \frac{(4 - x)(x - 2)}{x - 2}$$

$$xy - 2y = 4 - x$$

$$xy + x = 2y + 4$$

$$x(y + 1) = 2y + 4$$

$$x = \frac{2y + 4}{y + 1}$$

9. $3yz + 2 = 5x + z$ Solve for z.

$$3yz - z = 5x - 2$$

$$z(3y - 1) = 5x - 2$$

$$\frac{z(3y - 1)}{3y - 1} = \frac{5x - 2}{3y - 1}$$

$$z = \frac{5x - 2}{3y - 1}$$

11. $2xyz + 3yz = -6xy$ Solve for z.

$$z(2xy + 3y) = -6xy$$

$$z = \frac{-6xy}{2xy + 3y}$$

$$z = \frac{-6x}{2x + 3}$$

13. $\frac{1}{3}xy - 6y = 2y + 3$

Solve for y.

$$3\left[\frac{1}{3}xy - 6y\right] = 3(2y + 3)$$

$$xy - 18y = 6y + 9$$
$$xy - 24y = 9$$
$$y(x - 24) = 9$$

$$y = \frac{9}{x - 24}$$

15. $3rs - 2s = \frac{1}{2}(s + 2r)$

Solve for r.

$$2(3rs - 2s) = 2\cdot\frac{1}{2}(s + 2r)$$

$$6rs - 4s = s + 2r$$
$$6rs - 2r = 5s$$
$$r(6s - 2) = 5s$$

$$r = \frac{5s}{6s - 2}$$

17. $\frac{2}{3}x + ax = 2(x + 4) + 3$

Solve for x.

$$3\left[\frac{2}{3}x + ax\right] = 3(2(x + 4) + 3)$$

$$2x + 3ax = 6x + 24 + 9$$
$$3ax - 4x = 33$$
$$x(3a - 4) = 33$$

$$x = \frac{33}{3a - 4}$$

19. $\frac{3}{5}x - 2y = 6xy + \frac{4}{3}$

Solve for y.

$$15\left[\frac{3}{5}x - 2y\right] = 15\left[6xy + \frac{4}{3}\right]$$

$$9x - 30y = 90xy + 20$$
$$9x - 20 = 90xy + 30y$$

19. cont.

$$9x - 20 = y(90x + 30)$$

$$\frac{9x - 20}{90x + 30} = y$$

21. $I = P + PRT$ Solve for P.

$$I = P(1 + RT)$$

$$\frac{I}{1 + RT} = P$$

23. $x_2 - x_1 = \frac{y_2 - y_1}{m}$

Solve for m.

$$m\left[x_2 - x_1\right] = m\cdot\frac{y_2 - y_1}{m}$$

$$m\left[x_2 - x_1\right] = \left[y_2 - y_1\right]$$

$$m = \frac{y_2 - y_1}{x_2 - x_1}$$

25. $a_n = a_1 + nd - d$

Solve for d.

$$a_n - a_1 = nd - d$$

$$a_n - a_1 = d(n - 1)$$

$$\frac{a_n - a_1}{n - 1} = d$$

27. $Vr - R = O - Dr$
Solve for r.

$$Vr + Dr = O + R$$

$$r(V + D) = O + R$$

$$r = \frac{O + R}{V + D}$$

29.

$$S_n - S_n \cdot r = a_1 - a_1 \cdot r^n$$

Solve for $S_n$

$$S_n \cdot (1 - r) = a_1 - a_1 \cdot r^n$$

$$S_n = \frac{a_1 - a_1 \cdot r^n}{1 - r}$$

31.

$$S_n - S_n \cdot r = a_1 - a_1 \cdot r^n$$

Solve for $a_1$

$$S_n - S_n \cdot r = a_1 \cdot \left[1 - r^n\right]$$

$$\frac{S_n - S_n \cdot r}{1 - r^n} = a_1$$

33.

$$e = \frac{q_H + q_C}{q_H} \qquad \text{Solve for } q_H$$

$$e \cdot q_H = q_H \cdot \left[\frac{q_H + q_C}{q_H}\right]$$

$$e \cdot q_H = q_H + q_C$$

$$e \cdot q_H - q_H = q_C$$

$$q_H \cdot (e - 1) = q_C$$

$$q_H = \frac{q_C}{e - 1}$$

35.

$$b = \frac{3a + 4}{2a + 5}$$

The two equations are the same except that the variables are different.

37.

$$(4x - 3)(x + 2) - (x - 7)(x + 2)$$

$$(x + 2)[(4x - 3) - (x - 7)]$$

$$(x + 2)(4x - 3 - x + 7)$$

$$(x + 2)(3x + 4)$$

39.

$$2(3x - 2)^2 - 11(3x - 2) - 21$$

$$-14(3) = -42$$

and

$$-14 + 3 = -11$$

Therefore:

$$2(3x - 2)^2 - 14(3x - 2)$$

$$+ 3(3x - 2) - 21$$

$$2(3x - 2)[\ (3x - 2) - 7]$$

$$+ 3[\ (3x - 2) - 7]$$

$$(2(3x - 2) + 3)(\ (3x - 2) - 7)$$

$$(6x - 1)(3x - 9)$$

$$3(6x - 1)(x - 3)$$

JUST FOR FUN

1. $xy' + yy' = 1$ Solve for $y'$

$$y'(x + y) = 1$$

$$y' = \frac{1}{x + y}$$

3. $2xyy' - xy = x - 3y'$
Solve for $y'$.

$$2xyy' + 3y' = x + xy$$

$$y'(2xy + 3) = x + xy$$

$$y' = \frac{x + xy}{2xy + 3}$$

## REVIEW EXERCISES

1. $40x^2, 36x^3, 16x^5$

Factor each:

$40x^2 = 2^3 \cdot 5x^2$

$36x^3 = 2^2 \cdot 3^2 \cdot x^3$

$16x^5 = 2^4 \cdot x^5$

$GCF = 2^2 \cdot x^2 = 4x^2$

3. $15x^3 \cdot y^2 \cdot z, -6x^2 \cdot y^3, 30xy^4 \cdot z$

Factor each:

$15x^3 \cdot y^2 \cdot z = 3 \cdot 5 \cdot x^3 \cdot y^2 \cdot z$

$-6x^2 \cdot y^3 = -2 \cdot 3x^2 \cdot y^3$

$30xy^4 \cdot z = 2 \cdot 3 \cdot 5xy^4 \cdot z$

$GCF = 3xy^2$

5. $x(x + 5), x + 5, 2(x + 5)^2$

$GCF = (x + 5)$

7. $12x^2 + 4x - 8$

$= 4\left[3x^2 + x - 2\right]$

$= 4\left[3x^2 - 2x + 3x - 2\right]$

$= 4[x(3x - 2) + 1(3x - 2)]$

$= 4(x + 1)(3x - 2)$

9. $24x^6 - 13y^5 + 6z$

Cannot be factored.

11. $2x(4x - 3) + 4x - 3$

$= 2x(4x - 3) + 1(4x - 3)$

$= (2x + 1)(4x - 3)$

13. $4x(2x - 1) + 3(2x - 1)^2$

$= (2x - 1)[4x + 3(2x - 1)]$

$= (2x - 1)[4x + 6x - 3]$

$= (2x - 1)(10x - 3)$

15. $x^2 + 3x + 2x + 6$

$= x(x + 3) + 2(x + 3)$

$= (x + 2)(x + 3)$

17. $x^2 - 7x + 7x - 49$

$= x(x - 7) + 7(x - 7)$

$= (x + 7)(x - 7)$

19. $5x^2 + 20x - x - 4$

$= 5x(x + 4) - 1(x + 4)$

$= (5x - 1)(x + 4)$

21. $12x^2 - 8xy + 15xy - 10y^2$

$= 4x(3x - 2y) + 5y(3x - 2y)$

$= (4x + 5y)(3x - 2y)$

23. $3a^4 - 12a^2 \cdot b + 9a^2 \cdot b - 3ab^2$

$= 3a^2 \cdot \left[a^2 - 4b\right] + 9b\left[a^2 - 4b\right]$

$= \left[3a^2 + 9b\right]\left[a^2 - 4b\right]$

$= 3\left[a^2 + 3b\right]\left[a^2 - 4b\right]$

25. $x^2 - 8x + 15$

$= x^2 - 5x - 3x + 15$

$= x(x - 5) - 3(x - 5)$

$= (x - 3)(x - 5)$

27. $x^2 - 12x - 45$

$= x^2 - 15x + 3x - 45$

$= x(x - 15) + 3(x - 15)$

$= (x + 3)(x - 15)$

29. $x^2 - 15xy - 54y^2$

$= x^2 - 18xy + 3xy - 54y^2$

$= x(x - 18y) + 3y(x - 18y)$

$= (x + 3y)(x - 18y)$

31. $3x^2 - 18x + 27$

$= 3\left[x^2 - 6x + 9\right]$

$= 3(x - 3)^2$

33. $8x^3 + 10x^2 - 25x$

$= x\left[8x^2 + 10x - 25\right]$

$20 \times (-10) = -200$
and
$20 + (-10) = 10$
Therefore:

$= x\left[8x^2 - 10x + 20x - 25\right]$

$= x[2x(4x - 5) + 5(4x - 5)]$

$= x(2x + 5)(4x - 5)$

35. $4x^2 + 11xy - 3y^2$

$12 \times (-1) = -12$
and
$12 + (-1) = 11$
Therefore

$= 4x^2 - xy + 12xy - 3y^2$

$= x(4x - y) + 3y(4x - y)$

$= (x + 3y)(4x - y)$

37. $12x^3 + 61x^2 + 5x$

$= x\left[12x^2 + 61x + 5\right]$

$= x\left[12x^2 + 60x + x + 5\right]$

$= x[12x(x + 5) + 1(x + 5)]$

$= x(12x + 1)(x + 5)$

39. $x^4 - x^2 - 20$

$= x^4 - 5x^2 + 4x^2 - 20$

$= x^2 \cdot \left[x^2 - 5\right] + 4\left[x^2 - 5\right]$

$= \left[x^2 + 4\right]\left[x^2 - 5\right]$

41. $3(x + 2)^2 - 16(x + 2) - 12$

$-18 \times 2 = -36$
and
$-18 + (2) = -16$
Therefore:

$= 3(x - 2)^2 - 18(x + 2) + 2(x + 2) - 12$

$= 3(x+2)[(x+2) - 6] + 2[(x + 2) - 6]$

$= [3(x + 2) + 2][(x + 2) - 6]$

$= (3x + 8)(x - 4)$

43. $4x^2 - 16y^4 = 4\left[x^2 - 4y^4\right]$

$= 4\left[x - 2y^2\right]\left[x + 2y^2\right]$

45. $(x + 2)^2 - 9$

$= [(x + 2) - 3][(x + 2) + 3]$

$= (x - 1)(x + 5)$

47. $4x^2 - 12x + 9 = (2x - 3)^2$

49. $w^4 - 16w^2 + 64 = \left[w^2 - 8\right]^2$

51. $x^3 - 8 = (x - 2)\left[x^2 + 2x + 4\right]$

53. $27x^3 - 8y^3 = (3x - 2y)\left[9x^2 + 6xy + 4y^2\right]$

55. $8y^6 - 125x^3$

$= \left[2y^2 - 5x\right]\left[4y^4 + 10xy^2 + 25x^2\right]$

57. $x^2 y^2 - 2xy^2 - 15y^2$

$= y^2 \left[x^2 - 2x - 15\right]$

$= y^2 \left[x^2 - 5x + 3x - 15\right]$

$= y^2 (x(x - 5) + 3(x - 5))$

$= y^2 (x + 3)(x - 5)$

59. $3x^3 y^4 + 18x^2 y^4 - 6x^2 y^4 - 36xy^4$

$= 3xy^4 \left[x^2 + 6x - 2x - 12\right]$

$= 3xy^4 (x(x + 6) - 2(x + 6))$

$= 3xy^4 (x - 2)(x + 6)$

61. $2x^3 y + 16y$

$= 2y\left[x^3 + 8\right]$

$= 2y (x + 2)\left[x^2 - 2x + 4\right]$

63. $6x^3 - 21x^2 - 12x$

$= 3x\left[2x^2 - 7x - 4\right]$

$= 3x\left[2x^2 - 8x + x - 4\right]$

$= 3x[2x(x - 4) + 1(x - 4)]$

$= 3x(2x + 1)(x - 4)$

65. $3x^3 + 24y^3 = 3\left[x^3 + 8y^3\right]$

$= 3(x + 2y)\left[x^2 - 2xy + 4y^2\right]$

67. $4(2x + 3)^2 - 12(2x + 3) + 5$

$-2 \times (-10) = 20$
and
$-2 + (-10) = -12$
Therefore:

67. cont.

$= 4(2x + 3)^2 - 2(2x + 3) - 10(2x + 3) + 5$

$= 2(2x + 3)[2(2x + 3) - 1] - 5[2(2x + 3) -1]$

$= [2(2x + 3) - 5][2(2x + 3) - 1]$

$= (4x + 6 - 5)(4x + 6 - 1)$

$= (4x + 1)(4x + 5)$

69. $(x -1 )x^2 - (x - 1)x - 2(x - 1)$

$= (x -1 )\left[x^2 - 1x - 2\right]$

$= (x -1 )(x + 1)(x - 2)$

71. $9ax - 3bx + 12ay - 4by$

$= 3x(3a - b) + 4y(3a - b)$

$= (3x + 4y)(3a - b)$

73. $9x^4 - 12x^2 + 4$

$= (3x^2 - 2)^2$

75. $6(2a + 3)^2 - 7(2a + 3) - 3$

$-9 (2) = -18$
and
$-9 + 2 = -7$
Therefore:

$= 6(2a + 3)^2 - 9(2a + 3) + 2(2a + 3) - 3$

$= 3(2a + 3)[(2(2a + 3) - 3] + 1[2(2a + 3) - 3]$

$= [3(2a + 3) + 1][2(2a + 3) - 3]$

$= (6a + 9 + 1)(4a + 6 - 3$
$= (6a + 10)(4a + 3)$
$= 2(3a + 5)(4a + 3)$

77. $x^3 - \left[\frac{8}{27}\right] y^6$

$= \left[x - \frac{2}{3} y^2\right]\left[x^2 + \frac{2}{3} xy^2 + \frac{4}{9} y^4\right]$

79. $2x^2 = 3x$

$2x^2 - 3x = 0$

$x(2x - 3) = 0$

$x = 0$ or $2x - 3 = 0$
$x = 3/2$

Solution: {0, 3/2}

81. $x^2 - 2x - 24 = 0$

$(x - 6)(x + 4) = 0$

$x - 6 = 0$ or $x + 4 = 0$

$x = 6$ $x = -4$

Solution: {-4, 6}

83. $x^2 = -2x + 8$

$x^2 + 2x - 8 = 0$

$(x + 4)(x - 2) = 0$

$x + 4 = 0$ or $x - 2 = 0$
$x = -4$ $x = 2$

Solution: {-4, 2}

85. $x^3 - 6x^2 + 8x = 0$

$x\left[x^2 - 6x + 8\right] = 0$

$x(x - 4)(x - 2) = 0$

$x = 0$ or $x - 4 = 0$ or $x - 2 = 0$
$x = 4$ $x = 2$

Solution: {0,2,4}

87. $8x^2 - 3 = -10x$

$8x^2 + 10x - 3 = 0$

$8x^2 - 2x + 12x - 3 = 0$

$2x(4x - 1) + 3(4x - 1) = 0$

$(2x + 3)(4x - 1) = 0$

$2x + 3 = 0$ or $4x - 1 = 0$
$x = -3/2$ $x = 1/4$

Solution: {-3/2, 1/4}

89. $x(x + 3) = 2(x + 4) - 2$

$x^2 + 3x = 2x + 8 - 2$

$x^2 + x - 6 = 0$

$(x + 3)(x - 2) = 0$

$x + 3 = 0$ or $x - 2 = 0$
$x = -3$ $x = 2$

Solution: {-3,2}

91. Let w = the width and the length = w + 2.

$w(w + 2) = 63$

$w^2 + 2w - 63 = 0$

$(w + 9)(w - 7) = 0$

$w + 9 = 0$ or $w - 7 = 0$
$w = -9$ $w = 7$

Since width cannot be a negative value, the width equals 7 feet and the length = w + 2 = 9 feet.

93. Let x = the length of the sides of the second square and x + 4 = the length of the sides of the first squre.

$(x + 4)^2 = 81$

$(x + 4)^2 - 81 = 0$

$[(x + 4) - 9][(x + 4) + 9] = 0$

$(x - 5)(x + 13) = 0$

$x - 5 = 0$ or $x + 13 = 0$
$x = 5$ $x = -13$

Since the length cannot be a negative number, the length of the side of the second square is 5 inches and the length of the side on the first square is x + 4 = 9 inches.

95. $3x + 4y = xy + 6$ Solve for y.

$3x - 6 = xy - 4y$

$3x - 6 = y(x - 4)$

$$\frac{3x - 6}{x - 4} = y$$

97. $2(3xyz - x) = x - 2z$
Solve for z.

$6xyz - 2x = x - 2z$

$6xyz + 2z = x + 2x$

$z(6xy + 2) = 3x$

$$z = \frac{3x}{6xy + 2}$$

99. $$x = \frac{y + 2}{y - 3}$$

Solve for y .

$x(y - 3) = (y + 2)$

$xy - 3x = y + 2$

$xy - y = 3x + 2$

$y(x - 1) = 3x + 2$

$$y = \frac{3x + 2}{x - 1}$$

101. $\mu - 6\beta = 3\alpha\beta + 2\mu$

Solve for $\beta$.

$-\mu = 3\alpha\beta + 6\beta$

$-\mu = \beta(3\alpha + 6)$

$$\frac{-\mu}{3\alpha + 6} = \beta$$

PRACTICE TEST

1. $4x^2y - 4x$

$= 4x(xy - 1)$

2. $3x^2 + 12x + 2x + 8$

$= 3x(x + 4) + 2(x + 4)$

$= (3x + 2)(x + 4)$

3. $9x^3y^2 + 12x^2y^5 - 27xy^4$

$= 3xy^2\left[3x^2 + 4xy^3 - 9y^2\right]$

4. $5(x - 2)^2 + 15(x - 2)$

$= 5(x - 2)[(x - 2) + 3]$

$= 5(x - 2)(x + 1)$

5. $2x^2 + 4xy + 3xy + 6y^2$

$= 2x(x + 2y) + 3y(x + 2y)$

$= (2x + 3y)(x + 2y)$

6. $x^2 - 7xy + 12y^2$

$= x^2 - 4xy - 3xy + 12y^2$

$= x(x - 4y) - 3y(x - 4y)$

$= (x - 3y)(x - 4y)$

7. $3x^3 - 6x^2 - 9x$

$= 3x\left[x^2 - 2x - 3\right]$

$= 3x(x - 3)(x + 1)$

8. $6x^2 - 7x + 2$

$= 6x^2 - 4x - 3x + 2$

$= 2x(3x - 2) - 1(3x - 2)$

$= (2x - 1)(3x - 2)$

9. $5x^2 + 17x + 6$

$= 5x^2 + 15x + 2x + 6$

$= 5x(x + 3) + 2(x + 3)$

$= (5x + 2)(x + 3)$

10. $81x^2 - 16y^4$

$= \left[9x - 4y^2\right]\left[9x + 4y^2\right]$

11. $27x^3y^6 - 8y^6$

$= y^6\left[27x^3 - 8\right]$

11. cont.

$$= y^6(3x - 2)\left[9x^2 + 6x + 4\right]$$

12. $(x + 3)^2 + 2(x + 3) - 3$

$= (x + 3)^2 - 1(x + 3) + 3(x + 3) - 3$

$= (x + 3)[(x + 3) - 1] + 3[(x + 3) - 1]$

$= [(x + 3) + 3][(x + 3) - 1]$

$= (x + 6)(x + 2)$

13. $2x^4 + 5x^2 - 18$

$= 2x^4 - 4x^2 + 9x^2 - 18$

$= 2x^2\left[x^2 - 2\right] + 9\left[x^2 - 2\right]$

$= \left[2x^2 + 9\right]\left[x^2 - 2\right]$

14. $2(x - 5)(3x + 2) = 0$

$x - 5 = 0$ or $3x + 2 = 0$
$x = 5$ $\quad x = -2/3$

Solution: {-2/3, 5}

15. $4x^2 - 18 = 21x$

$4x^2 - 21x - 18 = 0$

$4x^2 - 24x + 3x - 18 = 0$

$4x(x - 6) + 3(x - 6) = 0$

$(4x + 3)(x - 6) = 0$

$(4x + 3)(x - 6) = 0$

$4x + 3 = 0$ or $x - 6 = 0$
$x = -3/4$ $\quad x = 6$

16. $x^3 + 4x^2 - 5x = 0$

$x\left[x^2 + 4x - 5\right] = 0$

16. cont.

$x(x - 1)(x + 5) = 0$

$x = 0$ or $x - 1 = 0$ or $x + 5 = 0$
$x = 1$ $\quad x = -5$

Solution: {-5,0,1}

17. $3(x + 2y) = x(y - 5)$
Solve for y.

$3x + 6y = xy - 5x$

$8x + 6y = xy$

$8x = xy - 6y$

$8x = y(x - 6)$

$$\frac{8x}{x - 6} = y$$

18. $\frac{2}{5}(w + p) = \frac{3}{4}(wp - 3)$

Solve for p.

Clear the fractions by multiplying through by the LCM 20.

$20 \cdot \frac{2}{5}(w + p) = 20 \cdot \frac{3}{4}(wp - 3)$

$8(w + p) = 15(wp - 3)$

$8w + 8p = 15wp - 45$

$8w + 45 = 15wp - 8p$

$8w + 45 = p(15w - 8)$

$$\frac{8w + 45}{15w - 8} = p$$

19. Let h = the height and
3h + 2 = the base.
The area of a triangle is found using the following formula.

$$A = \frac{1}{2} BH$$

Therefor:

$$28 = \frac{1}{2}(3h + 2)(h)$$

19. cont.

$$2 \cdot 28 = 2 \cdot \frac{1}{2} \cdot (3h + 2)(h)$$

$$56 = (3h + 2)h$$

$$56 = 3h^2 + 2h$$

$$0 = 3h^2 + 2y - 56$$

$$0 = (3h + 14)(h - 4)$$

$3h + 14 = 0$ or $h - 4 = 0$
$h = -14/3$ $\quad h = 4$

Since height cannot be negative the height equals 4m and the base = 3h + 2 = 14m.

20. $$s = -16t^2 + 48t + 448$$

The baseball strikes the ground when s = 0. Therefore:

$$0 = -16t^2 + 48t + 448$$

$$0 = -16\left[t^2 - 3t - 28\right]$$

$$0 = -16(t - 7)(t + 4)$$

$t - 7 = 0$ or $t + 4 = 0$
$t = 7$ $\quad t = -4$

Since time cannot be a negative number, t = 7 sec.

REVIEW TEST

1. $$\frac{\sqrt[3]{27} - \sqrt[3]{-8} + |-4|}{3^0 - 12 \div 3 \div 4 - 8}$$

$$= \frac{3 - (-2) + 4}{1 - 4 \div 4 - 8}$$

$$= \frac{5 + 4}{1 - 1 - 8} = \frac{-9}{8}$$

2. $$\frac{1}{3} \cdot (x - 6) = \frac{3}{4} \cdot (2x - 1)$$

$$12 \cdot \frac{1}{3} \cdot (x - 6) = 12 \cdot \frac{3}{4} \cdot (2x - 1)$$

2. cont.

$$4(x - 6) = 9(2x - 1)$$

$$4x - 24 = 18x - 9$$

$$-24 = 14x - 9$$

$$-15 = 14x$$

$$-15/14 = x$$

3. $$3p = \frac{2L - w}{4}$$

Solve for L.

$$12p = 2L - w$$

$$12p + w = 2L$$

$$\frac{12p + w}{2} = L$$

4. $4x - 3y = 9$

| x | y |
|---|---|
| 0 | -3 |
| 9/4 | 0 |

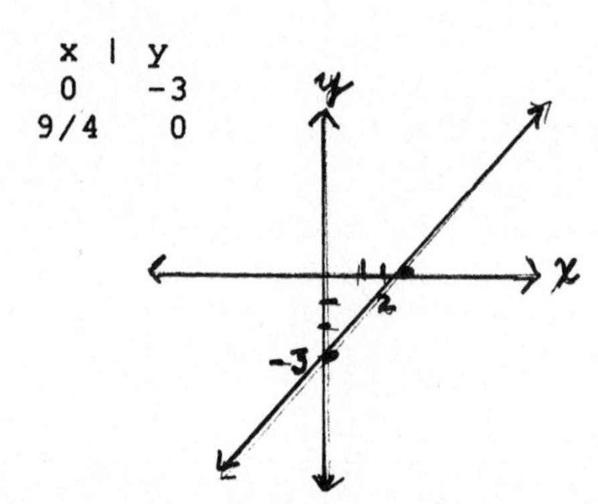


5. $2x - y \le 6$

Graph the line $2x - y = 6$ using a dashed line. Use (0,0) as a check point.
$2(0) - 0 \le 6$
Therefore, sheade the region containing the point (0,0).

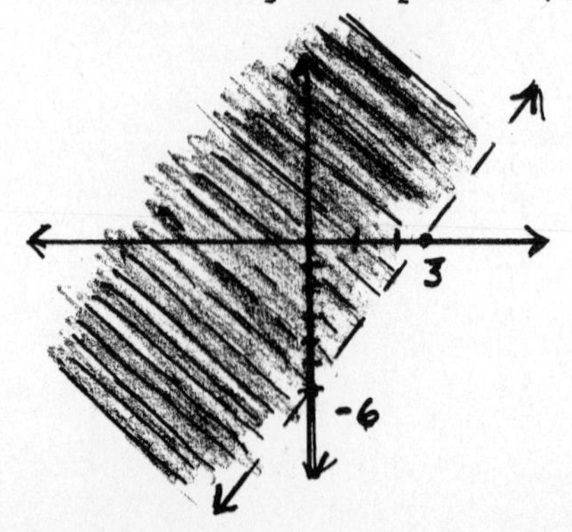

6. a) Yes, each x has a unique y.
b) No, (1,2) and (1,0) have the same x coordinate.

7. 1) $3x - 2y = 8$
2) $2x - 5y = 10$

Multiply equation 1 by 2 and equation 2 by -3. Add the results to eliminate x.

1) $6x - 4y = 16$
2) $-6x + 15y = -30$

---

$11y = -14$

$y = -14/11$

Substitute $y = -14/11$ into equation 1 or 2 and solve for x.

1) $3x - 2\left[\frac{-14}{11}\right] = 8$

$33x - 2(-14) = 88$

$33x + 28 = 88$

$33x = 60$

$x = 60/33 = 20/11$

Solution: $\left[\frac{20}{11}, \frac{-14}{11}\right]$

8. $\left[\frac{8x^{-2} \cdot y^{3}}{4xy^{-1}}\right]\left[\frac{2xy^{5}}{x^{-3}}\right]$

$= \left[2x^{-2-1} \cdot y^{3-(-1)}\right]\left[2x^{1-(-3)} \cdot y^{5}\right]$

$= \left[2x^{-3} \cdot y^{4}\right]\left[2x^{4} \cdot y^{5}\right]$

$= 4x^{-3+4} \cdot y^{4+5}$

$= 4xy^{9}$

9. $\dfrac{\left[2p^{4} \cdot q^{3}\right]\left[3pq^{4}\right]^{3}}{\left[4p^{-2} \cdot q^{3}\right]^{2}}$

$= \dfrac{2p^{4} \cdot q^{3} \cdot 27p^{3} \cdot q^{12}}{16p^{-4} \cdot q^{6}}$

$= \dfrac{27p^{7} \cdot q^{15}}{8p^{-4} \cdot q^{6}}$

$= \dfrac{27p^{7-(-4)} \cdot q^{15-6}}{8}$

$= \dfrac{27p^{11} \cdot q^{9}}{8}$

10. $3x^{2} - 4x - 6 - \left[5x - 4x^{2} - 6\right]$

$= 3x^{2} - 4x - 6 - 5x + 4x^{2} + 6$

$= 7x^{2} - 9x$

11.
$x^{2} - 3x - 6$
$2x - 5$

---

$2x^{3} - 6x^{2} - 12x$
$-5x^{2} + 15x + 30$

---

$2x^{3} - 11x^{2} + 3x + 30$

12. $\dfrac{9x^{3} \cdot y^{5} - 8x^{2} \cdot y^{4} - 12xy}{3x^{2} \cdot y}$

12. cont.

$$= \frac{9x^3 \cdot y^5}{3x^2 \cdot y} - \frac{8x^2 \cdot y^4}{3x^2 \cdot y} - \frac{12xy}{3x^2 \cdot y}$$

$$= 3xy^4 - \frac{8y^3}{3} - \frac{4}{x}$$

13.

$$f(2) = 3(2)^3 - 6(2)^2 - 4(2) + 3$$

$$= 3(8) - 6(4) - 8 + 3$$

$$= 24 - 24 - 8 + 3$$

$$= -5$$

14.

$$f(x) = x^2 - 6x + 8$$

a = 1 b = -6
Parabola opens up.
Axis of symmetry:

$$x = \frac{-(-6)}{2(1)} = 3$$

Vertex:

$$f(3) = 3^2 - 6(3) + 8$$
$$= -1$$

Vertex: (3,-1)

| x | y |
|---|---|
| 0 | 8 |
| 1 | 3 |
| 2 | 0 |
| 3 | -1 |
| 4 | 0 |
| 5 | 3 |
| 6 | 8 |

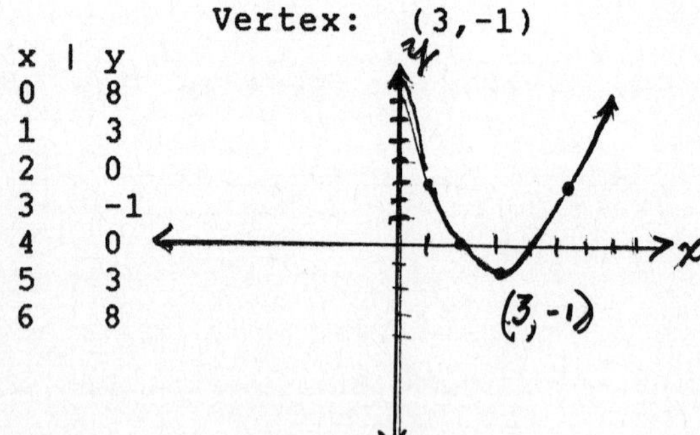


15. $x^4 - 3x^3 + 2x^2 - 6x$

$$= x\left[x^3 - 3x^2 + 2x - 6\right]$$

$$= x\left[x^2 \cdot (x - 3) + 2(x - 3)\right]$$

$$= x\left[x^2 + 2\right](x - 3)$$

16.

$$12x^2 \cdot y - 27xy + 6y$$

16. cont.

$$= 3y\left[4x^2 - 9x + 2\right]$$

$$= 3y\left[4x^2 - 8x - x + 2\right]$$

$$= 3y[4x(x - 2) - 1(x - 2)]$$

$$= 3y(4x - 1)(x - 2)$$

17. $y^4 - 2y^2 - 24$

$$= y^4 - 6y^2 + 4y^2 - 24$$

$$= y^2 \cdot \left[y^2 + 6\right] - 4\left[y^2 + 6\right]$$

$$= \left[y^2 - 4\right]\left[y^2 + 6\right] = (y + 2)(y - 2) \cdot (y^2 + 6)$$

18.

$$8x^3 - 27y^6$$

$$= \left[2x - 3y^2\right]\left[4x^2 + 6xy^2 + 9y^4\right]$$

19. Let x equal the number of pages in the manuscript.

$$.15x + 6(.05)x = 279$$
$$.15x + .3x = 279$$
$$.45x = 279$$
$$x = 620$$

There are 620 pages in the book.

20.

$$70 \le \frac{68 + 72 + 90 + 86 + x}{5} < 80$$

$$70 \le \frac{316 + x}{5} < 80$$

$$350 \le 316 + x < 400$$

$$34 \le x < 84$$

# CHAPTER 7

## EXERCISE SET 7.1

For problems 1-19, the domain is found by setting the denominator equal to zero and solving for the variable. The domain will be all reals except for those values.

1. $x = 0$

   Domain: $\{x \mid x \neq 0\}$

3. $2x - 6 = 0$

   $x = 3$

   Domain: $\{x \mid x \neq 3\}$

5. $x^2 + 4 = 0$

   There does not exist a value x such that

   $x^2 + 4 = 0$

   Therefore, the domain is all real numbers.

7. $(x - 2)^2 = 0$

   $x - 2 = 0$

   $x = 2$

   Domain: $\{x \mid x \neq 2\}$

9. $16 - r^2 = 0$

   $(4 - r)(4 + r) = 0$

   $4 - r = 0$ or $4 + r = 0$

   $r = 4 \qquad r = -4$

   Domain: $\{r \mid r \neq 4, r \neq -4\}$

11. $x^2 + 7x + 6 = 0$

    $(x + 6)(x + 1) = 0$

    $x + 6 = 0 \quad x + 1 = 0$

    $x = -6 \qquad x = -1$

    Domain: $\{x \mid x \neq -6, x \neq -1\}$

13. $y - 1 = 0$

    $y = 1$

    Domain: $\{y \mid y \neq 1\}$

15. $-8z + 15 = 0$

    $z = 15/8$

    Domain: $\{z \mid z \neq 15/8\}$

17. $x^2 - 16 = 0$

    $(x - 4)(x + 4) = 0$

    $x - 4 = 0 \quad x + 4 = 0$

    $x = 4 \qquad x = -4$

    Domain: $\{x \mid x \neq 4, x \neq -4\}$

19. $x^3 + 9x = 0$

    $x\left[x^2 + 9\right] = 0$

    $x = 0$

    There does not exist a value of x such that

    $x^2 + 9 = 0$

    Therefore, the domain is $\{x \mid x \neq 0\}$

21. $$\frac{x - xy}{x} = \frac{x}{x} - \frac{xy}{x}$$

    $$= 1 - y$$

23. $$\frac{5x^2 - 25x}{10} = \frac{5x(x - 5)}{5(2)}$$

    $$= \frac{x(x - 5)}{2}$$

25. $$\frac{x^2 - 4}{x + 2} = \frac{(x + 2)(x - 2)}{x + 2}$$

    $$= x - 2$$

27. $$\frac{5x^2 - 10xy}{25x} = \frac{5x(x - 2y)}{5x(5)}$$

    $$= \frac{x - 2y}{5}$$

29. $$\frac{4x^2 \cdot y + 12xy + 18x^3 \cdot y^3}{8xy^2}$$

    $$= \left[\frac{2xy\left[2x + 6 + 9x^2 \cdot y^2\right]}{8xy^2}\right]$$

    $$= \frac{2x + 6 + 9x^2 \cdot y^2}{4y}$$

31. $$\frac{3 - 2x}{2x - 3} = \frac{-1(2x - 3)}{2x - 3}$$

$$= -1$$

33. $$\frac{p + 4}{p^2 + 9p + 20} = \frac{p + 4}{(p + 4)(p + 5)}$$

$$= \frac{1}{p + 5}$$

35. $$\frac{x^2 - 2x - 24}{6 - x} = \frac{(x - 6)(x + 4)}{-1(x - 6)}$$

$$= -(x + 4)$$

37. $$\frac{4x^2 - 16x^4 + 6x^5 \cdot y}{8x^3 \cdot y}$$

$$= \frac{2x^2 \cdot \left[\ \ - 8x^2 + 3x^3 \cdot y\right]}{8x^3 \cdot y}$$

$$= \frac{2 - 8x^2 + 3x^3 \cdot y}{4xy}$$

39. $$\frac{y^2 - 10y + 24}{y^2 - 5y + 4}$$

$$= \frac{(y - 6)(y - 4)}{(y - 4)(y - 1)}$$

$$= \frac{y - 6}{y - 1}$$

41. $$\frac{x^5 + 4x^4 - 6x^3}{x^2 + 4x - 6}$$

41. cont.

$$= \left[x^3 \cdot \frac{x^2 + 4x - 6}{x^2 + 4x - 6}\right] = x^3$$

43. $$\frac{(x + 1)(x - 3) + (x + 1)(x - 2)}{2(x + 1)}$$

$$= \frac{(x + 1)(x - 3)}{2(x + 1)} + \frac{(x + 1)(x - 2)}{2(x + 1)}$$

$$= \frac{x - 3}{2} + \frac{x - 2}{2}$$

$$= \frac{2x - 5}{2}$$

45. $$\frac{x^2 - 8x + 5x - 40}{x^2 - 2x + 5x - 10}$$

$$= \frac{(x - 8)(x + 5)}{(x - 2)(x + 5)} = \frac{x - 8}{x - 2}$$

47. $$\frac{xy - yw + xz - zw}{xy + yw + xz + zw}$$

$$= \frac{(y + z)(x - w)}{(y + z)(x + w)} = \frac{x - w}{x + w}$$

49. $$\frac{a^3 - b^3}{a^2 - b^2} = \frac{(a - b)\left[a^2 + ab + b^2\right]}{(a + b)(a - b)}$$

$$= \frac{a^2 + ab + b^2}{a + b}$$

51. $$\frac{x^3 + 3x^2 - 4x - 12}{x^2 + 5x + 6}$$

$$= \frac{(x + 2)(x - 2)(x + 3)}{(x + 3)(x + 2)} = (x - 2)$$

53. $\frac{5}{x}\cdot\frac{x}{x} = \frac{5x}{x^2}$

55.

$$\frac{w+3}{w-3}\cdot\frac{w+3}{w+3} = \frac{w^2+6w+9}{w^2-9}$$

57.

$$\frac{1}{x^2y^3}\cdot\left[\frac{2xy^2z^2}{2xy^2z^2}\right] = \left[\frac{2xy^2z^2}{2x^3y^5z^2}\right]$$

59.

$$\left[\frac{x+1}{2(x-4)}\right]\cdot\left[\frac{4(x+2)}{4(x+2)}\right]$$

$$= \left[\frac{4(x+1)(x+2)}{8(x-4)(x+2)}\right]$$

61.

Factor:

$9y^2 + 15y + 4 = (3y + 4)(3y + 1)$
The factor missing from $\frac{y}{3y+4}$ is 3y +

$$\frac{y}{3y+4}\cdot\frac{3y+1}{3y+1} = \frac{y(3y+1)}{(3y+4)(3y+1)}$$

63.

$$\frac{\quad}{(x+5)(x-3)} = \frac{1}{x-3}$$

a) the missing numerator is x + 5.

b) Factor the denominator. For the reduced expression to equal 1/(x-3) the expression in the numerator should be divisable by x + 5. Therefore, the missing numerator is x + 5.

$$\frac{x+5}{(x+5)(x-3)} = \frac{1}{x-3}$$

65.

$$\frac{y^2 - y - 20}{\quad} = \frac{y+4}{y+1}$$

a) The missing denominator is (y - 5)(y + 1) or $y^2 - 4y + 5$

b) Factor the numerator.
$y^2 - y - 20 = (y - 5)(y + 5)$
For the expression to reduce to (y + 4)/(y + 1), the denominator must contain both the factors y + 1 and y - 5.

$$\frac{(y-5)(y+4)}{(y-5)(y+1)} = \frac{y+4}{y+1}$$

67. The expression is not a rational expression because $\sqrt{x}$ is not a polynomial.

69.a) $\frac{1}{x} = \frac{1}{1} \quad = 1$

b) $\frac{1}{x} = \frac{1}{10}$

c) $\frac{1}{x} = \frac{1}{100}$

d) As x increases in value, the value of the expression 1/x decreases.

71. a) Factor out -1 from either the numerator or denominator.

b)

$$\frac{3x^2 - 2x - 8}{-3x^2 + 2x + 8}$$

$$= \frac{3x^2 - 2x - 8}{(-1)\left[3x^2 - 2x - 8\right]} = -1$$

73. $3(y - 4) = -(x - 2)$
Write the equation in slope intercept form
$y = mx + b$
$3y - 12 = -x + 2$
$3y = -x + 14$

$$y = \frac{-1}{3}\cdot x + \frac{14}{3}$$

$m = -1/3, \quad b = 14/3$

75. $$\frac{\left[4x^{-3}\cdot y^{4}\right]\left[2x^{2}\cdot y^{-1}\right]}{12x^{-2}\cdot y^{3}}$$

$$= \frac{2x^{-3+2-(-2)}\cdot y^{4-1-3}}{3}$$

$$= \frac{2x}{3}$$

## JUST FOR FUN

1. $$f(x) = \frac{x^2 - 4}{x - 2}$$

a) $x - 2 = 0$
$x = 2$
Domain: $\{x \mid x \neq 2\}$

b) $$\frac{x^2 - 4}{x - 2} = \frac{(x + 2)\cancel{(x - 2)}}{\cancel{x - 2}}$$

$$= x + 2$$

Graph the line $f(x) = x + 2$. You must indicate the domain is $\{x|x = 2\}$ by breaking the line at $x = 2$.

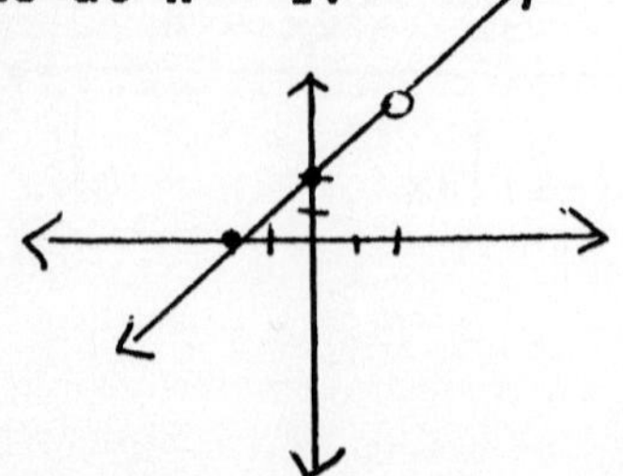

## EXERCISE SET 7.2

1. $$\frac{3x}{2y}\cdot\frac{y^2}{6} = \frac{xy}{4}$$

3. $$\frac{9x^3}{4} \div \frac{3}{16y^2} = \frac{9x^3}{4}\cdot\frac{16y^2}{3}$$

$$= 12x^3\cdot y^2$$

5. $$\frac{3y^2}{8x}\cdot\frac{9x^2}{y^3} = \frac{27x}{8y}$$

7. $$\frac{12a^2}{4bc} \div \frac{3a^2}{bc} = \frac{12a^2}{4bc}\cdot\frac{bc}{3a^2}$$

$$= \frac{12a^2\cdot bc}{12a^2\cdot bc} = 1$$

9. $$\frac{6x^5\cdot y^3}{5z^3}\cdot\frac{6x^4}{5yz^4} = \frac{36x^9\cdot y^2}{25z^7}$$

11. $$2xz \div \frac{4xy}{z} = \frac{2xz}{1}\cdot\frac{z}{4xy}$$

$$= \frac{z^2}{2y}$$

13. $$\frac{x - 3}{x + 5}\cdot\frac{2x^2 + 10x}{2x - 6}$$

$$= \frac{\cancel{(x - 3)}(\cancel{2}x)\cancel{(x + 5)}}{\cancel{(2)}\cancel{(x + 5)}\cancel{(x - 3)}} = x$$

15. $\frac{4-x}{x-4}\cdot\frac{x-3}{3-x} = \frac{-1(x-4)}{x-4}\cdot\frac{(-1)(x-3)}{x-3} = 1$

17. $\frac{x^2+7x+12}{x+4}\cdot\frac{1}{x+3} = \left[\frac{(x+3)(x+4)}{x+4}\cdot\frac{1}{x+3}\right] = 1$

19. $\frac{x^2+10x+21}{x+7} \div (x+3) = \frac{(x+7)(x+3)}{x+7}\cdot\frac{1}{x+3} = 1$

21. $\frac{x^2-9x+14}{x^2-5x+6} \div \frac{x^2-5x-14}{x+2} = \frac{(x-7)(x-2)}{(x-2)(x-3)}\cdot\frac{x+2}{(x-7)(x+2)}$

$= \frac{1}{x-3}$

23. $\frac{6x^2-14x-12}{6x+4}\cdot\frac{x+3}{2x^2-2x-12} = \frac{2(3x+2)(x-3)}{2(3x+2)}\cdot\frac{x+3}{2(x-3)(x+2)}$

$= \frac{x+3}{2(x+2)}$

25. $\frac{(x+2)^2}{x-2} \div \frac{x^2-4}{2x-4} = \frac{(x+2)(x+2)}{x-2}\cdot\frac{(2)(x-2)}{(x+2)(x-2)} = \frac{2(x+2)}{x-2}$

27. $\frac{x^2-y^2}{8x^2-16xy+8y^2}\cdot\frac{4x-4y}{x+y} = \frac{(x+y)(x-y)}{8(x-y)(x-y)}\cdot\frac{4(x-y)}{x+y} = \frac{1}{2}$

29. $\frac{x+2}{x^3-8}\cdot\frac{(x-2)^2}{x^2-4} = \frac{x+2}{(x-2)\left[x^2+2x+4\right]}\cdot\frac{(x-2)(x-2)}{(x+2)(x-2)} = \frac{1}{x^2+2x+4}$

31. $\frac{x^2-y^2}{x^2-2xy+y^2} \div \frac{x+y}{x-y} = \frac{(x+y)(x-y)}{(x-y)(x-y)}\cdot\frac{x-y}{x+y} = 1$

33.

$$\frac{\left[x^2-y^2\right]^2}{\left[x^2-y^2\right]^3} \div \frac{x^2+y^2}{x^4-y^4} = \frac{\left[x^2-y^2\right]\left[x^2-y^2\right]}{\left[x^2-y^2\right]\left[x^2-y^2\right]^2} \cdot \frac{\left[x^2-y^2\right]\left[x^2+y^2\right]}{x^2+y^2} = 1$$

35.

$$\frac{8a^3-1}{4a^2+2a+1} \div \frac{a-1}{(a-1)^2} = \frac{(2a-1)\left[4a^2+2a+1\right]}{4a^2+2a+1} \cdot \frac{(a-1)(a-1)}{a-1}$$

$$= (2a-1)(a-1)$$

37.

$$\frac{2x^3-7x^2+3x}{x^2+2x-3} \cdot \frac{x^2+3x}{(x-3)^2} = \frac{x(2x+1)(x-3)}{(x+3)(x-1)} \cdot \frac{x(x+3)}{(x-3)(x-3)}$$

$$= \frac{x^2(2x+1)}{(x-1)(x-3)}$$

39.

$$\frac{3r^2+17rs+10s^2}{6r^2+13rs-5s^2} \div \frac{6r^2+rs-2s^2}{6r^2-5rs+s^2} = \frac{(3r+2s)(r+5s)}{(2r+5s)(3r-s)} \cdot \frac{(3r-s)(2r-s)}{(2r-s)(3r+2s)}$$

$$= \frac{r+5s}{2r+5s}$$

41.

$$\frac{ac-ad+bc-bd}{ac+ad+bc+bd} \cdot \frac{pc+pd-qc-qd}{pcpd+qc-qd}$$

$$= \frac{(a+b)(c-d)}{(a+b)(c+d)} \cdot \frac{(p-q)(c+d)}{(p+q)(c-d)} = \frac{p-q}{p+q}$$

43.

$$\frac{2p^2+2pq-pq^2-q^3}{p^3+p^2+pq^2+q^2} \div \frac{p^3+p+p^2q+q}{p^3+p+p^2+1}$$

$$= \frac{\left[2p-q^2\right](p+q)}{\left[p^2+q^2\right](p+1)} \cdot \frac{(p+1)\left[p^2+1\right]}{(p+q)\left[p^2+1\right]} = \frac{2p-q^2}{p^2+q^2}$$

45. $$\frac{x^2 + 5x + 6}{x^2 - x - 20} \cdot \frac{2x^2 + 6x - 8}{x^2 - 9} \cdot \frac{x^2 - 3x}{x - 1}$$

$$= \frac{(x + 2)\cancel{(x + 3)}}{(x - 5)\cancel{(x + 4)}} \cdot \frac{2\cancel{(x + 4)}\cancel{(x - 1)}}{\cancel{(x - 3)}\cancel{(x + 3)}} \cdot \frac{x\cancel{(x - 3)}}{\cancel{x - 1}} = \frac{2x(x + 2)}{x - 5}$$

47.

$$\frac{x^3 + 64}{x - 2} \cdot \frac{x^2 - 4}{x + 4} \cdot \frac{x}{x + 2} = \frac{\cancel{(x + 4)}\left[x^2 + 4x + 16\right]}{\cancel{x - 2}} \cdot \frac{\cancel{(x - 2)}\cancel{(x + 2)}}{\cancel{x + 4}} \cdot \frac{x}{\cancel{x + 2}}$$

$$= x\left[x^2 + 4x + 16\right]$$

49.

$$\frac{a^2 - b^2}{2a^2 - 3ab + b^2} \cdot \frac{2a^2 - 7ab + 3b^2}{a^2 + ab} \div \frac{ab - 3b^2}{a^2 + 2ab + b^2}$$

$$= \frac{\cancel{(a + b)}\cancel{(a - b)}}{\cancel{(a - b)}\cancel{(2a - b)}} \cdot \frac{\cancel{(a - 3b)}\cancel{(2a - b)}}{a\cancel{(a + b)}} \cdot \frac{(a + b)(a + b)}{b\cancel{(a - 3b)}} = \frac{(a + b)^2}{ab}$$

51. $$\frac{5x^2(x - 1) - 3x(x - 1) - 2(x - 1)}{10x^2(x - 1) + 9x(x - 1) + 2(x - 1)} \cdot \frac{2x + 1}{x + 3}$$

$$= \frac{(x - 1)\left[5x^2 - 3x - 2\right]}{(x - 1)\left[10x^2 + 9x + 2\right]} \cdot \frac{2x + 1}{x + 3} = \frac{\cancel{(x - 1)}\cancel{(5x + 2)}(x - 1)}{\cancel{(x - 1)}\cancel{(5x + 2)}\cancel{(2x + 1)}} \cdot \frac{\cancel{2x + 1}}{x + 3}$$

$$= \frac{x - 1}{x + 3}$$

53. $$\frac{x^2 - x - 12}{x^2 + 2x - 3} \cdot \left[\frac{\qquad}{x^2 - 2x - 8}\right] = 1$$

$$\frac{(x - 4)(x + 3)}{(x + 3)(x - 1)} \cdot \left[\frac{\qquad}{(x - 4)(x + 2)}\right] = 1$$

53. cont.

$$\frac{\quad}{(x-1)(x+2)} = 1$$

For the rational expression to equal 1, the numerator must equal the denominator. Therefore, the missing numerator is $(x-1)(x+2)$

$$= x^2 + x - 2$$

55. $$\frac{x^2 - 9}{2x^2 + 3x - 2} \div \frac{2x^2 - 9x + 9}{\quad} = \frac{x+3}{2x-1}$$

$$\frac{(x-3)(x+3)}{(2x-1)(x+2)} \cdot \left[\frac{\quad}{(2x-3)(x-3)}\right] = \frac{x+3}{2x-1}$$

For this quotient to equal (x+3)/(2x-1), the missing numerator must be (x + 2)(2x - 3).

57. $$\frac{4x+2}{5} + 3 = 5$$

$4x + 2 + 15 = 25$
$4x = 8$
$x = 2$

59. $x + 2y = 4$
$-6x + 2y = 6$

$$D = \begin{bmatrix} 1 & 2 \\ -6 & 2 \end{bmatrix} = 14$$

$$D_x = \begin{bmatrix} 4 & 2 \\ 6 & 2 \end{bmatrix} = -4$$

$$D_y = \begin{bmatrix} 1 & 4 \\ -6 & 6 \end{bmatrix} = 30$$

$$x = \frac{D_x}{D} = \frac{30}{14} = \frac{15}{7}$$

$$y = \frac{D_y}{D} = \frac{-4}{14} = \frac{-2}{7}$$

EXERCISE SET 7.3

1. $\dfrac{2x - 7}{3} - \dfrac{4}{3} = \dfrac{2x - 11}{3}$

3. $\dfrac{x - 4}{x} - \dfrac{x + 4}{x}$

$= \dfrac{x - 4 - x - 4}{x} = \dfrac{-8}{x}$

5. $\dfrac{-t - 4}{t^2 - 16} + \dfrac{2(t + 4)}{t^2 - 16}$

$= \dfrac{-t - 4 + 2t + 8}{t^2 - 16} = \dfrac{t + 4}{(t + 4)(t - 4)}$

$= \dfrac{1}{t - 4}$

7. $\dfrac{4r + 12}{3 - r} - \dfrac{3r + 15}{3 - r}$

$= \dfrac{4r + 12 - 3r - 15}{3 - r}$

$= \dfrac{r - 3}{3 - r} = \dfrac{-1(r - 3)}{r - 3} = -1$

9. $\dfrac{-x^2}{x^2 + 5xy - 14y^2} + \dfrac{x^2 + xy + 7y^2}{x^2 + 5xy - 14y^2}$

$= \dfrac{-x^2 + x^2 + xy + 7y^2}{(x + 7y)(x - 2y)}$

$= \dfrac{y(x + 7y)}{(x + 7y)(x - 2y)} = \dfrac{y}{x - 2y}$

11. $\dfrac{x^3 - 10x^2 + 35x}{x(x - 6)} - \dfrac{x^2 + 5x}{x(x - 6)}$

$= \dfrac{x^3 - 10x^2 + 35x - x^2 - 5x}{x(x - 6)}$

$= \dfrac{x^3 - 11x^2 + 30x}{x(x - 6)}$

$= \dfrac{x(x - 5)(x - 6)}{x(x - 6)} = x - 5$

13. $\dfrac{3x^2 - x}{2x^2 - x - 21} + \dfrac{3x - 8}{2x^2 - x - 21} - \dfrac{x^2 - x + 27}{2x^2 - x - 21}$

$= \dfrac{3x^2 - x + 3x - 8 - x^2 + x - 27}{2x^2 - x - 21}$

$= \dfrac{2x^2 + 3x - 35}{2x^2 - x - 21}$

$= \dfrac{(2x - 7)(x + 5)}{(x + 3)(2x - 7)} = \dfrac{x + 5}{x + 3}$

15. $\dfrac{5x}{x + 1} + \dfrac{6}{x + 2}$

LCD $= (x + 1)(x + 2)$

17. $\dfrac{x + 3}{16x^2 \cdot y} - \dfrac{x^2}{3x^3}$

17. cont.

Factor the denominators:

$16x^2 y = 2^4 x^2 y$

$3x^3 = 3x^3$

$\text{LCD} = 2^4 \cdot (3)\left[x^3\right]y$

$= 48x^3 y$

19. $\dfrac{6z^2}{1} + \dfrac{9z}{z-3}$

$\text{LCD} = 1(z-3) = z - 3$

21. $\dfrac{a-2}{a^2 - 5a - 24} + \dfrac{3}{a^2 + 11a + 24}$

Factor each denominator:

$a^2 - 5a - 24 = (a-8)(a+3)$

$a^2 + 11a + 24 = (a+3)(a+8)$

$\text{LCD} = (a-8)(a+8)(a+3)$

23. $\dfrac{6}{x+3} - \dfrac{x+5}{x^2 - 4x + 3}$

Factor each denominator:

$x + 3 = x + 3$

$x^2 - 4x + 3 = (x-3)(x-1)$

$\text{LCD} = (x+3)(x-3)(x-1)$

25. $\dfrac{3x-5}{6x^2 + 13xy + 6y^2} + \dfrac{3}{3x^2 + 5xy + 2y^2}$

Factor each denominator:

$6x^2 + 13xy + 6y^2 = (3x+2y)(2x+3y)$

$3x^2 + 5xy + 2y^2 = (3x+2y)(x+y)$

$\text{LCD} = (3x+2y)(2x+3y)(x+y)$

27. $\dfrac{3}{x^2 + 3x - 4} - \dfrac{4}{4x^2 + 5x - 9} + \dfrac{x+2}{4x^2 + 25x + 36}$

Factor each denominator:

$x^2 + 3x - 4 = (x+4)(x-1)$

$4x^2 + 5x - 9 = (4x+9)(x-1)$

$4x^2 + 25x + 36 = (4x+9)(x+4)$

$\text{LCD} = (4x+9)(x+4)(x-1)$

29. $\dfrac{4}{3x} + \dfrac{2}{x} = \dfrac{4}{3x} + \dfrac{6}{3x} = \dfrac{10}{3x}$

31. $\dfrac{6}{x^2} + \dfrac{3}{2x} \quad \text{LCD} = 2x^2$

$\dfrac{12}{2x^2} + \dfrac{3x}{2x^2} = \dfrac{12 + 3x}{2x^2}$

33. $\dfrac{5}{6y} + \dfrac{3}{4y^2} \quad \text{LCD} = 3 \cdot 2^2 y^2 = 12y^2$

$\dfrac{10y}{12y^2} + \dfrac{9}{12y^2} = \dfrac{10y + 9}{12y^2}$

35. $\dfrac{5}{12x^4 y} - \dfrac{1}{5x^2 y^3}$

$\text{LCD} = 60x^4 y^3$

$\dfrac{25y^2}{60x^4 y^3} - \dfrac{12x^2}{60x^4 y^3}$

35. cont.

$$= \frac{25y^2 - 12x^2}{60x^4y^3}$$

37. $$\frac{4x}{3xy} + 2 = \frac{4}{3y} + \frac{2}{1}$$

LCD = 1(3y) = 3y

$$\frac{4}{3y} + \frac{6y}{3y} = \frac{4 + 6y}{3y}$$

39. $$\frac{5}{b - 2} + \frac{3x}{2 - b}$$

$$= \frac{5}{b - 2} - \frac{3x}{b - 2}$$

$$= \frac{5 - 3x}{b - 2}$$

41. $$\frac{b}{a - b} + \frac{a + b}{b}$$

LCD = b(a - b)

$$\frac{b^2}{b(a - b)} + \frac{(a + b)(a - b)}{b(a - b)}$$

$$= \frac{b^2}{b(a - b)} + \frac{a^2 - b^2}{b(a - b)}$$

$$= \frac{a^2}{b(a - b)}$$

43. $$\frac{z + 5}{z - 5} - \frac{z - 5}{z + 5}$$

LCD = (z - 5)(z + 5)

$$\frac{(z + 5)(z + 5)}{(z - 5)(z + 5)} - \frac{(z - 5)(z - 5)}{(z - 5)(z + 5)}$$

$$= \frac{z^2 + 10z + 25}{(z - 5)(z + 5)} - \frac{z^2 - 10z + 25}{(z - 5)(z + 5)}$$

43. cont.

$$= \frac{z^2 + 10z + 25 - z^2 + 10z - 25}{(z - 5)(z + 5)}$$

$$= \frac{20z}{(z - 5)(z + 5)}$$

45. $$\frac{x}{x^2 - 9} - \frac{4(x - 3)}{x + 3}$$

LCD = (x + 3)(x - 3)

$$\frac{x}{(x + 3)(x - 3)} - \frac{4(x - 3)(x - 3)}{(x + 3)(x - 3)}$$

$$= \frac{x - 4\left[x^2 - 6x + 9\right]}{(x - 3)(x + 3)}$$

$$= \frac{x - 4x^2 + 24x - 36}{(x - 3)(x + 3)}$$

$$= \frac{-4x^2 + 25x - 36}{(x + 3)(x - 3)}$$

47. $$\frac{2m + 1}{m - 5} - \frac{4}{m^2 - 3m - 10}$$

$$= \frac{2m + 1}{m - 5} - \frac{4}{(m - 5)(m + 2)}$$

LCD = (m - 5)(m + 2)

$$= \frac{(2m + 1)(m + 2)}{(m - 5)(m + 2)} - \frac{4}{(m - 5)(m + 2)}$$

$$= \frac{2m^2 + 5m + 2 - 4}{(m - 5)(m + 2)}$$

$$= \frac{2m^2 + 5m - 2}{(m - 5)(m + 2)}$$

49. $\dfrac{-x^2 + 5x}{(x-5)^2} + \dfrac{x+1}{x-5}$

LCD = $(x - 5)(x - 5)$

$= \dfrac{-x^2 + 5x}{(x-5)^2} + \dfrac{(x+1)(x-5)}{(x-5)(x-5)}$

$= \dfrac{-x^2 + 5x + x^2 - 4x - 5}{(x-5)^2}$

$= \dfrac{x-5}{(x-5)^2} = \dfrac{1}{x-5}$

51.

$\dfrac{x}{x^2 + 2x - 8} + \dfrac{x+2}{x^2 - 3x + 2}$

$= \dfrac{x}{(x+4)(x-2)} + \dfrac{x+2}{(x-2)(x-1)}$

LCD = $(x + 4)(x - 2)(x - 1)$

$= \dfrac{x(x-1)}{(x+4)(x-2)(x-1)}$

$+ \dfrac{(x+2)(x+4)}{(x-2)(x-1)(x+4)}$

$= \dfrac{x^2 - x + x^2 + 6x + 8}{(x+4)(x-2)(x-1)}$

$\dfrac{2x^2 + 5x + 8}{(x+4)(x-2)(x-1)}$

53.

$5 - \dfrac{x-1}{x^2 + 3x - 10}$

53. cont.

LCD = $x^2 + 3x - 10$

$= \dfrac{5\left[x^2 + 3x - 10\right]}{x^2 + 3x - 10} - \dfrac{x-1}{x^2 + 3x - 10}$

$= \dfrac{5x^2 + 15x - 50 - x + 1}{x^2 + 3x - 10}$

$= \dfrac{5x^2 + 14x - 49}{x^2 + 3x - 10}$

55. $\dfrac{3a-4}{4a+1} + \dfrac{3a+6}{4a^2 + 9a + 2}$

$= \dfrac{3a-4}{4a+1} + \dfrac{3a+6}{(4a+1)(a+2)}$

LCD = $(4a + 1)(a + 2)$

$= \dfrac{(3a-4)(a+2)}{(4a+1)(a+2)} + \dfrac{3a+6}{(4a+1)(a+2)}$

$= \dfrac{3a^2 + 2a - 8 + 3a + 6}{(4a+1)(a+2)}$

$= \dfrac{3a^2 + 5a - 2}{(4a+1)(a+2)}$

$= \dfrac{(3a-1)\cancel{(a+2)}}{(4a+1)\cancel{(a+2)}} = \dfrac{3a-1}{4a+1}$

57. $\dfrac{x+3}{3x^2 + 6x - 9} + \dfrac{x-3}{6x^2 - 15x + 9}$

$= \dfrac{x+3}{3(x+3)(x-1)} + \dfrac{x-3}{3(x-1)(2x-3)}$

$= \dfrac{1}{3(x-1)} + \dfrac{x-3}{3(x-1)(2x-3)}$

57. cont.

LCD = 3(x-1)(2x-3)

$$= \frac{1(2x - 3)}{3(x - 1)(2x - 3)} + \frac{x - 3}{3(x - 1)(2x - 3)}$$

$$= \frac{2x - 3 + x - 3}{3(x - 1)(2x - 3)}$$

$$= \frac{3x - 6}{3(x - 1)(2x - 3)}$$

$$= \frac{\cancel{3}(x - 2)}{\cancel{3}(x - 1)(2x - 3)} = \frac{x - 2}{(x - 1)(2x - 3)}$$

59. $$\frac{x - y}{x^2 - 4xy + 4y^2} + \frac{x - 3y}{x^2 - 4y^2}$$

$$= \frac{x - y}{(x - 2y)(x - 2y)} + \frac{x - 3y}{(x - 2y)(x + 2y)}$$

LCD = (x - 2y)(x - 2y)(x + 2y)

$$= \frac{(x - y)(x + 2y)}{(x - 2y)(x - 2y)(x + 2y)}$$

$$+ \frac{(x - 3y)(x - 2y)}{(x - 2y)(x - 2y)(x + 2y)}$$

$$= \frac{x^2 + xy - 2y^2 + x^2 - 5xy + 6y^2}{(x - 2y)(x - 2y)(x + 2y)}$$

$$= \frac{2x^2 - 4xy + 4y^2}{(x + 2y)(x - 2y)^2}$$

61. $$\frac{2x}{x - 3} - \frac{2x}{x + 3} + \frac{36}{x^2 - 9}$$

$$= \frac{2x}{x - 3} - \frac{2x}{x + 3} + \frac{36}{(x + 3)(x - 3)}$$

LCD = (x + 3)(x - 3)

61. cont.

$$= \frac{2x(x + 3)}{(x - 3)(x + 3)} - \frac{2x(x - 3)}{(x - 3)(x + 3)}$$

$$+ \frac{36}{(x - 3)(x + 3)}$$

$$= \frac{2x^2 + 6x - 2x^2 + 6x + 36}{(x - 3)(x + 3)}$$

$$= \frac{12x + 36}{(x - 3)(x + 3)} = \frac{12\cancel{(x + 3)}}{(x - 3)\cancel{(x + 3)}}$$

$$= \frac{12}{x - 3}$$

63. $$\frac{x^2 + 2}{x^2 - x - 2} + \frac{1}{x + 1} - \frac{x}{x - 2}$$

$$= \frac{x^2 + 2}{(x - 2)(x + 1)} + \frac{1}{x + 1} - \frac{x}{x - 2}$$

LCD = (x - 2)(x + 1)

$$= \frac{x^2 + 2}{(x - 2)(x + 1)} + \frac{1(x - 2)}{(x + 1)(x - 2)}$$

$$\frac{-x(x + 1)}{(x + 1)(x - 2)}$$

$$= \frac{x^2 + 2 + x - 2 - x^2 - x}{(x + 1)(x - 2)}$$

$$= \frac{0}{(x + 1)(x - 2)} = 0$$

65.

$$\frac{3x + 2}{x - 5} + \frac{x}{3x + 4} - \frac{7x^2 + 24x + 28}{3x^2 - 11x - 20}$$

$$= \frac{3x + 2}{x - 5} + \frac{x}{3x + 4} - \frac{7x^2 + 24x + 28}{(3x + 4)(x - 5)}$$

LCD = (3x + 4)(x − 5)

$$= \frac{(3x + 2)(3x + 4)}{(x - 5)(3x + 4)} + \frac{x(x - 5)}{(3x + 4)(x - 5)}$$

$$\frac{-\left[7x^2 + 24 + 28\right]}{(3x + 4)(x - 5)}$$

$$= \frac{9x^2 + 18x + 8 + x^2 - 5x - 7x^2 - 24x - 28}{(3x + 4)(x - 5)}$$

$$= \frac{3x^2 - 11x - 20}{(x - 5)(3x + 4)} = \frac{(3x + 4)(x - 5)}{(x - 5)(3x + 4)}$$

= 1

67. $$\frac{x}{x^2 - 10x + 24} - \frac{3}{x - 6} + 1$$

$$= \frac{x}{(x - 6)(x - 4)} - \frac{3}{x - 6} + \frac{1}{1}$$

LCD = (x − 6)(x − 4)

$$= \frac{x}{(x - 6)(x - 4)} - \frac{3(x - 4)}{(x - 6)(x - 4)}$$

$$+ \frac{(x - 6)(x - 4)}{(x - 6)(x - 4)}$$

$$= \frac{x - 3x + 12 + x^2 - 10x + 24}{(x - 6)(x - 4)}$$

$$= \frac{x^2 - 12x + 36}{(x - 6)(x - 4)}$$

67. cont.

$$= \frac{(x - 6)(x - 6)}{(x - 6)(x - 4)} = \frac{x - 6}{x - 4}$$

69.

$$\frac{3}{5x + 6} + \frac{x^2 - x}{5x^2 - 4x - 12} - \frac{4}{x - 2}$$

$$= \frac{3}{5x + 6} + \frac{x^2 - x}{(5x + 6)(x - 2)}$$

$$\frac{-4}{x - 2}$$

LCD = (5x + 6)(x − 2)

$$= \frac{3(x - 2)}{(5x + 6)(x - 2)}$$

$$+ \frac{x^2 - x}{(5x + 6)(x - 2)}$$

$$- \frac{4(5x + 6)}{(5x + 6)(x - 2)}$$

$$= \frac{3x - 6 + x^2 - x - 20x - 24}{(5x + 6)(x - 2)}$$

$$= \frac{x^2 - 18x - 30}{(5x + 6)(x - 2)}$$

71. $$\left[3 + \frac{1}{x + 3}\right]\left[\frac{x + 3}{x - 2}\right] = \frac{3(x + 3)}{x - 2} + \frac{x + 3}{(x + 3)(x - 2)}$$

$$= \frac{3x + 9}{x - 2} + \frac{1}{x - 2} = \frac{3x + 10}{x - 2}$$

73. $$\left[\frac{5}{a - 5} - \frac{2}{a + 3}\right] \div (3a + 25) = \left[\frac{5}{a - 5} - \frac{2}{a + 3}\right] \cdot \frac{1}{3a + 25}$$

$$= \frac{5(a + 3) - 2(a - 5)}{(a + 3)(a - 5)} \cdot \frac{1}{3a + 25} = \frac{3a + 25}{(a + 3)(a - 5)(3a + 25)} = \frac{1}{(a + 3)(a - 5)}$$

75. $$\left[\frac{x^2 + 4x - 5}{2x^2 + x - 3} \cdot \frac{2x + 3}{x + 1}\right] - \frac{2}{x + 2} = \left[\frac{(x + 5)(x - 1)(2x + 3)}{(2x + 3)(x - 1)(x + 1)}\right] - \frac{2}{x + 2}$$

$$= \frac{x + 5}{x + 1} - \frac{2}{x + 2} = \frac{(x + 5)(x + 2)}{(x + 1)(x + 2)} - \frac{2(x + 1)}{(x + 1)(x + 2)}$$

$$= \frac{x^2 + 7x + 10 - 2x - 2}{(x + 1)(x + 2)} = \frac{x^2 + 5x + 8}{(x + 1)(x + 2)}$$

77. $$\frac{2x^2 - 5x - 12}{x^2 + 3x - 10} - \frac{\qquad}{x^2 + 3x - 10} = \frac{-3x^2 - 4x - 6}{x^2 + 3x - 10}$$

Since the denominators, are the same, there is already a common denominator and the numerators do not have to be multiplied by another polynomial. We can assume that the missing numerator is of the form

$ax^2 + bx + c$ Therefore:

$$2x^2 - 5x - 12 + ax^2 + bx + c = -3x^2 - 4x - 6$$

| | | |
|---|---|---|
| $2x^2 + ax^2 = -3x^2$ | $-5x + bx = -4x$ | $-12 + c = -6$ |
| $2 + a = -3$ | $-5 + b = -4$ | $c = 6$ |
| $a = -5$ | $b = 1$ | |

The missing polynomial is: $-5x^2 + x + 6$

79. The denominators are the same in the sum but on the other side of the equal sign, the denominator is reduced to r-2. Therefore, before being reduced, the problem was:

$$\frac{r^2 - 6}{(r - 3)(r - 2)} - \frac{\quad}{(r - 3)(r - 2)} = \frac{r - 3}{(r - 3)(r - 2)}$$

The missing numerator is of the form $ar^2 + br + c$ □ and we can conclude:

$$\left[r^2 - 6\right] - \left[ar^2 + br + c\right] = r - 3$$

and

$r^2 - ar^2 = 0r^2$ $\qquad$ $-br = 1r$

$1 - a = 0$ $\qquad$ $b = -1$

$a = 1$

$-6 - c = -3$

$c = -3$

The missing polynomial is

$$r^2 - r - 3$$

81. Let t = the time on the fast speed and 14 - t = the time on the slower setting.

| | rate | x | time | = amount |
|---|---|---|---|---|
| fast | 80 | | t | 80t |
| slow | 60 | | (14-t) | 60(14-t) |

Since the amount of bottles filled and capped equals:

$80t = 60(14-t)$

$80t = 840 - 60t$

$140t = 840$

$t = 6$

a) The machine was used 6 minuets on the faster setting.

b) While the machine was on the faster setting, 80t = 480 bottles were filled. Since equal amounts were filled at each speed, the total amount filled equals 2(480) = 960.

83.

```
                 2x - 1
           ________________
                 2
(3x + 4) √ 6x  + 5x - 4
                 2
            6x  + 8x
          ----------------
                -3x - 4
                -3x - 4
                ------------
                      0
```

Therefore:

$$\frac{6x^2 + 5x - 4}{3x + 4} = 2x - 1$$

JUST FOR FUN

1. $$\frac{3}{x^2 - x - 6} - \frac{2}{x^2 + x - 6}$$

$$= \frac{3}{(x - 3)(x + 2)} - \frac{2}{(x + 3)(x - 2)}$$

LCD = $(x - 3)(x + 3)(x + 2)(x - 2)$

$$= \frac{3(x + 3)(x - 2) - 2(x - 3)(x + 2)}{(x - 3)(x + 3)(x + 2)(x - 2)}$$

$$= \frac{3x^2 + 3x - 18 - 2x^2 + 2x + 12}{(x + 3)(x - 3)(x + 2)(x - 2)}$$

$$= \frac{x^2 + 5x - 6}{(x + 3)(x - 3)(x + 2)(x - 2)}$$

or

$$\frac{(x + 6)(x - 1)}{(x + 3)(x - 3)(x + 2)(x - 2)}$$

## EXERCISE SET 7.4

1. Multiply by the LCM 5.

$$\frac{5(1) + 5\left[\frac{3}{5}\right]}{2(5) + 5\left[\frac{1}{5}\right]} = \frac{5 + 3}{10 + 1} = \frac{8}{11}$$

3. Multiply by the LCM 24.

$$\frac{2(24) + 24\left[\frac{3}{8}\right]}{1(24) + 24\left[\frac{1}{3}\right]} = \frac{48 + 9}{24 + 8} = \frac{57}{32}$$

5. Multiply by the LCM 360.

$$\frac{360\left[\frac{4}{9}\right] - 360\left[\frac{3}{8}\right]}{360(4) - 360\left[\frac{3}{5}\right]} = \frac{160 - 135}{1440 - 216} = \frac{25}{1224}$$

7. Multiply by the LCM 4x.

$$\frac{4x\left[x^2 \cdot \frac{y}{4}\right]}{4x\left[\frac{2}{x}\right]} = \frac{x^3 y}{8}$$

9. Multiply by the LCM $9z^5$

$$\frac{9z^5 \cdot \frac{8x^2 y}{3z^3}}{9z^5\left[\frac{4xy}{9z^5}\right]} = \frac{24x^2 y z^2}{4xy}$$

$$= 6xz^2$$

11. Multiply by the LCM y.

$$\frac{xy + y\frac{1}{y}}{y\frac{x}{y}} = \frac{xy + 1}{x}$$

13. Multiply by the LCM $x^2$ □ and simplify.

$$\frac{x^2\frac{9}{x} + x^2\left[\frac{3}{x^2}\right]}{3x^2 + x^2\frac{1}{x}} = \frac{9x + 3}{3x^2 + x}$$

$$= \frac{3(3x + 1)}{x(3x + 1)} = \frac{3}{x}$$

15. Multiply by the LCM y.

$$\frac{3y - y\left[\frac{1}{y}\right]}{2y - y\frac{1}{y}} = \frac{3y - 1}{2y - 1}$$

17. Multiply by the LCM xy.

$$\frac{xy\cdot\left[\frac{x}{y}\right] - xy\left[\frac{y}{x}\right]}{xy\cdot\frac{x+y}{x}} = \frac{x^2 - y^2}{y(x+y)}$$

$$= \frac{(x+y)(x-y)}{y(x+y)} = \frac{x-y}{y}$$

19. Multiply by the LCM ab.

$$\frac{ab\cdot\frac{a^2}{b} - ab\cdot b}{ab\cdot\frac{b^2}{a} - aba} = \frac{a^3 - ab^2}{b^3 - a^2\cdot b}$$

$$= \frac{a\left[a^2 - b^2\right]}{b\left[b^2 - a^2\right]} = \frac{a\left[a^2 - b^2\right]}{-b\left[a^2 - b^2\right]}$$

$$= \frac{-a}{b}$$

21. $$\frac{\frac{a}{b} - 2}{\frac{-a}{b} + 2} = \frac{\frac{a}{b} - 2}{-1\left[\left[\frac{a}{b}\right] - 2\right]} = -1$$

23. Multiply by the LCM $6x^2$

$$\frac{6x^2\cdot\left[\frac{4x+8}{3x^2}\right]}{6x^2\cdot\left[\frac{4x}{6}\right]} = \frac{2(4x+8)}{x^2\cdot(4x)}$$

$$= \frac{8x+16}{4x^3} = \frac{4(2x+4)}{4x^3}$$

23. cont.

$$= \frac{2(x+2)}{x^3}$$

25. Multiply by the LCM 4x.

$$\frac{4x\cdot\left[\frac{x}{4}\right] - 4x\left[\frac{1}{x}\right]}{4x(1) + \frac{4x(x+4)}{x}}$$

$$= \frac{x^2 - 4}{4x + 4x + 16} = \frac{x^2 - 4}{8x + 16}$$

$$= \frac{(x+2)(x-2)}{8(x+2)} = \frac{x-2}{8}$$

27. Multiply by the LCM $(x - 1)(x + 1)$.

$$\frac{(x-1)(x+1)\cdot\frac{1}{x-1} + 1(x-1)(x+1)}{\frac{1}{x+1}\cdot(x-1)(x+1) - 1(x-1)(x+1)}$$

$$= \frac{x + 1 + x^2 - 1}{(x-1) - \left[x^2 - 1\right]}$$

$$= \frac{x(x+1)}{x(1-x)} = \frac{x+1}{1-x}$$

29. Multiply by the LCM $(a - 2)(a + 2)$.

$$\frac{\dfrac{(a+2)(a-2)(a-2)}{a+2} - \dfrac{(a+2)(a-2)(a+2)}{a-2}}{\dfrac{(a+2)(a-2)(a-2)}{a+2} + \dfrac{(a+2)(a-2)(a+2)}{a-1}} = \frac{a^2 - 4a + 4 - (a^2 + 4a + 4)}{a^2 - 4a + 4 + a^2 + 4a + 4}$$

$$= \frac{-8a}{2a^2 + 8} = \frac{\not{2}(-4a)}{\not{2}\left[a^2 + 4\right]} = \frac{-4a}{a^2 + 4}$$

31. $$2a^{-2} + b = \frac{2}{a^2} + b$$

LCD $= a^2$

$$\frac{2}{a^2} + a^2 \cdot \frac{b}{a^2} = \frac{2 + a^2 b}{a^2}$$

33. $$\left[a^{-1} + b^{-1}\right]^{-1} = \frac{1}{\frac{1}{a} + \frac{1}{b}}$$

Multiply by the LCM ab.

$$\frac{ab(1)}{ab \cdot \frac{1}{a} + ab \cdot \frac{1}{b}} = \frac{ab}{b + a}$$

35. $$\frac{a^{-1} + b^{-1}}{\frac{1}{ab}} = \frac{\frac{1}{a} + \frac{1}{b}}{\frac{1}{ab}}$$

Multiply by the LCM ab.

35. cont.

$$\frac{ab \cdot \frac{1}{a} + ab \cdot \frac{1}{b}}{ab \cdot \frac{1}{ab}} = b + a$$

37. $$\frac{\frac{a}{b} + a^{-1}}{\frac{b}{a} + a^{-1}} = \frac{\frac{a}{b} + \frac{1}{a}}{\frac{b}{a} + \frac{1}{a}}$$

Multiply by the LCM ab.

$$\frac{ab \cdot \frac{a}{b} + ab \cdot \frac{1}{a}}{ab \cdot \frac{b}{a} + ab \cdot \frac{1}{a}} = \frac{a^2 + b}{b^2 + b}$$

39. $$\frac{x^{-1} - y^{-1}}{x^{-1} + y^{-1}} = \frac{\frac{1}{x} - \frac{1}{y}}{\frac{1}{x} + \frac{1}{y}}$$

Multiply by the LCM xy.

$$\frac{xy \cdot \frac{1}{x} - xy \cdot \frac{1}{y}}{xy \cdot \frac{1}{x} + xy \cdot \frac{1}{y}} = \frac{y - x}{y + x}$$

41. $$\frac{a^{-1} + b^{-1}}{(a + b)^{-1}} = \frac{\frac{1}{a} + \frac{1}{b}}{\frac{1}{a + b}}$$

Multiply by the LCM ab(a + b).

$$\frac{ab(a + b) \cdot \frac{1}{a} + ab(a + b) \cdot \frac{1}{b}}{ab(a + b) \cdot \frac{1}{a + b}}$$

$$= \frac{ab + b^2 + a^2 + ab}{ab}$$

$$= \frac{a^2 + 2ab + b^2}{ab} = \frac{(a + b)^2}{ab}$$

43. $$2x^{-1} - (3y)^{-1} = \frac{2}{x} - \frac{1}{3y}$$

The LCD = 3xy.

$$\frac{2(3y)}{3xy} - \frac{1x}{3xy} = \frac{6y - x}{3xy}$$

45. a)

$$E = \frac{\frac{1}{2} \cdot \frac{2}{3}}{\frac{2}{3} + \frac{1}{2}}$$

$$= \frac{\frac{1}{3}}{\frac{2}{3} + \frac{1}{2}}$$

Multiply by the LCM 6.

$$= \frac{6 \cdot \left[\frac{1}{3}\right]}{6 \cdot \frac{2}{3} + 6 \cdot \frac{1}{2}} = \frac{2}{4 + 3}$$

$$= \frac{2}{7}$$

b)

$$E = \frac{\frac{1}{2} \cdot \frac{4}{5}}{\frac{4}{5} + \frac{1}{2}}$$

$$= \frac{\frac{2}{5}}{\frac{4}{5} + \frac{1}{2}}$$

Multiply by the LCM 10.

$$= \frac{10 \cdot \frac{2}{5}}{10 \cdot \frac{4}{5} + 10 \cdot \frac{1}{2}}$$

45 b) cont.

$$= \frac{4}{8+5} = \frac{4}{13}$$

47.

$$R_T = \frac{R_1 \cdot R_2 \cdot R_3}{R_2 \cdot R_3 + R_1 \cdot R_3 + R_1 \cdot R_2}$$

Multiply by $\dfrac{1}{R_1 \cdot R_2 \cdot R_3}$

$$R_T = \frac{\left[R_1 \cdot R_2 \cdot R_3\right] \cdot \dfrac{1}{R_1 \cdot R_2 \cdot R_3}}{\left[R_2 \cdot R_3 + R_1 \cdot R_3 + R_1 \cdot R_2\right] \cdot \dfrac{1}{R_1 \cdot R_2 \cdot R_3}}$$

$$R_T = \frac{1}{\dfrac{R_2 \cdot R_3}{R_1 \cdot R_2 \cdot R_3} + \dfrac{R_1 \cdot R_3}{R_1 \cdot R_2 \cdot R_3} + \dfrac{R_1 \cdot R_2}{R_1 \cdot R_2 \cdot R_3}}$$

$$R_T = \frac{1}{\dfrac{1}{R_1} + \dfrac{1}{R_2} + \dfrac{1}{R_3}}$$

49. A complex fraction is one that has a fractional expression in its numerator or denominator or both.

51.
$$\frac{\left|\dfrac{-3}{9}\right| - \left[\dfrac{-5}{9}\right]\left|\dfrac{-3}{8}\right|}{|-5 - (-3)|}$$

$$= \frac{\dfrac{3}{9} + \dfrac{15}{72}}{2}$$

Multiply by the LCM 72.

51. cont.

$$\frac{72\left[\dfrac{3}{9}\right] + 72\left[\dfrac{15}{72}\right]}{2(72)}$$

$$= \frac{24 + 15}{144} = \frac{39}{144} = \frac{13}{48}$$

53. $6y - 3x < 12$
Graph the line $6y - 3x = 12$ (which is $2y - x = 4$) using a dashed line.

| x | y |
|---|---|
| 0 | 2 |
| -4 | 0 |

Use (0,0) as a check point.
Since $6(0) - 3(0) < 12$, shade the region which contains (0,0)

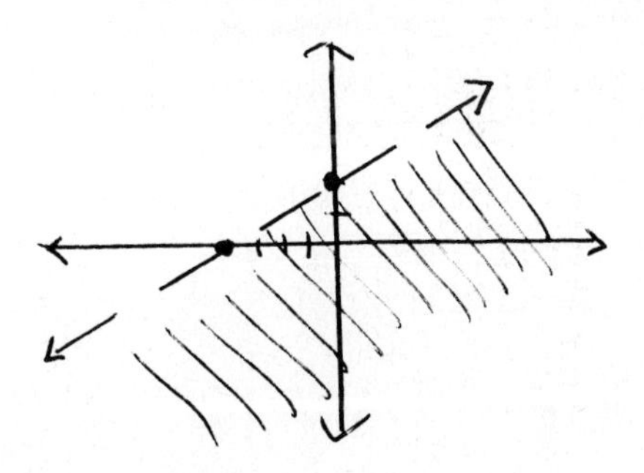

JUST FOR FUN

1.

$$\frac{1}{2a + \dfrac{1}{2a + \dfrac{1}{2a}}}$$

Beginning with the fraction, $\dfrac{1}{2a + \dfrac{1}{2a}}$ multiply by the LCM 2a.

$$\frac{1}{2a + \dfrac{1(2a)}{2a(2a) + \dfrac{1(2a)}{2a}}}$$

1. cont.

$$\frac{1}{2a + \dfrac{2a}{4a^2 + 1}}$$

Multiply again by the LCM $4a^2 + 1$

$$\frac{1\left[4a^2 + 1\right]}{2a\left[4a^2 + 1\right] + \dfrac{(2a)\left[4a^2 + 1\right]}{4a^2 + 1}}$$

$$= \frac{4a^2 + 1}{8a^3 + 4a}$$

EXERCISE SET 7.5

1. $\dfrac{2}{5} = \dfrac{x}{10}$

$\dfrac{20}{5} = x$

$x = 4$

3. $\dfrac{x}{8} = \dfrac{-15}{4}$

$x = \dfrac{-15(8)}{4}$

$x = -15(2) = -30$

5. $\dfrac{9}{3b} = \dfrac{-6}{2}$

$18 = -18b$
$b = -1$

7. $\dfrac{4x + 5}{6} = \dfrac{7}{2}$

7. cont.

$2(4x + 5) = 42$
$8x + 10 = 42$
$8x = 32$
$x = 4$

9. $\dfrac{6x + 7}{10} = \dfrac{2x + 9}{6}$

$6(6x + 7) = 10(2x + 9)$
$36x + 42 = 20x + 90$
$16x = 48$
$x = 3$

11. $\dfrac{x}{3} - \dfrac{3x}{4} = \dfrac{1}{12}$

Multiply by the LCM 12.

$12 \cdot \dfrac{x}{3} - 12 \cdot \dfrac{3x}{4} = 12 \cdot \dfrac{1}{12}$

$4x - 9x = 1$
$-5x = 1$
$x = -1/5$

13. $\dfrac{3}{4} - x = 2x$

Multiply by the LCM 4.

$4 \cdot \dfrac{3}{4} - 4x = 4 \cdot 2x$

$3 - 4x = 8x$
$3 = 12x$
$x = 1/4$

15. $\dfrac{5}{3x} + \dfrac{3}{x} = 1$

Multiply by the LCM 3x.

$3x \cdot \dfrac{5}{3x} + 3x \cdot \dfrac{3}{x} = 3x \cdot 1$

$5 + 9 = 3x$
$14 = 3x$
$x = 14/3$

17. $$\frac{x - 1}{x - 5} = \frac{4}{x - 5}$$

Since the denominators are the same, we can conclude that the numerators must also be equal.

$x - 1 = 4$
$x = 5$

But $x = 5$ would give a zero denominator. Therefore, there is no solution.

19. $$\frac{5y - 3}{7} = \frac{15y - 2}{28}$$

Cross multiply:

$28(5y - 3) = 7(15y - 2)$
$140y - 84 = 105y - 14$
$35y = 70$
$y = 2$

21. $$\frac{5}{-x - 6} = \frac{2}{x}$$

Cross multiply:

$5x = 2(-x-6)$
$5x = -2x - 12$
$7x = -12$
$x = -12/7$

23. $$\frac{x - 2}{x + 4} = \frac{x + 1}{x + 10}$$

Cross multiply:

$(x - 2)(x + 10) = (x + 1)(x + 4)$
$x^2 + 8x - 20 = x^2 + 5x + 4$
$8x - 20 = 5x + 4$
$3x - 20 = 4$
$3x = 24$
$x = 8$

25. $$x - \frac{4}{3x} = \frac{-1}{3}$$

Multiply by the LCM 3x.

$$3x \cdot x - 3x \cdot \frac{4}{3x} = 3x \cdot \frac{-1}{3}$$

$3x^2 + x - 4 = 0$

25. cont.

$(3x + 4)(x - 1) = 0$
$3x + 4 = 0 \quad x - 1 = 0$
$x = -4/3 \quad x = 1$

27. $$\frac{2x - 1}{3} - \frac{3x}{4} = \frac{5}{6}$$

Multiply by the LCM 12.

$$12 \cdot \frac{2x - 1}{3} - 12 \cdot \frac{3x}{4} = 12 \cdot \frac{5}{6}$$

$4(2x - 1) - 3(3x) = 2(5)$
$8x - 4 - 9x = 10$
$-x - 4 = 10$
$-x = 14$
$x = -14$

29. $$x + \frac{6}{x} = -5$$

Multiply by the LCM x.

$$x \cdot x + x \cdot \frac{6}{x} = x \cdot (-5)$$

$x^2 + 6 = -5x$

$x^2 + 5x + 6 = 0$

$(x + 3)(x + 2) = 0$
$x + 3 = 0 \quad x + 2 = 0$
$x = -3 \quad x = -2$

31. $$\frac{3y - 2}{y + 1} = 4 - \frac{y + 2}{y - 1}$$

Multiply by the LCM $(y + 1)(y - 1)$.

$$(y + 1)(y - 1)\left[\frac{3y - 2}{y + 1}\right]$$

$$= 4(y + 1)(y - 1) - (y + 1)(y - 1) \cdot \frac{y + 2}{y - 1}$$

$(y - 1)(3y - 2) = 4(y + 1)(y - 1)$
$\quad -(y + 1)(y + 2)$

$3y^2 - 5y + 2 = 4y^2 - 4 - y^2 - 3y - 2$

$3y^2 - 5y + 2 = 3y^2 - 3y - 6$

31. cont.

$-5y + 2 = -3y - 6$
$2 = 2y - 6$
$8 = 2y$
$y = 4$

33. $$\frac{1}{x + 3} + \frac{1}{x - 3} = \frac{-5}{x^2 - 9}$$

$$\frac{1}{x + 3} + \frac{1}{x - 3} = \frac{-5}{(x + 3)(x - 3)}$$

Multiply by the LCM $(x + 3)(x - 3)$

$$(x + 3)(x - 3) \cdot \frac{1}{x + 3} + (x + 3)(x - 3) \cdot \frac{1}{x - 3} = (x + 3)(x - 3) \cdot \frac{1}{(x + 3)(x - 3)}$$

$(x - 3) + (x + 3) = -5$
$2x = -5$
$x = -5/2$

35. $$\frac{2}{x - 3} - \frac{4}{x + 3} = \frac{8}{x^2 - 9}$$

$$\frac{2}{x - 3} - \frac{4}{x + 3} = \frac{8}{(x + 3)(x - 3)}$$

Multiply by the LCM $(x + 3)(x - 3)$

$$(x + 3)(x - 3) \cdot \frac{2}{x - 3} - (x + 3)(x - 3) \cdot \frac{4}{x + 3} = (x + 3)(x - 3) \cdot \frac{8}{(x + 3)(x - 3)}$$

$2(x + 3) - 4(x - 3) = 8$
$2x + 6 - 4x + 12 = 8$
$-2x + 18 = 8$
$-2x = -10$
$x = 5$

37. $$\frac{y}{2y + 2} + \frac{2y - 16}{4y + 4} = \frac{2y - 3}{y + 1}$$

$$\frac{y}{2y + 2} + \frac{2(y - 8)}{4(y + 1)} = \frac{2y - 3}{y + 1}$$

$$\frac{y}{2(y + 1)} + \frac{y - 8}{2(y + 1)} = \frac{2y - 3}{y + 1}$$

$$\frac{y + y - 8}{2(y + 1)} = \frac{2y - 3}{y + 1}$$

$$\frac{2y - 8}{2(y + 1)} = \frac{2y - 3}{y + 1}$$

$$\frac{y - 4}{y + 1} = \frac{2y - 3}{y + 1}$$

Since the denominators equal, we may assume the numerators are also equal.

$y - 4 = 2y - 3$
$-4 = y - 3$
$-1 = y$

But $y = -1$ would give zero denominators in the original equation. Therefore, there is no solution.

39. $$\frac{1}{x + 2} - \frac{1}{x - 2} = \frac{4}{x^2 - 4}$$

$$\frac{1}{x + 2} - \frac{1}{x - 2} = \frac{4}{(x - 2)(x + 2)}$$

Multiply by the LCM $(x - 2)(x + 2)$.

$$(x - 2)(x + 2) \cdot \frac{1}{x + 2} - (x - 2)(x + 2) \cdot \frac{1}{x - 2} = (x - 2)(x + 2) \cdot \frac{4}{(x - 2)(x + 2)}$$

$(x - 2) + (x + 2) = 4$
$2x = 4$
$x = 2$

There is no solution since $x = 2$ would result in zero denominators in the original equation.

41. Factor each of the denominators:

$$\frac{5}{x^2 + 4x + 3} + \frac{2}{x^2 + x - 6} = \frac{3}{x^2 - x - 2}$$

$$\frac{5}{(x + 3)(x + 1)} + \frac{2}{(x + 3)(x - 2)}$$

$$= \frac{3}{(x - 2)(x - 1)}$$

Multiply by the LCM $(x + 3)(x + 1)(x - 2)$

$$\frac{5(x + 3)(x + 1)(x - 2)}{(x + 3)(x + 1)}$$

$$+ \frac{2(x + 3)(x + 1)(x - 2)}{(x + 3)(x - 2)}$$

$$= \frac{3(x + 3)(x + 1)(x - 2)}{(x - 2)(x - 1)}$$

$5(x - 2) + 2(x + 1) = 3(x + 3)$
$5x - 10 + 2x + 2 = 3x + 9$
$7x - 8 = 3x + 9$
$4x = 17$
$x = 17/4$

43. $\frac{6x}{4} = \frac{6}{x}$

Cross multiply:
$6x^2 = 24$
$x^2 = 4$
$x = 2$
and
$6x = 12$
The missing lengths are 12 and 2.

45. $\frac{8}{2x + 10} = \frac{x + 3}{6}$

$\frac{4}{x + 5} = \frac{x + 3}{6}$

Cross multiply:

45. cont.

$24 = (x + 5)(x + 3)$

$24 = x^2 + 8x + 15$

$x^2 + 8x - 9 = 0$

$(x + 9)(x - 1) = 0$
$x + 9 = 0$ or $x - 1 = 0$
$x = -9$ $x = 1$
Since x cannot be a negative number, $x = 1$,
$x + 3 = 4$ and $2x + 10 = 12$
The lengths of the missing sides are 4 and 12.

47. $d = \frac{fL}{f + w}$ Solve for w.

Cross multiply

$d(f + w) = fL$
$df + dw = fL$
$dw = fL - df$
$w = \frac{fL - df}{d}$

49. $\frac{1}{p} + \frac{1}{q} = \frac{1}{f}$ ◻ Solve for p.

Multiply by the LCD pqf.

$pqf \cdot \frac{1}{p} + pqf \cdot \frac{1}{q} = pqf \cdot \frac{1}{f}$

$qf + pf = pq$
$qf = pq - pf$
$qf = p(q - f)$

$\frac{qf}{q - f} = p$

51. $\frac{1}{R_T} = \frac{1}{R_1} + \frac{1}{R_2}$ ◻ Solve for $R_1$

Multiply by the LCM $R_T \cdot R_1 \cdot R_2$

$$R_T \cdot R_1 \cdot R_2 \cdot \left[\frac{1}{R_T} = \frac{1}{R_1} + \frac{1}{R_2}\right]$$

51. cont.

$$R_1 \cdot R_2 = R_T \cdot R_2 + R_T \cdot R_1$$

$$R_1 \cdot R_2 - R_T \cdot R_1 = R_T \cdot R_2$$

$$R_1 \cdot \left[R_2 - R_T\right] = R_T \cdot R_2$$

$$R_1 = \frac{R_T \cdot R_2}{R_2 - R_T}$$

53.

$$z = \frac{\bar{x} - \mu}{\frac{\sigma}{\sqrt{n}}} \qquad \text{Solve for } \bar{x}.$$

$$z \cdot \left[\frac{\sigma}{\sqrt{n}}\right] = \left[\bar{x} - \mu\right]$$

$$z \cdot \left[\frac{\sigma}{\sqrt{n}}\right] + \mu = \bar{x}$$

55.

$$E = \frac{q}{\varepsilon_o \cdot A} - \frac{q'}{\varepsilon_o \cdot A}$$

Solve for $q'$.

$$E = \frac{q - q'}{\varepsilon_o \cdot A}$$

$$E\varepsilon_o \cdot A = q - q'$$

$$E\varepsilon_o \cdot A - q = -q'$$

$$q - E\varepsilon_o \cdot A = q'$$

57.

$$\frac{1}{C_T} = \frac{1}{C_1} + \frac{1}{C_2} + \frac{1}{C_3}$$

57. cont.

Solve for $C_T$

Multiply by the LCM $C_T \cdot C_1 \cdot C_2 \cdot C_3$ □

$$C_T \cdot C_1 \cdot C_2 \cdot C_3 \cdot \left[\frac{1}{C_T} = \frac{1}{C_1} + \frac{1}{C_2} + \frac{1}{C_3}\right]$$

$$C_1 \cdot C_2 \cdot C_3 = C_T \cdot C_2 \cdot C_3 + C_T \cdot C_1 \cdot C_3$$

$$+ C_T \cdot C_1 \cdot C_2$$

$$C_1 \cdot C_2 \cdot C_3 = C_T \cdot \left[C_2 \cdot C_3 + C_1 \cdot C_3 + C_1 \cdot C_2\right]$$

$$\frac{C_1 \cdot C_2 \cdot C_3}{C_2 \cdot C_3 + C_1 \cdot C_3 + C_1 \cdot C_2} = C_T$$

59.

$$\frac{1}{12} + \frac{1}{q} = \frac{1}{6} \qquad \text{Solve for } q.$$

Multiply by the LCM 12q.

$$12q \cdot \frac{1}{12} + 12q \cdot \frac{1}{q} = 12q \cdot \frac{1}{6}$$

$$q + 12 = 2q$$
$$12 = q$$

61.

$$\frac{1}{p} + \frac{1}{15} = \frac{1}{8}$$

Solve for p.
Multiply by the LCM 120p.

$$120p \cdot \frac{1}{p} + 120p \cdot \frac{1}{15} = 120p \cdot \frac{1}{8}$$

$$120 + 8p = 15p$$
$$120 = 7p$$
$$p = 17.14\text{cm}.$$

63. Let x = the object's distance and 3x = the image distance.

$$\frac{1}{x} + \frac{1}{3x} = \frac{1}{12}$$

Multiply by the LCM 12x.

$$12x \cdot \frac{1}{x} + 12x \cdot \frac{1}{3x} = 12x \cdot \frac{1}{12}$$

$12 + 4 = x$
$16 = x$
$3x = 48$
The object's distance is 16cm and the image distance is 48cm.

65. $$\frac{1}{R_T} = \frac{1}{200} + \frac{1}{700}$$

Multiply by the LCM $1400R_T$ □

$$1400R_T \cdot \frac{1}{R_T} = 1400R_T \cdot \frac{1}{200} + 1400R_T \cdot \frac{1}{700}$$

$1400 = 7R_T + 2R_T$

$1400 = 9R_T$ □

155.6OHMS $= R_T$

67. $$\frac{1}{R_T} = \frac{1}{300} + \frac{1}{500} + \frac{1}{3000}$$

Multiply by the LCM $3000R_T$ □

$$3000R_T \cdot \left[\frac{1}{R_T} = \frac{1}{300} + \frac{1}{500} + \frac{1}{3000}\right]$$

$3000 = 10R_T + 6R_T + R_T$

$3000 = 17R_T$

176.47ohms $= R_T$

69. An extraneous root is a number obtained when solving an equation that is not a true solution to the original equation.

71. a) Multiply both sides of the equation by the LCD 12. This removes the fractions.

b) $$12 \cdot \frac{x}{4} + 12 \cdot \frac{x}{3} = 2 \cdot 12$$

$3x - 4x = 24$
$-x = 24$
$x = -24$

c) Write each term with the LCM 12. This step allows the fractions to be added and subtracted.

$$\frac{x}{4} - \frac{x}{3} + 2 = \frac{3x}{12} - \frac{4x}{12} + \frac{24}{12}$$

$$= \frac{-x + 24}{12}$$

73. $$f(x) = \frac{1}{2} \cdot x^2 - 3x + 4$$

$$f(5) = \frac{1}{2} \cdot 5^2 - 3(5) + 4$$

$$= \frac{25}{2} - 15 + 4$$

$$= \frac{25}{2} - 11$$

$$= \frac{25}{2} - \frac{22}{2} = \frac{3}{2}$$

75. $$f(x) = x^2 - 4x - 6$$

a = 1. The parabola opens up.
Axis of symmetry: b = -4

$$x = \frac{-(-4)}{2(1)} = 2$$

Vertex:

$f(2) = 2^2 - 4(2) - 6 = -10$
Vertex: (2,-10)

75. cont.

Domain: All reals
Range: $\{y \mid y \geq -10\}$

| x | y |
|---|---|
| 0 | -6 |
| 1 | -9 |
| 2 | -10 |
| 3 | -9 |
| 4 | -6 |

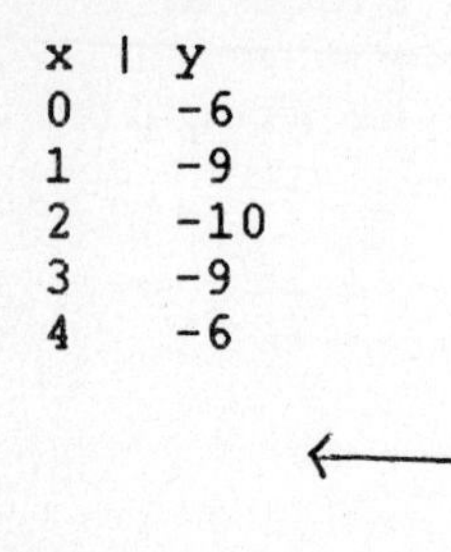

JUST FOR FUN

1. a)

$$T_a = \frac{.0601}{1 - (.33 + (.046 + .03)(1 - .33))}$$

$$T_a = \frac{.0601}{1 - (.33 + (.076)(.67))}$$

$$T_a = \frac{.0601}{1 - (.33 + .05092)}$$

$$T_a = \frac{.0601}{1 - .38092} = .0970795$$

$$T_a \approx 9.71\%$$

b) Mr. Levy should choose the Tax Free Money Market since 9.71% > 7.68%.

EXERCISE SET 7.6

1. Let x = the length of time it will take the computers to do the job if they work together.

$$\frac{x}{5} + \frac{x}{4} = 1$$

$$20 \cdot \frac{x}{5} + 20 \cdot \frac{x}{4} = 20 \cdot 1$$

1. cont.

$4x + 5x = 20$
$9x = 20$
$x = 2.22$hours.

3. Let x = the time needed to fill the pool if the pumps were working together.

$$\frac{x}{8} + \frac{x}{5} = 1$$

$$40 \cdot \frac{x}{8} + 40 \cdot \frac{x}{5} = 40 \cdot 1$$

$5x + 8x = 40$
$13x = 40$
$x = 3.08$ hours.

5. Let x = the length of time needed to fill the vat.

$$\frac{x}{20} - \frac{x}{25} = 1$$

$$100 \cdot \frac{x}{20} - 100 \cdot \frac{x}{25} = 100 \cdot 1$$

$5x - 4x = 100$
$x = 100$ hours

7. Let x = the length of time need for the smaller steam roller to do the job alone.

$$\frac{10}{16} + \frac{10}{x} = 1$$

$$16x \cdot \frac{10}{16} + 16x \cdot \frac{10}{x} = 16x \cdot 1$$

$10x + 160 = 16x$
$160 = 6x$
$x = 26.67$ hours

9. Let x = the amount of time needed for the student to do the job alone.

$$\frac{3.2}{5.7} + \frac{3.2}{x} = 1$$

9. cont.

$$5.7x \cdot \frac{3.2}{5.7} + 5.7x \cdot \frac{3.2}{x} = 5.7x \cdot 1$$

$3.2x + 18.24 = 5.7x$
$18.24 = 2.5x$
$x = 7.3$ hours

11. Let $x$ = the time to fill the tank.

$$\frac{x}{10} + \frac{x}{12} - \frac{x}{15} = 1$$

$$60 \cdot \frac{x}{10} + 60 \cdot \frac{x}{12} - 60 \cdot \frac{x}{15} = 60 \cdot 1$$

$6x + 5x - 4x = 60$
$7x = 60$
$x = 8.57$ hours

13. Let $x$ = the length of time it will take Nancy to do the job.

$$\frac{12}{x} + \frac{12}{2x} = 1$$

$$2x \cdot \frac{12}{x} + 2x \cdot \frac{12}{2x} = 2x \cdot 1$$

$24 + 12 = 2x$
$36 = 2x$
$x = 18$ hours

15. Let $x$ = the length of time needed for the smaller pipe to complete the job.

$$\frac{20}{60} + \frac{x}{80} = 1$$

$$\frac{1}{3} + \frac{x}{80} = 1$$

$$240 \cdot \frac{1}{3} + 240 \cdot \frac{x}{80} = 240 \cdot 1$$

15. cont.

$80 + 3x = 240$
$3x = 160$
$x = 53.33$ hours

17. Let $x$ = the unknown number.

$$\frac{3 + x}{2x} = \frac{1}{8}$$

$8(3 + x) = 2x$
$24 + 8x = 2x$
$24 = -6x$
$-4 = x$

19. Let $x$ and $(x + 1)$ represent the unknown numbers.

$$\frac{1}{x} + \frac{1}{x + 1} = \frac{11}{30}$$

$$30x(x + 1) \cdot \left[\frac{1}{x} + \frac{1}{x + 1} = \frac{11}{30}\right]$$

$$30(x + 1) + 30x = 11x(x + 1)$$

$$30x + 30 + 30x = 11x^2 + 11x$$

$$0 = 11x^2 - 49x - 30$$

$$0 = (11x + 6)(x - 5)$$

$11x + 6 = 0$ $\quad$ $x - 5 = 0$
$x = -6/11$ $\quad$ $x = 5$
Since $x$ is an integer, $x = 5$ and $x + 1 = 6$ are the two numbers.

21. Let $x$ = the unknown number.

$$\frac{5 + x}{7 + x} = \frac{4}{5}$$

$5(5 + x) = 4(7 + x)$
$25 + 5x = 28 + 4x$
$x = 3$

23. Let x = the unknown number.

$$\frac{1}{x - 3} = 2\left[\frac{1}{2x - 6}\right]$$

$$\frac{1}{x - 3} = \frac{1}{x - 3}$$

$x - 3 = x - 3$
$-3 = -3$

All reals except $x = 3$ becuase $x = 3$ would yield a zero denominator.

25. Let x = the number.

$$2 \cdot \left[\frac{1}{x^2}\right] - 3\left[\frac{1}{x}\right] = -1$$

$$2x^2 \cdot \left[\frac{1}{x^2}\right] - 3x^2 \cdot \left[\frac{1}{x}\right] = -x^2$$

$$2 - 3x = -x^2$$

$$x^2 - 3x + 2 = 0$$

$(x - 2)(x - 1) = 0$
$x - 2 = 0 \quad x - 1 = 0$
$x = 2 \quad x = 1$
Both 2 and 1 are solutions to this problem.

27. Let x = the speed of the current.

$$\frac{3}{20 + x} = \frac{2}{20 - x}$$

$3(20 - x) = 2(20 + x)$
$60 - 3x = 40 + 2x$
$20 = 5x$
4 mph = x

29. Let x = the distance between the starting and resting point.

$$\frac{x}{8} + \frac{1}{2} = \frac{x}{6}$$

29. cont.

$$24\left[\frac{x}{8}\right] + 24\left[\frac{1}{2}\right] = 24\left[\frac{x}{6}\right]$$

$3x + 12 = 4x$
12 miles = x

31. a) Let x = the distance Ray can go down stream.

$$\frac{x}{25} + \frac{x}{15} = 4$$

$$300 \cdot \frac{x}{25} + 300 \cdot \frac{x}{15} = 300 \cdot 4$$

$12x + 20x = 1200$
$32x = 1200$
x = 37.5 miles

b) Ray will need to turn back after traveling 37.5 miles. Therefore, since distance = rate x time
$37.5 = (20 + 5)t$
$1.5 = t$
Ray can travel the distance in 1.5 hours = 1 hour, 30 min. Ray must turn back at 9:30am.

33. Let x = the walking rate and 4x = the cycling rate.

$$\frac{30}{4x} + \frac{2}{x} = 6$$

$$4x \cdot \frac{30}{4x} + 4x \cdot \frac{2}{x} = 4x \cdot 6$$

$30 + 8 = 24x$
$38 = 24x$
1.58 mph= x
4x = 6.33 mph

The cycling rate is 6.33 mph.

35. Let x = the speed of the train and 5x = the speed of the plane.

$$\frac{900}{5x} + 12 = \frac{900}{x}$$

$$5x \cdot \frac{900}{5x} + 5x \cdot 12 = 5x \cdot \frac{900}{x}$$

35. cont.

$900 + 60x = 4500$
$60x = 3600$
$x = 60$ mph
$5x = 300$ mph

The speed of the train is 60 mph and the speed of the plane is 300 mph.

37. Let x = the distance between the space station and NASA headquarters.

$$\frac{x}{20000} + .6 = \frac{x}{18000}$$

$$4500000 \cdot \frac{x}{20000} + (4500000) \cdot .6 = 4500000 \cdot \frac{x}{18000}$$

$225x + 2700000 = 250x$
$2700000 = 25x$
$108000$ miles $= x$

39.

$$\begin{array}{r} 4x^2 - 6x - 1 \\ 3x - 4 \\ \hline -16x^2 + 24x + 4 \\ 12x^3 - 18x^2 - 3x \\ \hline 12x^3 - 34x^2 + 21x + 4 \end{array}$$

41. $8x^2 + 26x + 15$

$20(6) = 120$
and
$20 + 6 = 26$
Therefore:

$8x^2 + 26x + 15$

$= 8x^2 + 20x + 6x + 15$

$= 4x(2x + 5) + 3(2x + 5)$

$= (4x + 3)(2x + 5)$

EXERCISE SET 7.7

1. Direct
3. Direct
5. Direct
7. Inverse
9. Direct
11. Direct
13. Inverse
15. Direct

17. $C = kz^2$

$$= \frac{3}{4}\left[9^2\right] = 60.75$$

19. $x = \frac{k}{y}$

$$= \frac{5}{25} = 0.2$$

21.

$$L = \frac{k}{p^2}$$

$$= \frac{100}{4^2} = 6.25$$

23.

$$A = \frac{kR_1 \cdot R_2}{L^2}$$

$$= \left[\frac{3}{2}\right] \cdot (120) \cdot \frac{8}{5^2} = 57.6$$

25. $x = ky$
$9 = 18k$
$1/2 = k$

$$x = \frac{1}{2} \cdot 36$$

$= 18$

27. $y = kR^2$

$5 = k(5)^2$
$k = .2$

$y = .2(R)^2$

$y = .2(10)^2$
$y = 20$

29.

$$C = \frac{k}{J}$$

$$7 = \frac{k}{0.7}$$

$$4.9 = k$$

$$C = \frac{4.9}{J}$$

$$C = \frac{4.9}{12} = .41$$

31.

$$F = \frac{k \cdot m_1 \cdot m_2}{d}$$

$$20 = \frac{k(5)(10)}{.2}$$

$$k = .08$$

$$F = \frac{.08 \cdot m_1 \cdot m_2}{d}$$

$$F = \frac{.08 \cdot (10) \cdot (20)}{.4}$$

$$F = 40$$

33.

$$S = kIT^2$$

$$8 = k(20)4^2$$

$$k = .025$$

$$S = .025IT^2$$

$$S = (.025)(2)\left[2^2\right]$$

$$S = 0.2$$

35. $S = kF$
$1.4 = k(20)$
$k = .07$

$S = .07F$
$S = .07(10) = .7\text{in.}$

37.

$$I = \frac{k}{d^2}$$

$$20 = \frac{k}{15^2}$$

$$4500 = k$$

$$I = \frac{4500}{d^2}$$

$$I = \frac{4500}{(12)^2}$$

$I = 31.25$ foot candle

39.

$$w = \frac{k}{d^2}$$

$$140 = \frac{k}{(4000)^2}$$

$$2{,}240{,}000{,}000 = k$$

$$w = \frac{2240000000}{d^2}$$

$$w = \frac{2240000000}{(4100)^2}$$

$w = 133.25$ lbs

41.

$$R = \frac{kL}{A}$$

$$.2 = \frac{k(200)}{.05}$$

$$k = .00005$$

$$R = \frac{.00005L}{A}$$

$$R = \frac{.00005(5000)}{.01} = 25 \text{ ohms}$$

43. $\dfrac{d - 1}{-4 - 5} = \dfrac{2}{3}$

$\dfrac{d - 1}{-9} = \dfrac{2}{3}$

$3(d - 1) = -18$
$3d - 3 = -18$
$3d = -15$
$d = -5$

45. $|x - 2| < 4$

$-4 < x - 2 < 4$
$-2 < x < 6$

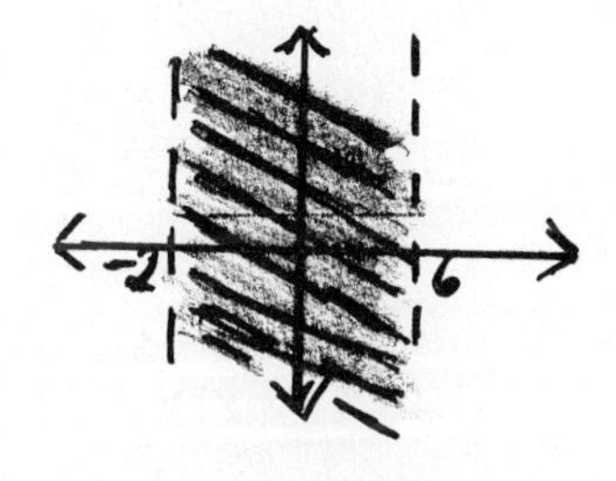


JUST FOR FUN

Let I = the intensity of illumination and d = the distance the subject is from the flash.

$I = \dfrac{k}{d^2}$

$\dfrac{1}{16} = \dfrac{k}{4^2}$

$1 = k$

$I = \dfrac{1}{d^2}$

$I = \dfrac{1}{3^2} = \dfrac{1}{9}$

REVIEW EXERCISES

1. $\dfrac{3}{x - 4}$

To find the domain, set the denominator equal to 0 and solve for x. The domain is all reals except that value.

$x - 4 = 0$
$x = 4$

Domain: $\{x|\ x \neq 4\}$

3. $\dfrac{-2x}{x^2 + 5}$

There is not a real value such that $x^2 + 5 = 0$ □ Therefore, the domain is all reals.

5. $\dfrac{x + 6}{x^2}$

Domain: $\{x|x \neq 0\}$

7. $\dfrac{x^2 + xy}{x + y} = \dfrac{x(x + y)}{x + y} = x$

9. $\dfrac{4 - 5x}{5x - 4} = \dfrac{-1(5x - 4)}{x - 4} = -1$

11. $\dfrac{2x^2 - 6x + 5x - 15}{2x^2 + 7x + 5}$

$= \dfrac{(2x + 5)(x - 3)}{(2x + 5)(x + 1)} = \dfrac{x - 3}{x + 1}$

13. $\dfrac{6x}{x + 1} - \dfrac{3}{x}$

The LCD is $x(x + 1)$

15. $\dfrac{19x - 5}{x^2 + 2x - 35} + \dfrac{3x - 2}{x^2 + 9x + 14}$

$= \dfrac{19x - 5}{(x + 7)(x - 5)} + \dfrac{3x - 2}{(x + 7)(x - 2)}$

The LCD = $(x + 7)(x - 5)(x + 2)$

17. $\dfrac{15x^2 \cdot y^3}{3z} \cdot \dfrac{6z^3}{5xy^3}$

$= \dfrac{90\, x^2 \cdot y^3}{15\, z} \cdot \dfrac{z^3}{xy^3}$

$= \dfrac{6x^2 \cdot y^3 \cdot z^2}{xy^3 \cdot z} = 6xz^2$

19. $\dfrac{8xy^2}{z} \div \dfrac{x^4 \cdot y^2}{4z^2}$

$= \dfrac{8xy^2}{z} \cdot \dfrac{4z^2}{x^4 \cdot y^2}$

$= \dfrac{32xy^2 \cdot z^2}{x^4 \cdot y^2 \cdot z} = \dfrac{32z}{x^3}$

21. $\dfrac{4x + 4y}{x^2 \cdot y} \cdot \dfrac{y^3}{8x}$

$= \dfrac{4(x + y)\left[y^3\right]}{8x^3 \cdot y}$

$= \dfrac{(x + y) \cdot y^3}{2x^3 \cdot y}$

$= \dfrac{(x + y)y^2}{2x^3}$

23. $\dfrac{a - 2}{a + 3} \cdot \dfrac{a^2 + 4a + 3}{a^2 - a - 2}$

$= \dfrac{(a - 2)(a + 3)(a + 1)}{(a + 3)(a - 2)(a + 1)} = 1$

25. $\dfrac{6x^2 - 4x}{2x - 3} - \dfrac{-3x + 12}{2x - 3} - \dfrac{2x + 4}{2x - 3}$

$= \dfrac{6x^2 - 4x + 3x - 12 - 2x - 4}{2x - 3}$

$= \dfrac{6x^2 - 3x - 16}{2x - 3}$

27. $6 + \dfrac{x}{x + 2}$

$= \dfrac{6(x + 2)}{x + 2} + \dfrac{x}{x + 2}$

$= \dfrac{6x + 12 + x}{x + 2}$

$= \dfrac{7x + 12}{x + 2}$

29. $\dfrac{x^2 - y^2}{x - y} \cdot \dfrac{x + y}{xy + x^2}$

$= \dfrac{(x + y)(x - y)(x + y)}{(x - y)(x)(y + x)}$

$= \dfrac{x + y}{x}$

31.

$\dfrac{6x^2 - 4x}{2x - 3} - \dfrac{-x + 4}{2x - 3} - \dfrac{6x^2 + x - 2}{2x - 3}$

$= \dfrac{6x^2 - 4x + x - 4 - 6x^2 - x + 2}{2x - 3}$

$= \dfrac{-4x - 2}{2x + 3}$

33. $\dfrac{4x^2 + 8x - 5}{2x + 5} \cdot \dfrac{x + 1}{4x^2 - 4x + 1}$

$= \dfrac{(2x + 5)(2x - 1)(x + 1)}{(2x + 5)(2x - 1)(2x - 1)}$

$= \dfrac{x + 1}{2x - 1}$

35.

$\dfrac{x^2 - 3xy - 10y^2}{6x} \div \dfrac{x + 2y}{12x^2}$

35. cont.

$= \dfrac{(x - 5y)(x + 2y)}{6x} \cdot \dfrac{12x^2}{x + 2y}$

$= 2x(x - 5y)$

37. $\dfrac{x - 4}{x - 5} - \dfrac{3}{x + 5}$

LCD = $(x - 5)(x + 5)$

$\dfrac{(x - 4)(x + 5)}{(x - 5)(x + 5)} - \left[\dfrac{3(x - 5)}{(x + 5)(x - 5)}\right]$

$= \dfrac{x^2 + x - 20 - 3x + 15}{(x - 5)(x + 5)}$

$= \dfrac{x^2 - 2x - 5}{(x - 5)(x + 5)}$

39. $\dfrac{x + 3}{x^2 - 9} + \dfrac{2}{x + 3}$

$= \dfrac{x + 3}{(x + 3)(x - 3)} + \dfrac{2}{x + 3}$

LCD = $(x + 3)(x - 3)$

$= \dfrac{x + 3}{(x + 3)(x - 3)} + \dfrac{2(x - 3)}{(x + 3)(x - 3)}$

$= \dfrac{x + 3 + 2x - 6}{(x + 3)(x - 3)}$

$= \dfrac{3x - 3}{(x + 3)(x - 3)} = \dfrac{3(x - 1)}{(x + 3)(x - 3)}$

41. $\dfrac{4x^2 - 16y^2}{9} \div \dfrac{(x + 2y)^2}{12}$

$= \dfrac{4(x - 2y)(x + 2y)}{9} \cdot \dfrac{12}{(x + 2y)(x + 2y}$

$= \dfrac{16(x - 2y)}{3(x + 2y)}$

43.

$$\frac{2x^2 + 10x + 12}{(x + 2)^2} \cdot \frac{x + 2}{x^3 + 5x^2 + 6x}$$

$$= \left[\frac{2(x + 3)(x + 2)}{(x + 2)(x + 2)}\right] \cdot \frac{x + 2}{x(x + 3)(x + 2)}$$

$$= \frac{2}{x(x + 2)}$$

45.

$$\frac{x + 5}{x^2 - 15x + 50} - \frac{x - 2}{x^2 - 25}$$

$$= \frac{x + 5}{(x - 5)(x - 10)} - \frac{x - 2}{(x - 5)(x + 5)}$$

LCD = $(x - 5)(x + 5)(x - 10)$

$$= \frac{(x + 5)(x + 5)}{(x - 5)(x + 5)(x - 10)}$$

$$- \frac{(x - 2)(x - 10)}{(x - 5)(x + 5)(x - 10)}$$

$$= \frac{\left[x^2 + 10x + 25\right] - \left[x^2 - 12x + 20\right]}{(x - 5)(x + 5)(x - 10)}$$

$$= \frac{x^2 + 10x + 25 - x^2 + 12x - 20}{(x - 5)(x + 5)(x - 10)}$$

$$= \frac{22x + 5}{(x + 5)(x - 5)(x - 10)}$$

47.

$$\frac{1}{x + 3} - \frac{2}{x - 3} + \frac{6}{x^2 - 9}$$

47. cont.

$$= \frac{1}{x + 3} - \frac{2}{x - 3} + \frac{6}{(x + 3)(x - 3)}$$

LCD = $(x + 3)(x - 3)$

$$= \frac{1(x - 3)}{(x + 3)(x - 3)} - \frac{2(x + 3)}{(x - 3)(x + 3)}$$

$$+ \frac{6}{(x + 3)(x - 3)}$$

$$= \frac{x - 3 - 2x - 6 + 6}{(x + 3)(x - 3)}$$

$$= \frac{-x - 3}{(x + 3)(x - 3)} = -1 \cdot \frac{x + 3}{(x + 3)(x - 3)}$$

$$= \frac{-1}{x - 3}$$

49.

$$\frac{x - 4}{x - 5} - \frac{3}{x + 5} - \frac{10}{x^2 - 25}$$

$$= \frac{x - 4}{x - 5} - \frac{3}{x + 5} - \frac{10}{(x + 5)(x - 5)}$$

LCD = $(x + 5)(x - 5)$

$$= \frac{(x - 4)(x + 5)}{(x - 5)(x + 5)} - \frac{3(x - 5)}{(x + 5)(x - 5)}$$

$$- \frac{10}{(x + 5)(x - 5)}$$

$$= \frac{x^2 + x - 20 - 3x + 15 - 10}{(x + 5)(x - 5)}$$

$$= \frac{x^2 - 2x - 15}{(x + 5)(x - 5)}$$

49. cont.

$$= \frac{(x - 5)(x + 3)}{(x - 5)(x + 5)}$$

$$= \frac{x + 3}{x + 5}$$

51. $$\frac{x^2 - x - 56}{x^2 + 14x + 49} \cdot \frac{x^2 + 4x - 21}{x^2 - 9x + 8} + \left[\frac{3}{x^2 + 8x - 9}\right]$$

$$= \frac{(x - 8)(x + 7)}{(x + 7)(x + 7)} \cdot \frac{(x + 7)(x - 3)}{(x - 1)(x - 8)} + \left[\frac{3}{x^2 + 8x - 9}\right]$$

$$= \frac{x - 3}{x - 1} + \frac{3}{(x + 9)(x - 1)}$$

LCD = (x - 1)(x + 9)

$$= \frac{(x - 3)(x + 9)}{(x - 1)(x + 9)} + \frac{3}{(x + 9)(x - 1)}$$

$$= \frac{x^2 + 6x - 27 + 3}{(x + 9)(x - 1)}$$

$$= \frac{x^2 + 6x - 24}{(x - 1)(x + 9)}$$

53. $$\left[\frac{\frac{15xy}{6z}}{\frac{3x}{z^2}}\right] = \frac{15xy}{6z} \cdot \frac{z^2}{3x} = \frac{5yz}{6}$$

57. $$\frac{\frac{4}{x} + \frac{2}{x^2}}{6 - \frac{1}{x}}$$ Multiply by the LCM $x^2$

$$= \frac{x^2 \cdot \frac{4}{x} + x^2 \cdot \frac{2}{x^2}}{x^2 \cdot 6 - x^2 \cdot \frac{1}{x}}$$

$$= \frac{4x + 2}{6x^2 - x} = \frac{2(2x + 1)}{x(6x - 1)}$$

59. $$\frac{x^{-2} + \frac{1}{x}}{\frac{1}{x^2} - \frac{1}{x}} = \frac{\frac{1}{x^2} + \frac{1}{x}}{\frac{1}{x^2} - \frac{1}{x}}$$

Multiply by the LCM $x^2$

$$\left[\frac{x^2 \cdot \frac{1}{x^2} + x^2 \cdot \frac{1}{x}}{x^2 \cdot \frac{1}{x^2} - x^2 \cdot \frac{1}{x}}\right] = \frac{1 + x}{1 - x}$$

61. $$\frac{4}{a} = \frac{16}{4}$$

Cross multiply:

16 = 16a
a = 1

63. $\frac{x}{6} = \frac{x-4}{2}$

Cross multiply

$2x = 6(x - 4)$
$2x = 6x - 24$
$0 = 4x - 24$
$24 = 4x$
$x = 6$

65. $\frac{x}{5} + \frac{x}{2} = 14$

Multiply by the LCM 10.

$10 \cdot \frac{x}{5} + 10 \cdot \frac{x}{2} = (10) \cdot 14$

$2x + 5x = 140$
$7x = 140$
$x = 20$

67. $\frac{1}{x-2} + \frac{1}{x+2} = \frac{1}{x^2 - 4}$

$\frac{1}{x-2} + \frac{1}{x+2} = \frac{1}{(x-2)(x+2)}$

Multiply by the LCD $(x + 2)(x - 2)$.

$\frac{1(x-2)(x+2)}{x-2} + \frac{1(x-2)(x+2)}{x+2} = \frac{1}{(x-2)(x+2)}$

$(x + 2) + (x - 2) = 1$
$2x = 1$
$x = 1/2$

69. $\frac{x}{x^2 - 9} + \frac{2}{x+3} = \frac{4}{x-3}$

$\frac{x}{(x-3)(x+3)} + \frac{2}{x+3} = \frac{4}{x-3}$

Multiply by the LCD $(x + 3)(x - 3)$

69. cont.

$\frac{x(x+3)(x-3)}{(x+3)(x-3)} + \frac{2(x+3)(x-3)}{x+3} = \frac{4(x+3)(x-3)}{x-3}$

$x + 2(x - 3) = 4(x + 3)$
$3x - 6 = 4x + 12$
$-18 = x$

71. $\frac{V_1 \cdot P_1}{T_1} = \frac{V_2 \cdot P_2}{T_2}$

Solve for $P_2$

Cross multiply:

$\left[V_1 \cdot P_1 \cdot T_2\right] = \left[V_2 \cdot P_2 \cdot T_1\right]$

$\frac{V_1 \cdot P_1 \cdot T_2}{V_2 \cdot T_1} = P_2$

73. $\frac{1}{R_T} = \frac{1}{R_1} + \frac{1}{R_2}$ Solve for $R_2$

Multiply by the LCD: $R_T \cdot R_1 \cdot R_2$

$R_T \cdot R_1 \cdot R_2 \cdot \left[\frac{1}{R_T} = \frac{1}{R_1} + \frac{1}{R_2}\right]$

$R_1 \cdot R_2 = R_T \cdot R_2 + R_T \cdot R_1$

$R_1 \cdot R_2 - R_T \cdot R_2 = R_T \cdot R_1$

$R_2 \cdot \left[R_1 - R_T\right] = \left[R_T \cdot R_1\right]$

$R_2 = \frac{R_T \cdot R_1}{R_1 - R_T}$

75. Let $x$ = the resistance of the first resistor and $2x$ = the resistance of the second.
Using the formula:

$$\frac{1}{R_T} = \frac{1}{R_1} + \frac{1}{R_2}$$

we have

$$\frac{1}{600} = \frac{1}{x} + \frac{1}{2x}$$

$$600x \cdot \frac{1}{600} = 600x \cdot \frac{1}{x} + 600x \cdot \frac{1}{2x}$$

$x = 600 + 300$
$x = 900$ ohms
$2x = 1800$ ohms

77. Let $x$ = the object distance and $2x$ = the image distance. Using the formula:

$$\frac{1}{f} = \frac{1}{p} + \frac{1}{q}$$

we have:

$$\frac{1}{10} = \frac{1}{x} + \frac{1}{2x}$$

$$10x \cdot \frac{1}{10} = 10x \cdot \frac{1}{x} + 10x \cdot \frac{1}{2x}$$

$x = 10 + 5$
$x = 15$cm

The object's distance is 15cm.

79. Let $x$ = the time it takes Pete to edit the manuscript.

$$\frac{1}{75} + \frac{1}{x} = \frac{1}{40}$$

$$600x \cdot \frac{1}{75} + 600x \cdot \frac{1}{x} = 600x \cdot \frac{1}{40}$$

$8x + 600 = 15x$
$600 = 7x$
$x = 85.71$ hours.

81. Let $x$ equal the unknown number.

$$1 - \frac{1}{2x} = \frac{1}{3x}$$

$$6x \cdot 1 - 6x \cdot \frac{1}{2x} = 6x \cdot \frac{1}{3x}$$

$6x - 3 = 2$
$6x = 5$
$x = 5/6$

83. Let $x$ = the speed of the car and $3x$ = the speed of the plane. Since

$\frac{d}{r} = t$ we have:

$\frac{450}{x}$ = the time it took the car to make the trip.

$\frac{450}{3x}$ = the time it took the plane to make the trip.

$\frac{450}{3x} + 6$ also describes the car's time.

Therefore:

$$\frac{450}{3x} + 6 = \frac{450}{x}$$

$$3x \cdot \frac{450}{3x} + 3x \cdot 6 = 3x \cdot \frac{450}{x}$$

$450 + 18x = 1350$
$18x = 900$
$x = 50$
$3x = 150$

The speed of the car is 50 mph and the speed of the plane is 150 mph.

85.

$$A = kC^2$$

$$5 = k(5)^2$$

$$k = .2$$

$$A = .2C^2$$

$$A = .2(10)^2$$
$$= 20$$

87.

$$w = \frac{kL}{A}$$

$$80 = \frac{k(100)}{20}$$

$$k = 16$$

$$w = \frac{16L}{A}$$

$$w = \frac{16(50)}{40}$$

$$w = 20$$

89. Let x = the map distance that represents 300 miles.

$$\frac{1}{60} = \frac{x}{300}$$

Cross multiply:

$$300 = 60x$$
$$x = 5$$

Five inches on the map represents 300 actual miles.

91.

$$d = kt^2$$

$$16 = k(1)^2$$

$$k = 16$$

$$d = 16t^2$$

$$d = 16(5)^2 = 400 \text{ feet.}$$

93. Let w = the temperature of the water.

$$t = \frac{k}{w}$$

$$1.7 = \frac{k}{70}$$

$$k = 119$$

$$t = \frac{119}{w}$$

$$t = \frac{119}{50}$$

$$t = 2.38 \text{ min.}$$

PRACTICE TEST

1. The domain is found by setting the denominator equal to zero and solving. The domain is all reals except those found when setting the denominator equal to zero.

$$x^2 - 3x - 28 = 0$$
$$(x - 7)(x + 4) = 9$$
$$x - 7 = 0 \qquad x + 4 = 0$$
$$x = 7 \qquad x = -4$$

Domain: $\{x \mid x \neq -4, x \neq 7\}$

2.

$$\frac{x^2 - 5x - 36}{9 - x} = \frac{(x - 9)(x + 4)}{-1(x - 9)}$$

$$= -(x + 4) \text{ or } -x - 4$$

3.

$$\frac{6x^2y^4}{4z^2} \cdot \frac{8xz^3}{9y^4}$$

$$= \frac{48x^3y^4z^3}{36y^4z^2} = \frac{4x^3z}{3}$$

4. $$\frac{a^2 - 9a + 14}{a - 2} \cdot \frac{a^2 - 4a - 21}{(a - 7)^2}$$

$$= \frac{(a - 7)(a - 2)}{a - 2} \cdot \frac{(a - 7)(a + 3)}{(a - 7)(a - 7)}$$

$$= a + 3$$

5. $$\frac{x^2 - 9y^2}{3x + 6y} \div \frac{x + 3y}{x + 2y}$$

$$= \frac{(x + 3y)(x - 3y)}{(3)(x + 2y)} \cdot \frac{x + 2y}{x + 3y}$$

$$= \frac{x - 3y}{3}$$

6. $$\frac{x^3 + y^3}{x + y} \div \frac{x^2 - xy + y^2}{x^2 + y^2}$$

$$= \frac{(x + y)\left[x^2 - xy + y^2\right]}{x + y} \cdot \frac{x^2 + y^2}{x^2 - xy + y^2}$$

$$= x^2 + y^2$$

7. $$\frac{5}{x} + \frac{3}{2x^2} = \frac{5(2x)}{x(2x)} + \frac{3}{2x^2}$$

$$= \frac{10x + 3}{2x^2}$$

8. $$\frac{x - 5}{x^2 - 16} - \frac{x - 2}{x^2 + 2x - 8}$$

$$= \frac{x - 5}{(x + 4)(x - 4)} - \frac{x - 2}{(x + 4)(x - 2)}$$

8. cont.

The LCD = $(x + 4)(x - 4)(x - 2)$

$$= \frac{(x - 5)(x - 2)}{(x + 4)(x - 4)(x - 2)} - \frac{(x - 2)(x - 4)}{(x + 4)(x - 4)(x - 2)}$$

$$= \frac{x^2 - 7x + 10 - x^2 + 6x - 8}{(x + 4)(x - 4)(x - 2)}$$

$$= \frac{-x + 2}{(x + 4)(x - 4)(x - 2)}$$

$$= \frac{-(x - 2)}{(x + 4)(x - 4)(x - 2)}$$

$$= \frac{-1}{(x + 4)(x - 4)}$$

9. $$\frac{x + 1}{4x^2 - 4x + 1} + \frac{3}{2x^2 + 5x - 3}$$

$$= \frac{x + 1}{(2x - 1)(2x - 1)} + \frac{3}{(2x - 1)(x + 3}$$

The LCD = $(2x - 1)(2x - 1)(x + 3)$

$$= \frac{(x + 1)(x + 3)}{(2x - 1)(2x - 1)(x + 3)} + \frac{3(2x - 1)}{(2x - 1)(2x - 1)(x + 3)}$$

$$= \frac{x^2 + 4x + 3 + 6x - 3}{(2x - 1)(2x - 1)(x + 3)}$$

$$= \frac{x^2 + 10x}{(2x - 1)^2(x + 3)}$$

10. $$\frac{\frac{1}{x}+\frac{1}{y}}{\frac{1}{x}-\frac{1}{y}} = \frac{xy\cdot\frac{1}{x}+xy\cdot\frac{1}{y}}{xy\cdot\frac{1}{x}-xy\cdot\frac{1}{y}}$$

$$= \frac{y+x}{y-x}$$

11. $$\frac{x+\frac{x}{y}}{x^{-1}+y^{-1}} = \frac{x+\frac{x}{y}}{\frac{1}{x}+\frac{1}{y}}$$

$$= \frac{xy\cdot x + xy\cdot\frac{x}{y}}{xy\cdot\frac{1}{x}+xy\cdot\frac{1}{y}}$$

$$= \frac{x^2y + x^2}{y+x}$$

12. $$\frac{x}{3}-\frac{x}{4} = 5$$

$$12\cdot\frac{x}{3} - 12\cdot\frac{x}{4} = 12\cdot 5$$

$$4x - 3x = 60$$
$$x = 60$$

13.
$$\frac{x}{x-8}+\frac{6}{x-2} = \frac{x^2}{x^2-10x+16}$$

$$\frac{x}{x-8}+\frac{6}{x-2} = \frac{x^2}{(x-8)(x-2)}$$

Multiply by the LCM $(x - 8)(x - 2)$.

13. cont.

$$\frac{x(x-8)(x-2)}{x-8}+\frac{6(x-8)(x-2)}{x-2}$$
$$= x^2\,\frac{(x-8)(x-2)}{(x-8)(x-2)}$$

$$x(x-2)+6(x-8) = x^2$$

$$x^2 - 2x + 6x - 48 = x^2$$

$$4x - 48 = 0$$
$$4x = 48$$
$$x = 12$$

14. $$P = \frac{kQ}{R}$$

$$8 = \frac{k(4)}{10}$$

$$k = 20$$

$$P = \frac{20Q}{R}$$

$$P = \frac{20(10)}{20} = 10$$

15. $$w = \frac{kPQ}{T^2}$$

$$6 = \frac{k(20)(8)}{4^2}$$

$$k = 3/5 = .6$$

$$w = \frac{.6PQ}{T^2}$$

$$w = \frac{.6(30)(4)}{(8)^2} = 1.125$$

16. Let x = the length of time Kris and Heather can level the field if they work together.

$$\frac{1}{8} + \frac{1}{5} = \frac{1}{x}$$

$$40x \cdot \frac{1}{8} + 40x \cdot \frac{1}{5} = 40x \cdot \frac{1}{x}$$

$5x + 8x = 40$
$13x = 40$
$x = 3.08$hrs.

## CHAPTER 8

### EXERCISE SET 8.1

1. $\sqrt{25} = 5$
   $5^2 = 5 \cdot 5 = 25$

3. $\sqrt[3]{-27} = -3$
   $(-3)^3 = (-3)(-3)(-3) = -27$

5. $\sqrt[3]{125} = 5$
   $5^3 = 5 \cdot 5 \cdot 5 = 125$

7. $\sqrt{-9}$ Not a real number.

9. $\sqrt[3]{-8} = -2$
   $(-2)^3 = (-2)(-2)(-2) = -8$

11. $\sqrt{144} = 12$
    $12^2 = 12 \cdot 12 = 144$

13. $\sqrt[5]{1} = 1$
    $1^5 = 1 \cdot 1 \cdot 1 \cdot 1 \cdot 1 = 1$

15. $\sqrt[3]{343} = 7$
    $7^3 = 7 \cdot 7 \cdot 7 = 343$

17. $\sqrt[4]{-16}$. Not a real number.

19. $-\sqrt{-25}$. Not a real number.

21. $-\sqrt{36} = -6$

23. $\sqrt{\frac{1}{9}} = \frac{1}{3}$

25. $\sqrt{-36}$ Not a real number.

27. $\sqrt[5]{-1} = -1$

29. $\sqrt{6^2} = |6| = 6$

31. $\sqrt{(-1)^2} = |-1| = 1$

33. $\sqrt{(43)^2} = |43| = 43$

35. $\sqrt{(147)^2} = |147| = 147$

37. $\sqrt{(-83)^2} = |-83| = 83$

39. $\sqrt{(179)^2} = |179| = 179$

41. $\sqrt{(y - 8)^2} = |y - 8|$

43. $\sqrt{(x - 3)^2} = |x - 3|$

45. $\sqrt{(3x + 5)^2} = |3x + 5|$

47. $\sqrt{(6 - 3x)^2} = |6 - 3x|$

49. $\sqrt{(y^2 - 4y + 3)^2} = |y^2 - 4y + 3|$

51. $\sqrt{(8a - b)^2} = |8a - b|$

53. (a) Every real number has two square roots; a positive or principle square root and a negative square root.

(b) $\sqrt{36} = \pm 6$

(c) When we say square root, we are referring to the positive root.

(d) $\sqrt{36} = 6$

55. There is not a real number such that when it is multiplied by itself, the answer is -49.

57. No. If the number under the radical is negative, the answer is not a real number.

59. If $\sqrt[n]{x} = a$, then $x = a^n$.

61. Choose a value of x that will make the expression 5x - 3 a negative number. For example, x = -2.

$\sqrt{(5(-2) - 3)^2} = \sqrt{(-13)^2} = |-13| = 13$. But $5(-2) - 3 = -13$. Therefore, $\sqrt{(5x - 3)^2} \neq 5x - 3$.

63. $\sqrt{(x + 4)^2} = x + 4$ for all values of x where $x + 4 \geq 0$. Therefore, solving for x, $x \geq -4$.

65. $\sqrt{(4x - 4)^2} = 4x - 4$ for all values of x where $4x - 4 \geq 0$. Therefore, solving for x, $x \geq 1$.

67. $\sqrt[n]{x}$ is not equal to a real number when n is an even integer and $x < 0$.

69. $\sqrt[n]{x^m}$ is not equal to a real number when n is an even integer, m is an odd integer and $x < 0$.

71. $x^3 + \frac{1}{27} = x^3 + (\frac{1}{3})^3$

$$= (x + \frac{1}{3})(x^2 - \frac{1}{3}x + \frac{1}{9})$$

73. $2x^4 - 3x^3 - 6x^2 + 9x$

$$= x[2x^3 - 3x^2 - 6x + 9]$$
$$= x[x^2(2x - 3) - 3(2x - 3)]$$
$$= x(2x - 3)(x^2 - 3)$$

**EXERCISE SET 8.2**

1. $\sqrt{x^3} = x^{3/2}$

3. $\sqrt{4^5} = 4^{5/2}$

5. $\sqrt[5]{x^4} = x^{4/5}$

7. $(\sqrt{x})^3 = x^{3/2}$

9. $(\sqrt[4]{5})^3 = 5^{3/4}$

11. $(\sqrt[7]{y^2}) = y^{2/7}$

13. $x^{1/2} = \sqrt{x}$

15. $z^{3/2} = \sqrt{z^3}$

17. $2^{1/4} = \sqrt[4]{2}$

19. $z^{9/4} = \sqrt[4]{z^9}$

21. $x^{4/5} = \sqrt[5]{x^4}$

23. $7^{1/3} = \sqrt[3]{7}$

25. $\sqrt{y^6} = y^{6/2} = y^3$

27. $\sqrt{z^8} = z^{8/2} = z^4$

29. $\sqrt[3]{x^9} = x^{9/3} = x^3$

31. $\sqrt[10]{z^5} = z^{5/10} = z^{1/2} = \sqrt{z}$

33. $(\sqrt{5})^2 = (5^{1/2})^2 = 5^{2/2} = 5 = 5$

35. $\sqrt[6]{y^6} = y^{6/6} = y^1 = y$

37. $(\sqrt[8]{x})^2 = x^{2/8} = x^{1/4} = \sqrt[4]{x}$

39. $(\sqrt[3]{x})^{15} = x^{15/3} = x^5$

41. $(\sqrt[10]{y})^5 = y^{5/10} = y^{1/2} = \sqrt{y}$

43. $\sqrt[18]{y^6} = y^{6/18} = y^{1/3} = \sqrt[3]{y}$

45. $4^{1/2} = \sqrt{4} = 2$

47. $27^{2/3} = (\sqrt[3]{27})^2 = 3^2 = 9$

49. $(-4)^{1/2} = \sqrt{-4}$. Not a real number.

51. $(\frac{9}{25})^{1/2} = \sqrt{\frac{9}{25}} = \frac{3}{5}$

53. $-16^{1/2} = -\sqrt{16} = -4$

55. $-27^{1/3} = -\sqrt[3]{27} = -3$

57. $27^{-1/3} = \frac{1}{27^{1/3}} = \frac{1}{\sqrt[3]{27}} = \frac{1}{3}$

59. $4^{-3/2} = \frac{1}{4^{3/2}} = \frac{1}{(\sqrt{4})^3} = \frac{1}{2^3} = \frac{1}{8}$

61. $(\frac{4}{49})^{-1/2} = \frac{1}{\sqrt{4/49}} = \frac{1}{\frac{2}{7}} = \frac{7}{2}$

63. $\left(\frac{8}{27}\right)^{-2/3} = \frac{1}{\left(\sqrt[3]{\frac{8}{27}}\right)^2} = \frac{1}{\left(\frac{2}{3}\right)^2}$

$= \frac{1}{\frac{4}{9}} = \frac{9}{4}$

65. $25^{1/2} + 36^{1/2}$

$= \sqrt{25} + \sqrt{36}$

$= 5 + 6 = 11$

67. $8^{-1/3} + 9^{-1/2}$

$= \frac{1}{8^{1/3}} + \frac{1}{9^{1/2}}$

$= \frac{1}{\sqrt[3]{8}} + \frac{1}{\sqrt{9}}$

$= \frac{1}{2} + \frac{1}{3} = \frac{5}{6}$

69. $x^5 \cdot x^{1/2} = x^{5 + 1/2}$

$= x^{11/2}$

71. $\frac{x^{1/2}}{x^{1/3}} = x^{1/2 - 1/3} = x^{1/6}$

73. $(x^{1/5})^{2/3} = x^{2/15}$

75. $(x^{1/2})^{-2} = x^{-1} = \frac{1}{x}$

77. $(6^{-1/3})^0 = 6^0 = 1$

79. $\frac{y^{-1/3}}{y^{-2}} = y^{-1/3-(-2)} = y^{5/3}$

81. $x^{5/3}\, x^{-7/2} = x^{5/3 - 7/2}$

$= x^{-11/6} = \frac{1}{x^{11/6}}$

83. $\left(\frac{64}{x}\right)^{1/3} = \frac{4}{x^{1/3}}$

85. $\left(\frac{x^{3/7}}{x^{1/2}}\right)^2 = \frac{x^{6/7}}{x^{2/2}}$

$= x^{6/7 - 1}$

$= x^{-1/7} = \frac{1}{x^{1/7}}$

87. $\left(\frac{y^4}{y^{-2/5}}\right)^{-3}$

$\frac{y^{-12}}{y^{6/5}} = y^{-12 - 6/5}$

$= y^{-66/5}$

$= \frac{1}{y^{66/5}}$

89. $\frac{x^{3/4}y^{-2}}{x^{1/2}y^2}$

$= x^{3/4 - 1/2}\, y^{-2-2}$

$= x^{1/4}y^{-4}$

$= \frac{x^{1/4}}{y^4}$

91. $\left(\frac{a^{1/2}\, b^{2/3}}{a^{-1/3}\, b^{3/5}}\right)^2$

$= \frac{a^1 b^{4/3}}{a^{-2/3} b^{6/5}}$

$a^{1-(-2/3)} b^{4/3 - 6/5}$

$a^{5/3}\, b^{2/15}$

93. $x^{3/2} + x^{1/2}$

$= x^{1/2}\, x^1 + x^{1/2}$

$= x^{1/2}(x + 1)$

**95.** $y^{1/3} - y^{4/3}$

$= y^{1/3} - y^{1/3}\, y^1$

$= y^{1/3}\,(1 - y)$

**97.** $y^{-3/5} + y^{2/5}$

$= y^{-3/5}\left(\frac{y^{-3/5}}{y^{-3/5}} + \frac{y^{2/5}}{y^{-3/5}}\right)$

$= y^{-3/5}\,(1 + y^{2/5 - (-3/5)})$

$= y^{-3/5}\,(1 + y)$

$= \frac{1 + y}{y^{3/5}}$

**99.** $y^{-1} - y$

$= y^{-1}\left(\frac{y^{-1}}{y^{-1}} - \frac{y}{y^{-1}}\right)$

$= y^{-1}\,(1 - y^{1-(-1)})$

$= y^{-1}\,(1 - y^2)$

$= \frac{1 - y^2}{y}$

**101.** $x^{-7} + x^{-5}$

$= x^{-7}\left(\frac{x^{-7}}{x^{-7}} + \frac{x^{-5}}{x^{-7}}\right)$

$= x^{-7}(1 + x^{-5-(-7)})$

$= \frac{1 + x^2}{x^{+7}}$

**103.** $x^{-1/2} + x^{-5/2}$

$= x^{-5/2}\left(\frac{x^{-1/2}}{x^{-5/2}} + \frac{x^{-5/2}}{x^{-5/2}}\right)$

$= x^{-5/2}(x^{-1/2-(-5/2)} + 1)$

$= \frac{x^2 + 1}{x^{5/2}}$

**105.** $2x^{-4} - 6x^{-5}$

$= 2x^{-5}\left(\frac{2x^{-4}}{2x^{-5}} - \frac{6x^{-5}}{2x^{-5}}\right)$

$= 2x^{-5}(x^{(-4-(5))} - 3)$

$= 2x^{-5}(x - 3)$

$= \frac{2(x - 3)}{x^5}$

**107.** $5x - 10x^{-1}$

$= 5x^{-1}\left(\frac{5x}{5x^{-1}} - \frac{10x^{-1}}{5x^{-1}}\right)$

$= 5x^{-1}(x^{1-(-1)} - 2)$

$= 5x^{-1}(x^2 - 2)$

$= \frac{5(x^2 - 2)}{x}$

**109.** Rewrite $x^{2/3} + 2x^{1/3} - 3$ as $(x^{1/3})^2 + 2(x^{1/3}) - 3$. Let $y=x^{1/3}$. $y^2 + 2y - 3 = (y+3)(y-1)$. Replacing y with $x^{1/3}$, $x^{2/3} + 2x^{1/3} - 3 = (x^{1/3} + 3)(x^{1/3} - 1)$.

**111.** Rewrite $x + 6x^{1/2} + 9$ as $(x^{1/2})^2 + 6(x^{1/2}) + 9$. Let $y = x^{1/2}$ $y^2 + 6y + 9 = (y + 3)^2$. Replacing y with $x^{1/2}$, $x + 6x^{1/2} + 9 = (x^{1/2}+3)^2$.

**113.** $2x^{2/7} - x^{1/7} - 3$

$= 2(x^{1/7})^2 - x^{1/7} - 3$

Let $y = x^{1/7}$

$2y^2 - y - 3 = (2y + 3)(y - 1)$

Replacing y with $x^{1/7}$, $2x^{2/7} - x^{1/7} - 3 = (2x^{1/7} + 3)(x^{1/7} - 1)$

**115.** $4x^{4/5} + 8x^{2/5} + 3$
$= 4(x^{1/5})^4 + 8(x^{1/5})^2 + 3.$
Let $y = x^{1/5}$
$4y^4 + 8y^2 + 3 = (2y^2+3)(2y^2+1).$
Replace y with $x^{1/5}$.
$4x^{4/5} + 8x^{2/5} + 3 = (2(x^{1/5})^2+3)(2(x^{1/5})^2+1)$
$= (2x^{2/5}+3)(2x^{2/5} + 1)$

**117.** $15x^{1/3} - 14x^{1/6} + 3$
$= 15x^{2/6} - 14x^{1/6} + 3$
$= 15(x^{1/2})^2 - 14x^{1/6} + 3$
Let $y = x^{1/6}$
$15y^2 - 14y + 3 = (5y - 3)(3y - 1)$
Replace y with $x^{1/6}$
$15x^{1/3} - 14x^{1/6} + 3$
$= (5x^{1/6}-3)(3x^{1/6}-1)$

**119.** $\sqrt[n]{a^n} = \sqrt[n]{a^n} = a$ when n is odd or when n is even and $a \geq 0$.

**121.** We can show $(a^{1/2} + b^{1/2})^2 \neq a + b$ by letting a = 9 and b = 16.
$(9^{1/2} + 16^{1/2})^2 = (3 + 4)^2$
$= 7^2$
$= 49$
But $a + b = 9 + 16$
$= 25$
Since $49 \neq 25$,
$(a^{1/2} + b^{1/2})^2 \neq a + b$

**123.** $\sqrt[3]{\sqrt{x}} = \sqrt{\sqrt[3]{x}}$ for $x \geq 0$.
Notice that both expressions equal $x^{1/6}$ when converted to exponential form.

**125.** $$\frac{a^{-2} + ab^{-1}}{ab^{-2} - a^{-2}b^{-1}}$$
$$= \frac{\frac{1}{a^2} + \frac{a}{b}}{\frac{a}{b^2} - \frac{1}{a^2b}}$$
$$= \frac{a^2b^2\left(\frac{1}{a^2}\right) + a^2b^2\left(\frac{a}{b}\right)}{a^2b^2\left(\frac{a}{b^2}\right) - a^2b^2\left(\frac{1}{a^2b}\right)}$$
$$= \frac{b^2 + a^3b}{a^3 - b}$$

**127.** Let t equal the time needed to travel either 500 miles or 560 miles and r equal the speed of the plane in still air.

$a = rt$
$500 = (r - 25)\ t$
or
$\frac{500}{r - 25} = t$
and
$560 = (r + 25)\ t$
or
$\frac{560}{r + 25} = t.$
Since both $\frac{500}{r - 25}$ and $\frac{560}{r + 25}$ equal t, we can conclude that $\frac{500}{r - 25} = \frac{560}{r + 25}$.

$500(r + 25) = 560(r - 25)$
$500r + 12500 = 560r - 14000$
$26500 = 60r$
$441.67 = r$
The speed of the plane in still air is 441.67 mph.

**JUST FOR FUN**

1. $(6x - 5)^{-3} + (6x - 5)^{-2}$.

   Let $y = 6x - 5$

   $$y^{-3} + y^{-2} = y^{-3}\left(\frac{y^{-3}}{y^{-3}} + \frac{y^{-2}}{y^{-3}}\right)$$

   $$= y^{-3}(1 + y^{1})$$

   Replacing y with 6x - 5,

   $$(6x - 5)^{-3} + (6x - 5)^{-2}$$

   $$= (6x - 5)^{-3}\,(1 + (6x - 5))$$

   $$= \frac{6x - 4}{(6x - 5)^{3}} = \frac{2(3x - 2)}{(6x - 5)^{3}}$$

## EXERCISE SET 8.3

1. $\sqrt{50} = \sqrt{25 \cdot 2} = \sqrt{25}\ \sqrt{2} = 5\sqrt{2}$

3. $\sqrt{32} = \sqrt{16} \cdot \sqrt{2} = \sqrt{16}\ \sqrt{2} = 4\sqrt{2}$

5. $\sqrt[3]{16} = \sqrt[3]{8 \cdot 2} = \sqrt[3]{8}\ \sqrt[3]{2} = 2\ \sqrt[3]{2}$

7. $\sqrt[3]{54} = \sqrt[3]{27}\ \sqrt[3]{2}$
$= 3\ \sqrt[3]{2}$

9. $\sqrt{x^3} = \sqrt{x^2 \cdot x} = \sqrt{x^2}\ \sqrt{x} = x\sqrt{x}$

11. $\sqrt{x^{11}} = \sqrt{x^{10}x} = \sqrt{x^{10}}\ \sqrt{x} = x^5\sqrt{x}$

13. $\sqrt{b^{27}} = \sqrt{b^{26}b} = \sqrt{b^{26}}\ \sqrt{b} = b^{13}\ \sqrt{b}$

15. $\sqrt[4]{y^9} = \sqrt[4]{y^8 \cdot y} = \sqrt[4]{y^8}\ \sqrt[4]{y}$
$= y^2\ \sqrt[4]{y}$

17. $\sqrt{24x^3} = \sqrt{4 \cdot 6x^2x} = \sqrt{4x^2}\ \sqrt{6x}$
$= 2x\sqrt{6x}$

19. $\sqrt[3]{24\ y^7} = \sqrt[3]{8}\ \sqrt[3]{3}\ \sqrt[3]{y^6}\ \sqrt[3]{y}$
$= 2y^2\ \sqrt[3]{3y}$

21. $\sqrt{x^3\ y^7} = \sqrt{x^2xy^6y}$
$= xy^3\ \sqrt{xy}$

23. $\sqrt[3]{81\ x^6y^2} = \sqrt[3]{27 \cdot 3\ x^6y^6y^2}$
$= 3x^2y^2\ \sqrt[3]{3y^2}$

25. $\sqrt[3]{54x^{12}y^{13}} = \sqrt[3]{27 \cdot 2x^{12}y^{12}y}$
$= \sqrt[3]{27x^{12}y^{12}}\ \sqrt[3]{2y}$
$= 3x^4y^4\ \sqrt[3]{2y}$

27. $\sqrt[5]{64x^{12}y^7} = \sqrt[5]{32 \cdot 2\ x^{10}x^2y^5y^2}$
$= 2x^2y\quad \sqrt[5]{2x^2y^2}$

29. $\sqrt[3]{32c^4w^9z} = \sqrt[3]{8 \cdot 4c^3cw^9z}$
$= 2cw^3\ \sqrt[3]{4cz}$

31. $\sqrt[3]{81x^7y^{21}z^{50}}$
$= \sqrt[3]{27 \cdot 3x^6xy^{21}z^{48}z^2}$
$= \sqrt[3]{27x^6y^{21}z^{48}}\ \sqrt[3]{3xz^2}$
$= 3x^2y^7z^{16}\ \sqrt[3]{3xz^2}$

33. $\sqrt{5}\ \sqrt{5} = \sqrt{25} = 5$

35. $\sqrt[3]{2}\ \sqrt[3]{4} = \sqrt[3]{8} = 2$

37. $\sqrt[3]{3}\ \sqrt[3]{54} = \sqrt[3]{162}$
$= \sqrt[3]{27 \cdot 6}$
$= 3\ \sqrt[3]{6}$

39. $\sqrt{15xy^4}\ \sqrt{6xy^3}$
$= \sqrt{90x^2y^7}$
$= \sqrt{9 \cdot 10x^2y^6y}$
$= 3xy^3\ \sqrt{10y}$

41. $(\sqrt{4x^3y^2})^2 = 4x^3y^2$

**43.** $\sqrt[3]{5xy^2}\ \sqrt[3]{25x^4y^{12}}$
$= \sqrt[3]{125x^5y^{14}}$
$= \sqrt[3]{125x^3x^2y^{12}y^2}$
$= 5xy^4\ \sqrt[3]{x^2y^2}$

**45.** $(\sqrt[3]{2x^3y^4})^2$
$= \sqrt[3]{4x^6y^8}$
$= \sqrt[3]{4x^6y^6y^2}$
$= x^2y^2\ \sqrt[3]{4y^2}$

**47.** $\sqrt[4]{12xy^4}\ \sqrt[4]{2x^3y^9z^7}$
$= \sqrt[4]{24x^4y^{13}z^7}$
$= \sqrt[4]{24x^4y^{12}yz^4z^3}$
$= xy^3z\ \sqrt[4]{24yz^3}$

**49.** $\sqrt[5]{x^{24}y^{30}z^9}\ \ \sqrt[5]{x^{13}y^8z^7}$
$= \sqrt[5]{x^{37}y^{38}z^{16}}$
$= \sqrt[5]{x^{35}x^2y^{35}y^3z^{15}z}$
$= x^7y^7z^3\ \sqrt[5]{x^2y^3z}$

**51.** $\sqrt{2}\ (\sqrt{6} + \sqrt{2})$
$= \sqrt{12} + \sqrt{4}$
$= 2\sqrt{3} + 2$

**53.** $\sqrt{3}\ (\sqrt{12} = \sqrt{6})$
$= \sqrt{36} - \sqrt{18}$
$= 6 - 3\sqrt{2}$

**55.** $\sqrt{2}\ (\sqrt{18} + \sqrt{8})$
$= \sqrt{36} + \sqrt{16}$
$= 6 + 4 = 10$

**57.** $\sqrt{3y}\ (\sqrt{27y^2} - \sqrt{y})$
$= \sqrt{81y^3} - \sqrt{3y^2}$
$= 9y\ \sqrt{y} - y\sqrt{3}$

**59.** $\sqrt[3]{2x^2y}\ (\sqrt[3]{4xy^5} + \sqrt[3]{12x^{10}y})$
$= \sqrt[3]{8x^3y^6} + \sqrt[3]{24x^{12}y^2}$
$= 2xy^2 + 2x^4\ \sqrt[3]{3y^2}$

**61.** $2\ \sqrt[3]{x^4y^5}\ (\sqrt[3]{8x^{12}y^4} + \sqrt[3]{16xy^9})$
$= 2\ \sqrt[3]{8x^{16}y^9} + 2\ \sqrt[3]{16x^5y^{14}}$
$= 2 \cdot 2\ x^5y^3\ \sqrt[3]{x}$
$+ 2 \cdot 2\ xy^4\ \sqrt[3]{2x^2y^2}$
$= 4x^5y^3\ \sqrt[3]{x} + 4xy^4\ \sqrt[3]{2x^2y^2}$

**63.** $3\sqrt{2xy^4}\ (\sqrt{20x^4y^8} - 2\sqrt{6xy^9})$
$= 3\sqrt{40x^5y^{12}} - 6\sqrt{12x^2y^{13}}$
$= 3 \cdot \sqrt{4 \cdot 10x^4xy^{12}}$
$- 6\sqrt{4 \cdot 3x^2y^{12}y}$
$= 3 \cdot 2x^2y^6\ \sqrt{10x} - 6 \cdot 2\ xy^6\ \sqrt{3y}$
$= 6x^2y^6\sqrt{10x} - 12xy^6\ \sqrt{3y}$

**65.** $\sqrt{24} = \sqrt{4}\ \sqrt{6} = 2\sqrt{6}$

**67.** $\sqrt[3]{32} = \sqrt[3]{8}\ \sqrt[3]{4} = 2\ \sqrt[3]{4}$

**69.** $\sqrt[3]{x^5} = x\sqrt[3]{x^2}$

71. $\sqrt{36x^5} = 6\sqrt{x^4x}$
$= 6x^2\sqrt{x}$

73. $\sqrt{x^5y^{12}} = \sqrt{x^4xy^{12}}$
$= x^2y^6\ \sqrt{x}$

75. $\sqrt[4]{16ab^{17}c^9}$
$= \sqrt[4]{16ab^{16}bc^8c}$
$= 4b^4c^2\ \sqrt{abc}$

77. $\sqrt{75}\ \sqrt{6}$
$= \sqrt{25\cdot 3}\ \sqrt{6}$
$= 5\sqrt{3}\ \sqrt{6}$
$= 5\sqrt{18}$
$= 5\sqrt{9}\ \sqrt{2}$
$= 5(3)\ \sqrt{2}$
$= 15\sqrt{2}$

79. $\sqrt{15x^2}\ \sqrt{6x^5}$
$= \sqrt{90x^7}$
$= \sqrt{9\cdot 10x^6x}$
$= 3x^3\ \sqrt{10x}$

81. $\sqrt{20xy^4}\ \sqrt{6x^5y^7}$
$= 2y^2\sqrt{5x}\cdot x^2y^3\sqrt{6xy}$
$= 2x^2y^5\ \sqrt{30x^2y}$
$= 2x^3y^5\ \sqrt{30y}$
$= 2x^3y^5\ \sqrt{30y}$

83. $\sqrt{x}\ (\sqrt{x} + 3)$
$= \sqrt{x^2} + 3\sqrt{x}$

$= x + 3\sqrt{x}$

85. $\sqrt[3]{4xy^2}\ \ \sqrt[3]{4xy^4}$
$= \sqrt[3]{16x^2y^6}$
$= \sqrt[3]{8\cdot 2\ x^2y^6}$
$= 2y^2\ \sqrt{2x^2}$

87. $\sqrt[3]{y}\ (2\ \sqrt[3]{y} - \sqrt[3]{y^8})$
$= 2\sqrt[3]{y^2} - \sqrt[3]{y^9}$
$= 2\sqrt[3]{y^2} - y^3$

89. $\sqrt[3]{3xy^2}\ (\sqrt[3]{4x^4y^3} - \sqrt[3]{8x^5y^4})$
$\sqrt[3]{12x^5y^5} - \sqrt[3]{8\cdot 3x^6y^6}$
$= xy\ \sqrt[3]{12x^2y^2} - 2x^2y^2\ \sqrt[3]{3}$

91. If n is even and a or b is negative, the numbers are not real numbers and the rule does not apply.

93. A real number is a number that can be represented on a number line.

95. $|a| = \begin{cases} a, & a \geq 0 \\ -a, & a < 0 \end{cases}$

97. $-4 < 2x - 3 \leq 5$
$-1 < 2x \leq 8$
$-1/2 < x \leq 4$

(a) $-\frac{1}{2}$    4

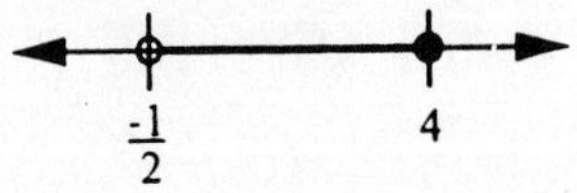

(b) $(-1/2,\ 4]$

(c) $\{x \mid -1/2 < x \leq 4\}$

**EXERCISE SET 8.4**

1. $\sqrt{\frac{27}{3}} = \sqrt{9} = 3$

3. $\frac{\sqrt{3}}{\sqrt{27}} = \sqrt{\frac{1}{9}} = \frac{1}{3}$

5. $\sqrt[3]{\frac{2}{16}} = \sqrt[3]{\frac{1}{8}} = \frac{1}{2}$

7. $\frac{\sqrt{24}}{\sqrt{3}} = \sqrt{8} = \sqrt{4}\ \sqrt{2} = 2\sqrt{2}$

9. $\sqrt{\frac{x^4}{25}} = \sqrt{\frac{x^4}{25}} = \frac{x^2}{5}$

11. $\sqrt{\frac{16x^4}{4}} = \sqrt{4x^4} = 2x^2$

13. $\sqrt{\frac{2x}{8x^5}} = \sqrt{\frac{1}{4x^4}} = \frac{1}{2x^2}$

15. $\sqrt{\frac{72x^2y^5}{8x^2y^7}} = \sqrt{\frac{9}{y^2}} = \frac{\sqrt{9}}{\sqrt{y^2}} = \frac{3}{y}$

17. $\frac{1}{\sqrt{3}} = \frac{1}{\sqrt{3}} \cdot \frac{\sqrt{3}}{\sqrt{3}} = \frac{\sqrt{3}}{\sqrt{9}} = \frac{\sqrt{3}}{3}$

19. $\frac{1}{\sqrt{2}} = \frac{1}{\sqrt{2}} \cdot \frac{\sqrt{2}}{\sqrt{2}} = \frac{\sqrt{2}}{\sqrt{4}} = \frac{\sqrt{2}}{2}$

21. $\frac{x}{\sqrt{5}} = \frac{x}{\sqrt{5}} \cdot \frac{\sqrt{5}}{\sqrt{5}} = \frac{x\sqrt{5}}{\sqrt{25}} = \frac{x\sqrt{5}}{5}$

23. $\frac{x}{\sqrt{y}} = \frac{x}{\sqrt{y}} \cdot \frac{\sqrt{y}}{\sqrt{y}} = \frac{x\sqrt{y}}{\sqrt{y^2}} = \frac{x\sqrt{y}}{y}$

25. $\sqrt{\frac{x}{2}} = \frac{\sqrt{x}}{\sqrt{2}} = \frac{\sqrt{x}\ \sqrt{2}}{\sqrt{2}\ \sqrt{2}} = \frac{\sqrt{2x}}{2}$

27. $\sqrt{\frac{5}{8}} = \frac{\sqrt{5}}{\sqrt{8}} \cdot \frac{\sqrt{8}}{\sqrt{8}} = \frac{\sqrt{40}}{\sqrt{64}} = \frac{\sqrt{4}\ \sqrt{10}}{8}$

$= \frac{2\sqrt{10}}{8} = \frac{\sqrt{10}}{4}$

29. $\frac{2\sqrt{3}}{\sqrt{5}} = \frac{2\sqrt{3}}{\sqrt{5}} \cdot \frac{\sqrt{5}}{\sqrt{5}} = \frac{2\sqrt{15}}{\sqrt{25}} = \frac{2\sqrt{15}}{5}$

31. $\frac{2\sqrt{3}}{\sqrt{32}} = \frac{2\sqrt{3}}{4\sqrt{2}} = \frac{2\sqrt{3}\ \sqrt{2}}{4\sqrt{2}\ \sqrt{2}}$

$= \frac{2\sqrt{6}}{8} = \frac{\sqrt{6}}{4}$

33. $\frac{1}{\sqrt[3]{2}}\ \frac{\sqrt[3]{4}}{\sqrt[3]{4}} = \frac{\sqrt[3]{4}}{\sqrt[3]{8}}$

$= \frac{\sqrt[3]{4}}{2}$

35. $\frac{1}{\sqrt[3]{3}} = \frac{1}{\sqrt[3]{3}}\ \frac{\sqrt[3]{9}}{\sqrt[3]{9}} = \frac{\sqrt[3]{9}}{3}$

37. $\sqrt[3]{\frac{5x}{y}} = \frac{\sqrt[3]{5x}}{\sqrt[3]{y}}$

$= \frac{\sqrt[3]{5x}\ \sqrt[3]{y^2}}{\sqrt[3]{y}\ \sqrt[3]{y^2}}$

$= \frac{\sqrt[3]{5xy^2}}{y}$

39. $\sqrt[3]{\frac{5x}{4y^2}} = \frac{\sqrt[3]{5x}}{\sqrt[3]{4y^2}} \cdot \frac{\sqrt[3]{4^2y}}{\sqrt[3]{4^2y}}$

$= \frac{\sqrt[3]{80xy}}{\sqrt[3]{4^3y^3}}$

$= \frac{\sqrt[3]{8}\ \sqrt[3]{10xy}}{4y}$

$= \frac{2\ \sqrt[3]{10xy}}{4y}$

$= \frac{\sqrt[3]{10xy}}{2y}$

41. $\frac{5x}{\sqrt[4]{2}} = \frac{5x}{\sqrt[4]{2}} \cdot \frac{\sqrt[4]{2^3}}{\sqrt[4]{2^3}}$

$= \frac{5x\ \sqrt[4]{8}}{\sqrt{2^4}}$

$= \frac{5x\ \sqrt[4]{8}}{2}$

43. $\sqrt[4]{\frac{2x}{4y^2}} = \sqrt[4]{\frac{x}{2y^2}} = \frac{\sqrt[4]{x}}{\sqrt[4]{2y^2}}$

$= \frac{\sqrt[4]{x}\ \sqrt[4]{8y^2}}{\sqrt[4]{2y^2}\ \sqrt[4]{8y^2}} = \frac{\sqrt[4]{8xy^2}}{2y}$

45. $\sqrt{\frac{8x^5y}{2z}} = \sqrt{\frac{4x^5y}{z}} = \frac{\sqrt{4x^4}\ \sqrt{xy}}{\sqrt{z}}$

$= \frac{2x^2\sqrt{xy}}{\sqrt{z}} \cdot \frac{\sqrt{z}}{\sqrt{z}}$

$= \frac{2x^2\sqrt{xyz}}{\sqrt{z^2}}$

$= \frac{2x^2\sqrt{xyz}}{z}$

47. $\sqrt{\frac{5xy^4}{2z}} = \frac{\sqrt{5xy^4}}{\sqrt{2z}} \cdot \frac{\sqrt{2z}}{\sqrt{2z}}$

$= \frac{\sqrt{10xy^4z}}{\sqrt{4z^2}}$

$= \frac{y^2\sqrt{10xz}}{2z}$

49. $\sqrt{\frac{5xy^6}{6z}} = \frac{\sqrt{5xy^6}\ \sqrt{6z}}{\sqrt{6z}\ \sqrt{6z}}$

$= \frac{\sqrt{30xy^6z}}{6z}$

$= \frac{y^3\sqrt{30xz}}{6z}$

51. $\sqrt{\frac{18x^4y^3}{2z}} = \sqrt{\frac{9x^4y^3}{z}}$

$= \frac{\sqrt{9x^4y^3}\ \sqrt{z}}{\sqrt{z}\ \sqrt{z}}$

$= \frac{\sqrt{9x^4y^3z}}{z}$

$= \frac{3x^2y\ \sqrt{yz}}{z}$

53. $\sqrt[3]{\frac{5x^6y^7}{z^2}} = \frac{\sqrt[3]{15x^6y^7}}{\sqrt[3]{z^2}} \cdot \frac{\sqrt[3]{z}}{\sqrt[3]{z}}$

$= \frac{\sqrt[3]{15x^6y^7z}}{\sqrt[3]{z^3}}$

$= \frac{\sqrt[3]{x^6y^6}\ \sqrt[3]{15yz}}{z}$

$= \frac{x^2y^2\ \sqrt[3]{15yz}}{z}$

**55.** $\sqrt[3]{\dfrac{32x^4y^9}{4x^5}} = \sqrt[3]{\dfrac{8y^9}{x}}$

$= \dfrac{\sqrt[3]{8y^9}\ \sqrt[3]{x^2}}{\sqrt[3]{x}\ \ \sqrt[3]{x^2}}$

$= \dfrac{\sqrt[3]{8x^2y^9}}{x} = \dfrac{2y^3\ \sqrt[3]{x^2}}{x}$

**57.** $(3 - \sqrt{3})(3 + \sqrt{3})$

$= 9 + 3\sqrt{3} - 3\sqrt{3} - \sqrt{3}\sqrt{3}$

$= 9 + 0 - \sqrt{9}$

$= 9 - 3$

$= 6$

**59.** $(6 - \sqrt{5})(6 + \sqrt{5})$

$= 6^2 - (\sqrt{5})^2$

$= 36 - 5$

$= 31$

**61.** $(\sqrt{x} + 5)(\sqrt{x} - 5)$

$= (\sqrt{x})^2 - 5^2$

$= x - 25$

**63.** $(\sqrt{x} + y)(\sqrt{x} - y)$

$= (\sqrt{x})^2 - y^2$

$= x - y^2$

**65.** $(x + \sqrt{y})(x - \sqrt{y})$

$= x^2\ (\sqrt{y})^2$

$= x^2 - y$

**67.** $(5 - \sqrt{y})(5 + \sqrt{y})$

$= 5^2 - (\sqrt{y})^2$

$= 25 - y$

**69.** $\dfrac{3}{1 + \sqrt{2}} = \dfrac{3}{1 + \sqrt{2}} \cdot \dfrac{(1 - \sqrt{2})}{(1 - \sqrt{2})}$

$= \dfrac{3 - 3\sqrt{2}}{1 - \sqrt{2} + \sqrt{2} - \sqrt{2}\sqrt{2}}$

$= \dfrac{3 - 3\sqrt{2}}{1 - \sqrt{4}} = \dfrac{3 - 3\sqrt{2}}{1 - 2}$

$= \dfrac{3 - 3\sqrt{2}}{-1} = 3\sqrt{2} - 3$

**71.** $\dfrac{3}{\sqrt{6} - 5} = \dfrac{3}{\sqrt{6} - 5} \cdot \dfrac{(\sqrt{6} + 5)}{(\sqrt{6} + 5)}$

$= \dfrac{3(\sqrt{6} + 5)}{\sqrt{6}\ \sqrt{6} + 5\sqrt{6} - 5\sqrt{6} - 25}$

$= \dfrac{3\sqrt{6} + 15}{\sqrt{36} - 25} = \dfrac{3\sqrt{6} + 15}{6 - 25}$

$= \dfrac{3\sqrt{6} + 15}{-19} = \dfrac{-3\sqrt{6} - 15}{19}$

**73.** $\dfrac{4}{\sqrt{2} - 7} = \dfrac{4}{\sqrt{2} - 7} \cdot \dfrac{(\sqrt{2} + 7)}{\sqrt{2} + 7)}$

$= \dfrac{4\sqrt{2} + 28}{\sqrt{2}\sqrt{2} + 7\sqrt{2} - 7\sqrt{2} - 49}$

$= \dfrac{4\sqrt{2} + 28}{\sqrt{4} - 49} = \dfrac{4\sqrt{2} + 28}{2 - 49}$

$= \dfrac{4\sqrt{2} + 28}{-47} = \dfrac{-4\sqrt{2} - 28}{47}$

**75.** $\dfrac{\sqrt{5}}{\sqrt{5} - \sqrt{6}} = \dfrac{\sqrt{5}}{\sqrt{5} - \sqrt{6}} \cdot \dfrac{(\sqrt{5} + \sqrt{6})}{(\sqrt{5} + \sqrt{6})}$

$= \dfrac{\sqrt{25} + \sqrt{30}}{\sqrt{25} + \sqrt{30} - \sqrt{30} - \sqrt{36}}$

$= \dfrac{5 + \sqrt{30}}{5 + 0 - 6} = \dfrac{5 + \sqrt{30}}{-1}$

$= -5 - \sqrt{30}$

77. $\dfrac{1}{\sqrt{17}-\sqrt{8}} = \dfrac{1}{\sqrt{17}-\sqrt{8}} \cdot \dfrac{(\sqrt{17}+\sqrt{8})}{(\sqrt{17}+\sqrt{8})}$

$= \dfrac{\sqrt{17}+\sqrt{8}}{\sqrt{17^2}+\sqrt{17}\sqrt{8}-\sqrt{17}\sqrt{8}-\sqrt{8^2}}$

$= \dfrac{\sqrt{17}+\sqrt{8}}{17-8} = \dfrac{\sqrt{17}+\sqrt{4}\sqrt{2}}{9}$

$= \dfrac{\sqrt{17}+2\sqrt{2}}{9}$

79. $\dfrac{5}{\sqrt{x}-3} = \dfrac{5}{\sqrt{x}-3} \cdot \dfrac{(\sqrt{x}+3)}{(\sqrt{x}+3)}$

$= \dfrac{5\sqrt{x}+15}{\sqrt{x^2}+3\sqrt{x}-3\sqrt{x}+9}$

$= \dfrac{5\sqrt{x}+15}{x-9}$

81. $-\dfrac{4 \quad (\sqrt{x}+y)}{\sqrt{x}-y \quad (\sqrt{x}+y)}$

$= \dfrac{4\sqrt{x}+4y}{(\sqrt{x})^2-y^2} = \dfrac{4\sqrt{x}+4y}{x-y^2}$

83. $\dfrac{(\sqrt{2}-1)(\sqrt{2}-1)}{(\sqrt{2}+1)(\sqrt{2}-1)}$

$= \dfrac{(\sqrt{2})^2-(2(1)\sqrt{2})+1^2}{(\sqrt{2})^2-1^2}$

$= \dfrac{2-2\sqrt{2}+1}{2-1} = 3-3\sqrt{2}$

85. $\dfrac{(\sqrt{x}-\sqrt{2y})(\sqrt{x}+\sqrt{y})}{(\sqrt{x}-\sqrt{y})(\sqrt{x}+\sqrt{y})}$

$= \dfrac{(\sqrt{x})^2+\sqrt{xy}-\sqrt{2xy}-\sqrt{2y^2}}{(\sqrt{x})^2-(\sqrt{y})^2}$

$= \dfrac{x+\sqrt{xy}-\sqrt{2xy}-y\sqrt{2}}{x-y}$

87. $\dfrac{2\sqrt{xy}-\sqrt{xy}}{\sqrt{x}+\sqrt{y}}$

$= \dfrac{\sqrt{xy}(\sqrt{x}-\sqrt{y})}{(\sqrt{x}+\sqrt{y})(\sqrt{x}-\sqrt{y})} = \dfrac{\sqrt{x^2y}-\sqrt{xy^2}}{x-y}$

$= \dfrac{x\sqrt{y}-y\sqrt{x}}{x-y}$

89. $\sqrt{\dfrac{x}{9}} = \dfrac{\sqrt{x}}{\sqrt{9}} = \dfrac{\sqrt{x}}{3}$

91. $\sqrt{\dfrac{2}{5}} = \dfrac{\sqrt{2}}{\sqrt{5}} = \dfrac{\sqrt{2}\,\sqrt{5}}{\sqrt{5}\,\sqrt{5}} = \dfrac{\sqrt{10}}{5}$

93. $= (\sqrt{5}+\sqrt{6})(\sqrt{5}-\sqrt{6})$
$= (\sqrt{5})^2-(\sqrt{6})^2 = 5-6 = -1$

95. $\sqrt{\dfrac{24x^3y^6}{5z}} = \dfrac{\sqrt{24x^3y^6}\,\sqrt{5z}}{\sqrt{5z}\,\sqrt{5z}}$

$= \dfrac{2xy^3\sqrt{30xz}}{5z}$

97. $\sqrt{\dfrac{12xy^4}{2x^3y^4}} = \sqrt{\dfrac{6}{x^2}} = \dfrac{\sqrt{6}}{x}$

**99.** $(\sqrt{x} + 3)(\sqrt{x} - 3) = (\sqrt{x})^2 - 3^2$
$= x - 9$

**101.** $\dfrac{7\sqrt{x}}{\sqrt{98}} = \dfrac{7\sqrt{x}}{7\sqrt{2}} = \dfrac{\sqrt{x}}{\sqrt{2}} = \dfrac{\sqrt{x}\ \sqrt{2}}{\sqrt{2}\ \sqrt{2}} = \dfrac{\sqrt{2x}}{2}$

**103.** $\sqrt[4]{\dfrac{3}{2x}} = \dfrac{\sqrt[4]{3}}{\sqrt[4]{2x}} = \dfrac{\sqrt[4]{3}\ \sqrt[4]{8x^3}}{\sqrt[4]{2x}\ \sqrt[4]{8x^3}} = \dfrac{\sqrt[4]{24x^3}}{2x}$

**105.** $\sqrt[3]{\dfrac{32y^{12}z^{10}}{2x}} = \sqrt[3]{\dfrac{16y^{12}z^{10}}{x}}$

$= \dfrac{\sqrt[3]{16y^{12}z^{10}}\ \sqrt[3]{x^2}}{\sqrt[3]{x}\ \sqrt[3]{x^2}}$

$= \dfrac{\sqrt[3]{16y^{12}z^{10}}}{x}$

$= 2y^4z\ \dfrac{\sqrt[3]{2x^2z}}{x}$

**107.** $\sqrt{\dfrac{2p}{q}} = \dfrac{\sqrt{2p}\ \sqrt{q}}{\sqrt{q}\ \sqrt{q}} = \dfrac{\sqrt{2pq}}{q}$

**109.** $(\sqrt{y} - 3)(\sqrt{y} + 3)$

$=(\sqrt{y})^2 - 3^2 = y - 9$

**111.** $\sqrt[4]{\dfrac{2x^7y^{12}z^4}{3x^9}}$

$= \dfrac{\sqrt[4]{2x^7y^{12}z^4}\ \sqrt[4]{27x^3}}{\sqrt[4]{3x^9}\ \sqrt[4]{27x^3}}$

$= \dfrac{\sqrt[4]{54\ x^{10}y^{12}z^4}}{3x^3}$

$= \dfrac{x^2y^3z\ \sqrt{54x^2}}{3x^3}$

$= \dfrac{y^3z\ \sqrt{54x^2}}{3x}$

**113.** (a) $\dfrac{1}{\sqrt{18}} = \dfrac{1}{3\sqrt{2}}$

$= \dfrac{1\ \sqrt{2}}{3\sqrt{2}\ \sqrt{2}} = \dfrac{\sqrt{2}}{3(2)} = \dfrac{\sqrt{2}}{6}$

(b) $\dfrac{1\ \sqrt{2}}{\sqrt{18}\ \sqrt{2}} = \dfrac{\sqrt{2}}{\sqrt{36}} = \dfrac{\sqrt{2}}{6}$

(c) $\dfrac{1\ \sqrt{18}}{\sqrt{18}\ \sqrt{18}} = \dfrac{\sqrt{18}}{18} = \dfrac{3\sqrt{2}}{18} = \dfrac{\sqrt{2}}{6}$

**115.** $\dfrac{2}{\sqrt{2}} = \dfrac{2\sqrt{2}}{\sqrt{2}\ \sqrt{2}} = \dfrac{2\sqrt{2}}{2} = \sqrt{2}$

$\dfrac{3}{\sqrt{3}} = \dfrac{3\sqrt{3}}{\sqrt{3}\ \sqrt{3}} = \dfrac{3\sqrt{3}}{3} = \sqrt{3}$

Since $3 > 2$, $\sqrt{3} > \sqrt{2}$ and $\dfrac{3}{\sqrt{3}} > \dfrac{2}{\sqrt{2}}$.

**117.** Yes $\sqrt[3]{\dfrac{2}{3}} = \dfrac{3\sqrt{18}}{3}$

**119.** (1) No perfect powers are factors of any radicand.

(2) No radicand contains fractions.

(3) No radicals are in the denominator.

**121.** Let r be the rate of the slower car and r + 10 be the rate of the faster.

Distance the 1st traveled and distance the second traveled = 270. Therefore

$$\begin{aligned} 3r + 3(r + 10) &= 270 \\ 3r + 3r + 30 &= 270 \\ 6r + 30 &= 270 \\ 6r &= 240 \\ r &= 40 \\ r + 10 &= 50 \end{aligned}$$

**123.**

$$\begin{array}{r} 4x^2 - 3x - 2 \\ 2x - 3 \\ \hline -12x^2 + 9x - 6 \\ 8x^3 - 6x^2 - 4x \quad\quad \\ \hline 8x^3 - 18x^2 + 5x - 6 \end{array}$$

**EXERCISE SET 8.5**

1. $4\sqrt{3} - 2\sqrt{3} = 2\sqrt{3}$

3. $4\sqrt{10} + 6\sqrt{10} - \sqrt{10} + 2$
$= 9\sqrt{10} + 2$

5. $12\sqrt[3]{15} + 5\sqrt[3]{15} - 8\sqrt[3]{15} = 9\sqrt[3]{15}$

7. $3\sqrt{y} - 6\sqrt{y} = -3\sqrt{y}$

9. $3\sqrt{5} - \sqrt[3]{x} + 4\sqrt{5} + 3\sqrt[3]{x}$
$= 7\sqrt{5} + 2\sqrt[3]{x}$

11. $5 + 4\sqrt[3]{x} - 8\sqrt[3]{x} = 5 - 4\sqrt[3]{x}$

13. $\sqrt{8} - \sqrt{12}$
$= 2\sqrt{2} - 2\sqrt{3}$
$= 2(\sqrt{2} - \sqrt{3})$

15. $-6\sqrt{75} + 4\sqrt{125}$
$= -6(5\sqrt{3}) + 4(5\sqrt{5})$
$= -30\sqrt{3} + 20\sqrt{5}$

17. $-4\sqrt{90} + 3\sqrt{40} + 2\sqrt{10}$
$= -4(3)\sqrt{10} + 3(2\sqrt{10}) + 2(\sqrt{10})$
$= -12\sqrt{10} + 6\sqrt{10} + 2\sqrt{10}$
$= -4\sqrt{10}$

19. $4\sqrt{32} - \sqrt{18} + 2\sqrt{128}$
$= 4(4\sqrt{2}) - 3\sqrt{2} + 2(8\sqrt{2})$
$= 16\sqrt{2} - 3\sqrt{2} + 16\sqrt{2}$
$= 29\sqrt{2}$

21. $2\sqrt{5x} - 3\sqrt{20x} - 4\sqrt{45x}$
$= 2\sqrt{5x} - 3(2\sqrt{5x}) - 4(3\sqrt{5x})$
$= 2\sqrt{5x} - 6\sqrt{5x} - 12\sqrt{5x}$
$= -16\sqrt{5x}$

23. $3\sqrt{50x^2} - 3\sqrt{72x^2} - 8x\sqrt{18}$
$= 3(5x\sqrt{2}) - 3(6x\sqrt{2}) - 8x(3\sqrt{2})$
$= 15x\sqrt{2} - 18x\sqrt{2} - 24x\sqrt{2}$
$= -27x\sqrt{2}$

25. $4\sqrt[3]{5} - 5\sqrt[3]{40}$
$= 4\sqrt[3]{5} - 5\sqrt[3]{8}\sqrt[3]{5}$
$= 4\sqrt[3]{5} - 5 \cdot 2\sqrt[3]{5}$
$= 4\sqrt[3]{5} - 10\sqrt[3]{5}$
$= -6\sqrt[3]{5}$

27. $2\sqrt[3]{16} + \sqrt[3]{54}$
$= 2\sqrt[3]{8}\sqrt[3]{2} + \sqrt[3]{27}\sqrt[3]{2}$
$= 2 \cdot 2\sqrt[3]{2} + 3\sqrt[3]{2}$
$= 4\sqrt[3]{2} + 3\sqrt[3]{2}$
$= 7\sqrt[3]{2}$

29. $3\sqrt{45x^3} + \sqrt{5x}$
$= 3(3x\sqrt{5x}) + \sqrt{5x}$
$= 9x\sqrt{5x} + \sqrt{5x}$
$= (9x+1)\sqrt{5x}$

31. $2a\sqrt{20a^3b^2} + 2b\sqrt{45a^5}$
$= 2a(2ab\sqrt{5a}) + 2b(3a^2)\sqrt{5a}$
$= 4a^2b\sqrt{5a} + 6a^2b\sqrt{5a}$
$= 10a^2b\sqrt{5a}$

**33.** $3y\sqrt[4]{48x^5} - x\sqrt[4]{3x^5y^4}$

$= 3y(2x\sqrt[4]{3x}) - x(xy)\sqrt{3x}$

$= 6xy\sqrt[4]{3x} - x^2y\sqrt{3x}$

$= (6xy - x^2y)\sqrt[4]{3x}$

**35.** $x\sqrt[3]{27x^5y^2} - x^2\sqrt[3]{x^2y^2}$

$+ 2\sqrt[3]{x^8y^2}$

$= x(3x\sqrt[3]{x^2y^2}) - x^2\sqrt[3]{x^2y^2}$

$+ 2x^2\sqrt[3]{x^2y^2}$

$= 3x^2\sqrt[3]{x^2y^2} - x^2\sqrt[3]{x^2y^2}$

$+ 2x^2\sqrt[3]{x^2y^2}$

$= 4x^2\sqrt[3]{x^2y^2}$

**37.** $\sqrt[3]{16x^9y^{10}} - 2x^2y\sqrt[3]{2x^3y^7}$

$= 2x^3y^3\sqrt[3]{2y} - 2x^2y(xy^2\sqrt[3]{2y})$

$= 2x^3y^3\sqrt[3]{2y} - 2x^3y^3\sqrt[3]{2y} = 0$

**39.** $\frac{1}{\sqrt{2}} + \frac{\sqrt{2}}{2} = \frac{1}{\sqrt{2}} \cdot \frac{\sqrt{2}}{\sqrt{2}} + \frac{\sqrt{2}}{2}$

$= \frac{\sqrt{2}}{\sqrt{4}} + \frac{\sqrt{2}}{2}$

$= \frac{\sqrt{2}}{2} + \frac{\sqrt{2}}{2}$

$= \frac{2\sqrt{2}}{2}$

$= \sqrt{2}$

**41.** $\sqrt{3} - \frac{1}{\sqrt{3}}$

$= \sqrt{3} - \frac{\sqrt{3}}{\sqrt{3}\cdot\sqrt{3}}$

$= \sqrt{3} - \frac{\sqrt{3}}{3}$

$= \frac{3\sqrt{3}}{3} - \frac{\sqrt{3}}{3}$

$= \frac{2\sqrt{3}}{3}$

**43.** $\sqrt{\frac{1}{6}} + \sqrt{24} = \frac{1}{\sqrt{6}} + 2\sqrt{6}$

$= \frac{\sqrt{6}}{\sqrt{6}\ \sqrt{6}} + 2\sqrt{6}$

$= \frac{\sqrt{6}}{6} + 2\sqrt{6}$

$= \frac{\sqrt{6}}{6} + \frac{(2\sqrt{6})}{6}\ 6$

$= \frac{\sqrt{6}}{6} + \frac{12\sqrt{6}}{6}$

$= \frac{13\sqrt{6}}{6}$

**45.** $\sqrt{\frac{1}{2}} + 3\sqrt{2} + \sqrt{18}$

$= \frac{\sqrt{1}}{\sqrt{2}} \cdot \frac{\sqrt{2}}{\sqrt{2}} + 3\sqrt{2} + \sqrt{9}\ \sqrt{2}$

$= \frac{\sqrt{2}}{\sqrt{4}} + 3\sqrt{2} + 3\sqrt{2}$

$= \frac{\sqrt{2}}{2} + 6\sqrt{2}$

$= \frac{\sqrt{2}}{2} + \frac{12\sqrt{2}}{2}$

$= \frac{13\sqrt{2}}{2}$

**47.** $4\sqrt{x} + \dfrac{1}{\sqrt{x}} + \sqrt{\dfrac{1}{x}} = 4\sqrt{x} + \dfrac{1}{\sqrt{x}}.\dfrac{\sqrt{x}}{\sqrt{x}} + \dfrac{\sqrt{1}}{\sqrt{x}}.\dfrac{\sqrt{x}}{\sqrt{x}}$

$= 4\sqrt{x} + \dfrac{\sqrt{x}}{\sqrt{x^2}} + \dfrac{\sqrt{x}}{\sqrt{x^2}}$

$= 4\sqrt{x} + \dfrac{\sqrt{x}}{x} + \dfrac{\sqrt{x}}{x}$

$= 4\sqrt{x} + \dfrac{2\sqrt{x}}{x}$

$= 2\sqrt{x}\ (2 + \dfrac{1}{x})$

**49.** $\dfrac{1}{2}\sqrt{18} - \dfrac{3}{\sqrt{12}} - 3\sqrt{50}$

$= \dfrac{1}{2}(3\sqrt{2}) - \dfrac{3}{\sqrt{2}} - 3(5\sqrt{2})$

$= \dfrac{3\sqrt{2}}{2} - \dfrac{3\ \sqrt{2}}{\sqrt{2}\ \sqrt{2}} - 15\sqrt{2}$

$= \dfrac{3\sqrt{2}}{2} - \dfrac{3\sqrt{2}}{2} - 15\sqrt{2}$

$= -\ 15\sqrt{2}$

**51.** $(\sqrt{3} + 4)(\sqrt{3} + 5)$

$= \sqrt{9} + 5\sqrt{3} + 4\sqrt{3} + 20$

$= 3 + 9\sqrt{3} + 20$

$= 23 + 9\sqrt{3}$

**53.** $(1+\sqrt{5})(6+\sqrt{5})$

$= 6 + \sqrt{5} + 6\sqrt{5} + \sqrt{5}^2$

$= 6 + 7\sqrt{5} + 5$

$= 11 + 7\sqrt{5}$

**55.** $(4 - \sqrt{2})(5 + \sqrt{2})$

$= 20 + 4\sqrt{2} - 5\sqrt{2} - (\sqrt{2})^2$

$= 20 - \sqrt{2} - 2$

$= 18 - \sqrt{2}$

**57.** $(\sqrt{5} + \sqrt{3})\ (\sqrt{5} + \sqrt{3})$

$= \sqrt{25} + \sqrt{15} + \sqrt{15} + \sqrt{9}$

$= 5 + 2\sqrt{15} + 3$

$= 8 + 2\sqrt{15}$

**59.** $(\sqrt{2} - \sqrt{3})(\sqrt{3} + \sqrt{8})$

$= \sqrt{6} + \sqrt{16} - \sqrt{9} - \sqrt{24}$

$= \sqrt{6} + 4 - 3 - \sqrt{4}\ \sqrt{6}$

$= \sqrt{6} - 2\sqrt{6} + 1$

$= -\sqrt{6} + 1$

$= 1 - \sqrt{6}$

**61.** $(2 - \sqrt{3})^2$

$= 2^2 - 2(2)(\sqrt{3}) + (\sqrt{3})^2$

$= 4 - 4\sqrt{3} + 3$

$= 7 - 4\sqrt{3}$

**63.** $(\sqrt{x} + \sqrt{3})(\sqrt{x} - \sqrt{12})$

$= (\sqrt{x} + \sqrt{3})(\sqrt{x} - 2\sqrt{3})$

$= (\sqrt{x})^2 - 2\sqrt{3}\sqrt{x} + \sqrt{3}\sqrt{x} - 2(\sqrt{3})^2$

$= x - 2\sqrt{3x} + \sqrt{3x} - 6$

$= x - \sqrt{3x} - 6$

**65.** $(2\sqrt{3x} - \sqrt{y})(3\sqrt{3x} + \sqrt{y})$

$= 6(\sqrt{3x})^2 + 2\sqrt{3x}\sqrt{y} - 3\sqrt{3x}\sqrt{y}$

$- (\sqrt{y})^2$

$= 18x - \sqrt{3xy} - y$

**67.** $(\sqrt[3]{4} - \sqrt[3]{6})(\sqrt[3]{2} - \sqrt[3]{36})$

$= \sqrt[3]{4}\ \sqrt[3]{2} - \sqrt[3]{4}\ \sqrt[3]{36} - \sqrt[3]{6}\ \sqrt[3]{2} + \sqrt[3]{6}\ \sqrt[3]{36}$

$= \sqrt[3]{8} - \sqrt[3]{144} - \sqrt[3]{12} + \sqrt[3]{216}$

$= 2 - 2\sqrt[3]{18} - \sqrt[3]{12} + 6$

$= 8 - 2\sqrt[3]{18} - \sqrt[3]{12}$

**69.** $\sqrt{5} + 2\sqrt{5} = 3\sqrt{5}$

**71.** $\sqrt{125} + \sqrt{20}$

$= 5\sqrt{5} + 2\sqrt{5}$

$= 7\sqrt{5}$

**73.** $\frac{\sqrt{6}}{2} + \frac{1}{\sqrt{6}}$

$= \frac{\sqrt{6}}{2} + \frac{\sqrt{6}}{\sqrt{6}\sqrt{6}}$

$= \frac{\sqrt{6}}{2} + \frac{\sqrt{6}}{6}$

$= \frac{3\sqrt{6}}{6} + \frac{\sqrt{6}}{6}$

$= \frac{4\sqrt{6}}{6} = \frac{2\sqrt{6}}{3}$

**75.** $-\sqrt[4]{x} + 6\sqrt[4]{x} - 2\sqrt[4]{x}$

$= 3\sqrt[4]{x}$

**77.** $2 + 3\sqrt{y} - 6\sqrt{y} + 5$

$= 7 - 3\sqrt{y}$

**79.** $(3\sqrt{2} - 4)(\sqrt{2} + 5)$

$= 3(\sqrt{2})^2 + 15\sqrt{2} - 4\sqrt{2} - 20$

$= 6 + 11\sqrt{2} - 20$

$= -14 + 11\sqrt{2}$

**81.** $4\sqrt{3x^3} - \sqrt{12x}$

$= 4x\sqrt{3x} - 2\sqrt{3x}$

$= (4x-2)\sqrt{3x}$

**83.** $\frac{3}{\sqrt{y}} - \sqrt{\frac{9}{y}} + \sqrt{y}$

$= \frac{3}{\sqrt{y}} - \frac{3}{\sqrt{y}} + \sqrt{y}$

$= \sqrt{y}$

**85.** $2\sqrt[3]{24a^3y^4} + 4a\sqrt[3]{81y^4}$

$= 2\sqrt[3]{8a^3y^3}\ \sqrt[3]{3y} + 4a\sqrt[3]{27y^3}\ \sqrt[3]{3y}$

$= 2(2ay)\ \sqrt[3]{3y} + 4a(3y)\ \sqrt[3]{3y}$

$= 4ay\ \sqrt[3]{3y} + 12ay\ \sqrt[3]{3y}$

$= 16ay\ \sqrt[3]{3y}$

**87.** $2x\sqrt[3]{xy} + 5y\sqrt[3]{x^4y^4}$

$2x\sqrt[3]{xy} + 5y\ (xy)\ \sqrt[3]{xy}$

$= 2x\ \sqrt[3]{xy} + 5xy^2\ \sqrt[3]{xy}$

$= (5xy^2 + 2x)\ \sqrt[3]{xy}$

**89.** $\frac{2}{\sqrt{50}} - 3\sqrt{50} - \frac{1}{\sqrt{8}}$

$= \frac{2}{5\sqrt{2}} - 3(5\sqrt{2}) - \frac{1}{2\sqrt{2}}$

$= \frac{2\sqrt{2}}{5(\sqrt{2})(\sqrt{2})} - 15\sqrt{2} - \frac{1(\sqrt{2})}{2\sqrt{2}(\sqrt{2})}$

$= \frac{2\sqrt{2}}{10} - 15\sqrt{2} - \frac{\sqrt{2}}{4}$

$= \frac{\sqrt{2}}{5} - 15\sqrt{2} - \frac{\sqrt{2}}{4}$

$= \frac{4\sqrt{2}}{20} - \frac{300\sqrt{2}}{20} - \frac{5\sqrt{2}}{20}$

$= \frac{-301\sqrt{2}}{20}$

91. $\sqrt{3} + 3\sqrt{2}$
$= 1.73 + 3(1.41)$
$= 1.73 + 4.24$
$= 5.97$

93. $\frac{1}{\sqrt{3} + 2} = \frac{1(\sqrt{3}-2)}{(\sqrt{3}+2)\sqrt{3}-2)}$

$= \frac{\sqrt{3} - 2}{3 - 4}$

$= \frac{\sqrt{3} - 2}{-1} = 2 - \sqrt{3}$

$2 + \sqrt{3}$ would be greater that $2 - \sqrt{3}$ which equals $\frac{1}{\sqrt{3} + 2}$

95. $\frac{(2x^{-2}y^{3})^{2}(x^{-1}y^{-3})}{(xy^{2})^{-2}}$

$= \frac{4x^{-4}y^{6}\ x^{-1}y^{-3}}{x^{-2}y^{-4}}$

$= 4x^{(-4-1-(-2))}y^{(6-3-(-4))}$

$= 4x^{(-5+2)}y^{(3+4)}$

$= 4x^{-3}y^{7}$

$= \frac{4y^{7}}{x^{3}}$

97. $(\frac{x^{3/4}y^{2/3}}{x^{1/2}y})^{2}$

$= \frac{x^{6/4}y^{4/3}}{xy^{2}}$

$= x^{(6/4-1)}\ y^{4/3-2}$

$= x^{2/4}\ y^{-2/3}$

$= \frac{x^{1/2}}{y^{2/3}}$

**JUST FOR FUN**

1. $\frac{1}{\sqrt[5]{3x^{7}y^{9}}} + \frac{2\sqrt[5]{81x^{3}y}}{3x^{2}y^{2}}$

$= \frac{1}{xy\ \sqrt[5]{3x^{2}y^{4}}} + \frac{2\sqrt[5]{81x^{3}y}}{3x^{2}y^{2}}$

$= \frac{1\sqrt[5]{3^{4}x^{3}y}}{xy(\sqrt[5]{3x^{2}y^{4}})(\sqrt[5]{3^{4}x^{3}y})}$

$+ \frac{2\sqrt[5]{81x^{3}y}}{3x^{2}y^{2}}$

$= \frac{\sqrt[5]{81x^{3}y}}{xy(3xy)} + \frac{2\sqrt[5]{81x^{3}y}}{3x^{2}y^{2}}$

$= \frac{\sqrt[5]{81x^{3}y}}{3x^{2}y^{2}} + \frac{2\sqrt[5]{81x^{3}y}}{3x^{2}y^{2}}$

$= \frac{3\sqrt[5]{81x^{3}y}}{3x^{2}y^{2}}$

$= \frac{\sqrt[5]{81x^{3}y}}{x^{2}y^{2}}$

**EXERCISE SET 8.6**

Some checks have not been shown to conserve space.

1. $$\sqrt{x} = 5$$ $$(\sqrt{x})^2 = 5^2$$ $$x = 25$$

3. $$\sqrt[3]{x} = 2$$ $$(\sqrt[3]{x})^3 = 2^3$$ $$x = 8$$

5. $$\sqrt[4]{x} = 3$$ $$(\sqrt[4]{x})^4 = 3^4$$ $$x = 81$$

7. $$-\sqrt{2x + 4} = -6$$ $$(-\sqrt{2x + 4})^2 = (-6)^2$$ $$2x + 4 = 36$$ $$2x = 32$$ $$x = 16$$

Check: $-\sqrt{2x + 4} = -6$
$$-\sqrt{2(16) + 4} = -6$$
$$-\sqrt{32 + 4} = -6$$
$$-\sqrt{36} = -6$$
$$-6 = -6$$

9. $$\sqrt[3]{2x + 11} = 3$$ $$(\sqrt[3]{2x + 11})^3 = 3^3$$ $$2x + 11 = 27$$ $$2x = 16$$ $$x = 8$$

11. $$\sqrt[3]{3x} + 4 = 7$$ $$\sqrt[3]{3x} = 3$$ $$(\sqrt[3]{3x})^3 = 3^3$$ $$3x = 27$$ $$x = 9$$

13. $$\sqrt{2x - 3} = 2\sqrt{3x - 2}$$ $$(\sqrt{2x - 3})^2 = (2\sqrt{3x - 2})^2$$ $$2x - 3 = 12x - 8$$ $$-10x = -5$$ $$x = \frac{5}{10} = \frac{1}{2}$$

Check: $\sqrt{2x - 3} = 2\sqrt{3x - 2}$
$$\sqrt{2(1/2) - 3} = \sqrt{3(1/2) - 2}$$
$$\sqrt{1 - 3} = \sqrt{3/2 - 2}$$
$$\sqrt{-2} \neq 2\sqrt{-1/2}$$

False

$\frac{1}{2}$ is not a solution to this equation.

No solution.

15. $$\sqrt{5x + 10} = -\sqrt{3x + 8}$$ $$(\sqrt{5x + 10}) = (-\sqrt{3x + 8})^2$$ $$5x + 10 = 3x + 8$$ $$2x = -2$$ $$x = -1$$

Check: $\sqrt{5x + 10} = -\sqrt{3x + 8}$
$$\sqrt{5(-1) + 10} = -\sqrt{3(-1) + 8}$$
$$\sqrt{-5 + 10} = -\sqrt{-3 + 8}$$
$$\sqrt{5} \neq -\sqrt{5}$$

False.

Thus, -1 is not a solution to this equation.

17. $$\sqrt[3]{6x + 1} = \sqrt[3]{2x + 5}$$ $$(\sqrt[3]{6x + 1})^3 = (\sqrt[3]{2x + 5})^3$$ $$6x + 1 = 2x + 5$$ $$4x = 4$$ $$x = 1$$

**19.**

$$\sqrt{x^2 + 9x + 3} = -x$$
$$(\sqrt{x^2 + 9x + 3})^2 = (-x)^2$$
$$x^2 + 9x + 3 = x^2$$
$$9x + 3 = 0$$
$$9x = -3$$
$$x = \frac{-3}{9} = \frac{-1}{3}$$

Check: $\sqrt{x^2 + 9x + 3} = -x$

$$\sqrt{(-\tfrac{1}{3})^2 + 9(-\tfrac{1}{3}) + 3} = -(-\tfrac{1}{3})$$
$$\sqrt{\tfrac{1}{9} - 3 + 3} = \frac{1}{3}$$
$$\sqrt{\tfrac{1}{9}} = \frac{1}{3}$$
$$\frac{1}{3} = \frac{1}{3}$$

**21.**

$$\sqrt{m^2 + 4m - 20} = m$$
$$(\sqrt{m^2 + 4m - 20})^2 = (m)^2$$
$$m^2 + 4m - 20 = m^2$$
$$4m - 20 = 0$$
$$4m = 20$$
$$m = 5$$

**23.**

$$\sqrt{2y + 3} + y = 0$$
$$\sqrt{2y + 3} = -y$$
$$(\sqrt{2y + 3})^2 = (-y)^2$$
$$2y + 3 = y^2$$
$$y^2 - 2y - 3 = 0$$
$$(y - 3)(y + 1) = 0$$
$$y - 3 = 0 \text{ or } y + 1 = 0$$
$$y = 3 \qquad y = -1$$

Number 23 (continued)

Check:

$$\sqrt{2y + 3} + y = 0$$
$$\sqrt{2(3) + 3} + 3 = 0$$
$$\sqrt{9} + 3 = 0$$
$$3 + 3 = 0$$
$$6 \neq 0$$

False.
3 is not a solution.

Check:

$$\sqrt{2y + 3} + y = 0$$
$$\sqrt{2(-1) + 3} + (-1) = 0$$
$$\sqrt{1} - 1 = 0$$
$$1 - 1 = 0$$
$$0 = 0$$

True.
-1 is a solution.

**25.**

$$-\sqrt{x} = 2x - 1$$
$$(-\sqrt{x})^2 = (2x - 1)^2$$
$$x = 4x^2 - 4x + 1$$
$$0 = 4x^2 - 5x + 1$$
$$0 = (x - 1)(4x - 1)$$
$$x - 1 = 0 \qquad 4x - 1 = 0$$
$$x = 1 \qquad 4x = 1$$
$$x = 1/4$$

Check:

$-\sqrt{x} = 2x - 1$

$-\sqrt{x} = 2(1) - 1$

$-1 \neq 1$

False

Check:

$-\sqrt{x} = 2x - 1$

$-\sqrt{1/4} = 2(1/4) - 1$

$-1/2 = 1/2 - 1$

$-1/2 = 1/2 - 2/2$

$-1/2 = -1/2$

Only 1/4 is a solution. 1 is an extraneous root.

**27.**
$$\sqrt{x^2 + 8} = x + 2$$
$$(\sqrt{x^2 + 8})^2 = (x + 2)^2$$
$$x^2 + 8 = x^2 + 4x + 4$$
$$8 = 4x + 4$$
$$4 = 4x$$
$$1 = x$$

**29.**
$$\sqrt[3]{x - 12} = \sqrt[3]{5x+16}$$
$$(\sqrt[3]{x - 12})^3 = (\sqrt[3]{5x+16})^3$$
$$x - 12 = 5x + 16$$
$$-28 = 4x$$
$$-7 = x$$

Check:
$$\sqrt[3]{x - 12} = \sqrt[3]{5x+16}$$
$$\sqrt[3]{-7 - 12} = \sqrt[3]{5(-7)+16}$$
$$\sqrt[3]{-19} = \sqrt[3]{-19}$$

**31.**
$$\sqrt{x + 7} = 2x - 1$$
$$(\sqrt{x + 7})^2 = (2x - 1)^2$$
$$x + 7 = 4x^2 - 4x + 1$$
$$0 = 4x^2 - 5x - 6$$
$$0 = (4x + 3)(x - 2)$$
$$4x + 3 = 0 \quad \text{or} \quad x - 2 = 0$$
$$4x = -3 \qquad x = 2$$
$$x = -3/4$$

A check shows that 2 is a solution and that -3/4 is not a solution.

**33.**
$$\sqrt{2a - 3} = \sqrt{2a} - 1$$
$$(\sqrt{2a - 3})^2 = (\sqrt{2a} - 1)^2$$
$$2a - 3 = 2a - 2\sqrt{2a} + 1$$
$$-4 = -2\sqrt{2a}$$
$$-2 = -\sqrt{2a}$$
$$(-2)^2 = (-\sqrt{2a})^2$$
$$4 = 2a$$
$$2 = a$$

**35.**
$$\sqrt{x + 1} = 2 - \sqrt{x}$$
$$(\sqrt{x + 1})^2 = (2 - \sqrt{x})^2$$
$$x + 1 = 4 - 4\sqrt{x} + x$$
$$-3 = -4\sqrt{x}$$
$$(-3)^2 = (-4\sqrt{x})^2$$
$$9 = 16x$$
$$\frac{9}{16} = x$$

Check:
$$\sqrt{x + 1} = 2 = \sqrt{x}$$
$$\sqrt{\frac{9}{16} + 1} = 2 - \sqrt{\frac{9}{16}}$$
$$\sqrt{\frac{9}{16} + \frac{16}{16}} = 2 - \frac{3}{4}$$
$$\sqrt{\frac{25}{16}} = \frac{8}{4} - \frac{3}{4}$$
$$\frac{5}{4} = \frac{5}{4}$$

9/16 is a solution.

**37.**
$$\sqrt{x + 7} = 5 - \sqrt{x - 8}$$
$$(\sqrt{x + 7})^2 = (5 - \sqrt{x - 8})^2$$
$$x + 7 = 25 - 10\sqrt{x - 8} + x - 8$$
$$x + 7 = 17 - 10\sqrt{x - 8} + x$$
$$-10 = -10\sqrt{x - 8}$$
$$1 = \sqrt{x - 8}$$
$$(1)^2 = (\sqrt{x - 8})^2$$
$$1 = x - 8$$
$$9 = x$$

**39.**
$$\sqrt{b-3} = 4 - \sqrt{b+5}$$
$$(\sqrt{b-3})^2 = (4 - \sqrt{b+5})^2$$
$$b - 3 = 16 - 8\sqrt{b+5} + b + 5$$
$$b - 3 = 21 - 8\sqrt{b+5} + b$$
$$-24 = -8\sqrt{b+5}$$
$$3 = \sqrt{b+5}$$
$$(3)^2 = (\sqrt{b+5})^2$$
$$9 = b + 5$$
$$4 = b$$

**41.**
$$\sqrt{2x+8} - \sqrt{2x-4} = 2$$
$$(\sqrt{2x+8})^2 = (2 + \sqrt{2x-4})^2$$
$$2x+8 = 4+4\sqrt{2x-4} + (2x-4)$$
$$2x+8 = 2x+ 4\sqrt{2x-4}$$
$$8 = 4\sqrt{2x-4}$$
$$2 = \sqrt{2x-4}$$
$$2^2 = (\sqrt{2x-4})^2$$
$$4 = 2x-4$$
$$8 = 2x$$
$$x = 4$$

**43.**
$$\sqrt{2x+4} - \sqrt{x+3} - 1 = 0$$
$$(\sqrt{2x+4})^2 = (\sqrt{x+3}+1)^2$$
$$2x+4 = (x+3)+2\sqrt{x+3}+1$$
$$2x+4 = x+4+2\sqrt{x+3}$$
$$x = 2\sqrt{x+3}$$
$$x^2 = (2\sqrt{x+3})^2$$
$$x^2 = 4(x+3)$$
$$x^2 = 4x+12$$
$$x^2-4x-12 = 0$$
$$(x-6)(x+2) = 0$$
$$x - 6 = 0 \qquad x + 2 = 0$$
$$x = 6 \qquad x = -2$$

A check shows that 2 is not a solution and that 6 is a solution.

**45.** $p = \sqrt{2v}$ solve for v
$$p^2 = (\sqrt{2v})^2$$
$$p^2 = 2v$$
$$\frac{p^2}{2} = v$$

**47.** $V = \sqrt{2gh}$. Solve for g.
$$V = (\sqrt{2gh})^2$$
$$V^2 = 2gh$$
$$\frac{V^2}{2h} = g$$

**49.** $V = \sqrt{\frac{FR}{m}}$. Solve for F.
$$V^2 = \left(\sqrt{\frac{FR}{m}}\right)^2$$
$$V^2 = \frac{FR}{m}$$
$$V^2m = FR$$
$$\frac{V^2m}{R} = F$$

**51.** $x = \sqrt{\frac{m}{k}} \cdot V_0$.
Solve for m.
$$x^2 = \left(\sqrt{\frac{m}{k}}\, V_0\right)^2$$
$$x^2 = \frac{mV_0^2}{k}$$
$$x^2k = mV_0^2$$
$$\frac{x^2K}{V_0^2} = m$$

**53.** The equation $\sqrt{x+3} = -\sqrt{2x+1}$ can have no real solution because $\sqrt{x+3}$ cannot be equal to a negative number.

**55.** 0 is the only solution to the equation:

$\sqrt[3]{x^2} = -\sqrt[3]{x^2}$.

For any non-zero value, the left side of the equation is positive and the right side is negative. When x = 0, however, the left side will equal the right side, since 0 is neither positive nor negative.

**57.**

$$\frac{\frac{4a^2 - 9b^2}{4a^2 + 12ab + 9b^2}}{\frac{6a^2b}{8a^2b^2 - 12ab^3}}$$

$$= \frac{\frac{(2a - 3b)(2a + 3b)}{(2a + 3b)(2a + 3b)}}{\frac{6a^2b}{4ab^2(2a - 3b)}}$$

$$= \frac{6a^2b}{4ab^2(2a - 3b)} = \frac{3a}{2b(2a + 3b)}$$

**59.**

$$\frac{2}{x + 3} - \frac{1}{x - 3} + \frac{2x}{x^2 - 9}$$

$$= \frac{2}{x + 3} - \frac{1}{x - 3} + \frac{2x}{(x + 3)(x - 3)}$$

$$= \frac{2(x-3)}{(x + 3)(x - 3)} - \frac{(x + 3)}{(x + 3)(x - 3)} + \frac{2x}{(x + 3)(x - 3)}$$

$$= \frac{2x - 6 - x - 3 + 2x}{(x + 3)(x - 3)}$$

$$= \frac{3x - 9}{(x + 3)(x - 3)}$$

$$= \frac{3(x - 3)}{(x + 3)(x - 3)} = \frac{3}{x + 3}$$

**JUST FOR FUN**

**1.**

$$\sqrt{4x + 1} - \sqrt{3x - 2} = \sqrt{x - 5}$$

$$(\sqrt{4x + 1} - \sqrt{3x - 2})^2 = (\sqrt{x - 5})^2$$

$$4x + 1 - 2\sqrt{(4x + 1)(3x - 2)} + 3x - 2 = x - 5$$

$$7x - 1 - 2\sqrt{12x^2 - 5x - 2} = x - 5$$

$$-2\sqrt{12x^2 - 5x - 2} = -6x - 4$$

$$\sqrt{12x^2 - 5x - 2} = 3x + 2$$

$$(\sqrt{12x^2 - 5x - 2})^2 = (3x + 2)^2$$

$$12x^2 - 5x - 2 = 9x^2 + 12x + 4$$

$$3x^2 - 17x - 6 = 0$$

$$(x - 6)(3x + 1) = 0$$

$$x - 6 = 0 \quad \text{or} \quad 3x + 1 = 0$$

$$x = 6 \qquad x = -1/3$$

A check will show that 6 is a solution but -1/3 is not a solution.

**3.**

$$\sqrt{\sqrt{x + 25} - \sqrt{x}} = 5$$

$$(\sqrt{\sqrt{x + 25} - \sqrt{x}})^2 = 5^2$$

$$\sqrt{x + 25} - \sqrt{x} = 25$$

$$\sqrt{x + 25} = 25 + \sqrt{x}$$

$$(\sqrt{x + 25})^2 = (25 + \sqrt{x})^2$$

$$x + 25 = 625 + 50\sqrt{x} + x$$

$$25 = 625 + 50\sqrt{x}$$

$$-600 = 50\sqrt{x}$$

$$-12 = \sqrt{x}$$

$$(-12)^2 = (\sqrt{x})^2$$

$$144 = x$$

A check will show that 144 is NOT a solution.

**EXERCISE SET 8.7**

1. 5 inches

3. $\sqrt{45} \approx 6.71$ inches

5. $4^2 + 1.5^2 = x^2$
   $x = \sqrt{16 + 2.25}$
   $x = \sqrt{18.25} \approx 4.27$ meters

7. Let x = the length of the diagonal
   $x = \sqrt{12^2 + 5^2}$
   $x = \sqrt{169}$
   $x = 13$ in

9. Let x = the distance between home and second base which is the diagonal of the rectangle.

   $x = \sqrt{90^2 + 90^2}$
   $x = \sqrt{16200}$
   $x \approx 127.28$ feet

11. Let x = the unknown side of the square
    $A = L^2$
    $60 = L^2$
    $L = \sqrt{60} \approx 7.75$ in

13. $V = \sqrt{2gh}$ Recall from Example 2 that $g = 32$

    $V = \sqrt{2(32)(80)}$
    $= \sqrt{5120} \approx 71.55$ ft/sec

15. $T = 2\pi\sqrt{\frac{L}{32}}$
    $T = 2\pi\sqrt{\frac{8}{32}}$
    $= 2\pi\sqrt{\frac{1}{4}}$
    $= 2\pi(\frac{1}{2})$
    $= \pi \approx 3.14$ sec.

17. $A = \sqrt{s(s-a)(s-b)(s-c)}$
    Recall from Example 4 that

    $s = \frac{a + b + c}{2} = \frac{6 + 8 + 10}{2} = 12$

    Therefore

    $A = \sqrt{12(12-6)(12-8)(12-10)}$
    $= \sqrt{12(6)(4)(2)}$
    $= \sqrt{576}$
    $= 24$ sq. ft.

19. $N = .2(\sqrt{R})^3$
    $= .2(\sqrt{149.4})^3$
    $= .2\ (12.223)^3$
    $= .2(1826.106)$
    $= 365.2$ days

21. $R = \sqrt{F_1^2 + F_2^2}$
    $= \sqrt{600^2 + 800^2}$
    $= \sqrt{1,000,000}$
    $= 1000$ lb

23. $C = \sqrt{gh}$
    $= \sqrt{32(10)}$
    $= \sqrt{320}$
    $= 17.89$ ft/sec.

**25.** $x = \frac{-b \pm \sqrt{b^2 - 4ac}}{2a}$

(a) $x = \frac{-0 \pm \sqrt{0^2 - 4(1)(-4)}}{2(1)}$

$= \frac{\pm\sqrt{16}}{2}$

$x = \frac{\sqrt{16}}{2} = \frac{4}{2} = 2 \quad x = -\frac{\sqrt{16}}{2} = \frac{4}{2} = -2$

(b) $x = \frac{-1 \pm \sqrt{1^2 - 4(1)(-12)}}{2(1)}$

$= \frac{-1 \pm \sqrt{49}}{2}$

$x = \frac{-1 + \sqrt{49}}{2} = \frac{-1 + 7}{2} = \frac{6}{2} = 3$

or

$x = \frac{-1 - \sqrt{49}}{2} = \frac{-1 - 7}{2} = \frac{-8}{2} = -4$

(c) $x = \frac{-5 \pm \sqrt{5^2 - 4(2)(-12)}}{2(2)}$

$x = \frac{-5 \pm \sqrt{121}}{4}$

$x = \frac{-5 \pm 11}{4}$

$x = \frac{-5 + 11}{4} = \frac{6}{4} = \frac{3}{2}$

$X = \frac{-5 - 11}{4} = \frac{-16}{4} = -4$

(d) $x = \frac{-4 \pm \sqrt{4^2 - 4(-1)(5)}}{2(-1)}$

$x = \frac{-4 \pm \sqrt{36}}{-2}$

$x = \frac{-4 \pm 6}{-2}$

$x = \frac{-4 + 6}{-2} = \frac{2}{-2} = -1$

or

$x = \frac{-4 - 6}{-2} = \frac{-10}{-2} = 5$

**27.** To find the domain, set the denominator $3x^2 - 10x - 8 = 0$ and solve. The domain will be all real numbers except the solutions.

$3x^2 - 10x - 8 = -0$
$\quad -\underline{12} \times \underline{2} = -24$
$\quad -\underline{12} + \underline{2} = -10$

$3x^2 - 12x + 2x - 8 = 0$
$3x(x - 4) + 2(x - 4) = 0$
$(3x + 2)(x - 4) = 0$
$3x + 2 = 0 \qquad x - 4 = 0$
$x = -2/3 \qquad x = 4$

Domain = $\{x \mid x \neq -2/3 \text{ and } x \neq 4\}$

**29.** $\frac{4b}{b+1} = 4 - \frac{5}{b}$

Multiply each side of the equation by the LCD $b(b+1)$

$\frac{4b(b)(b+1)}{b+1} = 4b(b+1) - \frac{5b(b+1)}{b}$

$4b^2 = 4b^2 + 4b - 5b - 5$
$0 = -b - 5$
$b = -5$

**JUST FOR FUN**

1. Since the gravity on the moon is $\frac{1}{6}$ that of the earth, $g = \frac{1}{6}(32)$

$v = \sqrt{2(\frac{1}{6})(32)(10)}$
$= \sqrt{1066.67}$
$\approx 32.66$ ft/sec

**EXERCISE SET 8.8**

**1.** $3 = 3 + 0i$

**3.** $3 + \sqrt{-4} = 3 + \sqrt{4}\,\sqrt{-1}$
$= 3 + 2i$

**5.** $6 + \sqrt{3} = (6 + \sqrt{3}) + 0i$

**7.** $\sqrt{-25} = 0 + \sqrt{25}\,\sqrt{-1}$
$= 0 + 5i$

**9.** $4 + \sqrt{-12} = 4 + \sqrt{12}\,\sqrt{-1}$
$= 4 + \sqrt{4}\,\sqrt{3}\,\sqrt{-1}$
$= 4 + 2i\sqrt{3}$

**11.** $\sqrt{-25} = 2i = 0 + \sqrt{25}\,\sqrt{-1} - 2i$
$= 0 + 5i - 2i$
$= 0 + 3i$

**13.** $9 - \sqrt{-9} = 9 - \sqrt{9}\,\sqrt{-1}$
$= 9 - 3i$

**15.** $2i - \sqrt{-80} = 0 + 2i - \sqrt{16}\,\sqrt{5}\,\sqrt{-1}$
$= 0 + 2i - 4i\sqrt{5}$
$= 0 + (2 - 4\sqrt{5})i$

**17.** $(12 - 6i) + (3 + 2i)$
$= 12 - 6i + 3 + 2i$
$= 15 - 4i$

**19.** $(12 + \frac{5}{9}i) - (4 - \frac{3}{4}i)$
$= 12 + \frac{5}{9}i - 4 + \frac{3}{4}i$
$= 8 + \frac{5}{9}i + \frac{3}{4}i$
$= 8 + \frac{20i}{36} + \frac{27}{36}i$
$= 8 + \frac{47}{36}i$

**21.** $(13 - \sqrt{-4}) - (-5 + \sqrt{-9})$
$= 13 - \sqrt{-4} + 5 - \sqrt{-9}$
$= 13 - \sqrt{4}\,\sqrt{-1} + 5 - \sqrt{9}\,\sqrt{-1}$
$= 13 - 2i + 5 - 3i$
$= 18 - 5i$

**23.** $(\sqrt{3} + \sqrt{2}) + (3\sqrt{2} - \sqrt{-8})$
$= \sqrt{3} + \sqrt{2} + 3\sqrt{2} - \sqrt{8}\,\sqrt{-1}$
$= \sqrt{3} + \sqrt{2} + 3\sqrt{2} - \sqrt{4}\,\sqrt{2}\,\sqrt{-1}$
$= \sqrt{3} + \sqrt{2} + 3\sqrt{2} - 2i\sqrt{2}$
$= (4\sqrt{2} + \sqrt{3}) - 2i\sqrt{2}$

**25.** $(19 + \sqrt{-147}) + (\sqrt{-75})$
$= 19 + \sqrt{-147}\,\sqrt{-1} + \sqrt{75}\,\sqrt{-1}$
$= 19 + \sqrt{49}\,\sqrt{3}\,\sqrt{-1} + \sqrt{25}\,\sqrt{3}\,\sqrt{-1}$
$= 19 + 7i\sqrt{3} + 5i\sqrt{3}$
$= 19 + 12i\sqrt{3}$

**27.** $(\sqrt{12} + \sqrt{-49}) - (\sqrt{49} - \sqrt{-12})$
$= \sqrt{12} + \sqrt{49}\,\sqrt{-1} - \sqrt{49} + \sqrt{12}\,\sqrt{-1}$
$= \sqrt{4}\,\sqrt{3} + \sqrt{49}\,\sqrt{-1} - \sqrt{49} + \sqrt{4}\,\sqrt{3}\,\sqrt{-1}$
$= 2\sqrt{3} + 7i - 7 + 2i\sqrt{3}$
$= 2\sqrt{3} - 7 + (7 + 2\sqrt{3})i$

**29.** $2(-3 - 2i) = -6 - 4i$

**31.** $i(6 + i) = 6(i) + i(i)$
$= 6i + i^2$
$= 6i - 1$ or $-1 + 6i$

**33.** $-3.5i\,(6.4 - 1.8i)$
$= -3.5i(6.4) - 3.5i(-1.8i)$
$= -22.4i + 6.3i^2$
$= -22.4i + 6.3(-1)$
$= -6.3 - 22.4i$

**35.** $\sqrt{-4}(\sqrt{3} + 2i) = \sqrt{4}\ \sqrt{-1}\ (\sqrt{3} + 2i)$

$= 2i(\sqrt{3} + 2i)$

$= 2i\sqrt{3} + 4i^2$

$= 2i\sqrt{3} + 4(-1)$

$= 2i\sqrt{3} - 4$ or $-4 + 2i\sqrt{3}$

**37.** $\sqrt{-6}\ (\sqrt{3} + \sqrt{-6})$

$= \sqrt{6}\ \sqrt{-1}\ (\sqrt{3} + \sqrt{6}\ \sqrt{-1})$

$= i\sqrt{6}\ (\sqrt{3} + i\sqrt{6})$

$= i\sqrt{18} + i^2\sqrt{36}$

$= i\sqrt{9}\cdot\sqrt{2} - 1\sqrt{36}$

$= 3i\sqrt{2} - 6$ or $-6 + 3i\sqrt{2}$

**39.** $(3 + 2i)\ (1 + i)$

$= 3(1) + 3(i) + 2i(1) + 2i(i)$

$= 3 + 3i + 2i + 2i^2$

$= 3 + 3i + 2i + 2\ (-1)$

$= 3 + 3i + 2i - 2$

$= 1 + 5i$

**41.** $(4 - 6i)\ (3 - i)$

$= 4(3) - 4(i) - 6i(3) - 6i(-i)$

$= 12 - 4i - 18i + 6i^2$

$= 12 - 4i - 18i + 6(-1)$

$= 12 - 4i - 18i - 6$

$= 6 - 22i$

**43.** $(\frac{1}{4} + \sqrt{-3})\ (2 + \sqrt{3})$

$= \frac{1}{4}(2) + \frac{1}{4}\sqrt{3} + 2\sqrt{-3} + \sqrt{-3}\ \sqrt{3}$

$= \frac{1}{2} + \frac{\sqrt{3}}{4} + 2i\sqrt{3} + i\sqrt{3}\ \sqrt{3}$

$= \frac{1}{2} + \frac{\sqrt{3}}{4} + 2i\sqrt{3} + 3i$

$= \frac{1}{2} + \frac{\sqrt{3}}{4} + 2i\sqrt{3} + 3i$

$= (\frac{1}{2} + \frac{\sqrt{3}}{4}) + (2\sqrt{3} + 3)i$

**45.** $(5 - \sqrt{-8})(\frac{1}{4} + \sqrt{-2})$

$= 5(\frac{1}{4}) + 5\sqrt{-2} - \sqrt{-8}(\frac{1}{4}) - \sqrt{-8}\ \sqrt{-2}$

$= \frac{5}{4} + 5i\sqrt{2} - \frac{2i\sqrt{2}}{2} - i^2\sqrt{8}\ \sqrt{2}$

$= \frac{5}{4} + 5i\sqrt{2} - \frac{i\sqrt{2}}{2} - i^2\sqrt{16}$

$= \frac{5}{4} + 5i\sqrt{2} - \frac{i\sqrt{2}}{2} - 4(-1)$

$= \frac{5}{4} + 5i\sqrt{2} - \frac{i\sqrt{2}}{2} + 4$

$= (\frac{5}{4} + \frac{4}{1}) + (5\ \sqrt{2} - \frac{\sqrt{2}}{2})i$

$= \frac{21}{4} + \frac{9\sqrt{2}}{2}\ i$

**47.** $\frac{-5}{-3i} = \frac{-5}{-3i} \cdot \frac{i}{i} = \frac{-5i}{-3i^2} = \frac{-5i}{-3(-1)}$

$= -\frac{5i}{3}$

**49.** $\frac{2 + 3i}{2i} = \frac{2 + 3i}{2i} \cdot \frac{-i}{-i}$

$= \frac{(2 + 3i)(-i)}{-2i^2}$

$= \frac{-2i - 3i^2}{-2i^2}$

$= \frac{-2i - 3(-1)}{-2(-1)}$

$= \frac{-2i + 3}{2}$ or $\frac{3 - 2i}{2}$

**51.** $\frac{2 + 5i}{5i} = \frac{2 + 5i}{5i} \cdot \frac{-i}{-i}$

$= \frac{(2 + 5i)(-1)}{-5i^2}$

$= \frac{-2i - 5i^2}{-5i^2}$

$= \frac{-2i - 5(-1)}{-5(-1)}$

$= \frac{-2i + 5}{5}$ or $\frac{5 - 2i}{5}$

**53.** $$\frac{7}{7-2i}=\frac{7}{7-2i}\cdot\frac{7+2i}{7+2i}$$
$$=\frac{7(7+2i)}{49-4i^2}$$
$$=\frac{49+14i}{49-4(-1)}$$
$$=\frac{49+14i}{49+4}$$
$$=\frac{49+14i}{53}$$

**55.** $$\frac{6-3i}{4+2i}=\frac{6-3i}{4+2i}\cdot\frac{4-2i}{4-2i}$$
$$=\frac{(6-3i)(4-2i)}{16-4i^2}$$
$$=\frac{24-12i-12i+6i^2}{16-4i^2}$$
$$=\frac{24-12i-12i+6(-1)}{16-4(-1)}$$
$$=\frac{24-12i-12i-6}{16+4}$$
$$=\frac{18-24i}{20}$$
$$=\frac{\overset{1}{\cancel{2}}(9-12i)}{\underset{10}{\cancel{20}}}$$
$$=\frac{9-12i}{10}$$

**57.** $$\frac{4}{6-\sqrt{-4}}=\frac{4}{6-\sqrt{4}\sqrt{-1}}$$
$$=\frac{4}{6-2i}\cdot\frac{6+2i}{6+2i}$$
$$=\frac{4(6+2i)}{36-4i^2}$$
$$=\frac{24+8i}{36-4(-1)}$$
$$=\frac{24+8i}{36+4}$$
$$=\frac{\overset{1}{\cancel{8}}(3+i)}{\underset{5}{\cancel{40}}}$$
$$=\frac{3+i}{5}$$

**59.** $$\frac{\sqrt{6}}{\sqrt{3}-\sqrt{-9}}=\frac{\sqrt{6}}{\sqrt{3}-\sqrt{-9}\sqrt{-1}}$$
$$=\frac{\sqrt{6}}{\sqrt{3}-3i}\cdot\frac{\sqrt{3}+3i}{\sqrt{3}+3i}$$
$$=\frac{\sqrt{6}(\sqrt{3}+3i)}{\sqrt{9}-9i^2}$$
$$=\frac{\sqrt{18}+3i\sqrt{6}}{3-9(-1)}$$
$$=\frac{\sqrt{9}\sqrt{2}+3i\sqrt{6}}{3+9}$$
$$=\frac{3\sqrt{2}+3i\sqrt{6}}{12}$$
$$=\frac{\overset{1}{\cancel{3}}(\sqrt{2}+i\sqrt{6})}{\underset{4}{\cancel{12}}}$$
$$=\frac{\sqrt{2}+i\sqrt{6}}{4}$$

**61.** $$\frac{\sqrt{10}+\sqrt{-3}}{5-\sqrt{-20}}=\frac{\sqrt{10}+\sqrt{3}\sqrt{-1}}{5-\sqrt{4}\sqrt{5}\sqrt{-1}}$$
$$=\frac{\sqrt{10}+\sqrt{3}i}{5-2i\sqrt{5}}\cdot\frac{5+2i\sqrt{5}}{5+2i\sqrt{5}}$$
$$=\frac{(\sqrt{10}+i\sqrt{3})(5+2i\sqrt{5})}{5^2-4i^2\sqrt{5}^2}$$
$$=\frac{5\sqrt{10}+2i\sqrt{50}+5i\sqrt{3}+2i^2\sqrt{15}}{5^2-4(-1)(5)}$$
$$=\frac{5\sqrt{10}+2i\sqrt{25}\sqrt{2}+5i\sqrt{3}+2(-1)\sqrt{15}}{25+20}$$
$$=\frac{(5\sqrt{10}-2\sqrt{15})+(10\sqrt{2}+5\sqrt{3})i}{45}$$

**63.** $$(4-2i)+(3-5i)$$
$$=7-7i$$

**65.** $(8 - \sqrt{-6}) - (2 - \sqrt{-24})$
$= (8 - i\sqrt{6}) - (2 - 2i\sqrt{6})$
$= 8 - i\sqrt{6} - 2 + 2i\sqrt{6}$
$= 6 + i\sqrt{6}$

**67.** $5.2(4 - 3.2i)$
$= 5.2(4) - 5.2(3.2i)$
$= 20.8 - 16.64i$

**69.** $\sqrt{-6}\,(\sqrt{3} - \sqrt{-10})$
$= i\sqrt{6}(\sqrt{3} - i\sqrt{10})$
$= i\sqrt{6}\,\sqrt{3} - i^2\sqrt{6}\,\sqrt{10}$
$= i\sqrt{18} - i^2\sqrt{60}$
$= 3i\sqrt{2} - (-1)2\,\sqrt{15}$
$= 3i\sqrt{2} + 2\sqrt{15}$
$= 2\sqrt{15} + 3i\sqrt{2}$

**71.** $(\sqrt{3} + 2i)\,(\sqrt{6} - \sqrt{-8})$
$= (\sqrt{3} + 2i)\,(\sqrt{6} - 2i\sqrt{2})$
$= \sqrt{3}\,\sqrt{6} - 2i\sqrt{3}\,\sqrt{2} + 2i\sqrt{6}$
$\quad - (2i)(2i\sqrt{2})$
$= \sqrt{18} - 2i\sqrt{6} + 2i\sqrt{6} - 4i^2\sqrt{2}$
$= 3\sqrt{2} - 4(-1)\sqrt{2}$
$= 3\sqrt{2} + 4\sqrt{2}$
$= 7\sqrt{2}$

**73.** $\dfrac{5 - 4i}{2i}\,\dfrac{(-2i)}{(-2i)}$
$= \dfrac{-10i + 8i^2}{-4(i)^2}$
$= \dfrac{-10i + 8(-1)}{-4(-1)}$
$= \dfrac{-8 - 10i}{4} = \dfrac{-4 - 5i}{2}$

**75.** $\dfrac{4}{\sqrt{3} - \sqrt{-4}} = \dfrac{4}{\sqrt{3} - 2i}$
$= \dfrac{4(\sqrt{3} + 2i)}{(\sqrt{3} - 2i)(\sqrt{3} + 2i)}$
$= \dfrac{4\sqrt{3} + 8i}{3 - 4i^2}$
$= \dfrac{4\sqrt{3} + 8i}{3 - 4(-1)}$
$= \dfrac{4\sqrt{3} + 8i}{7}$

**77.** $(5 - \frac{5}{9}i) - (2 - \frac{3}{5}i)$
$= 5 - \frac{5}{9}i - 2 + \frac{3}{5}i$
$= 3 - \frac{5}{9}i + \frac{3}{5}i$
$= 3 - \frac{25}{45}i + \frac{27}{45}i$
$= 3 + \frac{2}{45}i$

**79.** $(\frac{2}{3} - \frac{1}{5}i)(\frac{3}{5} - \frac{3}{4}i)$
$= (\frac{2}{3})(\frac{3}{5}) - \frac{2}{3}(\frac{3}{4}i) - (\frac{1}{5}i)(\frac{3}{5})$
$\quad + (\frac{1}{5}i)(\frac{3}{4}i)$
$= \frac{2}{5} - \frac{1}{2}i - \frac{3}{25}i + \frac{3}{20}i^2$
$= \frac{2}{5} - \frac{1}{2}i - \frac{3}{25}i + \frac{3}{20}(-1)$
$= (\frac{2}{5} - \frac{3}{20}) + (-\frac{1}{2} - \frac{3}{25})i$
$= (\frac{8}{20} - \frac{3}{20}) + (-\frac{25}{50} - \frac{6}{50})i$
$= \frac{5}{20} - \frac{31}{50}i$
$= \frac{1}{4} - \frac{31}{50}i$

**81.** $\frac{(5-3i)(5i)}{-5i\ (5i)} = \frac{25i - 15i^2}{-25i^2}$

$= \frac{25i - 15(-1)}{-25(-1)}$

$= \frac{15 + 25i}{25} = \frac{3 + 5i}{5}$

**83.** $(5.23-6.41i) - (8.56-4.5i)-7.1i$

$= 5.23-6.41i-8.56+4.5i-7.1i$

$= -3.33-9.01i$

**85.** $z = \frac{V}{I}$

$z = \frac{1.8 + .5i}{.6i}$

$= \frac{(1.8 + .5i)(-.6i)}{(.6i)(-.6i)}$

$= \frac{-1.08i - .3i^2}{-.36i^2}$

$= \frac{-1.08i - .3(-1)}{-.36(-1)}$

$= \frac{.3 - 1.08i}{.36}$

$= .83 - 3i$

**87.** $z_1 = \frac{z_1\ z_2}{z_1 + z_2}$

$= \frac{(2 - i)(4 + i)}{(2 - i) + (4 + i)}$

$= \frac{8 + 2i - 4i - i^2}{6}$

$= \frac{8 - 2i - (-1)}{6}$

$= \frac{9 - 2i}{6}$

$= 1.5 - .33i$

**89.** $i^{23} = (i^4)^5 \cdot i^3 = 1^5 i^3$

$= 1i^3$

$= i^3$

$= -i$

**91.** $i^{-1} = \frac{1}{i}$

$= \frac{-i}{i(-i)}$

$= \frac{-i}{-i^2}$

$= \frac{-i}{-(-1)}$

$= \frac{-i}{1}$

$= -i$

Therefore, $i^{-1} = -i$

**93.** False.
0 + 2i is a complex number but
0 + 2i is not a real number.

**95.** False.
2i + (-2i) = 0
and 0 is not an imaginary number.

**97.** True

**99.** Answers varies from student to student.

**101.** $f(x) = x^2 - 2x + 1$
$a = 1 \quad b = -2 \quad c = 1$

Axis of symmetry

$x = -\frac{b}{2a} = -\frac{(-2)}{2(a)} = 1$

Vertex

$f(1) = 1^2 - 2(1) + 1$

$= 0$

Vertex (1,0)
The parabola opens up

Number 101 (Continued)

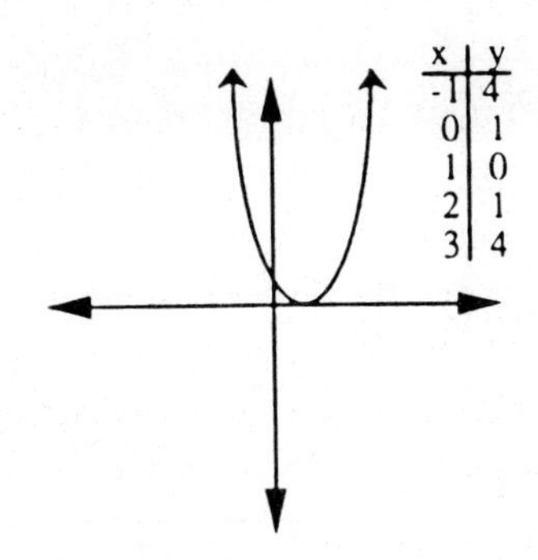


**103.** $15r^2 s^2 + rs - 6$

$\underline{10} \times (\underline{-9}) = (-6)(15) = -90$

$\underline{10} + (\underline{-9}) = 1$

$= 15r^2 s^2 - 9rs + 10rs - 6$
$= 3rs(5rs - 3) + 2(5rs - 3)$
$= (3rs + 2)(5rs - 3)$

## EXERCISE SET 8.9

1. The value under the radical must be greater than or equal to zero. Therefore:
$x + 2 \geq 0$
$x \geq -2$
Domain: = $\{x | x \geq -2\}$

3. The value under the radical must be greater than or equal to zero. Therefore:
$4x \geq 0$
$x \geq 0$
Domain = $\{x | x \geq 0\}$

5. $f(x) = \sqrt{4 - x}$
$4 - x \geq 0$
$4 \geq x$
$x \leq 4$

Domain = $\{x | x \leq 4\}$

7. $f(x) = -\sqrt{x + 4}$
$x + 4 \geq 0$
$x \geq -4$

Domain = $\{x | x \geq -4\}$

9. $f(x) = -\sqrt{3x + 5}$
$3x + 5 \geq 0$
$3x \geq -5$
$x \geq -5/3$
Domain = $\{x | x \geq -5/3\}$

11. $f(x) = \sqrt{7x - \frac{1}{2}}$

$7x - \frac{1}{2} \geq 0$

$7x \geq \frac{1}{2}$

$\frac{1}{7} \cdot 7x \geq \frac{1}{2} \cdot \frac{1}{7}$

$x \geq \frac{1}{14}$

Domain = $\{x \mid x \geq \frac{1}{14}\}$

13. $f(x) = \sqrt{\frac{5}{3} - 4x}$

$\frac{5}{3} - 4x \geq 0$

$\frac{5}{3} \geq 4x$

$\frac{1}{4}(\frac{5}{3}) \geq 4x\ (\frac{1}{4})$

$\frac{5}{12} \geq x$

$X \leq 5/12$
Domain = $\{x | x \leq 5/12\}$

15. $f(x) = \sqrt{x + 4}$
$x + 4 \geq 0$
$x \geq -4$

Domain = $\{x | x \geq -4\}$

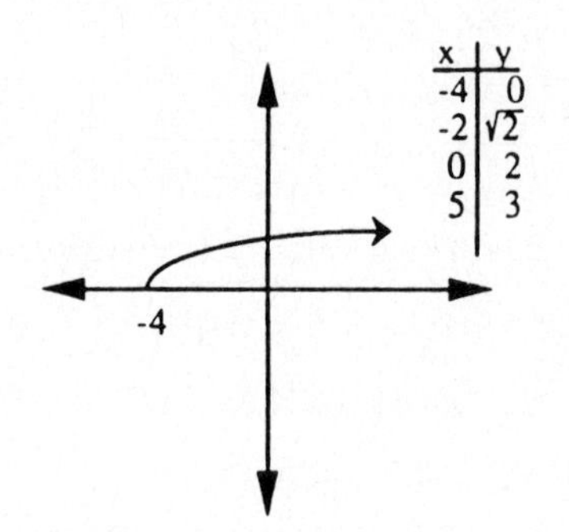


Range = $\{y | y \geq 0\}$

17. $f(x) = \sqrt{x - 2}$
$x - 2 \geq 0$
$x \geq 2$

Domain = $\{x | x \geq 2\}$

| x | y |
|---|---|
| 2 | 0 |
| 3 | 1 |
| 6 | 2 |

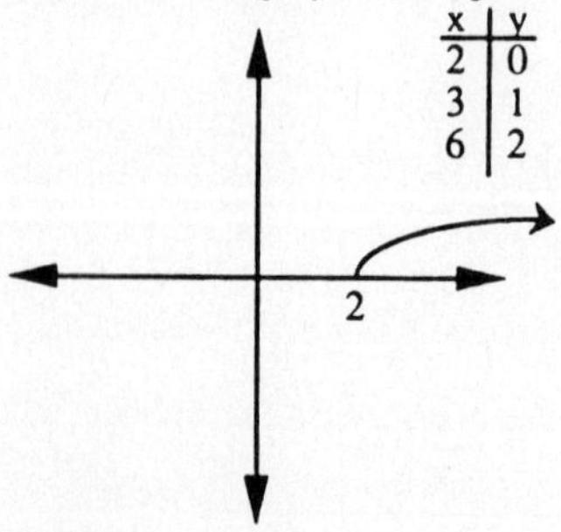

2

Range = $\{y | y \geq 0\}$

**19.** $f(x) = -\sqrt{x+3}$

$x + 3 \geq 0$

$x \geq -3$

Domain = $\{x \mid x \geq -3\}$

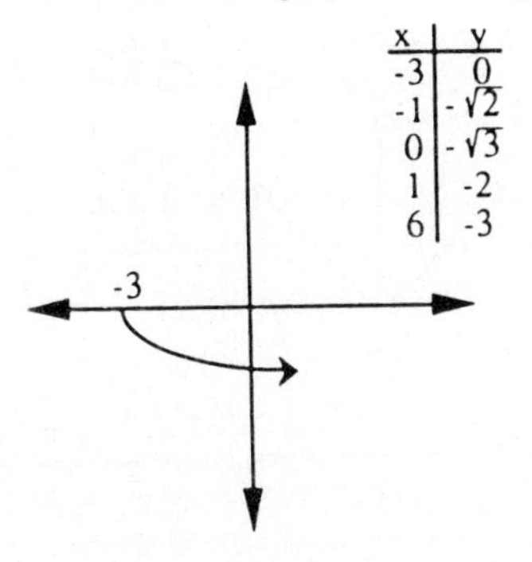

| x | y |
|---|---|
| -3 | 0 |
| -1 | $-\sqrt{2}$ |
| 0 | $-\sqrt{3}$ |
| 1 | -2 |
| 6 | -3 |



Range = $\{y \mid y \leq 0\}$

**21.** $f(x) = \sqrt{3-x}$

$3 - x \geq 0$

$3 \geq x \qquad x \leq 3$

Domain = $\{x \mid x \leq 3\}$

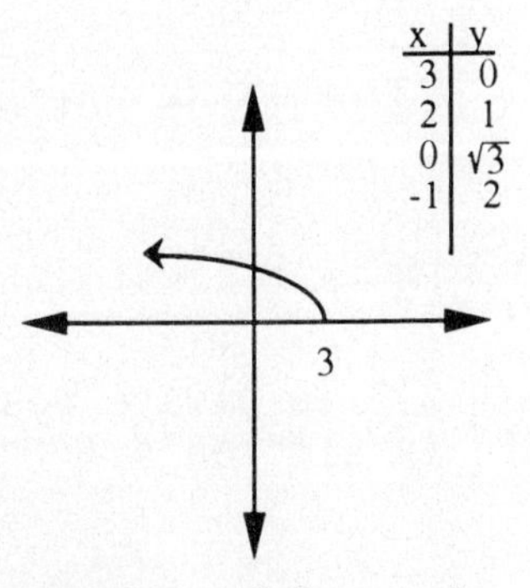

| x | y |
|---|---|
| 3 | 0 |
| 2 | 1 |
| 0 | $\sqrt{3}$ |
| -1 | 2 |



Range = $\{y \mid y \geq 0\}$

**23.** $f(x) = -\sqrt{2-x}$

$2 - x \geq 0$

$2 \geq x$

$x \leq 2$

Domain = $\{x \mid x \leq 2\}$

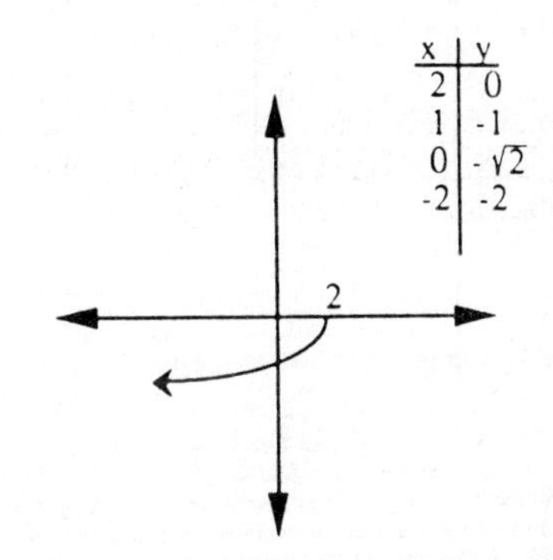

| x | y |
|---|---|
| 2 | 0 |
| 1 | -1 |
| 0 | $-\sqrt{2}$ |
| -2 | -2 |



Range = $\{y \mid y \leq 0\}$

**25.** $f(x) = \sqrt{2x}$

$2x \geq 0$

$x \geq 0$

Domain = $\{x \mid x \geq 0\}$

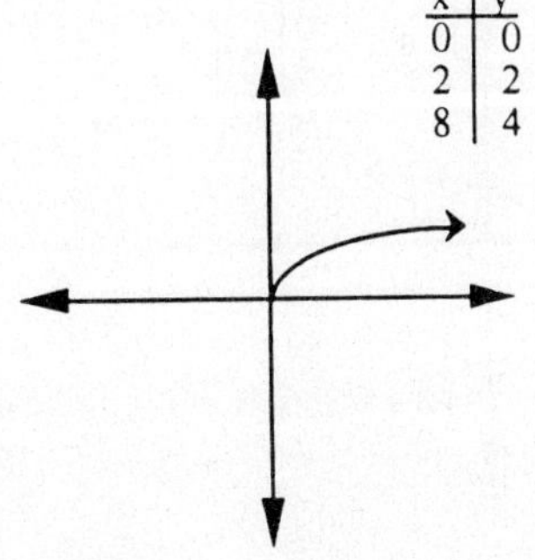

| x | y |
|---|---|
| 0 | 0 |
| 2 | 2 |
| 8 | 4 |

Range = $\{y \mid y \geq 0\}$

**27.** $f(x) = \sqrt{2x + 1}$

$2x + 1 \geq 0$

$x \geq -1/2$

Domain $= \{x | x \geq -1/2\}$

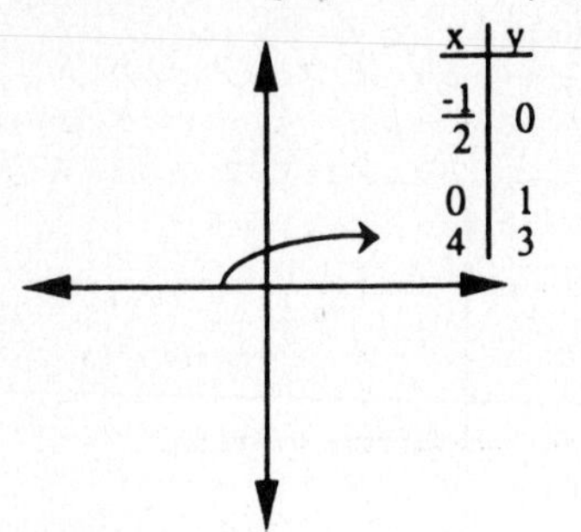

| x | y |
|---|---|
| $\frac{-1}{2}$ | 0 |
| 0 | 1 |
| 4 | 3 |

Range $= \{y | y \geq 0\}$

**29.** $f(x) = \sqrt{2x + 4}$

$2x + 4 \geq 0$

$2x \geq -4$

$x \geq -2$

Domain $= \{x | x \geq -2\}$

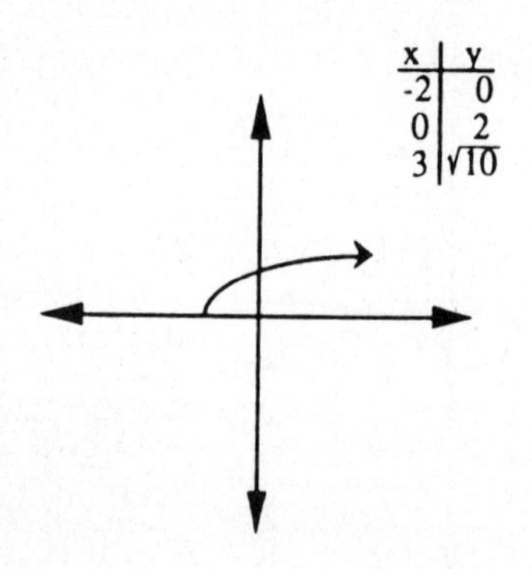

| x | y |
|---|---|
| -2 | 0 |
| 0 | 2 |
| 3 | $\sqrt{10}$ |

Range $= \{y | y \geq 0\}$

**31.** $y = \sqrt{x^2}$

(a) Since $x^2$ is always positive so the domain is all real numbers.

(b)

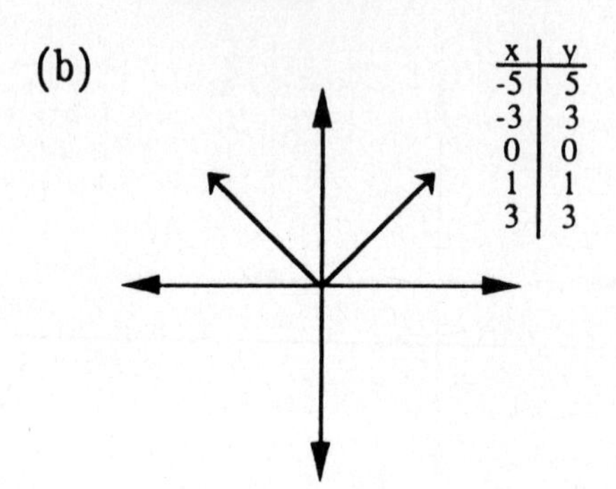

| x | y |
|---|---|
| -5 | 5 |
| -3 | 3 |
| 0 | 0 |
| 1 | 1 |
| 3 | 3 |

(c) Range $= \{y | y \geq 0\}$

**33.** $y = \sqrt{x^2 - 4}$

The value 0 is not in the domain of the function since

$\sqrt{0^2 - 4} = \sqrt{-4}$ which is not a real number.

**35.** $y = \sqrt{(x + 2)^2}$

Since $(x + 2)^2$ is always positive, the domain is all reals.

**37.** $y = \sqrt[3]{x}$

No, this is a cube root function.
The domain is all reals.

The range is all reals.

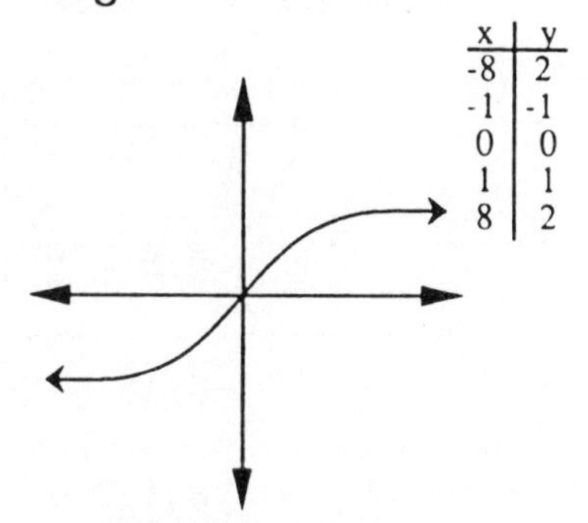


The range is all reals.

**39.** $\dfrac{\frac{2}{a} + \frac{1}{2a}}{\frac{a}{2} - a}$

$= \dfrac{2a(\frac{2}{a}) + 2a(\frac{1}{2a})}{2a(\frac{a}{2}) - 2a(a)}$

$= \dfrac{4 + 1}{a^2 - 2a^2}$

$= -\dfrac{5}{a^2}$

**41.** $\sqrt{x + 6} + \sqrt{x + 2} - 1 = 1$

$(\sqrt{x + 6}) = (2 - \sqrt{x + 2})^2$

$x + 6 = 4 - 4\sqrt{x + 2} + x + 2$

$x + 6 = 6 + x - 4\sqrt{x + 2}$

$0 = -4\sqrt{x + 2}$

$0 = \sqrt{x + 2}$

$0 = x + 2$

$-2 = x$

The solution $x = -2$ checks in the original equation.

**JUST FOR FUN**

**1.** $f(x) = \sqrt{4 - x^2}$

(a) $4 - x^2 \geq 0$

$(2 - x)(2 + x) \geq 0$

$2 - x = 0 \quad 2 + x = 0$

$x = 2 \quad x = -2$

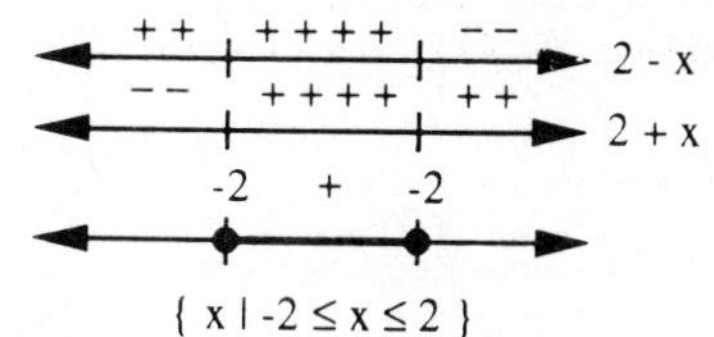


Domain $= \{x | -2 \leq x \leq 2\}$

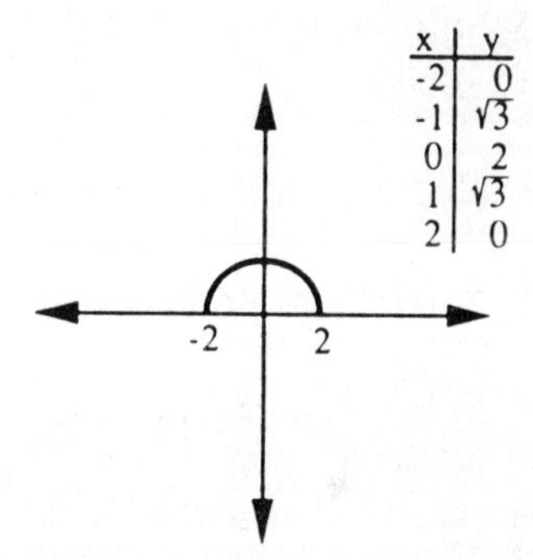


Range $= \{y | 0 \leq y \leq 2\}$

**3.** $f(x) = \sqrt[3]{x - 2}$

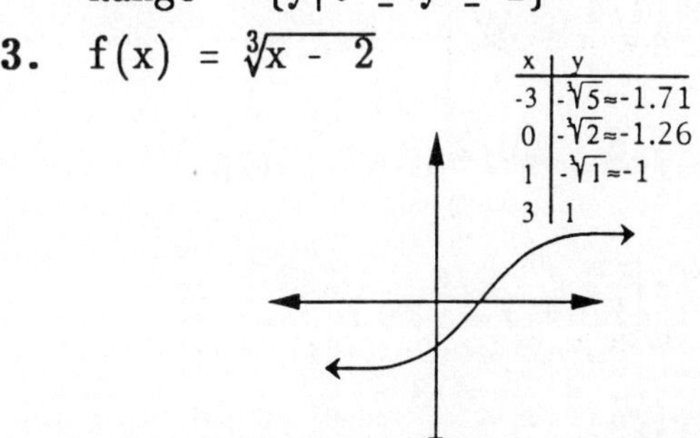

x | y
-3 | -∛5≈-1.71
0 | -∛2≈-1.26
1 | -∛1≈-1
3 | 1


The domain is all real numbers.
The range is all real numbers.

**REVIEW EXERCISES**

1. $\sqrt{9} = 3$

3. $\sqrt[3]{-8} = -2$

5. $\sqrt[3]{27} = 3$

7. $\sqrt{6^2} = 6$

9. $\sqrt{(-7)^2} = \sqrt{49} = 7$

11. $\sqrt{x^2} = |x|$

13. $\sqrt{(x-y)^2} = |x-y|$

15. $\sqrt{x^5} = x^{5/2}$

17. $\sqrt[4]{y^3} = y^{3/4}$

19. $x^{1/2} = \sqrt{x}$

21. $2^{1/4} = \sqrt[4]{2}$

23. $\sqrt{5^6} = 5^{6/2} = 5^3 = 125$

25. $\sqrt[6]{y^2} = y^{2/6} = y^{1/3} = \sqrt[3]{y}$

27. $(\sqrt[3]{4})^6 = 4^{6/3} = 4^2 = 16$

29. $\sqrt[20]{x^4} = x^{4/20} = x^{1/5} = \sqrt[5]{x}$

31. $8^{1/3} = \sqrt[3]{8} = 2$

33. $(-25)^{1/2} = \sqrt{-25}$
Not a real number.

35. $x^{1/2}\, x^{2/5} = x^{1/2 + 2/5} = x^{9/10}$

37. $$\left(\frac{y^{-3/5}}{y^{1/5}}\right)^{2/3} = \frac{y^{-3/5\cdot 2/3}}{y^{1/5\cdot 2/3}}$$
$$= \frac{y^{-2/5}}{y^{2/15}}$$
$$= y^{-2/5 - 2/15}$$
$$= y^{-6/15 - 2/15}$$
$$= y^{-8/15}$$
$$= \frac{1}{y^{8/15}}$$

39. $$x^{1/5} + x^{6/5} = x^{1/5}\left(\frac{x^{1/5}}{x^{1/5}} + \frac{x^{6/5}}{x^{1/5}}\right)$$
$$= x^{1/5}(1 + x^{6/5 - 1/5})$$
$$= x^{1/5}\ (1 + x^{5/5})$$
$$= x^{1/5}(1 + x)$$

41. $$x^{-1/2} + x^{-2/3}$$
$$= x^{-3/6} + x^{-4/6}$$
$$= x^{-4/6}\ \left(\frac{x^{-3/6}}{x^{-4/6}} + \frac{x^{-4/6}}{x^{-4/6}}\right)$$
$$= x^{-2/3}\ (x^{-3/6 - (-4/6)} + 1)$$
$$= x^{-2/3}(x^{1/6} + 1)$$
$$= \frac{x^{1/6} + 1}{x^{2/3}}$$

**43.** Let $a = x^{1/4}$
Substitute a for
$x^{1/4}$ in $x^{1/2} - 5x^{1/4} + 4$
$= (x^{1/4})^2 - 5x^{1/4} + 4$
$= a^2 - 5a - 4$
$= (a - 4)(a - 1)$
Replace a with $x^{1/4}$:
$= (x^{1/4} - 4)(x^{1/4} - 1)$

**45.** Let $a = x^{1/2}$
Substitute $a = x^{1/2}$ in
$6x - 5x^{1/2} - 6$
$= 6(x^{1/2})^2 - 5x^{1/2} - 6$
$= 6a^2 - 5a - 6$
$= (3a + 2)(2a - 3)$
Replace a with $x^{1/2}$:
$= (3x^{1/2} + 2)(2x^{1/2} - 3)$

**47.** $\sqrt{24} = \sqrt{4}\,\sqrt{6}$
$= 2\sqrt{6}$

**49.** $\sqrt[3]{16} = \sqrt[3]{8}\,\sqrt[3]{2}$
$= 2\,\sqrt[3]{2}$

**51.** $\sqrt{50x^3y^2} = \sqrt{25}\,\sqrt{2}\,\sqrt{x^2}\sqrt{x}\,\sqrt{y^6}\,\sqrt{y}$
$= 5xy^3\,\sqrt{2xy}$

**53.** $\sqrt[4]{16x^9y^{12}} = \sqrt[4]{16}\,\sqrt[4]{x^8}\,\sqrt[4]{x}\,\sqrt[4]{y^{12}}$
$= 2x^2y^3\,\sqrt[4]{x}$

**55.** $\sqrt[5]{32x^{12}y^7z^{17}}$
$= \sqrt[5]{32}\,\sqrt[5]{x^{10}}\,\sqrt[5]{x^2}\,\sqrt[5]{y^5}\,\sqrt[5]{y^2}$
$\sqrt[5]{z^{15}}\,\sqrt[5]{z^2}$
$= 2x^2\,yz^3\,\sqrt[5]{x^2y^2z^2}$

**57.** $\sqrt{5x}\,\sqrt{8x5} = \sqrt{40x^6}$
$= \sqrt{4}\,\sqrt{10}\,\sqrt{x^6}$
$= 2x^3\,\sqrt{10}$

**59.** $(\sqrt[3]{5x^2y^3})^2 = \sqrt[3]{(5x^2y^3)^2}$
$= \sqrt[3]{25x^4y^6}$
$= \sqrt[3]{25}\,\sqrt[3]{x^3}\,\sqrt[3]{x}\,\sqrt[3]{y^6}$
$= xy^2\,\sqrt[3]{25x}$

**61.** $\sqrt{3x}\,(\sqrt{12x} - \sqrt{20})$
$= \sqrt{36x^2} - \sqrt{60x}$
$= 6x - 2\sqrt{15x}$

**63.** $\sqrt[3]{2x^2y^3}\,(\sqrt[3]{4x^4y^2} + \sqrt[3]{9xy^{12}})$
$= \sqrt[3]{8x^6y^{10}} + \sqrt[3]{18x^3y^{15}}$
$= 2x^2y^3\,\sqrt[3]{y} + xy^5\,\sqrt[3]{18}$

**65.** $\sqrt{\frac{1}{4}} = \frac{\sqrt{1}}{\sqrt{4}} = \frac{1}{2}$

**67.** $\sqrt[3]{\frac{3}{81}} = \sqrt[3]{\frac{1}{27}} = \frac{\sqrt[3]{1}}{\sqrt[3]{27}} = \frac{1}{3}$

**69.** $\dfrac{\sqrt[3]{2x^9}}{\sqrt[3]{16x^6}} = \sqrt[3]{\dfrac{2x^9}{16x^6}}$

$= \sqrt[3]{\dfrac{x^3}{8}} = \dfrac{\sqrt[3]{x^3}}{\sqrt[3]{8}} = \dfrac{x}{2}$

**71.** $\sqrt[3]{\dfrac{54x^3y^6}{2y^3}} = \sqrt[3]{27x^3y^3} = 3xy$

**73.** $\dfrac{1}{\sqrt{2}} = \dfrac{1}{\sqrt{2}}\dfrac{\sqrt{2}}{\sqrt{2}} = \dfrac{\sqrt{2}}{2}$

**75.** $\dfrac{x}{\sqrt{7}} = \dfrac{x}{\sqrt{7}}\dfrac{\sqrt{7}}{\sqrt{7}} = \dfrac{x\sqrt{7}}{7}$

**77.** $\sqrt{\dfrac{y}{x}} = \dfrac{\sqrt{y}}{\sqrt{x}}$

$= \dfrac{\sqrt{y}}{\sqrt{x}}\dfrac{\sqrt{x}}{\sqrt{x}} = \dfrac{\sqrt{xy}}{x}$

**79.** $\dfrac{2}{\sqrt[3]{x}} = \dfrac{2\sqrt[3]{x^2}}{\sqrt[3]{x}\,\sqrt[3]{x^2}}$

$= \dfrac{2\sqrt[3]{x^2}}{\sqrt[3]{x^3}} = \dfrac{2\sqrt[3]{x^2}}{x}$

**81.** $\sqrt{\dfrac{3x^2}{y}} = \dfrac{\sqrt[3]{x^2}}{\sqrt{y}} = \dfrac{x\sqrt{3}\,\sqrt{y}}{\sqrt{y}\,\sqrt{y}} = \dfrac{x\sqrt{3y}}{y}$

**83.** $\sqrt{\dfrac{25x^2y^5}{3z}} = \dfrac{\sqrt{25x^2y^5}}{3z} = \dfrac{5xy^2\sqrt{y}}{\sqrt{3z}}$

$= \dfrac{5xy^2\sqrt{y}}{\sqrt{3z}}\dfrac{\sqrt{3z}}{\sqrt{3z}}$

$= \dfrac{5xy^2\sqrt{3y\,z}}{3z}$

**85.** $\sqrt{\dfrac{27x^4z^5}{5y}} = \dfrac{\sqrt{27x^4z^5}}{\sqrt{5y}}$

$= 3x^2z^2\dfrac{\sqrt{3}}{\sqrt{5y}}$

$= \dfrac{3x^2z^2\sqrt{3z}\,\sqrt{5y}}{\sqrt{5y}\,\sqrt{5y}}$

$= \dfrac{3x^2z^2\sqrt{15yz}}{5y} = \dfrac{3x^2y\sqrt{15yz}}{5}$

**87.** $\sqrt[3]{\dfrac{4x^5y^3}{x^6}} = \sqrt[3]{\dfrac{4y^3}{x}} = \dfrac{\sqrt[3]{4y^3}}{\sqrt[3]{x}}$

$= \dfrac{y\sqrt[3]{4}\,\sqrt[3]{x^2}}{\sqrt[3]{4}\,\sqrt[3]{x^2}} = \dfrac{y\sqrt[3]{4x^2}}{x}$

**89.** $(3 - \sqrt{2})(3 + \sqrt{2})$
$= 3^2 - (\sqrt{2})^2$
$= 9 - 2 = 7$

**91.** $(\sqrt{x} + y)\ (\sqrt{x} - y)$
$= (\sqrt{x})^2 - y^2$
$= x - y^2$

93. $(\sqrt{3} + 5)^2 = (\sqrt{3})^2 + 2(5)\sqrt{3} + 5^2$
$= 3 + 10\sqrt{3} + 25$
$= 28 + 10\sqrt{3}$

95. $(\sqrt{x} - \sqrt{3y})(\sqrt{x} + \sqrt{5y})$
$= (\sqrt{x})^2 + \sqrt{x}\sqrt{5y} - \sqrt{x}\,\sqrt{3y} - \sqrt{3y}\,\sqrt{5y}$
$= x + \sqrt{5xy} - \sqrt{3xy} - \sqrt{15y^2}$
$= x + \sqrt{5xy} - \sqrt{3xy} - y\sqrt{15}$

97. $(\sqrt[3]{2x} - \sqrt[3]{3y})(\sqrt[3]{3x} - \sqrt[3]{2y})$
$= \sqrt[3]{2x}\,(\sqrt[3]{3x}) - (\sqrt[3]{2x})(\sqrt[3]{2y})$
$- \sqrt[3]{3y}(\sqrt[3]{3x}) + \sqrt[3]{3y}\,\sqrt[3]{2y}$
$= \sqrt[3]{6x^2} - \sqrt[3]{4xy} - \sqrt[3]{9xy} + \sqrt[3]{6y^2}$

99. $\dfrac{3}{4-\sqrt{2}} = \dfrac{3(4 + \sqrt{2})}{(4-\sqrt{2})(4+\sqrt{2})}$

$= \dfrac{12 + 3\sqrt{2}}{16 - 2} = \dfrac{12 + 3\sqrt{2}}{14}$

101. $\dfrac{\sqrt{x}\,(\sqrt{x} - \sqrt{y})}{(\sqrt{x} + \sqrt{y})(\sqrt{x} - \sqrt{y})} = \dfrac{x - \sqrt{xy}}{x - y}$

103. $\dfrac{(\sqrt{x} - 3)(\sqrt{x} - 3)}{(\sqrt{x} + 3)(\sqrt{x} - 3)} = \dfrac{x - 6\sqrt{x} + 9}{x - 9}$

105. $\dfrac{4 \qquad (\sqrt{y+2} + 3)}{(\sqrt{y+2} - 3)(\sqrt{y+2} + 3)}$

$= \dfrac{4\sqrt{y+2} + 12}{(y + 2) - 9} = \dfrac{4\sqrt{y+2} + 12}{y - 7}$

107. $\sqrt[3]{x} + 3\sqrt[3]{x} - 2\sqrt[3]{x} = 2\sqrt[3]{x}$

109. $\sqrt{3} + \sqrt{27} - \sqrt{192}$
$\sqrt{3} + 3\sqrt{3} - \sqrt{64}\,\sqrt{3}$
$\sqrt{3} + 3\sqrt{3} - 8\sqrt{3} = -\,4\sqrt{3}$

111. $4\sqrt{2} - \dfrac{3}{\sqrt{32}} + \sqrt{50}$
$= 4\sqrt{2} - \dfrac{3}{4\sqrt{2}} + 5\sqrt{2}$
$= 4\sqrt{2} - \dfrac{3\sqrt{2}}{4\sqrt{2}\sqrt{2}} + 5\sqrt{2}$
$= 4\sqrt{2} - \dfrac{3\sqrt{2}}{8} + 5\sqrt{2}$
$= \frac{8}{8}(4\sqrt{2}) - \dfrac{3\sqrt{2}}{8} + (\frac{8}{8})5\sqrt{2}$
$= \dfrac{32\sqrt{2}}{8} - \dfrac{3\sqrt{2}}{8} + \dfrac{40\sqrt{2}}{8} = \dfrac{69\sqrt{2}}{8}$

113. $2\sqrt[3]{x^7y^8} - \sqrt[3]{x^4y^2} + 3\sqrt[3]{x^{10}y^2}$
$= 2x^2y^2\,\sqrt{xy^2} - x\,\sqrt[3]{xy^2} + 3x^3\,\sqrt{xy^2}$
$= (2x^2y^2 - x + 3x^3)\,\sqrt[3]{xy^2}$

115. $\sqrt{3x + 4} = 5$
$(\sqrt{3x + 4})^2 = 5^2$
$3x + 4 = 25$
$3x = 21$
$x = 7$
7 checks in the original equation.

117. $\sqrt{x^2 + 2x - 4} = x$
$(\sqrt{x^2 + 2x - 4})^2 = (x)^2$
$x^2 + 2x - 4 = x^2$
$2x - 4 = 0$
$x = 2$
2 checks in the original equation.

119. $\sqrt{x^2 + 5} = x + 1$

$(\sqrt{x^2 + 5})^2 = (x + 1)^2$

$x^2 + 5 = x^2 + 2x + 1$

$5 = 2x + 1$

$4 = 2x$

$x = 2$

2 checks in the original equation.

121. $\sqrt{6x - 5} - \sqrt{2x + 6} - 1 = 0$

$\sqrt{6x - 5} = \sqrt{2x + 6} + 1$

$(\sqrt{6x - 5})^2 = (\sqrt{2x + 6} + 1)^2$

$6x - 5 = 2x + 6 + 2\sqrt{2x + 6} + 1$

$6x - 5 = 2x + 7 + 2\sqrt{2x + 6}$

$4x - 12 = 2\sqrt{2x + 6}$

$\frac{4x}{2} - \frac{12}{12} = \frac{2}{2}\sqrt{2x + 6}$

$2x - 6 = \sqrt{2x + 6}$

$(2x - 6)^2 = (\sqrt{2x + 6})^2$

$4x^2 - 24x + 36 = 2x + 6$

$\frac{4x^2}{2} - \frac{26x}{2} + \frac{30}{2} = \frac{0}{2}$

$2x^2 - 13x + 15 = 0$

Factor $2x^2 - 14x + 15$:

$-\underline{10} \times \underline{-3} = 30$

$-\underline{10} + \underline{-3} = -13$

$2x^2 - 10x - 3x + 15 = 0$

$2x(x - 5) - 3(x - 5) = 0$

$(2x - 3)(x - 5) = 0$

$2x - 3 = 0 \qquad x - 5 = 0$

$x = 3/2 \qquad x = 5$

The solution x = 5 checks in the original equation but x = 3/2 does not check. Therefore the only solution is x = 5.

123. $r = \sqrt{\frac{A}{\pi}}$

Solve for A

$r^2 = (\sqrt{\frac{A}{\pi}})^2$

$r^2 = \frac{A}{\pi}$

$\pi r^2 = A$

125. $\nu = \sqrt{2gh}$

$= \sqrt{2(32)(20)}$

$= \sqrt{1280} \approx 35.78$

127. $5 = 5 + 0i$

129. $2 - \sqrt{-4}$

$= 2 - \sqrt{-1}\sqrt{4}$

$= 2 - 2i$

131. $(3 + 2i) + (4 - i) = 7 + i$

133. $(5 + \sqrt{-9}) - (3 - \sqrt{-4})$

$= (5 + \sqrt{-1}\sqrt{9}) - (3 - \sqrt{-1}\sqrt{4})$

$= (5 + 3i) - (3 - 2i)$

$= 5 + 3i - 3 + 2i$

$= 2 + 5i$

135. $4(3 + 2i)$

$= 12 + 8i$

137. $\sqrt{8}(\sqrt{-2} + 3)$

$= \sqrt{8}(\sqrt{-1}\sqrt{2} + 3)$

$= \sqrt{8}(i\sqrt{2} + 3)$

$= i\sqrt{16} + 3\sqrt{8}$

$= 4i + 3(2\sqrt{2})$

$= 4i + 6\sqrt{2}$

or $6\sqrt{2} + 4i$

**139.** $(4 + 3i)(2 - 3i)$

$= 8 - 12i + 6i - 9i^2$
$= 8 - 6i - 9(-1)$
$= 8 - 6i + 9$
$= 17 - 6i$

**141.** $\frac{2}{3i} = \frac{2(-3i)}{3i(-3i)}$

$= \frac{-6i}{-9(i^2)}$

$= \frac{-6i}{-9(-1)}$

$= \frac{-6i}{9}$

$= \frac{-2i}{3}$

**143.** $\frac{2+\sqrt{3}}{2i} = \frac{(2+\sqrt{3})(-2i)}{2i(-2i)}$

$= \frac{-4i - 2i\sqrt{3}}{-4i^2}$

$= \frac{-4i - 2i\sqrt{3}}{4}$

$= \frac{2(-2i - i\sqrt{3})}{4}$

$= \frac{-2i - i\sqrt{3}}{2}$

$= \frac{(-2 - \sqrt{3})i}{2}$

**145.** $\frac{\sqrt{3}}{5 - \sqrt{-6}} = \frac{\sqrt{3}}{5 - i\sqrt{6}}$

$= \frac{\sqrt{3}\,(5+i\sqrt{6})}{(5-i\sqrt{6})(5+i\sqrt{6})}$

$= \frac{5\sqrt{3} + i\sqrt{18}}{25 + 6}$

$= \frac{5\sqrt{3} + 3i\sqrt{2}}{31}$

**147.** $f(x) = \sqrt{x-6}$

$x - 6 \geq 0$
$x \geq 0$

Domain $= \{x | x \geq 6\}$

**149.** $f(x) = \sqrt{10 - 4x}$

$10 - 4x \geq 0$
$-4x \geq -10$
$x \leq 5/2$

Domain $= \{x | x \leq 5/2\}$

**151.** $f(x) = \sqrt{x}$

$x \geq 0$

Domain $= \{x | x \geq 0\}$

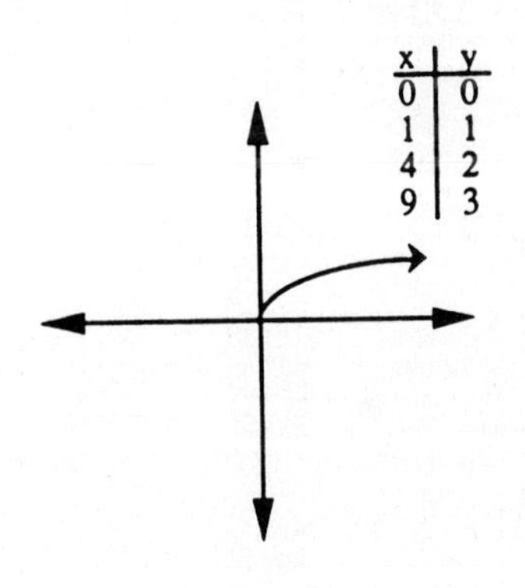

| x | y |
|---|---|
| 0 | 0 |
| 1 | 1 |
| 4 | 2 |
| 9 | 3 |

Range $= \{y | y \geq 0\}$

**153.** $f(x) = \sqrt{5 - x}$

$5 - x \geq 0$

$5 \geq x$ or $x \leq 5$

Domain = $\{x | x \leq 5\}$

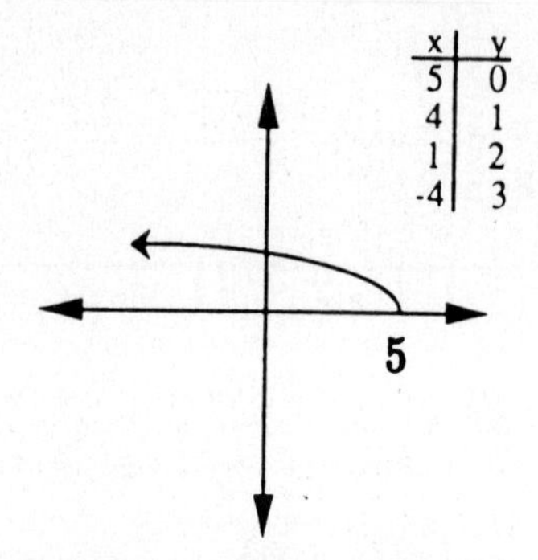


Range = $\{y | y \geq 0\}$

**PRACTICE TEST**

1. $\sqrt{(-26)^2} = |-26| = 26$

2. $\sqrt{(3x - 4)^2} = |3x - 4|$

3. $\left(\frac{y^{2/3}\ y^{-1}}{y^{1/4}}\right)^2 = \frac{y^{4/3}\ y^{-2}}{y^{2/4}}$

$= y^{4/3 - 2 - 1/2}$

$= y^{(8/6 - 12/6 - 3/6)}$

$= y^{-7/6}$

$= \frac{1}{y^{7/6}}$

4. $2x^{2/3} + x^{1/3} - 10$

Let $a = x^{1/3}$.

Substitute $a = x^{1/3}$ in the equation.

$2x^{2/3} + x^{1/3} - 10$

$= 2(x^{1/3})^2 + x^{1/3} - 10$

$= 2a^2 + a - 10$

$= (2a + 5)(a - 2)$

Replace a with $x^{1/3}$

$= (2x^{1/3} + 5)(x^{1/3} - 2)$

5. $\sqrt{50x^5y^8} = \sqrt{25x^4y^8}\ \sqrt{2x}$

$= 5x^2y^4\sqrt{2x}$

6. $\sqrt[3]{4x^5y^2}\ \sqrt[3]{10x^6y^8} = \sqrt[3]{40x^{11}y^{10}}$

$= \sqrt[3]{8x^9y^9}\cdot\sqrt[3]{5x^2y} = 2x^3y^3\ \sqrt[3]{5x^2y}$

7. $\sqrt{\frac{2x^4y^5}{8z}} = \frac{\sqrt{2x^4y^5}}{\sqrt{8z}} = \frac{\sqrt{x^4y^4}\sqrt{2y}}{\sqrt{4}\ \sqrt{2z}}$

$= \frac{x^2y^2\sqrt{2y}}{2\sqrt{2z}}\cdot\frac{\sqrt{2z}}{\sqrt{2z}} = \frac{x^2y^2\sqrt{4yz}}{2\sqrt{(2z)^2}}$

$= \frac{x^2y^2\sqrt{4}\ \sqrt{yz}}{2(2z)} = \frac{2x^2y^2\sqrt{yz}}{4z}$

$= \frac{x^2y^2\sqrt{yz}}{2z}$

8. $\sqrt[3]{\frac{1}{x}} = \frac{\sqrt[3]{1}}{\sqrt[3]{x}}\cdot\frac{\sqrt[3]{x^2}}{\sqrt[3]{x^2}} = \frac{\sqrt[3]{x^2}}{\sqrt[3]{x^3}} = \frac{\sqrt[3]{x^2}}{x}$

9. $\frac{\sqrt{2}}{2+\sqrt{8}}\cdot\frac{2-\sqrt{8}}{2-\sqrt{8}} = \frac{\sqrt{2}(2-\sqrt{8})}{4-8} = \frac{2\sqrt{2}-\sqrt{16}}{-4}$

$= \frac{2\sqrt{2} - 4}{-4} = \frac{2(\sqrt{2} - 2)}{-4} = \frac{\sqrt{2} - 2}{-2}$

$= \frac{-\sqrt{2} + 2}{2}$

10. $\sqrt{27} + 2\sqrt{3} - 5\sqrt{75}$

$= \sqrt{9}\ \sqrt{3} + 2\sqrt{3} - 5\sqrt{25}\ \sqrt{3}$

$= 3\sqrt{3} + 2\sqrt{3} - 25\sqrt{3}$

$= -20\sqrt{3}$

11. $\sqrt[3]{8x^3y^5} + 2\sqrt[3]{x^6y^8}$

$= 2xy\ \sqrt[3]{y^2} + 2x^2y^2\ \sqrt[3]{y^2}$

$= (2xy + 2x^2y^2)\ \sqrt[3]{y^2}$

12. $(\sqrt{5} - 3)(2 - \sqrt{8})$

$= \sqrt{5}(2) - \sqrt{5}\ \sqrt{8} - 3(2) + 3\sqrt{8}$

$= 2\sqrt{5} - \sqrt{40} - 6 + 3\sqrt{8}$

$= 2\sqrt{5} - \sqrt{4}\ \sqrt{10} - 6 + 3\sqrt{4}\ \sqrt{2}$

$= 2\sqrt{5} - 2\sqrt{10} - 6 + 6\sqrt{2}$

**13.** $\sqrt{4x - 3} = 7$

$(\sqrt{4x - 3})^2 = 7^2$

$4x - 3 = 49$

$4x = 52$

$x = 13$

The solution x = 13 checks in the original equation.

**14.** $\sqrt{x^2 - x - 12} = x + 3$

$(\sqrt{x^2 - x - 12})^2 = (x + 3)^2$

$x^2 - x - 12 = x^2 + 6x + 9$

$-x - 12 = 6x + 9$

$-12 = 7x + 9$

$-21 = 7x$

$x = -3$

The solution x = - 3 checks in the original equation.

**15.** $\sqrt{x - 15} = \sqrt{x} - 3$

$(\sqrt{x - 15})^2 = (\sqrt{x} - 3)^2$

$x - 15 = x - 6\sqrt{x} + 9$

$-15 = -6\sqrt{x} + 9$

$\frac{-24}{-6} = \frac{-6\sqrt{x}}{-6}$

$4 = \sqrt{x}$

Take the equation $4 = \sqrt{x}$ and then square each side to obtain 16 = x. This value checks in the original solution.

**16.** $w = \frac{\sqrt{2gh}}{4}$. Solve for h

$4w = \sqrt{2gh}$

$(4w)^2 = (\sqrt{2gh})^2$

$16w^2 = 2gh$

$\frac{16w^2}{2g} = \frac{2gh}{2g}$

$\frac{8w^2}{g} = h$

**17.** Let x be the length of the ladder.

$x = \sqrt{12^2 + 5^2}$

$= \sqrt{169}$

$= 13$ feet

**18.** $(6 - \sqrt{-4})(3 + \sqrt{-2})$

$= (6 - 2i)\ (3 + i\sqrt{2})$

$= 18 + 6i\sqrt{2} - 6i - 2i^2\sqrt{2}$

$= 18 + 6i\sqrt{2} - 6i - 2(-i)\sqrt{2}$

$= 18 + 6i\sqrt{2} - 6i + 2\sqrt{2}$

$= (18 + 2\sqrt{2}) + (6\sqrt{2} - 6)i$

**19.** $\frac{\sqrt{5}}{2 - \sqrt{-8}}$

$= \frac{\sqrt{5}}{2 - 2i\sqrt{2}}$

$= \frac{\sqrt{5}\ (2 + 2i\sqrt{2})}{(2 - 2i\sqrt{2})(2 + 2i\sqrt{2})}$

$= \frac{2\sqrt{5} + 2i\sqrt{10}}{4 - 4i^2(\sqrt{2})^2}$

$= \frac{2\sqrt{5} + 2i\sqrt{10}}{4 + 8}$

$= \frac{2\sqrt{5} + 2i\sqrt{10}}{12}$

$= \frac{\sqrt{5} + i\sqrt{10}}{6}$

20. $f(x) = \sqrt{x + 2}$

(a) $x + 2 \geq 0$

$x \geq -2$

Domain = $\{x | x \geq -2\}$

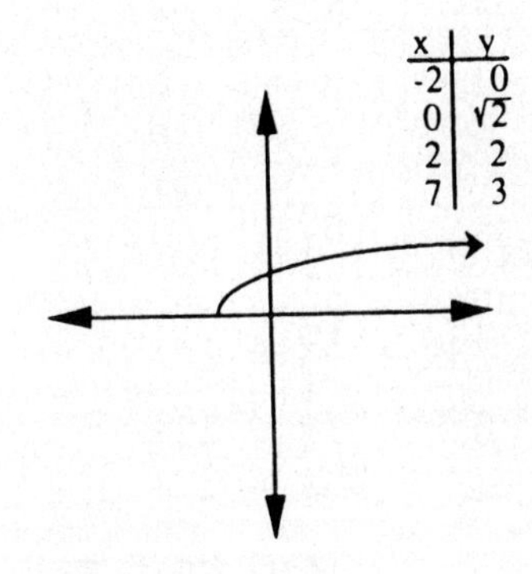


Range = $\{y | y \geq 0\}$

**CUMULATIVE REVIEW TEST**

**1.**

$$\frac{1}{5}(x - 3) = \frac{3}{4}(x + 3) - x$$

$$20\left(\frac{1}{5}(x - 3)\right) = 20\left(\frac{3}{4}(x + 3)\right) - 20x$$

$$4(x - 3) = 5(3)(x + 3) - 20x$$

$$4x - 12 = 15x + 45 - 20x$$

$$4x - 12 = -5x + 45$$

$$9x - 12 = 45$$

$$9x = 57$$

$$x = \frac{57}{9}$$

**2.** $5x - 2y = 6$
The graph will be a straight line

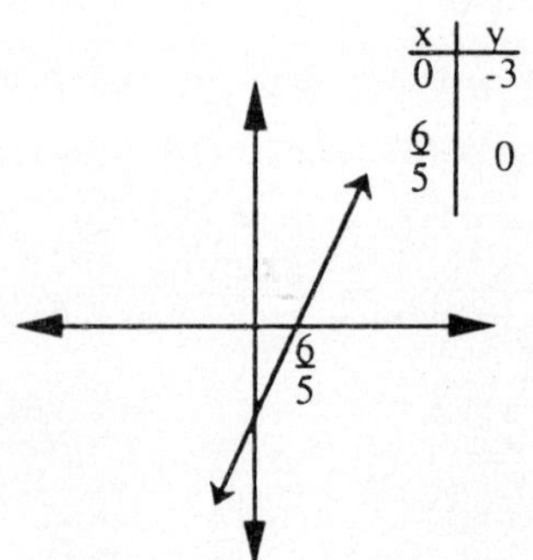


**3.** (a) Relation: A set of ordered pairs.

(b) Function: A set of ordered pairs no two of which have the same first coordinate.

**4.** (1) $x - 4y = 6$
(2) $3x - y = 2$
Multiply equation (1) by -3. Add the result to equation (2).

$$\begin{aligned} (1)\quad -3x + 12y &= -18 \\ (2)\quad 3x - y &= 2 \\ \hline 11y &= -6 \\ y &= -\frac{16}{11} \end{aligned}$$

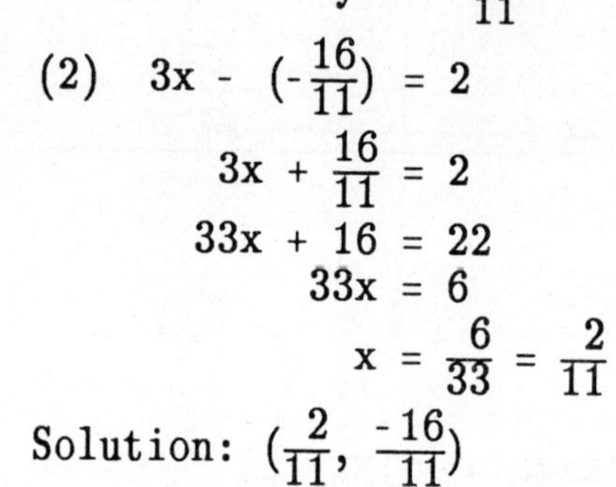

$$(2)\quad 3x - \left(-\frac{16}{11}\right) = 2$$

$$3x + \frac{16}{11} = 2$$

$$33x + 16 = 22$$

$$33x = 6$$

$$x = \frac{6}{33} = \frac{2}{11}$$

Solution: $\left(\frac{2}{11}, \frac{-16}{11}\right)$

**5.**

$$\left(\frac{3x^2y^{-2}}{x^4y^{-5}}\right)\left(\frac{2xy^3}{x^2y^{-3}}\right)^{-1}$$

$$= \left(\frac{3x^2y^{-2}}{x^4y^{-5}} \cdot \frac{2^{-1}x^{-1}y^{-3}}{x^{-2}y^3}\right)$$

$$= \frac{3x^{2-1}y^{-2-3}}{2x^{4-2}y^{-5+3}} = \frac{3xy^{-5}}{2x^2y^{-2}}$$

$$= \frac{3y^{-5-(-2)}}{2x} = \frac{3y^{-3}}{2x} = \frac{3}{2xy^3}$$

**6.**

$$\begin{array}{rrrr} & 3x^2 & -\ 4x & -\ 6 \\ & & 2x & -\ 5 \\ \hline 6x^3 & -\ 8x^2 & -\ 12x & \\ & -15x^2 & +\ 20x & +\ 30 \\ \hline 6x^3 & -23x^2 & +\ 8x & +\ 30 \end{array}$$

**7.** Using synthetic division:

$$\begin{array}{r|rrr} -2 & 3 & 10 & 10 \\ & & -6 & -8 \\ \hline & 3 & 4 & 2 \end{array}$$

$$\frac{3x^2 + 10x + 10}{x + 2} = 3x + 4 + \frac{2}{x + 2}$$

**8.** $f(x) = x^3 + 4x - 1$

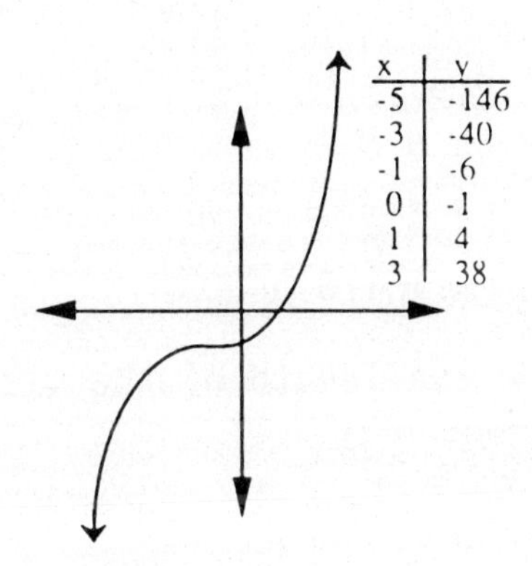

x | y
-5 | -146
-3 | -40
-1 | -6
0 | -1
1 | 4
3 | 38


**9.** $2x^2 - 12x + 18 - 2y^2$

$= 2(x^2 - 6x + 9 - y^2)$

$= 2[(x^2 - 6x + 9) - y^2]$

$= 2[(x-3)^2 - y^2]$

$= 2[(x-3) - y][(x-3) + y]$

$= 2[x - y - 3][x + y - 3]$

**10.** $\dfrac{x - 4}{5x - 3}$

The domain is found by setting 5x - 3 equal to zero and solving. The domain will be all reals except that solution.

$5x - 3 = 0$

$x = 3/5$

Domain: $\{x | x \neq \frac{3}{5}\}$

**11.** $\dfrac{(x+2)(x-4) + (x-1)(x-4)}{3(x-4)}$

$\dfrac{(x+2)(x-4)}{3(x-4)} + \dfrac{(x-1)(x-4)}{3(x-4)}$

$= \dfrac{x + 2}{3} + \dfrac{x - 1}{3} = \dfrac{2x + 1}{3}$

**12.** $\dfrac{4x^2 + 8x + 3}{2x^2 - x - 1} \cdot \dfrac{x^2 - 1}{4x^2 + 12x + 9}$

$= \dfrac{(2x+1)(2x+3)}{(2x+1)(x-1)} \cdot \dfrac{(x-1)(x+1)}{(2x+3)(2x+3)}$

$= \dfrac{x + 1}{2x + 3}$

**13.** $\dfrac{x + 1}{x^2 + 2x - 3} - \dfrac{x}{2x^2 + 11 + 15}$

$= \dfrac{x + 1}{(x+3)(x-1)} - \dfrac{x}{(2x+5)(x+3)}$

LCD $= (x+3)(2x+5)(x-3)$

$= \dfrac{(x+1)(2x+5)}{(x+3)(x-1)(2x+5)}$

$- \dfrac{x(x-1)}{(2x+5)(x+3)(x-1)}$

$= \dfrac{2x^2 + 7x + 5 - x^2 + x}{(x+3)(x-1)(2x+5)}$

$= \dfrac{x^2 + 8x + 5}{(x+3)(x-1)(2x+5)}$

14. $4 - \frac{5}{y} = \frac{4y}{y+1}$

$4y(y+1) - \frac{5y(y+1)}{y} = \frac{4y(y)(y)}{y+1}$

$4y^2 + 4y - 5y - 5 = 4y^2$

$4y - 5y - 5 = 0$

$-y - 5 = 0$

$-y = 5$

$y = -5$

15. $\left(\frac{x^2 y^{1/2}}{x^{1/4}}\right)^2 = \frac{x^4 y}{x^{2/4}} = x^{4-1/2} y$

$= x^{8/2 - 1/2} y$

$= x^{7/2} y$

16. $\sqrt[3]{4x^{10}y^{20}}\ \sqrt[3]{4x^3y^9}$

$= \sqrt[3]{16x^{13}y^{29}} = \sqrt[3]{16}\ \sqrt[3]{x^{13}}\ \sqrt[3]{y^{29}}$

$= 2\ \sqrt[3]{2}\ x^4\ \sqrt[3]{x}\ y^9\ \sqrt[3]{y^2}$

$= 2x^4\ y^9\ \sqrt[3]{2xy^2}$

17. $\sqrt{2x^2+7} + 3 = 8$

$\sqrt{2x^2 + 7} = 5$

$\left(\sqrt{2x^2 + 7}\right)^2 = 5^2$

$2x^2 + 7 = 25$

$2x^2 - 18 = 0$

$2(x^2 - 9) = 0$

$2(x+3)(x-3) = 0$

$x + 3 = 0 \quad x - 3 = 0$

$x = -3 \qquad x = 3$

Both $x = -3$ and $x = 3$. Check in the original equation.

18. $\frac{2}{3 + \sqrt{-6}} = \frac{2}{3 + i\sqrt{6}}$

$= \frac{2(3 - i\sqrt{6})}{(3+i\sqrt{6})(3 - i\sqrt{6})}$

$= \frac{6 - 2i\sqrt{6}}{9 + 6} = \frac{6 - 2i\sqrt{6}}{15}$

19. Let x = the time needed to paint the room together.

$\frac{x}{2} + \frac{x}{3} = 1$

$6\left(\frac{x}{2}\right) + 6\left(\frac{x}{3}\right) = 6$

$3x + 2x = 6$

$5x = 6$

$x = 6/5$

$= 1\ 1/5$ hours

20. Let x = the length of the wire.

$x = \sqrt{30^2 + 20^2}$

$= \sqrt{1300}$

$\approx 36.06$ feet.

**EXERCISE SET 9.1**

1. $x^2 = 25$

$\sqrt{x^2} = \sqrt{25}$

$x = \pm 5$

3. $y^2 = 75$

$\sqrt{y^2} = \sqrt{75}$

$y = \pm \sqrt{75}$

$= \pm 5\sqrt{3}$

5. $z^2 + 12 = 40$

$z^2 = 28$

$\sqrt{z^2} = \sqrt{28}$

$z = \pm \sqrt{28}$

$z = \pm 2\sqrt{7}$

7. $(x-4)^2 = 16$

$\sqrt{(x-4)^2} = \sqrt{16}$

$x - 4 = 4$ or $x - 4 = -4$

$x = 8$ or $x = 0$

9. $(z+\frac{1}{3})^2 = \frac{4}{9}$

$\sqrt{(z+1/3)^2} = \sqrt{4/9}$

$z + \frac{1}{3} = \frac{2}{3}$ or $z + \frac{1}{3} = -\frac{2}{3}$

$z = \frac{1}{3}$ or $z = -1$

11. $(x + 1.8)^2 = 0.81$

$\sqrt{(x+1.8)^2} = \sqrt{0.81}$

$x + 1.8 = .9$ or $x + 1.8 = -.9$

$x = -.9$ or $x = -2.7$

13. $(2x - 5)^2 = 12$

$\sqrt{(2x-5)^2} = \sqrt{12}$

$2x - 5 = \sqrt{12}$ or $2x - 5 = -\sqrt{12}$

$x = \frac{\sqrt{12}+5}{2}$ $x = \frac{-\sqrt{12}+5}{2}$

$= \frac{2\sqrt{3}+5}{2}$ $= \frac{-2\sqrt{3}+5}{2}$

15. $(2y + \frac{1}{2})^2 = \frac{4}{25}$

$\sqrt{(2y + \frac{1}{2})^2} = \sqrt{\frac{4}{25}}$

$2y + \frac{1}{2} = \frac{2}{5}$ or $2y + \frac{1}{2} = \frac{-2}{5}$

$10(2y + \frac{1}{2} = \frac{2}{5})$ $10(2y + \frac{1}{2} = \frac{-2}{5})$

$20y + 5 = 4$ $20y + 5 = -4$

$y = -\frac{1}{20}$ $y = \frac{-9}{20}$

**17.** Solve for S

$$A = S^2$$
$$\sqrt{A} = \sqrt{S^2}$$
$$S = \sqrt{A}$$

**19.** Solve for r

$$A = \pi r^2$$
$$\frac{A}{\pi} = r^2$$
$$\sqrt{\frac{A}{\pi}} = \sqrt{r^2}$$
$$\sqrt{\frac{A}{\pi}} = r$$

**21.** Solve for F

$$F_x^2 + F_y^2 = F^2$$
$$\sqrt{F_x^2 + F_y^2} = \sqrt{F^2}$$
$$F = \sqrt{F_x^2 + F_y^2}$$

**23.** Solve for $V_x$

$$V^2 = V_x^2 + V_y^2$$
$$V^2 - V_y^2 = V_x^2$$
$$\sqrt{V^2 - V_y^2} = \sqrt{V_x^2}$$
$$\sqrt{V^2 - V_y^2} = V_x$$

**25.** Solve for b

$$L = a^2 - b^2$$
$$b^2 = a^2 - L$$
$$\sqrt{b^2} = \sqrt{a^2 - L}$$
$$b = \sqrt{a^2 - L}$$

**27.** Solve for t

$$v = m + nt^2$$
$$v - m = nt^2$$
$$\frac{v-m}{n} = t^2$$
$$\sqrt{\frac{v-m}{n}} = \sqrt{t^2}$$
$$\sqrt{\frac{v-m}{n}} = t$$

**29.** Solve for b: $d = b^2 - 4ac$

$$d + 4ac = b^2$$
$$\sqrt{d + 4ac} = \sqrt{b^2}$$
$$\sqrt{d + 4ac} = b$$

**31.** Solve for a: $w = 3L + 2d^2$

$$w - 3L + 2d^2$$
$$\frac{w-3L}{2} = d^2$$
$$\sqrt{\frac{w-3L}{2}} = \sqrt{d^2}$$
$$\sqrt{\frac{w-3L}{2}} = d$$

In solving this problem, capital L was used since the lower case is l which looks more like the "number 1".

**33.** $3^2 + x^2 = 5^2$

$x^2 = 16$

$x = 4$

**35.** $x^2 + 15^2 = 20^2$

$x^2 = 175$

$x = \sqrt{175} \approx 13.23$

**37.** $6^2 + (\sqrt{5})^2 = x^2$

$36 + 5 = x^2$

$41 = x^2$

$x = \sqrt{41} \approx 6.40$

**39.** $x^2 + 4^2 = 12^2$

$x^2 = 128$

$x = \sqrt{128} \approx 11.31$

**41.** $8^2 + 6^2 = x^2$

$100 = x^2$

$x = \sqrt{100} = 10$

**43.** $x^2 + 2x - 3 = 0$

$x^2 + 2x = 3$

$x^2 + 2x + 1 = 3 + 1$

$(x+1)^2 = 4$

$x + 1 = \pm 2$

$x = \pm 2 - 1$

$x = 2 - 1$ or $x = -2 - 1$

$x = 1$ or $x = -3$

**45.** $x^2 - 4x - 5 = 0$

$x^2 - 4x = 5$

$x^2 - 4x + 4 = 5 + 4$

$(x-2)^2 = 9$

$x - 2 = \pm 3$

$x = \pm 3 + 2$

$x = 3 + 2$ or $x = -3 + 2$

$x = 5$ or $x = -1$

**47.** $x^2 + 3x + 2 = 0$

$x^2 + 3x = -2$

$x^2 + 3x + \frac{9}{4} = -2 + \frac{9}{4}$

$(x + \frac{3}{2})^2 = \frac{1}{4}$

$x + \frac{3}{2} = \pm \frac{1}{2}$

$x = \pm \frac{1}{2} - \frac{3}{2}$

$x = -\frac{1}{2} - \frac{3}{2}$ or $x = \frac{1}{2} - \frac{3}{2}$

$x = -\frac{4}{2}$ or $x = -\frac{2}{2}$

$x = -2$ or $x = -1$

**49.** $x^2 - 8x + 15 = 0$

$x^2 - 8x = -15$

$x^2 - 8x + 16 = -15 + 16$

$(x-4)^2 = 1$

$x - 4 = \pm 1$

$x = \pm 1 + 4$

$x = 1 + 4$ or $x = -1 + 4$

$x = 5$ or $x = 3$

**51.** $x^2 + 2x + 15 = 0$

$$x^2 + 2x = -15$$

$$x^2 + 2x + 1 = -15 + 1$$

$$(x + 1)^2 = -14$$

$$x + 1 = \pm i\sqrt{14}$$

$$x = \pm i\sqrt{14} - 1$$

$$x = -1 + i\sqrt{14} \text{ or } x -1 \ \ -i\sqrt{14}$$

**53.** $x^2 = -5x - 6.$

$$x^2 + 5x = -6$$

$$x^2 + 5x + \frac{25}{4} = -6 + \frac{25}{4}$$

$$\left[x + \frac{5}{2}\right]^2 = \frac{-24}{4} + \frac{25}{4}$$

$$\left[x + \frac{5}{2}\right]^2 = \frac{1}{4}$$

$$x + \frac{5}{2} = \pm \frac{1}{2}$$

$$x = \pm \frac{1}{2} - \frac{5}{2}$$

$$x = \frac{1}{2} - \frac{5}{2} \text{ or } x = -\frac{1}{2} - \frac{5}{2}$$

$$x = -\frac{4}{2} \text{ or } x = -\frac{6}{2}$$

$$x = -2 \text{ or } x = -3$$

**55.** $x^2 + 9x + 18 = 0$

$$x^2 + 9x = -18$$

$$x^2 + 9x + \frac{81}{4} = -18 + \frac{81}{4}$$

$$\left[x + \frac{9}{2}\right]^2 = \frac{-72}{4} + \frac{81}{4}$$

$$\left[x + \frac{9}{2}\right]^2 = \frac{9}{4}$$

$$x + \frac{9}{2} = \pm \frac{3}{2}$$

$$x = \pm \frac{3}{2} - \frac{9}{2}$$

$$x = \frac{3}{2} - \frac{9}{2} \text{ or } x = -\frac{3}{2} - \frac{9}{2}$$

$$x = -\frac{6}{2} \text{ or } x = -\frac{12}{2}$$

$$x = -3 \text{ or } x = -6$$

**57.** $x^2 = 15x - 56$

$$x^2 - 15x = -56$$

$$x^2 - 15x + \frac{225}{4} = -56 + \frac{225}{4}$$

$$\left[x - \frac{15}{2}\right]^2 = \frac{-224}{4} + \frac{225}{4}$$

$$\left[x - \frac{15}{2}\right]^2 = \frac{1}{4}$$

$$x - \frac{15}{2} = \pm \frac{1}{2}$$

$$x = \pm \frac{1}{2} + \frac{15}{2}$$

$$x = -\frac{1}{2} + \frac{15}{2} \text{ or } x = \frac{1}{2} + \frac{15}{2}$$

$$x = \frac{14}{2} \text{ or } x = \frac{16}{2}$$

$$x = 7 \text{ or } x = 8$$

**59.**

$$-4x = -x^2 + 12$$

$$x^2 - 4x = 12$$

$$x^2 - 4x + 4 = 12 + 4$$

$$(x - 2)^2 = 16$$

$$x - 2 = \pm 4$$

$$x = \pm 4 + 2$$

$x = 4 + 2$ or $x = -4 + 2$
$x = 6$ or $x = -2$

**61.**

$$\frac{1}{2}x^2 + x - 3 = 0$$

$$2\left[\frac{1}{2}x^2 + x - 3 = 0\right]$$

$$x^2 + 2x - 6 = 0$$

$$x^2 + 2x = 6$$

$$x^2 + 2x + 1 = 7$$

$$(x+ 1)^2 = 7$$

$x + 1 = \sqrt{7}$ or $x + 1 = -\sqrt{7}$
$x = \sqrt{7} - 1$ or $x = -\sqrt{7} - 1$

**63.**

$$6x + 6 = -x^2$$

$$x^2 + 6x = -6$$

$$x^2 + 6x + 9 = -6 + 9$$

$$(x + 3)^2 = 3$$

$$x + 3 = \pm \sqrt{3}$$

$$x = -3 \pm \sqrt{3}$$

$x = -3 + \sqrt{3}$ or $x = -3 - \sqrt{3}$

**65.**

$$-x^2 + 5x = -8$$

$$x^2 - 5x = 8$$

$$x^2 - 5x + \frac{25}{4} = 8 + \frac{25}{4}$$

$$\left[x - \frac{5}{2}\right]^2 = \frac{32}{4} + \frac{25}{4}$$

$$\left[x - \frac{5}{2}\right]^2 = \frac{57}{4}$$

$$x - \frac{5}{2} = \pm \frac{\sqrt{57}}{4}$$

$$x = \pm \frac{\sqrt{57}}{2} + \frac{5}{2}$$

$x = \dfrac{5 + \sqrt{57}}{2}$ or $x = \dfrac{5 - \sqrt{57}}{2}$

**67.**

$$-\frac{1}{4}x^2 - \frac{1}{2}x = 0$$

$$-4\left[-\frac{1}{4}x^2 - \frac{1}{2}x = 0\right]$$

$$x^2 + 2x = 0$$

$$x^2 + 2x + 1 = 1$$

$$(x + 1)^2 = 1$$

$x + 1 = 1$ or $x + 1 = -1$
$x = 0$ or $x = -2$

**69.**

$$12x^2 - 4x = 0$$

$$x^2 - \frac{1}{3}x = 0$$

$$x^2 - \frac{1}{3}x + \frac{1}{36} = \frac{1}{36}$$

$$\left[x - \frac{1}{6}\right]^2 = \frac{1}{36}$$

$$x - \frac{1}{6} \pm \frac{1}{6}$$

$$x = \pm \frac{1}{6} + \frac{1}{6}$$

$x = -\frac{1}{6} + \frac{1}{6}$ or $x = \frac{1}{6} + \frac{1}{6}$

$x = 0$ or $x = \frac{2}{6}$

$x = \frac{1}{3}$

**71.**

$$-\frac{1}{2}x^2 - x + \frac{3}{2} = 0$$
$$-2\left[-\frac{1}{2}x^2 - x + \frac{3}{2} = 0\right]$$
$$x^2 + 2x - 3 = 0$$
$$x^2 + 2x = 3$$
$$x^2 + 2x + 1 = 4$$
$$(x + 1)^2 = 4$$
$$x + 1 = 2 \quad \text{or} \quad x + 1 = -2$$
$$x = 1 \quad \text{or} \quad x = -3$$

**73.**

$$2x^2 + 18x + 4 = 0$$
$$x^2 + 9x + 2 = 0$$
$$x^2 + 9x = -2$$
$$x^2 + 9x + \frac{81}{4} = -2 + \frac{81}{4}$$
$$\left[x + \frac{9}{2}\right]^2 = -\frac{8}{4} + \frac{81}{4}$$
$$\left[x + \frac{9}{2}\right]^2 = \frac{73}{4}$$
$$x + \frac{9}{2} = \pm \frac{\sqrt{73}}{\sqrt{4}}$$
$$x = \frac{\pm\sqrt{73} - 9}{2}$$
$$x = \frac{-9 + \sqrt{73}}{2} \quad \text{or} \quad x = \frac{-9 - \sqrt{73}}{2}$$

**75.**

$$3x^2 + 33x + 72 = 0$$
$$x^2 + 11x + 24 = 0$$
$$x^2 + 11x = -24$$
$$x^2 + 11x + \frac{121}{4} = -24 + \frac{121}{4}$$
$$\left[x + \frac{11}{2}\right]^2 = -\frac{96}{4} + \frac{121}{4}$$
$$\left[x - \frac{11}{2}\right]^2 = \frac{25}{4}$$
$$x + \frac{11}{2} = \pm \frac{5}{2}$$
$$x = \pm \frac{5}{2} - \frac{11}{2}$$
$$x = -\frac{5}{2} - \frac{11}{2} \quad \text{or} \quad x = \frac{5}{2} - \frac{11}{2}$$
$$x = -\frac{16}{2} \quad \text{or} \quad x = -\frac{6}{2}$$
$$x = -8 \quad \text{or} \quad x = -3$$

**77.**

$$\frac{2}{3}x^2 + \frac{4}{3}x + 1 = 0$$
$$3\left[\frac{2}{3}x^2 + \frac{4}{3}x + 1\right] = 0$$
$$2x^2 + 4x + 3 = 0$$
$$x^2 + 2x + \frac{3}{2} = 0$$
$$x^2 + 2x = -\frac{3}{2}$$
$$x^2 + 2x + 1 = -\frac{3}{2} + 1$$
$$(x+1)^2 = -\frac{1}{2}$$
$$x + 1 = \sqrt{\frac{-1}{2}} \quad \text{or} \quad x + 1 = -\sqrt{\frac{-1}{2}}$$
$$x + 1 = \frac{i\sqrt{2}}{2} \quad \text{or} \quad x + 1 = -\frac{i\sqrt{2}}{2}$$
$$x = \frac{i\sqrt{2}}{2} + 1 \quad \text{or} \quad x = -\frac{-i\sqrt{2}}{2} - 1$$
$$x = \frac{i\sqrt{2} + 2}{2} \quad \text{or} \quad x = \frac{-i\sqrt{2} - 2}{2}$$

**79.**

$$-3x^2 + 6x = 6$$
$$x^2 - 2x = -2$$
$$x^2 - 2x + 1 = -2 + 1$$
$$(x - 1)^2 = -1$$
$$x - 1 = \pm i$$
$$x = 1 \pm i$$
$$x = 1 + i \quad \text{or} \quad x = 1 - i$$

**81.**
$$\frac{5}{2}x^2 + \frac{3}{2}x - \frac{5}{4} = 0$$
$$4\left[\frac{5}{2}x^2 + \frac{3}{2}x - \frac{5}{4} = 0\right]$$
$$10x^2 + 6x - 5 = 0$$
$$x^2 + \frac{3}{5}x - \frac{1}{2} = 0$$
$$x^2 + \frac{3}{5}x = 1/2$$
$$x^2 + \frac{3}{5}x + \frac{9}{100} = \frac{1}{2} + \frac{9}{100}$$
$$\left[x + \frac{3}{10}\right]^2 = \frac{59}{100}$$
$$x + \frac{3}{10} = \frac{\sqrt{59}}{10} \text{ or } x + \frac{3}{10} = \frac{-\sqrt{59}}{10}$$
$$x = \frac{-3}{10} + \frac{\sqrt{59}}{10} \text{ or } x = \frac{-3}{10} - \frac{\sqrt{59}}{10}$$
$$x = \frac{-3 + \sqrt{59}}{10} \text{ or } x = \frac{-3 - \sqrt{59}}{10}$$

**83.**
$$3x^2 + \frac{1}{2}x = -4$$
$$x^2 + \frac{1}{6}x = -\frac{4}{3}$$
$$x^2 + \frac{1}{6}x + \frac{1}{144} = -\frac{4}{3} + \frac{1}{144}$$
$$\left[x - \frac{1}{12}\right]^2 = \frac{-192 + 1}{144}$$
$$\left[x - \frac{1}{12}\right]^2 = \frac{-191}{144}$$
$$x + \frac{1}{12} = \pm \frac{i\sqrt{191}}{\sqrt{144}}$$
$$x = \frac{\pm i\sqrt{191}}{12} - \frac{1}{12}$$
$$x = \frac{-1 + i\sqrt{191}}{12} \text{ or } x = \frac{-1 - i\sqrt{191}}{12}$$

**85.**
$$(x+a)^2 - k = 0$$
$$(x+a)^2 = k$$
$$x+a = \sqrt{k} \quad \text{or} \quad x + a = -\sqrt{k}$$
$$x = -a + \sqrt{k} \quad \text{or} \quad x = -a - \sqrt{k}$$

**87.** Let x = the first integer.
Then x + 2 = the next integer

$$x(x+2) = 63$$
$$x^2 + 2x + 1 = 63 + 1$$
$$(x + 1)^2 = 64$$
$$x + 1 = \pm 8$$
$$x = \pm 8 - 1$$
$$x = 8 - 1$$
$$x = 7$$

Since the integers are positive and x + 2 = 9.

**89.** Let x = width of rectangle.
Then 2x + 2 = length
A = length · width
$$60 = (2x + 2)(x)$$
$$2x^2 + 2x = 60$$
$$x^2 + x = 30$$
$$x^2 + x + \frac{1}{4} = 30 + \frac{1}{4}$$
$$\left[x - \frac{1}{2}\right]^2 = \frac{120}{4} + \frac{1}{4}$$
$$\left[x - \frac{1}{2}\right]^2 = \frac{121}{4}$$
$$x + \frac{1}{2} = \pm \frac{11}{2}$$
$$x = \pm \frac{11}{2} - \frac{1}{2}$$
$$x = \frac{11}{2} - \frac{1}{2}$$
$$x = \frac{10}{2} = 5$$

Length = 2(5) + 2 = 10 + 2 = 12.
The rectangle is 5 x 12.

**91.** Let x = the length of the side. Then

$$d = x + 12$$
$$2x^2 = (x + 12)^2$$
$$2x^2 = x^2 + 24x + 144$$
$$2x^2 - x^2 - 24x = 144$$
$$x^2 - 24x = 144$$
$$x^2 - 24x + 144 = 144 + 144$$
$$(x - 12)^2 = 288$$
$$x - 12 = \pm 12\sqrt{2}$$
$$x = 12 + 12\sqrt{2}$$
$$\approx 28.97 \text{ ft.}$$

Since the length is positive.

**93.** Make the coefficient of the squared ten equal to 1.

**95.** Varies from Student to Student.

**97.** $\sqrt{(x^2 - 4x)^2} = |x^2 - 4x|$

**99.** $\dfrac{x^{\frac{3}{4}} y^{\frac{1}{2}}}{x^{\frac{1}{4}} y^{2}} = x^{\frac{3}{4} - \frac{1}{4}} y^{\frac{1}{2} - 2}$

$$= x^{\frac{2}{4}} y^{-\frac{3}{2}}$$

$$= \frac{x^{\frac{1}{2}}}{y^{\frac{3}{2}}}$$

**JUST FOR FUN**

**1.** (a) To find the surface area, we must first determine the radius.

$$v = \pi r^2 h$$
$$160 = \pi r^2 (10)$$
$$16 = \pi r^2$$
$$\frac{16}{\pi} = r^2$$
$$\frac{4}{\sqrt{\pi}} = r$$

Now that we know the radius equals $\frac{4}{\sqrt{\pi}}$, we can use the formula

$$s = 2\pi r^2 + 2\pi rh$$

to calculate the surface area.

$$S = 2\pi \left(\frac{4}{\sqrt{\pi}}\right)^2 + 2\pi\left(\frac{4}{\sqrt{\pi}}\right) 10$$
$$= 2\pi \left(\frac{16}{\pi}\right) + \frac{80\pi}{\sqrt{\pi}}$$
$$= 32 + \frac{80\pi}{\sqrt{\pi}} \approx 173.80 \text{ in}$$

(b)
$$S = 2\pi r^2 + 2\pi rh$$
$$160 = 2\pi r^2 + 2\ \pi r(10)$$
$$\frac{160}{2\pi} = \frac{2\pi r^2}{2\pi} + \frac{20\pi r}{2\pi}$$
$$\frac{80}{\pi} = r^2 + 10r$$
$$\frac{80}{\pi} + 25 = r^2 + 10r + 25$$
$$\frac{80}{\pi} + 25 = (r+5)^2$$
$$\frac{80 + 25\pi}{\pi} = (r+5)^2$$
$$\sqrt{\frac{80 + 25\pi}{\pi}} = r+5$$
$$\sqrt{\frac{80 + 25\pi}{\pi}} - 5 = r$$
$$r \approx 2.1$$

**3.**
$$x^2 + 4x + y^2 - 6y = 3$$
$$(x^2 + 4x + 4) + (y^2 - 6y + 9)$$
$$= 3 + 4 + 9$$
$$(x + 2)^2 + (y - 3)^2 = 16$$

**5.** $x^2 - 4y^2 + 4x - 16y - 28 = 0$
$$x^2 + 4x - 4y^2 - 16y = 28$$
$$x^2 + 4x + 4 - 4y^2 - 16y = 28 + 4$$
$$(x+ 2)^2 - 4(y^2 + 4y) = 32$$
$$(x + 2)^2 - 4(y^2 + 4y + 4) = 32 - 16$$
$$(x + 2)^2 - 4(y + 2)^2 = 16$$

## EXERCISE SET 9.2

1. $x^2 + 4x - 3 = 0$

$$b^2 - 4ac = 4^2 - 4(1)(-3) = 16 + 12 = 28$$

Since $28 > 0$ there are two real solutions.

3. $2x^2 - 4x + 7 = 0$

$$b^2 - 4ac = (-4)^2 - 4(2)(7) = 16 - 56 = -40$$

Since $-40 < 0$ there are no real solutions.

5. $5x^2 + 3x - 7 = 0$

$$b^2 - 4ac = 3^2 - 4(5)(-7) = 9 + 140 = 149$$

Since $149 > 0$ there are two real solutions.

7. $4x^2 - 24x = -36$

$4x^2 - 24x + 36 = 0$

$$b^2 - 4ac = (-24)^2 - 4(4)(36) = 576 - 576 = 0$$

Since the discriminant is 0 there is one real solution.

9. $2x^2 - 8x + 5 = 0$

$$b^2 - 4ac = (-8)^2 - 4(2)(5) = 64 - 40 = 14$$

Since $14 > 0$ there are two real solutions.

11. $-3x^2 + 5x - 8 = 0$

$$b^2 - 4ac = 5^2 - 4(-3)(-8) = 25 - 96 = -71$$

Since $-71 < 0$ there are no real solutions.

13. $x^2 + 7x - 3 = 0$

$$b^2 - 4ac = 7^2 - 4(1)(-3) = 49 + 12 = 61$$

Since $61 > 0$ there are two real solutions.

15. $4x^2 - 9 = 0$

$$b^2 - 4ac = 0^2 - 4(4)(-9) = 144$$

Since $144 > 0$ there are two real solutions.

17. $x^2 - 3x + 2 = 0$

$$x = \frac{-b \pm \sqrt{b^2 - 4ac}}{2a}$$

$$x = \frac{-(-3) \pm \sqrt{(-3)^2 - 4(1)(2)}}{2(1)}$$

$$x = \frac{3 \pm \sqrt{9 - 8}}{2}$$

$$x = \frac{3 + \sqrt{1}}{2}$$

$$x = \frac{3 - 1}{2} = \frac{2}{2} = 1$$

or

$$x = \frac{3 + 1}{2} = \frac{4}{2} = 2$$

The solutions are 1 and 2.

For equations 19 - 59 the quadratic formula is used.

$x = \frac{-b \pm \sqrt{b^2 - 4ac}}{2a}$ is not written out in the following problems.

**19.** $x^2 - 9x + 20$

$$x = \frac{-(-9) \pm \sqrt{(-9)^2 - 4(1)(20)}}{2(1)}$$

$$x = \frac{9 \pm \sqrt{81 - 80}}{2}$$

$$x = \frac{9 \pm \sqrt{1}}{2}$$

$$x = \frac{9 \pm 1}{2}$$

$$x = \frac{9 - 1}{2} = \frac{8}{2} = 4$$

or

$$x = \frac{9 + 1}{2} = \frac{10}{2} = 5.$$

The solutions are 4 and 5.

**21.** $x^2 + 5x - 24 = 0$

$$x = \frac{-5 \pm \sqrt{5^2 - 4(1)(-24)}}{2}$$

$$x = \frac{-5 \pm \sqrt{25 + 96}}{2}$$

$$x = \frac{-5 \pm \sqrt{121}}{2}$$

$$x = \frac{-5 \pm 11}{2}$$

$$x = \frac{-5 + 11}{2} = \frac{6}{2} = 3$$

or

$$x = \frac{-5 - 11}{2} = \frac{-16}{2} = -8$$

The solutions are 3 and -8.

**23.** $x^2 = 13x - 36$

$x^2 - 13x + 36 = 0$

$$x = \frac{-(-13) \pm \sqrt{(-13)^2 - 4(1)(36)}}{2(1)}$$

$$x = \frac{13 \pm \sqrt{169 - 144}}{2}$$

$$x = \frac{13 \pm \sqrt{25}}{2}$$

$$x = \frac{13 \pm 5}{2}$$

$$x = \frac{13 - 5}{2} = \frac{8}{2} = 4$$

or

$$x = \frac{13 + 5}{2} = \frac{18}{2} = 9$$

The solutions are 4 and 9.

**25.** $x^2 - 25 = 0$

$$x = \frac{0 \pm \sqrt{0^2 - 4(1)(-25)}}{2(1)}$$

$$x = \frac{\pm \sqrt{100}}{2}$$

$$x = \frac{\pm 10}{2}$$

$$x = \frac{10}{2} = 5$$

or

$$x = \frac{-10}{2} = -5$$

The solutions are 5 and -5.

**27.** $x^2 - 3x = 0$

$$x = \frac{-(-3) \pm \sqrt{(-3)^2 - 4(1)(0)}}{2(1)}$$

$$x = \frac{3 \pm \sqrt{9}}{2}$$

$$x = \frac{3 \pm 3}{2}$$

$$x = \frac{3 - 3}{2} = 0$$

or

$$x = \frac{3 + 3}{2} = \frac{6}{2} = 3$$

The solutions are 0 and 3.

**29.** $4p^2 - 7p + 13 = 0$

$$x = \frac{-(-7) \pm \sqrt{(-7)^2 - 4(4)(13)}}{2(4)}$$

$$x = \frac{7 \pm \sqrt{49 - 208}}{8}$$

$$x = \frac{7 \pm \sqrt{-159}}{8}$$

$$x = \frac{7 \pm i\sqrt{159}}{8}$$

or

$$x = \frac{7 - i\sqrt{159}}{8}$$

The solutions are $\frac{7 + i\sqrt{159}}{8}$ and $\frac{7 - i\sqrt{159}}{8}$.

**31.** $2y^2 - 7y + 4 = 0$

$$x = \frac{-(-7) \pm \sqrt{(-7)^2 - 4(2)(4)}}{2(2)}$$

$$x = \frac{7 \pm \sqrt{49 - 32}}{4}$$

$$x = \frac{7 \pm \sqrt{17}}{4}$$

$$x = \frac{7 + \sqrt{17}}{4}$$

or

$$x = \frac{7 - \sqrt{17}}{4}$$

The solutions are $\frac{7 + \sqrt{17}}{4}$ and $\frac{7 - \sqrt{17}}{4}$.

**33.** $6x^2 = -x + 1$

$6x^2 + x - 1 = 0$

$$x = \frac{-1 \pm \sqrt{(1)^2 - 4(6)(-1)}}{2(6)}$$

$$x = \frac{-1 \pm \sqrt{1 + 24}}{12}$$

$$x = \frac{-1 \pm \sqrt{25}}{12}$$

$$x = \frac{-1 \pm 5}{12}$$

$$x = \frac{-1 + 5}{12} = \frac{4}{12} = \frac{1}{3}$$

or

$$x = \frac{-1 - 5}{12} = \frac{6}{12} = -\frac{1}{2}$$

The solutions are 1/3 and -1/2.

**35.** $2x^2 - 4x - 1 = 0$

$$x = \frac{-(-4) \pm \sqrt{(-4)^2 - 4(2)(-1)}}{(2)}$$

$$x = \frac{4 \pm \sqrt{16 + 8}}{4}$$

$$x = \frac{4 \pm \sqrt{24}}{4}$$

$$x = \frac{4 \pm 2\sqrt{6}}{4}$$

$$x = \frac{4 + 2\sqrt{6}}{4} = \frac{2 + \sqrt{6}}{2}$$

or

$$x = \frac{4 - 2\sqrt{6}}{4} = \frac{2 - \sqrt{6}}{2}$$

The solutions are $\frac{2 + \sqrt{6}}{2}$ and $\frac{2 - \sqrt{6}}{2}$.

**37.**
$$4s^2 - 8s + 6 = 0$$
$$\frac{1}{2}(4s^2 - 8s + 6 = 0)$$
$$2s^2 - 4s + 3 = 0$$
$$s = \frac{-(-4) \pm \sqrt{(-4)^2 - 4(2)(3)}}{2(2)}$$
$$s = \frac{4 \pm \sqrt{16 - 24}}{4}$$
$$s = \frac{4 \pm \sqrt{-8}}{4}$$
$$s = \frac{4 \pm 2i\sqrt{2}}{4}$$
$$s = \frac{4 + 2i\sqrt{2}}{4} = \frac{2 + i\sqrt{2}}{2}$$
or
$$s = \frac{4 - 2i\sqrt{2}}{4} = \frac{2 - 1\sqrt{2}}{2}$$
The solutions are $\frac{2 - i\sqrt{2}}{2}$ and

$\frac{2 - i\sqrt{2}}{2}$.

**39.**
$$4x^2 = x + 5$$
$$4x^2 - x - 5 = 0$$
$$x = \frac{-(-1) \pm \sqrt{(-1)^2 - 4(4)(-5)}}{2(4)}$$
$$x = \frac{1 \pm \sqrt{1 + 80}}{8}$$
$$x = \frac{1 \pm \sqrt{81}}{8}$$
$$x = \frac{1 \pm 9}{8}$$
$$x = \frac{1 \pm 9}{8} = \frac{10}{8} = \frac{5}{4}$$
or
$$x = \frac{1 - 9}{8} = \frac{-8}{8} = -1.$$
The solutions are 5/4 and -1.

**41.**
$$6x^2 - 21x = 27$$
$$\frac{6x^2}{3} - \frac{21}{3}x = \frac{27}{3}$$
$$2x^2 - 7x = 9$$
$$2x^2 - 7x - 9 = 0$$
$$x = \frac{-(-7) \pm \sqrt{(-7)^2 - 4(2)(-9)}}{2(2)}$$
$$x = \frac{7 \pm \sqrt{49 + 72}}{4}$$
$$x = \frac{7 \pm \sqrt{121}}{4}$$
$$x = \frac{7 \pm 11}{4}$$
$$x = \frac{7 + 11}{4} = \frac{18}{4} = \frac{9}{2}$$
or
$$x = \frac{7 - 11}{4} = \frac{-4}{4} = -1$$
The solutions are 9/2 and -1.

**43.**
$$-2x^2 + 11x - 15 = 0$$
$$x = \frac{-11 \pm \sqrt{(11)^2 - 4(-2)(-15)}}{2(-2)}$$
$$x = \frac{-11 \pm \sqrt{121 - 120}}{-4}$$
$$x = \frac{-11 \pm \sqrt{1}}{-4}$$
$$x = \frac{-11 \pm 1}{-4}$$
$$x = \frac{11 + 1}{4} = \frac{12}{4} = 3$$
or
$$x = \frac{11 - 1}{4} = \frac{10}{4} = \frac{5}{2}.$$
The solutions are 3 and 5/2.

45. $-2x^2 - x - 3 = 0$

$$\frac{-2x^2}{-1} - \frac{x}{-1} - \frac{3}{-1} = \frac{0}{-1}$$

$$2x^2 + x + 3 = 0$$

$$x = \frac{-1 \pm \sqrt{(1)^2 - 4(2)(3)}}{2(2)}$$

$$x = \frac{-1 \pm \sqrt{1 - 24}}{4}$$

$$x = \frac{-1 \pm i\sqrt{23}}{4}$$

$$x = \frac{-1 + i\sqrt{23}}{4}$$

or

$$x = \frac{-1 - i\sqrt{23}}{4}.$$

The solutions are $\frac{-1 + i\sqrt{23}}{4}$ and $\frac{-1 - i\sqrt{23}}{4}$.

47. $2x^2 + 6x = 0$

$$x = \frac{-6 \pm \sqrt{6^2 - 4(2)(0)}}{2(2)}$$

$$x = \frac{-6 \pm \sqrt{36 - 0}}{4}$$

$$x = \frac{-6 \pm 6}{4}$$

$$x = \frac{-6 + 6}{4} = \frac{0}{4} = 0$$

or

$$x = \frac{-6 - 6}{4} = \frac{-12}{4} = -3.$$

The solutions are 0 and -3.

49. $4x^2 - 7 = 0$

$$x = \frac{-0 \pm \sqrt{(0)^2 - 4(4)(-7)}}{2(4)}$$

$$x = \frac{\pm \sqrt{112}}{8}$$

$$x = \frac{\pm 4\sqrt{7}}{8}$$

$$x = \frac{4\sqrt{7}}{8} = \frac{\sqrt{7}}{2} \qquad \text{or}$$

$$x = \frac{-4\sqrt{7}}{8} = \frac{-\sqrt{7}}{2}.$$

The solutions are $\frac{\sqrt{7}}{2}$ and $\frac{-\sqrt{7}}{2}$.

51. $\frac{x^2}{2} + 2x = \frac{2}{3} = 0$

$$x = \frac{-2 \pm \sqrt{2^2 - 4(\frac{1}{2})(\frac{2}{3})}}{2(\frac{1}{2})}$$

$$x = \frac{-2 \pm \sqrt{4 - \frac{4}{3}}}{1}$$

$$x = -2 \pm \sqrt{\frac{12}{3} - \frac{4}{3}}$$

$$x = -2 \pm \sqrt{\frac{8}{3}}$$

$$x = -2 \pm \frac{\sqrt{8} \cdot \sqrt{3}}{\sqrt{3} \cdot \sqrt{3}}$$

$$x = -2 \pm \frac{\sqrt{24}}{3}$$

$$x = -2 \pm \frac{2\sqrt{6}}{3}$$

$$x = \frac{-6 \pm 2\sqrt{6}}{3}$$

$$x = \frac{-6 + 2\sqrt{6}}{3} \qquad \text{or}$$

$$x = \frac{-6 - 2\sqrt{6}}{3}$$

The solutions are $\frac{-6 + 2\sqrt{6}}{3}$ and $\frac{-6 - 2\sqrt{3}}{3}$.

**53.** $-x^2 + \frac{11}{33}x + \frac{10}{3} = 0$

$-1\ (-x^2 + 11/3\ x + \frac{10}{3} = 0)$

$x^2 - \frac{11}{3}x - \frac{10}{3} = 0$

$3x^2 - 11x - 10 = 0$

$x = \frac{-(-11) \pm \sqrt{(-11)^2 - 4(3)(-10)}}{2(3)}$

$x = \frac{11 \pm \sqrt{121 + 120}}{6}$

$x = \frac{11 \pm \sqrt{241}}{6}$

$x = \frac{11 + \sqrt{241}}{6}$

or

$x = \frac{11 - \sqrt{241}}{6}$.

The solutions are $\frac{11 + \sqrt{241}}{6}$ and $\frac{11 - \sqrt{241}}{6}$.

**55.** $.1x^2 + .6x - 1.2 = 0$

$10(.1x^2 + .6x - 1.2 = 0)$

$x^2 + 6x - 12 = 0$

$x = \frac{-6 \pm \sqrt{6^2 - 4(1)(-12)}}{2(1)}$

$x = \frac{-6 \pm \sqrt{84}}{2}$

$x = \frac{-6 \pm 2\sqrt{21}}{2}$

$x = -3 + \sqrt{21}$ or $x = -3 - \sqrt{21}$

**57.** $-1.62x^2 - .94x + 4.85 = 0$

$100\ (-1.62\ x^2 - .94x + 4.85 = 0)$

$162\ x^2 + 94x - 485 = 0$

$x = \frac{-94 \pm \sqrt{(-94)^2 - 4(162)(-485)}}{2(162)}$

$x = \frac{-94 \pm \sqrt{323,116}}{324} = \frac{-94 \pm 2\sqrt{80,779}}{324}$

$= \frac{-47 \pm \sqrt{80,779}}{162}$

**59.** Let x = the number

Then $2x^2 + 3x = 14$

$2x^2 + 3x - 14 = 0$

$x = \frac{-3 \pm \sqrt{3^2 - 4(2)(-14)}}{2(2)}$

$x = \frac{-3 \pm \sqrt{9 + 112}}{4}$

$x = \frac{-3 \pm \sqrt{121}}{4}$

$x = \frac{-3 + 11}{4}$

$x = \frac{-3 + 11}{4}$ since x is positivee

$x = \frac{8}{4}$

$x = 2$

**61.** Let x = width.

Then $3x - 2$ = length

A = (length)(width)

$21\ ft.^2 = (3x^2 - 2)(x)$

$21 = 3x^2 - 2x$

$3x^2 - 2x - 21 = 0$

$x = \frac{-(-2) \pm \sqrt{(-2)^2 - 4(3)(-21)}}{2(3)}$

$x = \frac{2 \pm \sqrt{4 + 252}}{6}$

$x = \frac{2 \pm \sqrt{256}}{6}$

$x = \frac{2 \pm 16}{6}$

$x = \frac{2 + 16}{6}$. Since width is positive

$x = \frac{18}{6} = 3$ ft.

width = 3 ft.

length = 3(3 ft.) - 2 ft.

= 9 ft. - 2 ft.

= 7 ft.

**63.** Let x = amount each side is to be reduced by
Then 6 - x = New Width
and 8 - x = New Length
New Area = 1/2 (Old Area)
= 1/2 (6 · 8)
= 1/2 (48)
= 24
New Area = (New Width)(New Length)
24 = (6-x)(8-x)

$0 = 48 - 14x + x^2 - 24$

$0 = x^2 - 14x + 24$

$$x = \frac{-(-14) \pm \sqrt{(-14)^2 - 4(1)(24)}}{2(1)}$$

$$x = \frac{14 \pm \sqrt{196 - 96}}{2}$$

$$x = \frac{14 \pm \sqrt{100}}{2}$$

$$x = \frac{14 \pm 10}{2}$$

$$x = \frac{14 + 10}{2} = \frac{24}{2} = 12$$

$x \neq 12$ since this would give a negative value for width and length.

or

$$x = \frac{14 - 10}{2} = \frac{4}{2} = 2$$

x = 2 inches since this gives a positive value for width and length.

**65.** $h = \frac{1}{2} a t^2 + v_o t + h_0$

$h = 0,\ a_0 = -32,\ v_o = -20$

$$0 = \frac{1}{2}(-32)t^2 + 20t + 120$$

$$0 = -16t^2 - 20t + 120$$

$$\frac{0}{4} = -\frac{16t^2}{4} + \frac{20t}{4} + \frac{120}{4}$$

$$0 = -4t^2 - 5t + 30$$

$$t = \frac{5 \pm \sqrt{25 - 4(-4)(30)}}{2(-4)}$$

$$= \frac{5 \pm \sqrt{505}}{-8}$$

$$t = \frac{5 + \sqrt{505}}{-8} \text{ or } t = \frac{5 - \sqrt{505}}{-8}$$

$t = -3.43$ $t = 2.18$

The sand bags will hit the ground in 2.18 sec.

**67.** Since gravity is $\frac{1}{6}$th that of earths, a in this problem will equal $\frac{1}{6}(-32) = \frac{-16}{3}$

$$0 = \frac{1}{2}(-\frac{16}{3})\ t^2 + 40t + 60$$

$$0 = -\frac{8}{3}t^2 + 40t + 60$$

$$3(0 = -\frac{8}{3}t^2 + 40t + 60)$$

$$0 = -8t^2 + 120t + 180$$

$$\frac{0}{4} = \frac{-8t^2}{4} + \frac{120t}{4} + \frac{180}{4}$$

$$0 = -2t^2 + 30t + 45$$

$$t = \frac{-30 \pm \sqrt{30^2 - 4(-2)(45)}}{2(-2)}$$

$$t = \frac{-30 \pm \sqrt{1260}}{-4}$$

$$t = \frac{-30 + \sqrt{1260}}{-4} \text{ or}$$

$$t = \frac{-30 - \sqrt{1260}}{-4}$$

$t = -1.37$ or $t = 16.37$.

Neil Armstrong will land on the ground in 16.37 sec.

**69.** In this problem, we will use the relationship

$$\frac{\text{distance}}{\text{rate}} = \text{time}$$

| | distance | rate | time |
|---|---|---|---|
| 1st trip | 300 | r | $\frac{300}{r}$ |
| 2nd trip | 300 | r+10 | $\frac{300}{r+10}$ |

Time for both trips is 11 hours

Number 69 (Continued)

Therefore, $\frac{300}{r} + \frac{300}{r+10} = 11$

$$r(r+10)\ [\ \frac{300}{r} + \frac{300}{r+10} = 11]$$

$$300(r + 10) + 300\ r = 11(r)\ (r+10)$$

$$300r + 3000 + 300r = 11r^2 + 110r$$

$$0 = 11r^2 - 490r - 3000$$

$$r = \frac{-(-490) \pm \sqrt{(-490)^2 - 4(11)(-3000)}}{2(11)}$$

$$r = \frac{490 + \sqrt{372100}}{22} \text{ or}$$

$$r = \frac{490 - \sqrt{372100}}{22}$$

$r = 50 \quad r = -5$

The teacher's speed going was 50 mph while his return speed was r+10 or 60 mph.

**71.**

| | Rate of work | Time worked | Part of task comp. |
|---|---|---|---|
| small heater | $\frac{1}{t+6}$ | 42 | $\frac{42}{t+6}$ |
| large heater | $\frac{1}{t}$ | 42 | $\frac{42}{t}$ |

$$\frac{42}{t+6} + \frac{42}{t} = 1$$

$$(t+6) + [\ \frac{42}{t+6} + \frac{42}{t} = 1]$$

$$42t + 42(t+6) = t(t+6)$$

$$42t + 42t + 252 = t^2 + 6t$$

$$0 = t^2 - 78t - 252$$

$$t = \frac{-(-78) \pm \sqrt{(-78)^2 - 4(1)(-25)}}{2(1)}$$

$$t = \frac{78 \pm \sqrt{7092}}{2}$$

$$t = \frac{78 + \sqrt{7092}}{2} \text{ or } t = \frac{78 - \sqrt{7092}}{2}$$

$t = 81.11 \qquad t = -3.11$

Number 71 (Continued)

It will take the large heater 81.11 minutes to heat the garage and t + 6 or 87.11 min for the smaller heater to heat the garage.

**73.** (a) $P = .2(1)^2 + 1.5(1) - 1.2$
$= .5$ thousands

(b) $P = .2(5)^2 + 1.5(5) - 1.2$
$= 11.3$ thousand

(c) $1.6 = .2t + 1.5t - 1.2$
$16 = 2t + 15t - 12$
$0 = 2t + 15t - 28$

$$t = \frac{-15 \pm \sqrt{15^2 - 4(2)(-28)}}{2(2)}$$

$$t = \frac{-15 \pm \sqrt{449}}{4}$$

$$t = \frac{-15 + \sqrt{449}}{4} \text{ or } t = \frac{-15 - \sqrt{449}}{4}$$

$t = 1.55$ or $t = -9.05$

The company should charge $1.55 per tape.

**75.** (a) $6 = .01(0)^2 + .2(0) + 1.2$
$= 1.2$

(b) $6 = .01(4)^2 + 2(4) + 1.2$
$= 2.16$

(c) $3.2 = .01h^2 + .2h + 1.2$
$0 = .01h^2 + 2h - 2$

$$h = \frac{-.2 \pm \sqrt{(-.2)^2 - 4(.01)(-2)}}{2(.01)}$$

$$h = \frac{-.2 + \sqrt{.12}}{.02} \text{ or}$$

$$h = \frac{-.2 - \sqrt{.12}}{.02}$$

$h = 7.3$ $h = -27.3$

The student must study 7.3 hours per week to achieve a GPA of 3.2.

**77.** Yes, the solutions are the same. If

$$-6x^2 + \frac{1}{2}x - 5 = 0$$

is multiplied by -1, the result is

$$6x^2 - \frac{1}{2}x + 5 = 0.$$

Since the equations are the same, the solutions will also be the same.

**79.** $\sqrt[5]{64\, x^9\, y^{12}\, z^{20}}$

$= \sqrt[5]{32 \cdot 2\, x^9\, y^{12}\, z^{20}}$

$= 2xy^2\, z^4\, \sqrt{2x^4y^2}$

**81.** $\dfrac{x + \sqrt{y}}{x - \sqrt{y}}$

$$= \frac{x + \sqrt{y}}{(x - \sqrt{y})} \frac{(x + \sqrt{y})}{(x + \sqrt{z})}$$

$$= \frac{x^2 + 2x\sqrt{y + y}}{x^2 - y}$$

**JUST FOR FUN**

**1.** $x^2 - \sqrt{5x} - 10 = 0$

$$x = \frac{-(-\sqrt{5}) \pm \sqrt{(-\sqrt{5})^2 - 4(1)(-10)}}{2(1)}$$

$$x = \frac{\sqrt{5} \pm \sqrt{5 + 40}}{2}$$

$$x = \frac{\sqrt{5} \pm \sqrt{45}}{2}$$

$$x = \frac{\sqrt{5} \pm 3\sqrt{5}}{2}$$

$$x = \frac{\sqrt{5} + 3\sqrt{5}}{2} = \frac{4\sqrt{5}}{2} = 2\sqrt{5}$$

or

Number 1 (Continued)

$$x = \frac{\sqrt{5} - 3\sqrt{5}}{2} = \frac{-2\sqrt{5}}{2} = -\sqrt{5}$$

The solutions are $2\sqrt{5}$ and $-\sqrt{5}$.

3. $a^4 - 5a^2 + 4 = 0$

$$a^2 = \frac{-(-5) \pm \sqrt{(-5)^2 - 4(1)(4)}}{2(1)}$$

$$a^2 = \frac{5 \pm \sqrt{25 - 16}}{2}$$

$$a^2 = \frac{5 \pm \sqrt{9}}{2}$$

$$a^2 = \frac{5 \pm 3}{2}$$

$$a^2 = \frac{5 - 3}{2} = \frac{2}{2} = 1$$

$a = \pm \sqrt{1}$
$a = 1$ or $-1$
or

$$a^2 = \frac{5 + 3}{2} = \frac{8}{2} = 4$$

$a = \pm \sqrt{4}$
$a = 2$ or $-2$

$\therefore$ The solutions are 1, -1, 2, and -2.

5. Let s = side of cube. Then s + .2 = length of cube and

$(s+.2)^3 = s^3 + 6$
$(s+.2)^3 = (s+.2)(s+.2)(s+.2)$
$(s+.2)^3 = (s^2+.4s+.04)(s+.2)$
$(s+.2)^3 = s^3+.2s^2+.4s^2$
$\quad + 0.8s+.04s+.008$
$(s+.2)^3 = s^3+.6x^2+.12s+.008$
$s^3+.6s^2+.12s+.008 = s^3+6$
$.6s^2+.12s+5.992 = 0$

$$s = \frac{-.12 \pm \sqrt{(.12)^2 - 4(.6)(-5.992)}}{2(.6)}$$

$$s = \frac{-.12 \pm \sqrt{.0144 + 14.3808}}{1.2}$$

$$s = \frac{-.12 \pm \sqrt{14.3952}}{1.2}$$

Since length is positive

$$s = \frac{-.12 + \sqrt{14.3952}}{1.2}$$

$$s \approx \frac{-.12 + 3.794}{1.2}$$

$s \approx 3.0617$

7. $x^2 + 3xy + 4 = 0$

$$x = \frac{-3y \pm \sqrt{(3y)^2 - 4(1)(4)}}{2(1)}$$

$$x = \frac{-3y \pm \sqrt{9y^2 - 16}}{2}$$

## EXERCISE SET 9.3

1. $x^2 - 3x - 10 \geq 0$
$(x + 2)(x - 5) \geq 0$

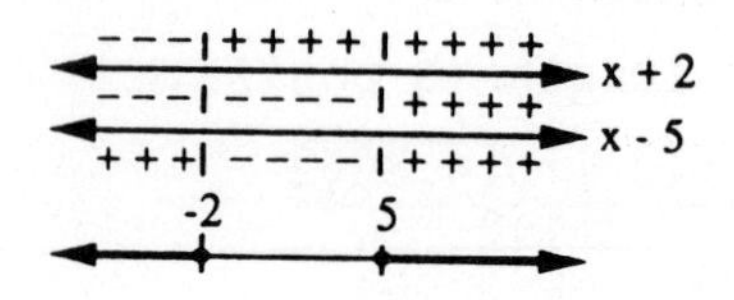


3. $x^2 + 4x > 0$
$x(x + 4) > 0$

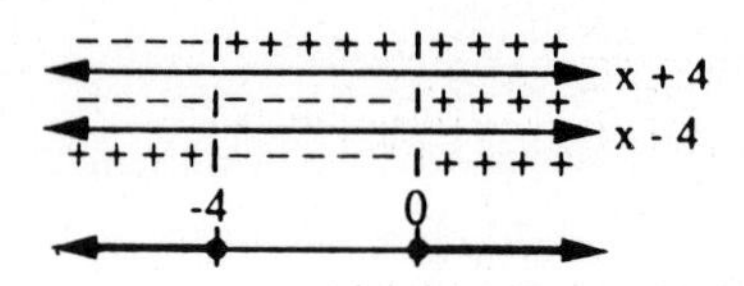


5. $x^2 - 16 < 0$
$(x + 4)(x - 4) < 0$

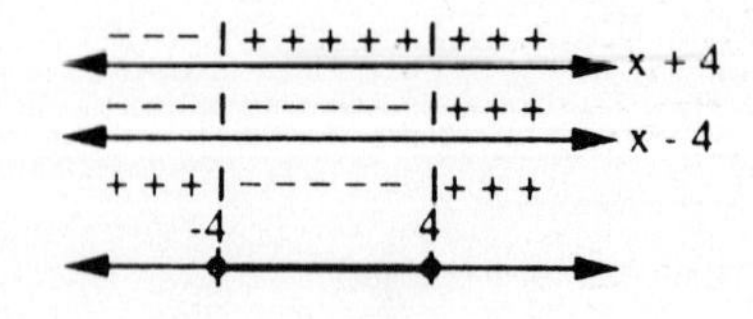


7. $x^2 + 9x + 20 \geq 0$
$(x + 5)(x + 4) \geq 0$

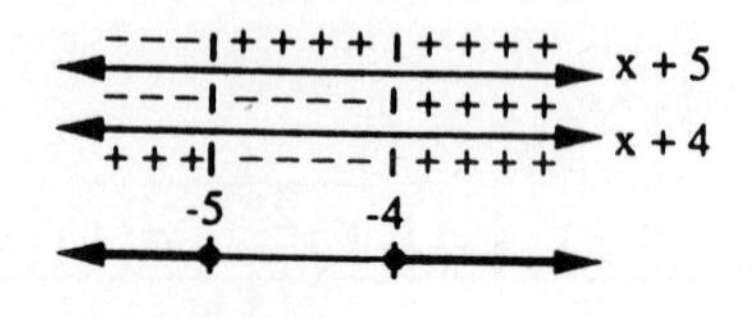


9. $x^2 - 6x \leq 27$
$x^2 - 6x - 27 \leq 0$
$(x + 3)(x - 9) \leq 0$

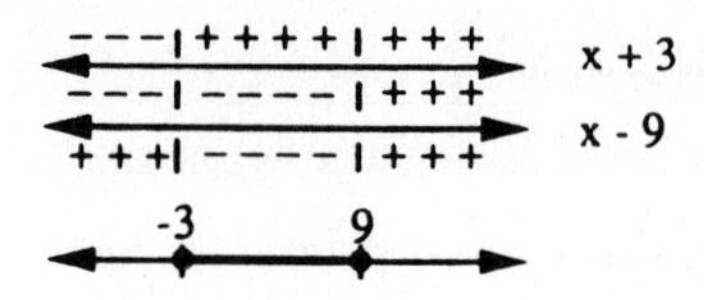


11. $x^2 \leq 2x + 35$
$x^2 - 2x - 35 \leq 0$
$(x + 5)(x - 7) \leq 0$

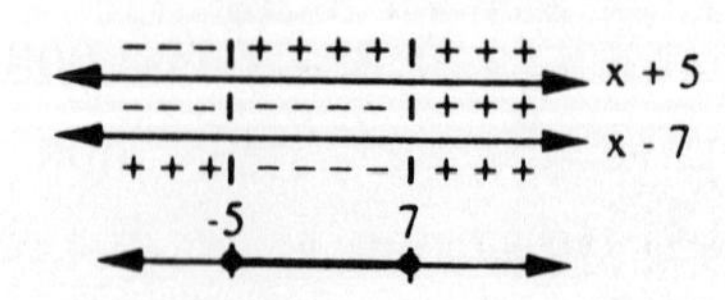

**13.** $x^2 < 36$

$x^2 - 36 < 0$

$(x + 6)(x - 6) < 0$

- - - |+ + + + | + + + + x + 6
- - - | - - - - | + + + + x - 6
+ + + | - - - - | + + + +
-6 6

**15.** $2x^2 + 10x + 12 > 0$

$2(x^2 + 5x + 6) > 0$

$2(x + 3)(x + 2) > 0$

- - - |+ + + + | + + + + x + 3
- - - | - - - - | + + + + x + 2
+ + + | - - - - | + + + +
-3 -2

**17.** $2x^2 - 7x - 15 \le 0$

$(2x + 3)(x - 5) \le 0$

- - - |+ + + + | + + + + 2x +3
- - - | - - - - | + + + + x - 5
+ + + | - - - - | + + + +
$-\frac{3}{2}$ 5

**19.** $4x^2 - 11x \le 20$

$4x^2 - 11x - 20 \le 0$

$(4x + 5)(x - 4) \le 0$

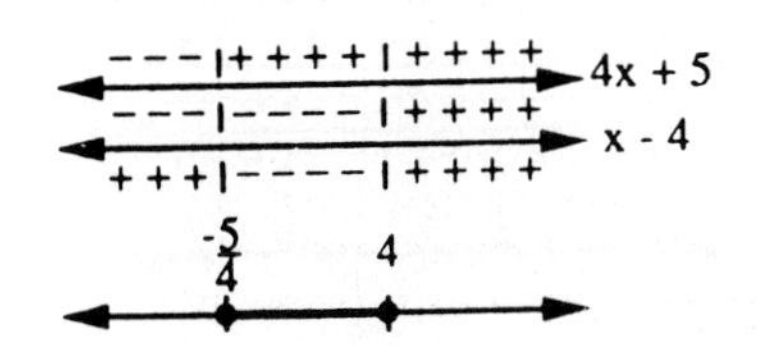


**21.** $2y^2 + y < 15$

$2y^2 + y - 15 < 0$

$(2y - 5)(y + 3) < 0$

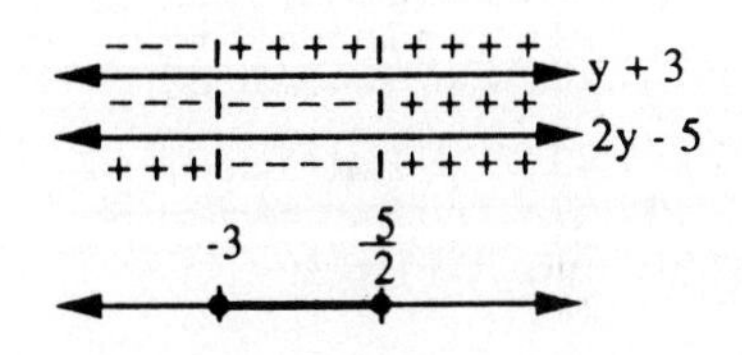


**23.** $3x^2 + x - 10 \ge 0$

$(3x - 5)(x + 2) \ge 0$

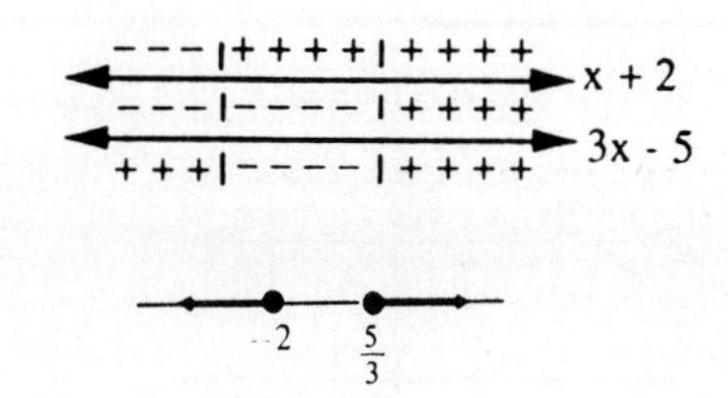

**25.** $(x - 1)(x + 1)(x + 4) > 0$

$x - 1 = 0 \quad x + 1 = 0 \quad x + 4 = 0$

$x = 1 \quad x = -1 \quad x = -4$

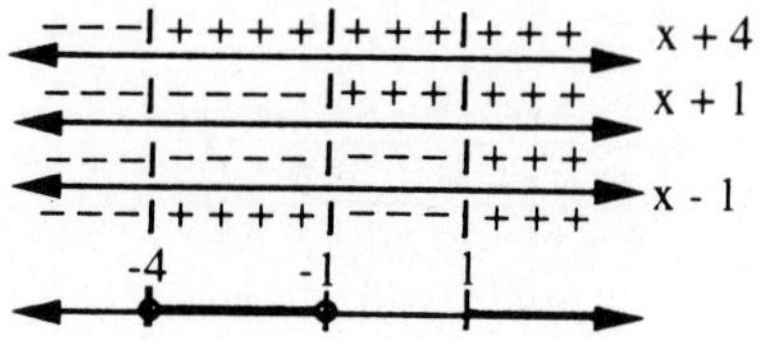

$(-4,-1) \cup (1, \infty)$

**27.** $(x - 4)(x - 1)(x + 3) \leq 0$

$x - 4 = 0 \quad x - 1 = 0 \quad x + 3 = 0$

$x = 1 \quad x = 1 \quad x = -3$

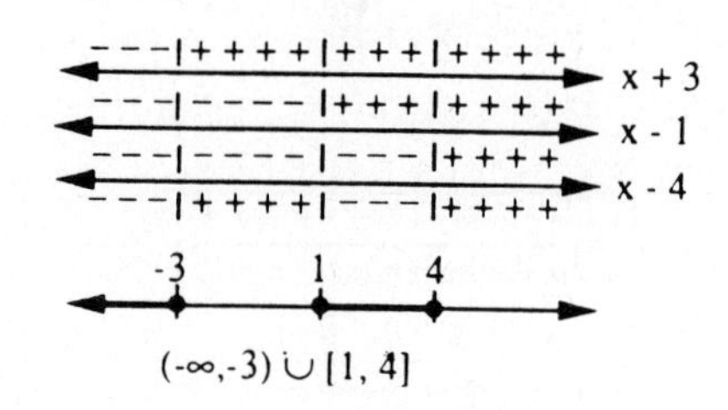


$(-\infty,-3) \cup [1, 4]$

**29.** $x(x - 3)(2x + 6) \geq 0$

$x = 0 \quad x - 3 = 0 \quad 2x + 6 = 0$

$x = 0 \quad x = 3 \quad 2x = -6$

$x = 0 \quad x = 3 \quad x = -3$

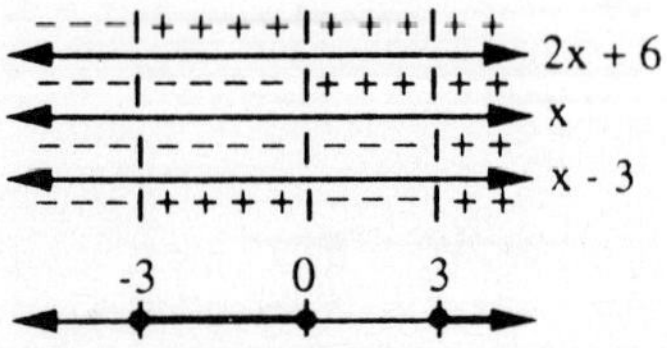

$[-3, 0] \cup [3, \infty)$

**31.** $(2x + 6)(3x - 6)(x+6) > 0$

$2x + 6 = 0 \quad 3x - 6 = 0 \quad x + 6 = 0$

$2x = -6 \quad 3x = 6 \quad x = -6$

$x = -3 \quad x = 2 \quad x = -6$

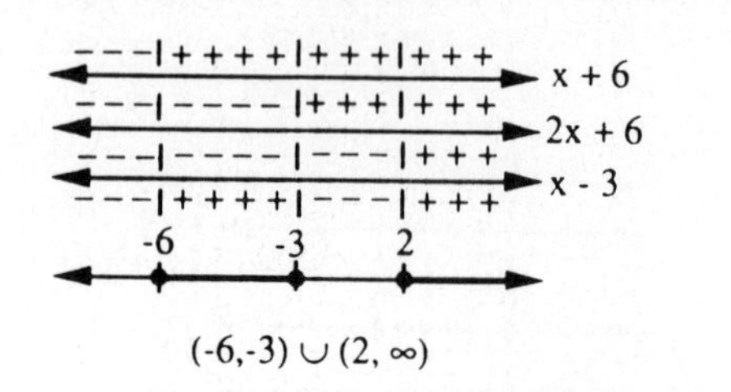


$(-6,-3) \cup (2, \infty)$

**33.** $(x + 2)(x + 2)(x - 3) \geq 0$

$x + 2 = 0 \quad x - 3 = 0$

$x = -2 \quad x = 3$

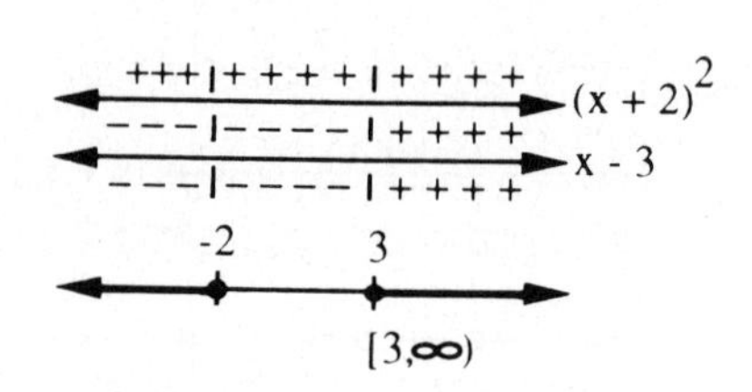


$[3,\infty)$

**35.** $(x + 3)(x + 3)(x - 4) < 0$

$x + 3 = 0 \quad x - 4 = 0$

$x = -3 \quad x = 4$

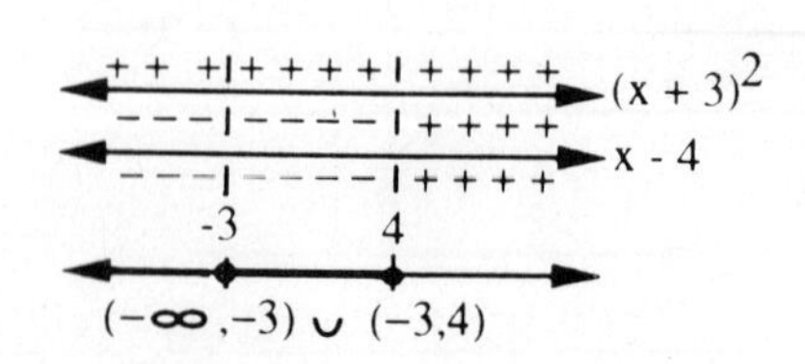


$(-\infty,-3) \cup (-3,4)$

**37.** $\dfrac{x + 3}{x - 1} > 0$

$x \neq 1$

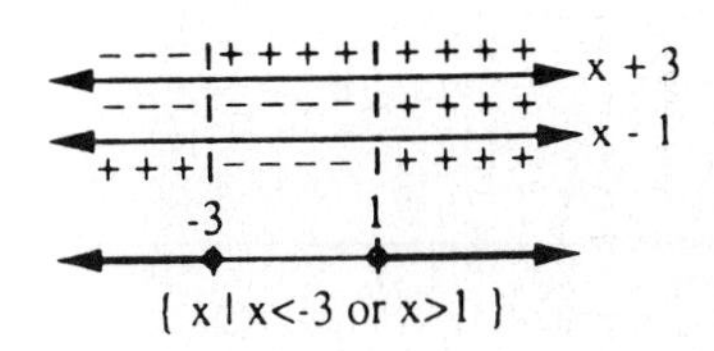


**39.** $\dfrac{x - 4}{x - 1} \leq 0$

$x \neq 1$

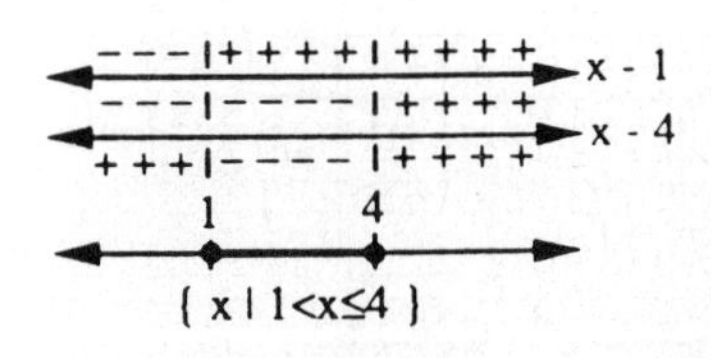


**41.** $\dfrac{2x - 4}{x - 1} < 0$

$x \neq 1$

- - - | + + + + | + + + +   x - 1
- - - | - - - - | + + + +   2x - 4
+ + + | - - - - | + + + +

1   2

{ x | 1<x<2 }

**43.** $\dfrac{3x + 6}{2x - 1} \geq 0$

$x \neq 1/2$

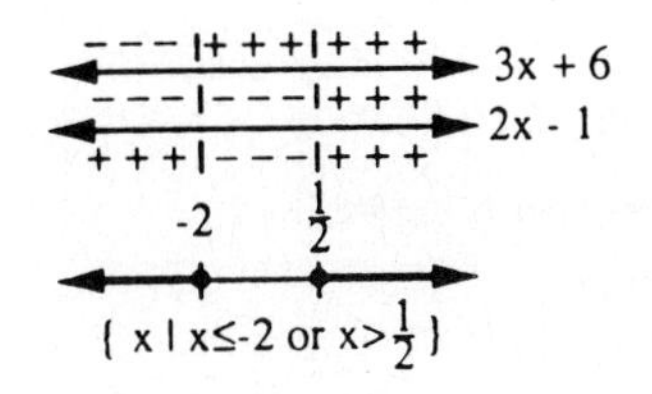


**45.** $\dfrac{x + 4}{x - 4} \leq 0$

$x \neq 4$

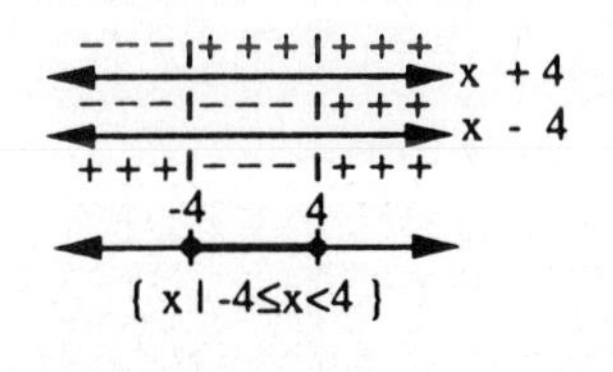


**47.** $\dfrac{4x - 2}{2x - 4} > 0$

$x \neq 2$

- - - | + + + | + + +   4x - 2
- - - | - - - | + + +   2x - 4
+ + + | - - - | + + +

$\frac{1}{2}$   2

{ x | x<$\frac{1}{2}$ or x>2 }

**49.** $\dfrac{(x + 2)(x - 4)}{x + 6} < 0$

$x \neq -6$

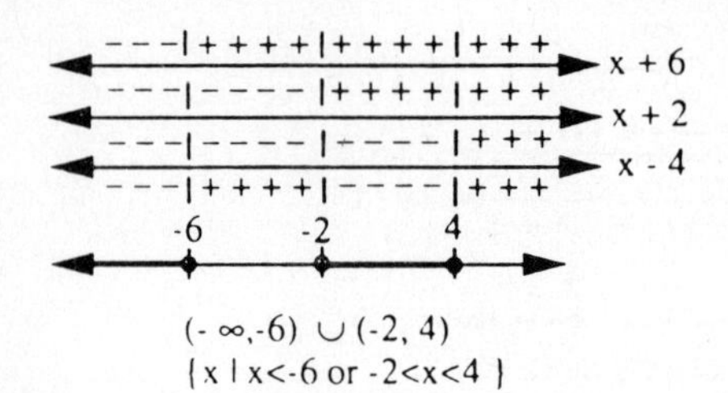


**51.** $\dfrac{(x - 6)(x - 1)}{x - 3} \geq 0$

$x \neq 3$

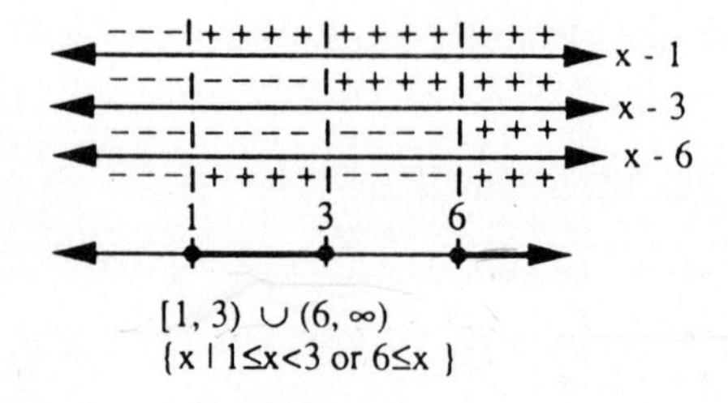


**53.** $\dfrac{x - 6}{(x + 4)(x - 1)} \leq 0$

$x \neq -4 \quad x \neq 1$

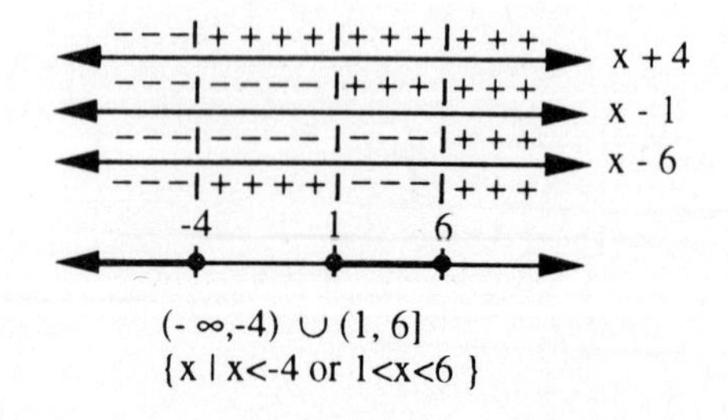


**55.** $\dfrac{(x - 3)(2x + 5)}{(x - 6)} > 0$

$x \neq 6$

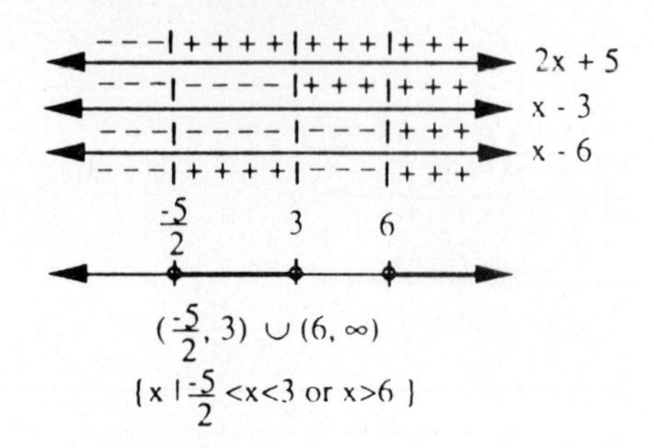


**57.** $\dfrac{(x + 2)(2x - 3)}{x} \geq 0$

$x \neq 0$

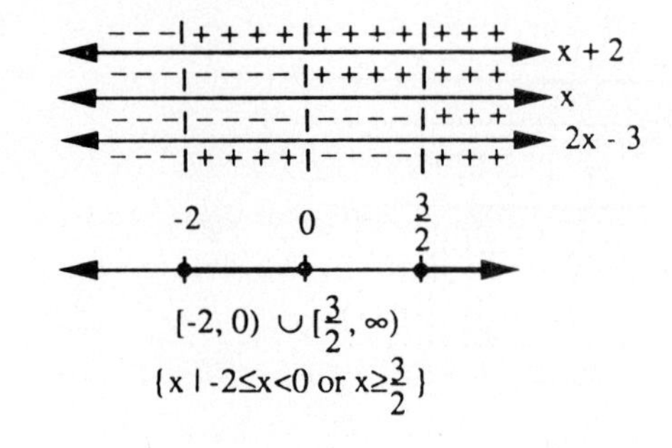

**59.**

$$\frac{2}{x - 3} \geq -1$$

$$\frac{2}{x - 3} + 1 \geq 0$$

$$\frac{2}{x - 3} + \frac{1(x - 3)}{x - 3} \geq 0$$

$$\frac{2 + x - 3}{x - 3} \geq 0$$

$$\frac{x - 1}{x - 3} \geq 0$$

$$x \neq 3$$

- - - | + + + | + + + x - 1
- - - | - - - | + + + x - 3
+ + + | - - - | + + +
1 3

**61.**

$$\frac{4}{x - 2} - 2 \geq 0$$

$$\frac{4}{4 - 2} - \frac{2(x - 2)}{x - 2} \geq 0$$

$$\frac{4 - 2x + 4}{x - 2} \geq 0$$

$$\frac{8 - 2x}{x - 2} \geq 0$$

$$x \neq 2$$

- - - | + + + + | + + + + x - 2
+ + + | + + + + | - - - - 8 - 2x
- - - | + + + + | - - - -
2 4

**63.**

$$\frac{2x - 5}{x - 4} \leq 1$$

$$\frac{2x - 5}{x - 4} - 1 \leq 0$$

$$\frac{2x - 5}{x - 4} - \frac{1(x - 4)}{x - 4} \leq 0$$

$$\frac{2x - 5 - x + 4}{x - 4} \leq 0$$

$$\frac{x - 1}{x - 4} \leq 0$$

$$x \neq 4$$

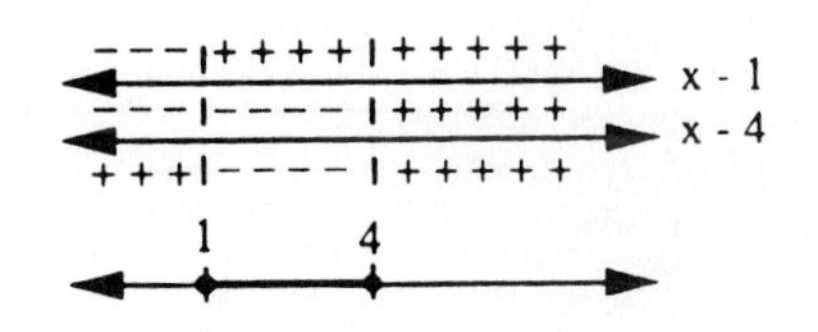


**65.**

$$\frac{w}{3w - 2} > -2$$

$$\frac{w}{3w - 2} + 2 > 0$$

$$\frac{w}{3w - 2} + \frac{2(3w - 2)}{3w - 2} > 0$$

$$\frac{w + 6w - 4}{3w - 2} > 0$$

$$\frac{7w - 4}{3w - 2} > 0$$

$$w \neq \frac{2}{3}$$

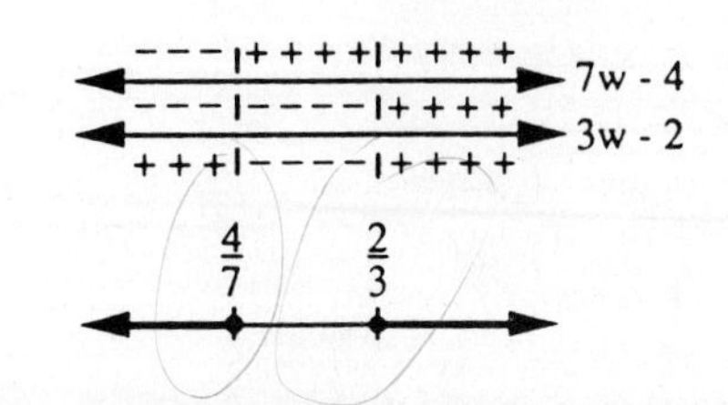

**67.** $$\frac{x}{3x - 1} < -1$$
$$\frac{x}{3x - 1} + 1 < 0$$
$$\frac{x}{3x - 1} + \frac{1(3x - 1)}{3x - 1} < 0$$
$$\frac{x + 3x - 1}{3x - 1} < 0$$
$$\frac{4x - 1}{3x - 1} < 0$$
$$x \neq \frac{1}{3}$$

- - - | + + + | + + +  4x - 1
- - - | - - - | + + +  3x - 1
+ + + | - - - | + + +

$\frac{1}{4}$ $\frac{1}{3}$

**69.** $$\frac{x + 8}{x + 2} > 1$$
$$\frac{x + 8}{x + 2} - 1 > 0$$
$$\frac{x + 8}{x + 2} - \frac{1(x + 2)}{x + 2} > 0$$
$$\frac{x + 8 - x - 2}{x + 2} > 0$$
$$\frac{6}{x + 2} > 0$$
$$x \neq -2$$

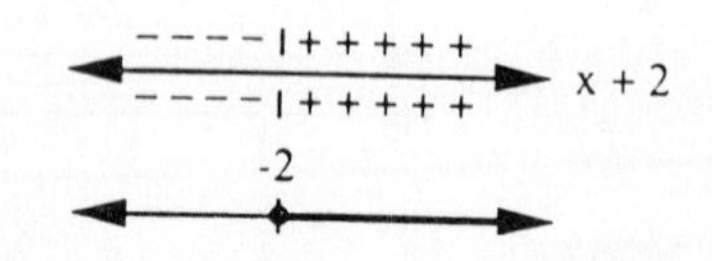


**71.** $(x + 3)^2(x - 1)^2 \geq 0$

The solution is all real numbers since any nonzero number squared is positive and zero squared is zero.

**73.** $x^2(x - 3)^2(x + 4)^2 < 0$

There is no real solution to this inequality since any nonzero real number squared is positive and zero squared is zero. Therefore, for any real number x,

$$x^2(x - 3)^2(x + 4)^2 \geq 0$$

**75.** $$\frac{x^2}{(x - 3)^2} > 0.$$

The solution is all real numbers, except 0 and 3 since x = 0 gives us 0 > 0 which is false, and x = 3 gives us a zero in the denominator, which is undefined.

**77.** $f(x) = \sqrt{4 - x}$.
The domain of f(x) can be found where $4 - x \geq 0$ or $4 \geq x$. Therefore, the domain is all $x \leq 4$.

**79.** $(\sqrt{-8} + \sqrt{2})(\sqrt{-2} - \sqrt{8})$
$= \sqrt{-8}\sqrt{-2} - \sqrt{-8}\sqrt{8} + \sqrt{2}\sqrt{-2} - \sqrt{8}\sqrt{2}$
$= i^2\sqrt{16} - \sqrt{-64} + \sqrt{-4} - \sqrt{16}$
$= -4 - 8i + 2i - 4$
$= -8 - 6i$

**JUST FOR FUN**

1. $(x + 1)(x - 3)(x + 5)(x + 9) \geq 0$

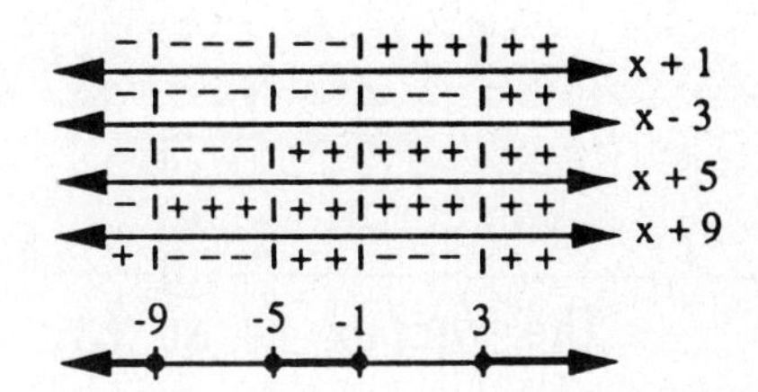


3. $x^2 - x + 1 > 0$

$$x = \frac{-(-1) \pm \sqrt{(-1)^2 - 4(1)(1)}}{2}$$

$$= \frac{1 \pm \sqrt{-3}}{2}.$$

Since the critical point is an imaginary number, the solution is all reals.

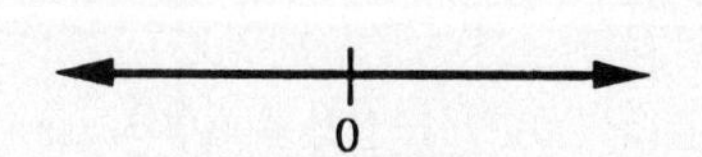

**EXERCISE 9.4**

1. $y = x^2 + 8x + 15$

(a) Since a = 1 the parabola opens upward

(b) $y = 0^2 + 8(0) + 15 = 15$
The y intercept is 15.

(c) $x = \frac{-b}{2a}$  $y = \frac{4ac - b^2}{4a}$

$= \frac{-8}{2(1)} = \frac{4(1)(15) - 8^2}{4(1)}$

$= \frac{-8}{2} = \frac{60 - 64}{4}$

$= -4 = \frac{-4}{4} = -1$

The vertex is at (-4, -1).

(d) $0 = x^2 + 8x + 15$
$= (x + 5)(x + 3)$
$x + 5 = 0$ or $x + 3 = 0$
$x = -5$   $x = -3$
The roots are -3 and -5.

(e)

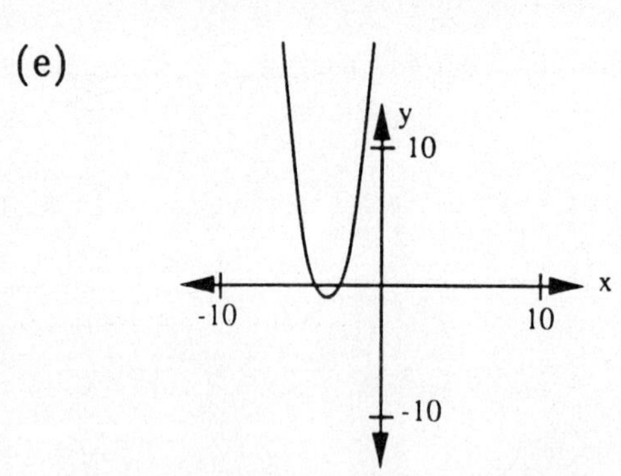


3. $f(x) = x^2 - 6x + 4$

(a) Since a = 1 the parabola opens upward

(b) $y = 0^2 - 6(0) + 4 = 4$
The y intercept is 4.

Number 3 (Continued)

(c) $x = \frac{-b}{2a}$  $y = \frac{4ac - b^2}{4a}$

$= \frac{-(-6)}{2(1)} = \frac{4(1)(4) - (-6)^2}{4(1)}$

$= \frac{6}{2} = \frac{16 - 36}{4}$

$= 3 = \frac{-20}{4} = -5.$

The vertex is at (3, -5).

(d) $0 = x^2 - 6x + 4.$

Since $x^2 - 6x + 4$ is not factorable check the discriminant.

$b^2 - 4ac = (-6)^2 - 4(1)(4)$
$= 36 - 16$
$= 20$

Since 20 > 0 there are two real roots.

$x = \frac{-b \pm \sqrt{b^2 - 4ac}}{2a}$

$= \frac{-(-6) \pm \sqrt{20}}{2(1)}$

$= \frac{6 \pm 2\sqrt{5}}{2}$

$x = \frac{6 + 2\sqrt{5}}{2} = 3 + \sqrt{5}$ or

$x = \frac{6 - 2\sqrt{5}}{2} = 3 - 5$

The roots are $3 + \sqrt{5}$ and $3 - \sqrt{5}$.

Number 3 (Continued)

(e)

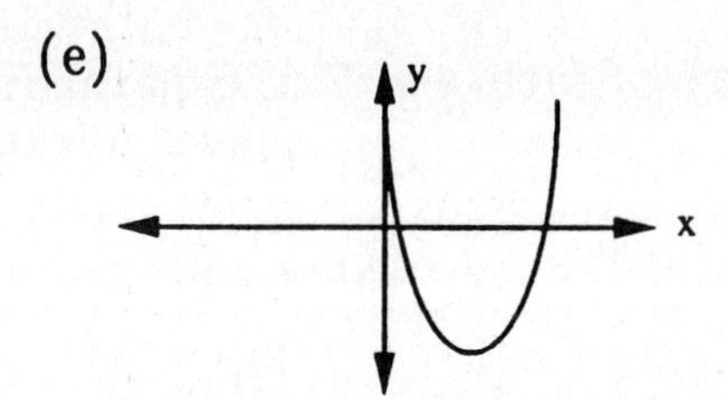


**5.** $y = x^2 + 6x + 9.$

(a) Since a = 1 the parabola opens upward

(b) $y = 0^2 + 6(0) + 9 = 9.$
The y intercept is 9.

(c) Note: In solving these problems, for part (c), the first column of calculations is for "x" and the second column is for "y".

$$x = \frac{-b}{2a} \quad y = \frac{4ac - b^2}{4a}$$

$$= \frac{-b}{2(1)} \quad = \frac{4(1)(9) - 6^2}{4(1)}$$

$$= \frac{-6}{2} \quad = \frac{36 - 36}{0}$$

$$= -3 \quad = 0$$

The vertex is at (-3, 0).

(d) $0 = x^2 + 6x + 9$
$= (x + 3)(x + 3)$
$= (x + 3)^2$
$x = -3$
The double root is -3.

(e)

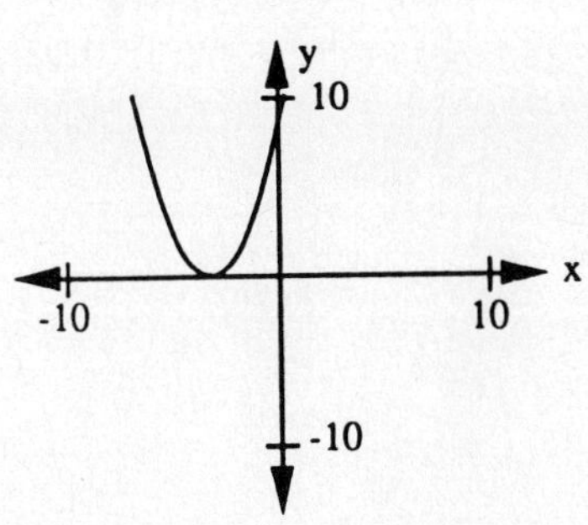


**7.** $f(x) = x^2 - 4x + 4$

(a) Since a = 1 the curve opens upward

(b) $y = 0^2 - 4(0) + 4 = 4$
The y intercept is 4.

(c) $x = \frac{-b}{2a} \quad y = \frac{4ac - b^2}{4a}$

$$= \frac{-(-4)}{2(1)} \quad = \frac{4(1)(4) - (-4)^2}{4(1)}$$

$$= \frac{4}{2} \quad = \frac{16 - 16}{4}$$

$$= 2 \quad = 0$$

The vertex is at (2, 0).

(d) $0 = x^2 - 4x + 4$
$= (x - 2)(x - 2)$
$= (x - 2)^2$
$= x - 2$
$x = 2$

The double root is 2.

(e)

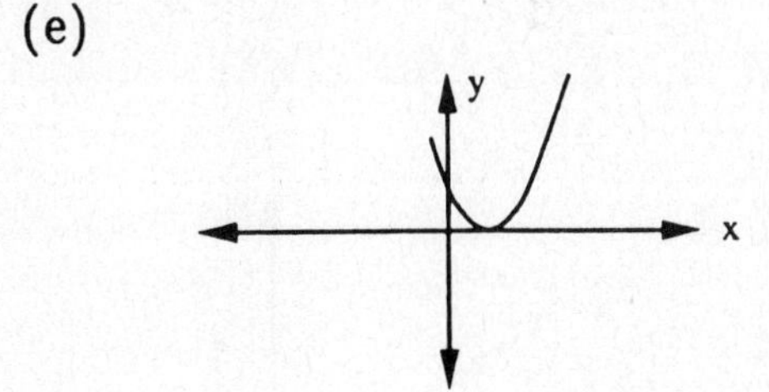

**9.** $y = 2x^2 - x - 6$

(a) Since $a = 2$ the parabola opens upward.

(b) $y = 2(0)^2 - 0 - 6 = -6$.
The y intercept is $-6$.

(c) $x = \frac{-b}{2a}$ $\quad y = \frac{4ac - b^2}{4a}$

$= \frac{-(-1)}{2(2)}$ $\quad = \frac{4(2)(-6) - (-1)^2}{4(2)}$

$= \frac{1}{4}$ $\quad = \frac{-48 - 1}{8}$

$= \frac{-49}{8}$

The vertex is at $(\frac{1}{4}, \frac{-49}{8})$.

(d) $0 = 2x^2 - x - 6$

$= (2x + 3)(x - 2)$

$2x + 3 = 0$ or $x - 2 = 0$

$x = \frac{-3}{2}$ $\quad x = 2$.

The roots are $\frac{-3}{2}$ and 2.

(e)

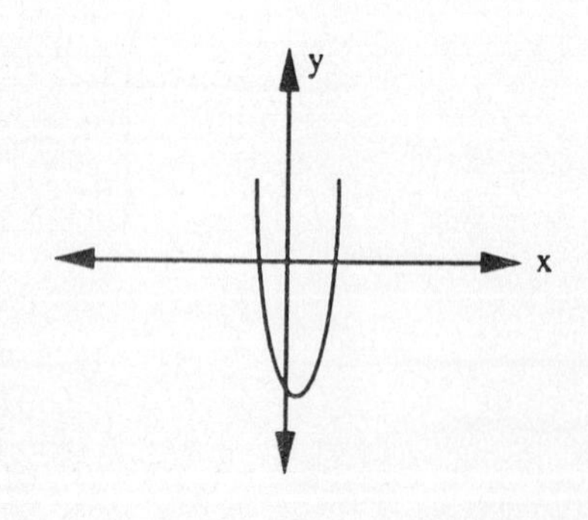


**11.** $y = 2x^2 - 3x - 9$.

(a) Since $a = 2$ the parabola opens upward.

(b) $y = 2(0)^2 - 3(0) - 9 = -9$.
The y intercept is $-9$.

(c) $x = \frac{-b}{2a}$ $\quad y = \frac{4ac - b^2}{4a}$

$= \frac{-(-3)}{2(2)}$ $\quad = \frac{4(2)(-9) - (-3)^2}{4(2)}$

$= \frac{3}{4}$ $\quad = \frac{-72 - 9}{8}$

$= \frac{-81}{8}$.

The vertex is at $(\frac{3}{4}, \frac{-81}{8})$.

(d) $0 = 2x^2 - 3x - 9$

$= (2x + 3)(x - 3)$

$2x + 3 = 0$ or $x - 3 = 0$

$x = \frac{-3}{2}$ $\quad x = 3$.

The roots are $\frac{-3}{2}$ and 3.

(e)

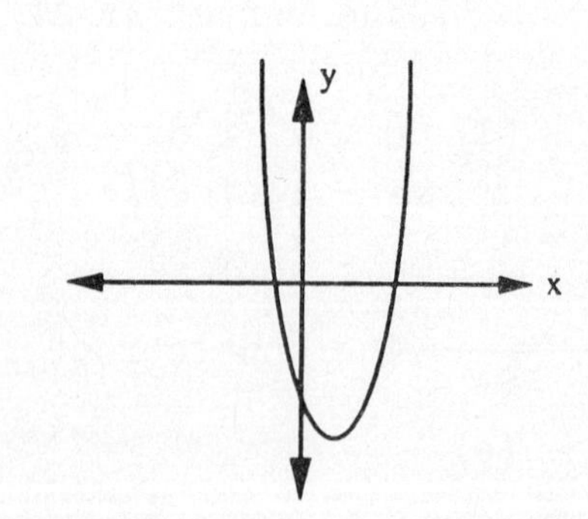

**13.** $y = 3x^2 + 4x + 3$

(a) Since $a = 3$ the parabola opens upward.

(b) $y = 3(0) + 4(0) + 3 = 3$
The y intercept is 3.

(c) $x = \frac{-b}{2a}$ $\quad y = \frac{4ac - b^2}{4a}$

$= \frac{-4}{2(3)} \quad = \frac{4(3)(3) - 4^2}{4(3)}$

$= \frac{-4}{6} \quad = \frac{36 - 16}{12}$

$= \frac{-2}{3} \quad = \frac{20}{12} = \frac{5}{3}$

The vertex is at $(\frac{-2}{3}, \frac{5}{3})$.

(d) $0 = 3x^2 + 4x + 3$

Since this is not factorable, check the discriminant.

$b^2 - 4ac = 4^2 - 4(3)(3)$
$= 16 - 36$
$= -20.$

Since $-20 < 0$ there are no real roots.

(e)

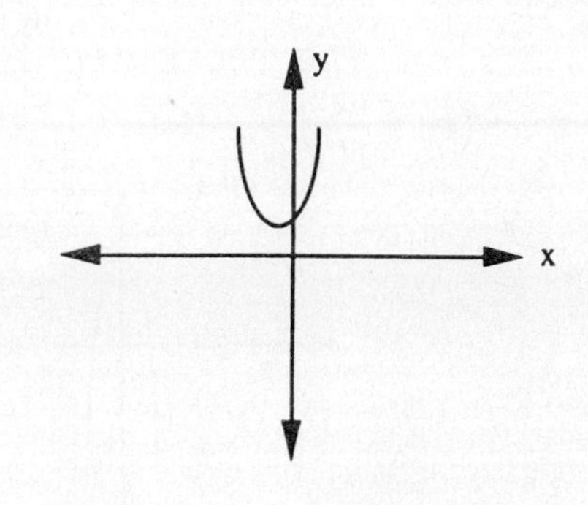


**15.** $f(x) = -2x^2 - 6x + 4$

(a) Since $a = -2$ the parabola opens downward.

(b) $y = -2(0)^2 - 6(0) + 4 = 4.$
The y intercept is 4.

(c) $x = \frac{-b}{2a}$ $\quad y = \frac{4ac - b^2}{4a}$

$= \frac{-(-6)}{2(-2)} \quad = \frac{4(-2)(4) - (-6)^2}{4(-2)}$

$= \frac{6}{-4} \quad = \frac{-32 - 36}{-8}$

$= \frac{-3}{2} \quad = \frac{-68}{-8} = \frac{17}{2}.$

The vertex is at $(\frac{-3}{2}, \frac{17}{2})$.

(d) $0 = -2x^2 - 6x + 4.$
Since this is not factorable, check the discriminant.

$b^2 - 4ac = (-6)^2 - 4(-2)(4)$
$= 36 + 32$
$= 68$

$\therefore$ There are two real roots.

$\frac{-b \pm \sqrt{b^2 - 4ac}}{2a} = \frac{-(-6) \pm \sqrt{68}}{2(-2)}$

$= \frac{6 \pm 2\sqrt{17}}{-4}$

$= \frac{3 \pm \sqrt{17}}{-2}$

$x = \frac{-3 + \sqrt{17}}{2}$ or $x = \frac{-3 - \sqrt{17}}{2}$

Number 15 (Continued)

(e)

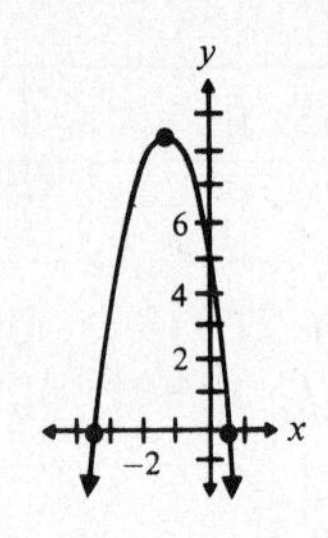


17. $y = x^2 + 4$

(a) Since a = 1 the parabola opens upward.

(b) $y = 0^2 + 4 = 4$

(c) $x = \frac{-b}{2a}$ $y = \frac{4ac - b^2}{4a}$

$= \frac{-0}{2(1)}$ $= \frac{4(1)(4) - 0^2}{4(1)}$

$= 0$ $= \frac{16}{4}$

$= 4$

The vertex is at (0,4).

(d) $0 = x^2 = 4.$
Since this is not factorable, check the discriminant.

$b^2 - 4ac = 0 - 4(1)4 = -16$

$\therefore$ There are no real roots.

Number 17 (Continued)

(e)

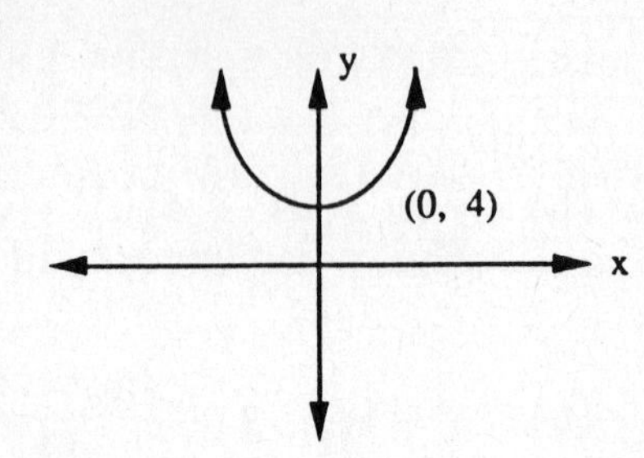


19. $y = -9x^2 + 4$

(a) Since a = -9 the parabola opens downward.

(b) $y = -9(0)^2 + 4 = 4.$
The y intercept is 4.

(c) $x = \frac{-b}{2a}$ $y = \frac{4ac - b^2}{4a}$

$= \frac{-0}{2(-9)}$ $= \frac{4(-9)(4) - 0^2}{4(-9)}$

$= 0$ $= \frac{-144 - 0}{36} = 4.$

The vertex is at (0,4).

(d) $0 = -9x^2 + 4$

$= -1(9x^2 - 4)$

$= -1(3x - 2)(3x + 2)$

$3x - 2 = 0$ or $3x + 2 = 0$

$3x = 2$ $3x = -2$

$x = \frac{2}{3}$ $x = \frac{-2}{3}.$

The roots are $\frac{2}{3}$ and $\frac{-2}{3}$.

(e)

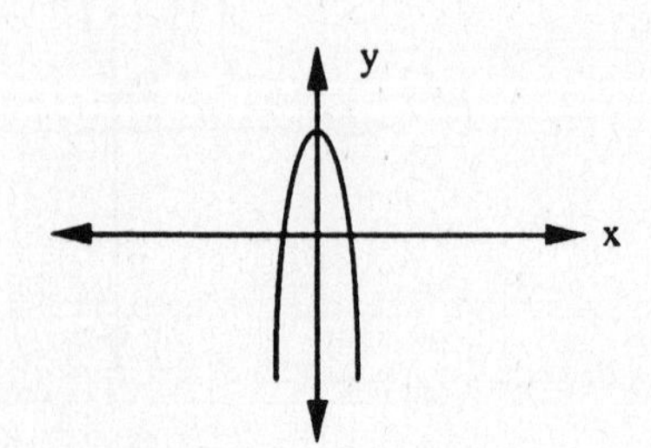

21. $y = -x^2 + 6x$

(a) Since a = -1 the parabola opens downward.

(b) $y = -0^2 + 6(0) = 0.$
The y intercept is 0.

(c) $x = \frac{-b}{2a}$ $y = \frac{4ac - b^2}{4a}$

$= \frac{-6}{2(-1)}$ $= \frac{4(-1)(0) - 6^2}{4(-1)}$

$= \frac{-6}{-2}$ $= \frac{0 - 36}{-4}$

$= 3$ $= \frac{-36}{-4} = 9$

The vertex is at (3,9).

(d) $0 = -x^2 + 6x$
$= -x(x - 6)$
$-x = 0$ or $x - 6 = 0$
$x = 0$ $x = 6$
The roots are 0 and 6.

(e)

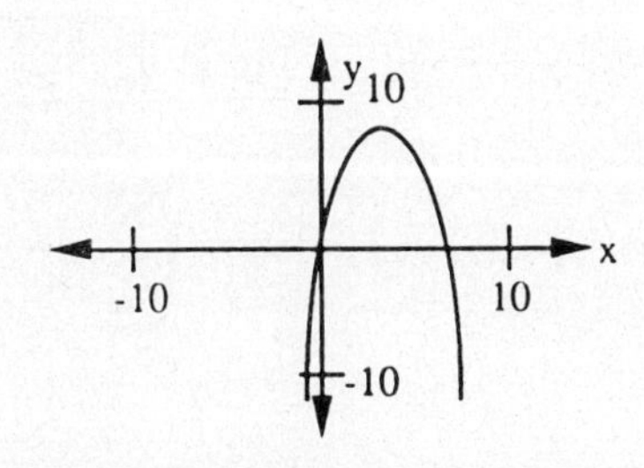


23. $f(x) = -5x^2 + 5$

(a) Since a = -5 the parabola opens downward.

(b) $y = -5(0)^2 + 5 = 5.$
The y intercept is 5.

(c) $x = \frac{-b}{2a}$ $y = \frac{4ac - b^2}{4a}$

$= \frac{-0}{2(-5)}$ $= \frac{4(-5)(5) - 0^2}{4(-5)}$

$= 0$ $= \frac{-100}{-20} = 5.$

The vertex is at (0,5).

Number 23 (Continued)

(d) $0 = -5x^2 + 5$
$= -5(x^2 - 1)$
$= -5(x + 1)(x - 1)$
$x + 1 = 0$ or $x - 1 = 0$
$x = -1$ $x = 1$
The roots are -1 and 1.

(e)

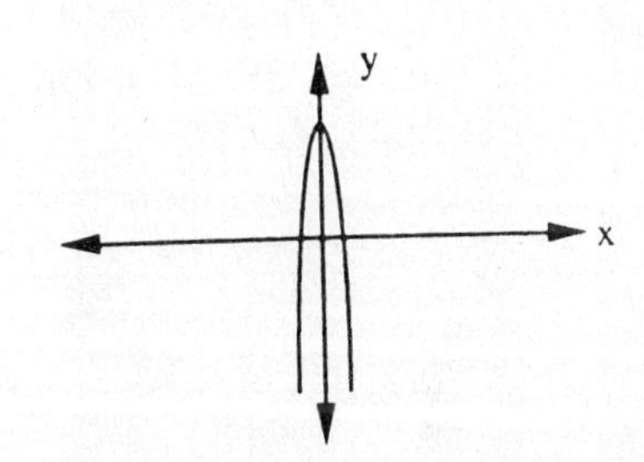


25. $y = 3x^2 + 4x - 6.$

(a) Since a = 3 the parabola opens upward.

(b) $y = 3(0)^2 + 4(0) - 6 = -6$
The y intercept is -6.

(c) $x = \frac{-b}{2a}$ $y = \frac{4ac - b^2}{4a}$

$= \frac{-4}{2(3)}$ $= \frac{4(3)(-6) - 4^2}{4(3)}$

$= \frac{-4}{6}$ $= \frac{-72-16}{12}$

$= \frac{-2}{3}$ $= \frac{-88}{12} = \frac{-22}{3}$

The vertex is at $(\frac{-2}{3}, \frac{-22}{3})$.

Number 25 (Continued)

(d) $0 = 3x^2 + 4x - 6.$
Since this is not factorable, check the discriminant.

$$b^2 - 4ac = 4^2 - 4(3)(-6) = 16 + 72 = 88$$

Since $88 > 0$ there are two real roots.

$$x = \frac{-b \pm \sqrt{b^2 - 4ac}}{2a} = \frac{-4 \pm \sqrt{88}}{2(3)} = \frac{-4 \pm 2\sqrt{22}}{6}$$

$$x = \frac{-4 - 2\sqrt{22}}{6} = \frac{-2 - \sqrt{22}}{3} \text{ or}$$

$$x = \frac{-4 + 2\sqrt{22}}{6} = \frac{-2 + \sqrt{22}}{3}.$$

The solutions are $\frac{-2 - \sqrt{22}}{3}$ and $\frac{-2 + \sqrt{22}}{3}$.

(e)

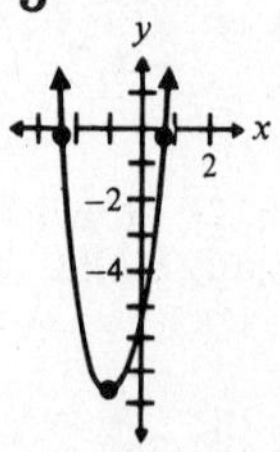

27. $y = -x^2 + 3x + 6$

(a) Since $a = -1$ the parabola opens downward.

(b) $y = -0^2 + 3(0) + 6 = 6.$
The y intercept is 6.

Number 27 (Continued)

(c) $x = \frac{-b}{2a}$ $y = \frac{4ac - b^2}{4a}$

$$= \frac{-3}{2(-1)} = \frac{4(-1)(6) - 3^2}{4(-1)}$$

$$= \frac{3}{2} = \frac{-24 - 9}{-4}$$

$$= \frac{-33}{-4} = \frac{33}{4}.$$

The vertex is at $(\frac{3}{2}, \frac{33}{4})$.

(d) $0 = -x^2 + 3x + 6.$
Since this is not factorable, check the discriminant.

$$b^2 - 4ac = 3^2 - 4(-1)(6) = 9 + 24 = 33$$

Since $33 > 0$ there are two real roots.

$$x = \frac{-b \pm \sqrt{b^2 - 4ac}}{2a} = \frac{-3 \pm \sqrt{33}}{2(-1)} = \frac{-3 \pm \sqrt{33}}{-2}$$

$$x = \frac{3 - \sqrt{33}}{2} \text{ or } x = \frac{3 + \sqrt{33}}{2}$$

The roots are $\frac{3 - \sqrt{33}}{2}$ and $\frac{3 + \sqrt{33}}{2}$.

(e)

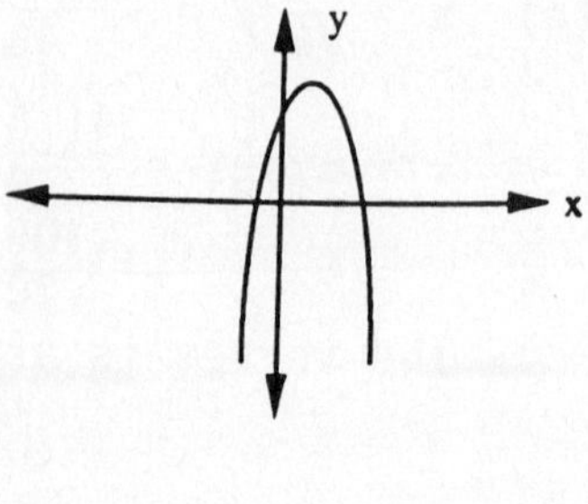

**29.** $f(x) = -4x^2 + 6x - 9$

(a) Since a = -4 the parabola opens downward.

(b) $y = -4(0)^2 + 6(0) - 9 = -9$
The y intercept is -9.

(c) $x = \frac{-b}{2a}$ $y = \frac{4ac - b^2}{4a}$

$= \frac{-6}{2(-4)}$ $= \frac{4(-4)(-9) - 6^2}{4(-4)}$

$= \frac{-6}{-8}$ $= \frac{144 - 36}{-16}$

$= \frac{3}{4}$ $= \frac{108}{-16} = \frac{-27}{4}$

The vertex is at $(\frac{3}{4}$ and $\frac{-27}{4})$.

(d) $0 = -4x^2 + 6x - 9.$
Since this is not factorable, check the discriminant.

$b^2 - 4ac = 6^2 - 4(-4)(-9)$
$= 36 - 144$
$= -108.$

Since $-108 < 0$ there are no real roots.

(e)

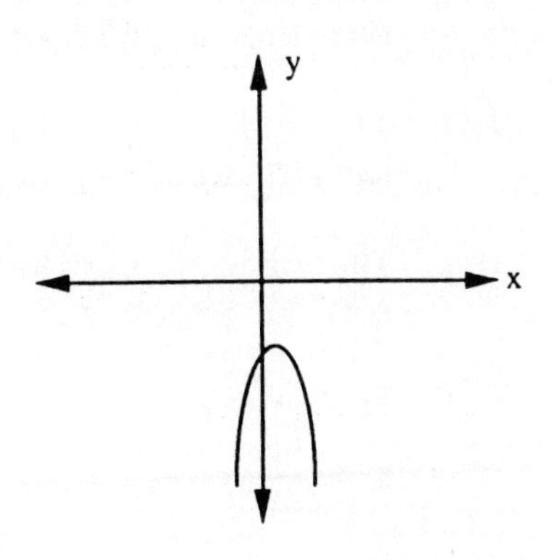


**31.** If 4 and 6 are the roots, the factors must be (x - 4) and (x - 6).

Therefore the equation is:
$y = (x - 4)(x - 6)$
$= x^2 - 6x - 4x + 24$
$= x^2 - 10x + 24$

**33.** If 3 and -4 are the roots, the factors must be (x - 3) and (x + 4).

Therefore, the equation is:
$y = (x - 3)(x + 4)$
$= x^2 + 4x - 3x - 12$
$= x^2 + x - 12$

**35.** If 2 and -3 are the roots, the factors must be (x - 2) and (x + 3).

Therefore, the equation is:
$y = (x - 2)(x + 3)$
$= x^2 + 3x - 2x - 6$
$= x^2 + x - 6$

**37.** If 2 and 2 are the roots, the factors must be (x - 2) and (x - 2).

Therefore, the equation is:
$y = (x - 2)(x - 2)$
$= x^2 - 2x - 2x + 4$
$= x^2 - 4x + 4$

**39.** If -2 and 2/3 are the roots, the factors must be (x + 2) and (3x - 2).

Therefore, the equation is:

$y = (x + 2)(3x - 2)$
$= 3x^2 - 2x + 6x - 4$
$= 3x^2 + 4x - 4$

**41.** If -1/2 and 2/3 are the roots, the factors must be $(2x + 1)$ and $(3x - 2)$.

Therefore, the equation is:

$$y = (2x + 1)(3x - 2)$$
$$= 6x^2 - 4x + 3x - 2$$
$$= 6x^2 - x - 2$$

**43.** $C = 1/2(N^2 - N)$
$= 1/2(8^2 - 8)$
$= 1/2(64 - 8)$
$= 1/2(56)$
$= 28$

**45.** $h = -0.3t^2 + 6.5t + 6,\ t \leq 10$
$t = 8$
$h = -0.3(8)^2 + 6.5(8) + 6$
$= -0.3(64) + 52 + 6$
$= -19.2 + 58$
$= 38.8$ feet

**47.** $d = 0.0003w^2 + 0.05w \quad w = 200$ lb.
$d = 0.0003(200)^2 + 0.05(200)$
$= 0.0003(40{,}000) + 10$
$= 1.2 + 10$
$= 11.2$ lbs.

**49.** $d = \nu_0 t - 1/2gt^2$

(a) $\nu_0 = 192$ ft/sec $t = 3$ sec
$d = 192$ ft/sec (3 sec)
$- 1/2$ (32 ft/sec)(3 sec)$^2$
$= 576$ ft$- 1/2(32$ ft/sec$^2)$ (9 sec)$^2$
$= 576$ ft $- 144$ ft
$= 432$ ft.

Number 49 (Continued)

(b) $t = \frac{-b}{2a} = \frac{-192}{-2(16)} = 6$

When $t = 6$
$d = 192(6) - 1/2(32)(6)^2$
$= 1152 - 576$
$= 576$

The vertex is at (6, 576).
To find roots, set $d = 0$.

$0 = -16t^2 + 192t$
$= -16t(t - 12)$
$-16t = 0$ or $t - 12 = 0$
$t = 0 \qquad t = 12$

The roots are 0 and 12.

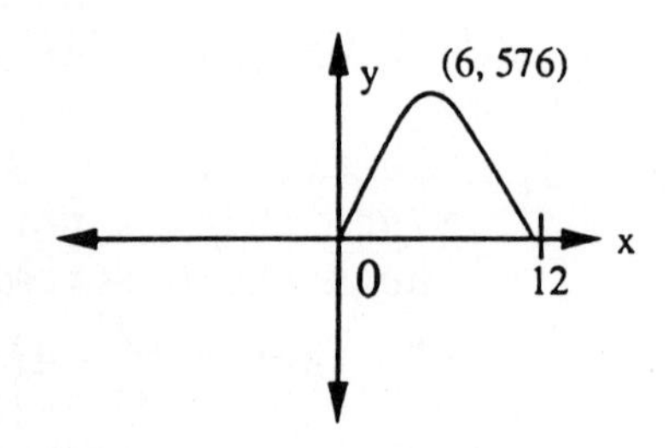


(c) The object will reach its maximum height at 576 ft.

(d) It will reach its maximum height at 6 seconds.

(e) The object strikes at $t = 12$ seconds.

**51.** (a) The vertex
$x = \frac{-b}{2a} = \frac{-24}{2} = 12$

$I = -(12)^2 + 24(12) - 44$
$= -144 + 288 - 44$
$= 100$

Number 51 (Continued)

The vertex is at (12,100).

To find the root set I = 0.

$0 = -x^2 + 24x - 44$

$= -1(x^2 - 24x + 44)$

$= -1(x - 2)(x - 22)$

$x - 2 = 0$ or $x - 22 = 0$

$x = 2$ $\quad x = 22.$

The roots are 2 and 22.

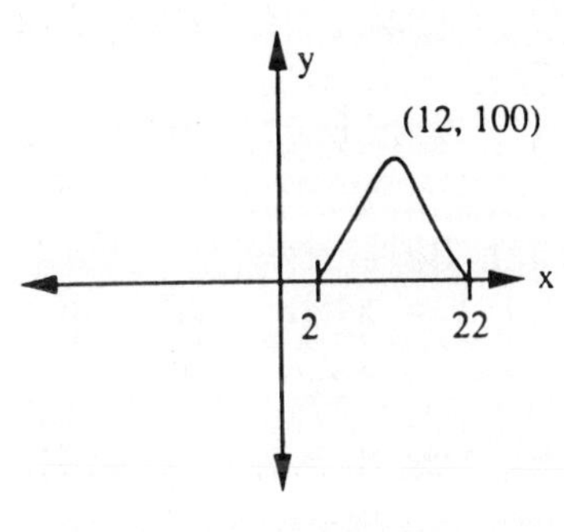


(b) The minimum cost will be \$2 since the smaller root is 2.

(c) The maximum cost is \$22 since the larger root is 22.

(d) maximum $x = \frac{-b}{2a}$

$= \frac{-24}{2(-1)} = \frac{-24}{-2} = 12$

Therefore, they should charge \$12.

(e) max $I = -(12)^2 + 24(12) - 44$

$= -144 + 288 - 44$

$= 100$

Since I is in hundreds of dollars the maximum income is 100(\$100) = \$10,000.

**53.** $h = 32t - 16t^2$

maximum height occurs at:

$x = \frac{-b}{2a} = \frac{-32}{2(-16)}$

$= \frac{-32}{-32} = 1$ second

The maximum height is:

$y = \frac{4ac - b^2}{4a}$

$= \frac{4(-16)(0) - (32)^2}{4(-16)}$

$= \frac{0 - 1024}{-64} = \frac{-1024}{-64}$

$y = 16$ feet

**55.** $y = ax^2 + bx + c$

(a) No real roots means no x-intercepts

(b) One real root gives one x-intercept

(c) Two real roots which are distinct give two x-intercepts

**57.** If the vertex is in the first or second quadrants and $a > 0$, there are no $x$ intercepts. If the vertex is in the third or fourth quadrant and $a < 0$, there are no $x$ intercepts.

If the vertex is on the $x$ axis, there is a single $x$ intercept. In all other cases, there are two distinct $x$ intercepts.

**59.** (a) The graphs will have the same x-intercepts but $y = x^2 - 8x + 12$ will open up and $y = -x^2 + 8x - 12$ will open down.

(b) Yes, because the x intercepts are located by setting $x^2 - 8x + 12$ and $-x^2 + 8x - 12$ equal to zero. They have the same solution set, therefore the x-intercepts are equal.

(c) No. The vertex for $y = x^2 - 4x + 4$ have the same x-coordinate as $y = -x^2 + 4x - 4$ but the y coordinate will differ by sign.

(d)

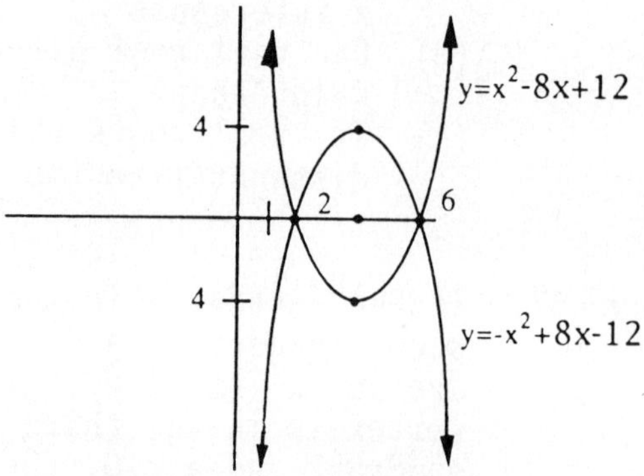

y=x²-8x+12
4
2
6
4
y=-x²+8x-12

**61.** Let x equal the amount of money in the 12% account an 15000 - x equal the amount in the 5.5% account.

$900 = .12(x) + .055(15000 - x)$
$900 = .12x + 825 - .055x$
$75 = .065x$
$x = 1153.85$

$1,153.85 should be invested in the 12% account and 15000 - x = $13,846.15 should be invested in the 5.5% account.

**63.** $6x^2 + 17x - 45$
$\underline{27} \times (\underline{-10}) = -270$
$\underline{27} + (\underline{-10}) = 17$

$6x^2 - 10x + 27x - 45$
$2x(3x-5) + 9(3x-5)$
$(2x+9)(3x-5)$

**JUST FOR FUN**

1. $P = 2(L + w)$
$\frac{64}{2} = L + w$
$\frac{64}{2} - w = L$
Let $x = w$
Then $32 - x = 1$
and $A = x(32 - x)$
$= 32x - x^2$
maximum $x = \frac{-b}{2a} = \frac{-32}{2(-1)}$
$= \frac{-32}{-2} = 16$

$\therefore$ width = 16
and length = 32 - 16 = 16.

## EXERCISE SET 9.5

1. (a) $(f+g)(x) = (x^2+2x-8)+(x-2)$
$= x^2 + 3x - 10$
(b) $(f+g)(2) = 2^2 + 3(2) - 10$
$= 0$

3. (a) $(fg)(x) = (x^2+2x-8)(x-2)$
$= x^3 - 12x + 16$
(b) $(fg)(-1) = (-1)^3 - 12(-1)+16$
$= 27.$

5. (a) $(f\circ g)(x) = f(g(x))$
$= (x-2)^2 + 2(x-2) - 8$
$= x^2 - 4x + 4 + 2x - 4 - 8$
$= x^2 - 2x - 8$
(b) $(f\circ g)(3) = 3^2 - 2(3) - 8$
$= -5$

7. (a) $(g+f)(x) = (x+2) + (x^2-4)$
$= x^2 + x - 2$
(b) $(g+f)(-2) = (-2)^2 + (-2)-2$
$= 0$

9. (a) $(gf)(x) = (x+2)(x^2-4)$
$= x^3 + 2x^2 - 4x - 8$
(b) $(gf)(-4) = (-4)^3 + 2(-4)^2$
$- 4(-4) - 8$
$= -24$

11. (a) $(g\circ f)(x) = (x^2-4) + 2$
$= x^2 - 2$
(b) $(g\circ f)(4) = 4^2 - 2 = 14$

13. (a) $(f+g)(x) = (x-4) + \sqrt{x+6}$
(b) $(f+g)(3) = (3-4) + \sqrt{3+6}$
$= -1 + \sqrt{9}$
$= 2$

15. (a) $(fg)(x) = (x-4)(\sqrt{x+6})$
(b) $(fg)(10) = (10-4)(\sqrt{10+6})$
$= 6(\sqrt{16})$
$= 24$

17. (a) $(f\circ g)(x) = (\sqrt{x+6}) - 4$
(b) $(f\circ g)(7) = \sqrt{7+6} - 4$
$= \sqrt{13} - 4$

19. Yes. Addition is commutative.

21. Yes. Multiplication is also commutative.

23. No. Composition is not commutative.

25. (a) $(f\circ g)(x) = \sqrt{(x^2-5)+5}$
$= \sqrt{x^2}$
$= x$
$(g\circ f)(x) = (\sqrt{x+5})^2 - 5$
$= x + 5 - 5$
$= x$
(b) Since values of $x < 0$ are not in the domain of $g(x)$, values of $x < 0$ are not in the domain of $(f\circ g)(x)$.

27. Yes, both linear and quadratic functions are polynomial functions.

29. All polynomial functions are rational functions but not all rational functions are polynomial functions.

## EXERCISE SET 9.6

1. The graph fails the horizontal line test and is therefore not one-to-one.

3. The graph passes the horizontal line test and is therefore one-to-one.

5. The graph fails the horizontal line test and is therefore not one-to-one.

7. {(-2,4), (3,-7), (5,3) (-6,0)}. The function is one-to-one since there is a unique y for each distinct x.

9. {(-4,2), (5,3) (0,2), (3,7)} The function is not one-to-one since there is not a unique y for each distinct x.

11. f(x):
Domain: {-2, -1, 2, 4, 9}
Range: {0, 3, 4, 6, 7}
$f^{-1}(x)$:
Domain: {0, 3, 4, 6, 7}
Range: {-2, -1, 2, 4, 9}

13. f(x):
Domain: {-2.9, 0, 1.7, 5.7}
Range: {-3.4, 3, 4, 9.76}
$f^{-1}(x)$:
Domain: {-3.4, 3, 4, 9.76}
Range: {-2.9, 0, 1.7, 5.7}

15. $f(x) = 2x + 8$
$f^{-1}(x) = \dfrac{x - 8}{2}$

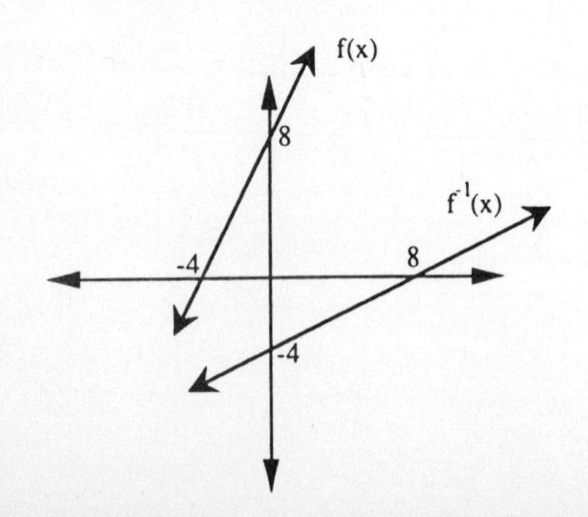


17. $y = f(x) = -3x - 10$
$f^{-1}(x) = \dfrac{-(x+10)}{3}$

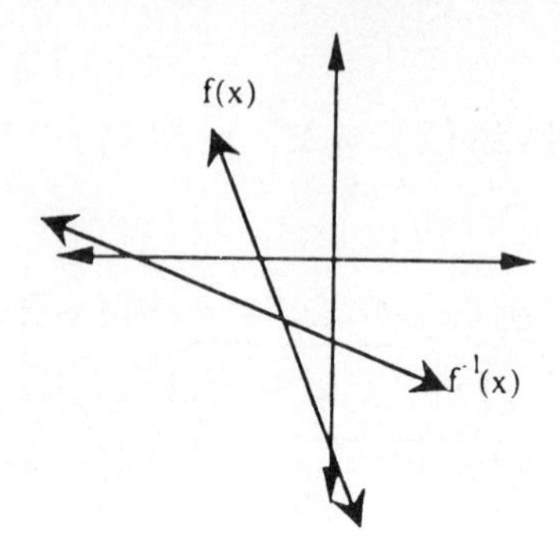


19. $y = f(x) = 2x - \dfrac{3}{5}$
$f^{-1}(x) = (5x+3)/10$

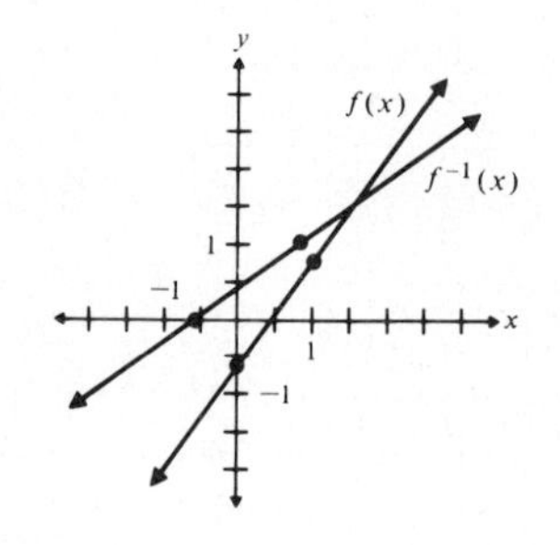


21. $f(x) = 6-3x$
$f^{-1}(x) = \dfrac{(-x+6)}{3}$

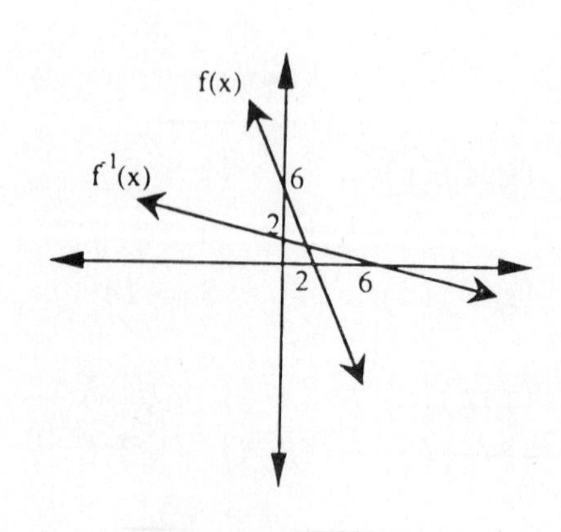

**23.** $f(x) = \frac{6+4x}{3}$

$f^{-1}(x) = \frac{(3x-6)}{4}$

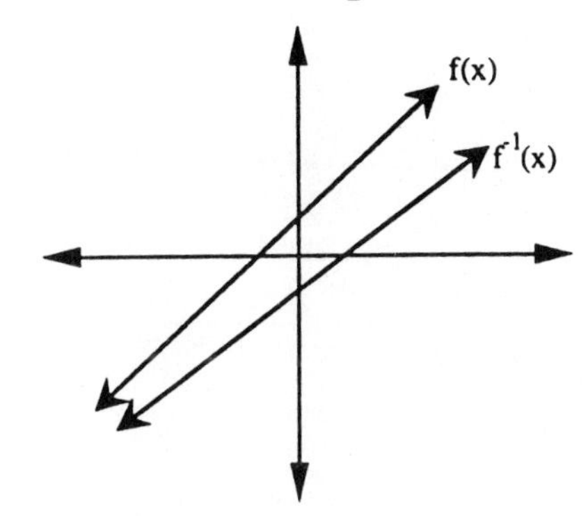


**25.** $y = f(x) = -\frac{2}{3} + \frac{5}{8}x$

$f^{-1}(x) = (24x + 16)/15$

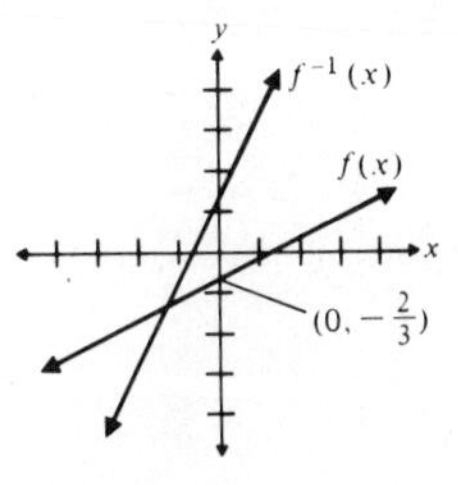


**27.** $(f \circ f^{-1})(x) = (x+4) - 4$
$= x$

$(f^{-1} \circ f)(x) = (x-4) + 4$
$= x$

**29.** $(f \circ f^{-1})(x) = \frac{(5x+4)-4}{5}$
$= \frac{5x}{5} = x$

$(f^{-1} \circ f)(x) = 5(\frac{x-4}{5}) + 4$
$= x - 4 + 4$
$= x$

**31.** $(f \circ f^{-1})(x) = \sqrt[3]{(x^3+2) - 2}$
$= \sqrt[3]{x^3}$
$= x$

$(f^{-1} \circ f)(x) = (\sqrt[3]{x-2})^3 + 2$
$= x - 2 + 2$
$= x$

**33.** Functions for which each value in the range has a unique value in the domain.

**35.** The domain of a function is the range of the inverse function. The range of the function is the domain of the (inverse) function.

**37.**
(1) $2x + 3y - 4z = 18$
(2) $x - y - z = 3$
(3) $x - 2y - 2z = 2$

Multiply Equation (2) by -2. Add the result to equation (1) and eliminate x.

(1) $2x + 3y - 4z = 18$
(2) $-2x + 2y + 2z = -6$

(4) $5y - 2z = 12$

Multiply equation (2) by -1. Add the result to equation (3) and eliminate x.

(2) $-x + y + z = -3$
(3) $x - 2y - 2z = 2$

(5) $-y - z = -1$

Multiply equation (5) by 5. Add the result to equation (4) to eliminate y and solve for z.

(4) $5y - 2z = 12$
(5) $-5y - 5z = -5$

$-7z = 7$
$z = -1$

Number 37 (Continued)

Substitute $z = -1$ into equation (4) or (5) and solve for y.

(5) $-(y) - (-1) = -1$

$-y + 1 = -1$

$-y = -2$

$y = 2$

Substitute $z = -1$ and $y = 2$ into equation (1), (2) or (3) and solve for x.

(2) $x - 2 - (-1) = 3$

$x - 2 + 1 = 3$

$x - 1 = 3$

$x = 4$

Solution: (4, 2, -1)

39. $\sqrt{\dfrac{24x^3y^2}{3xy^3}} = \sqrt{8x^{(3-1)}y^{(2-3)}}$

$= \sqrt{8x^2y^{-1}} = 2x\sqrt{2y^{-1}}$

$= \dfrac{2x\sqrt{2}}{\sqrt{y}} = \dfrac{2x\sqrt{2}\ \sqrt{y}}{\sqrt{y}\ \sqrt{y}}$

$= \dfrac{2x\sqrt{2y}}{y}$

**JUST FOR FUN**

1. $y = \sqrt{x^2 - 9}$ $\quad x \geq 3$

$y^2 = x^2 - 9$

$y^2 + 9 = x^2$

$\sqrt{y^2 + 9} = x$

Therefore $f^{-1}(x) = \sqrt{x^2 + 9}$

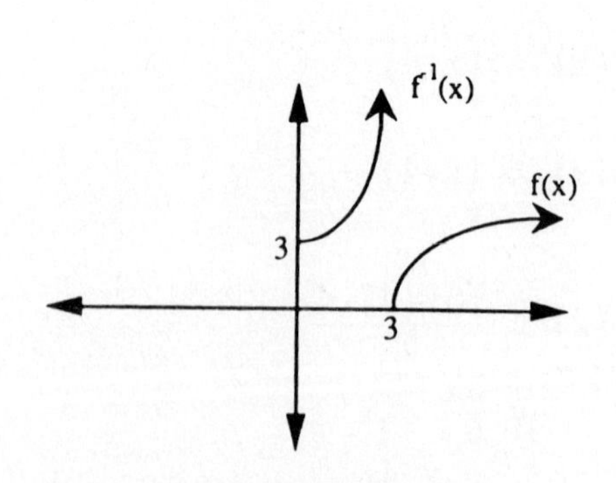


f(x):

Domain: $\{x | x > 3\}$

Range: $\{y | y \geq 0\}$

$f^{-1}(x)$

Domain: $\{x | x \geq 0\}$

Range: $\{y | y \geq 3\}$

## REVIEW EXERCISES

1. $(x - 4)^2 = 20$

$\sqrt{(x-4)^2} = \sqrt{20}$

$x - 4 = \pm 2\sqrt{5}$

$x = 4 \pm 2\sqrt{5}$

3. $(x - \frac{2}{3})^2 = \frac{1}{9}$

$\sqrt{(x - \frac{2}{3})} = \sqrt{\frac{1}{9}}$

$x - \frac{2}{3} = \pm \frac{1}{3}$

$x - \frac{2}{3} = \frac{1}{3} \quad x - \frac{2}{3} = - \frac{1}{3}$

$x = 1 \quad\quad x = \frac{1}{3}$

5. Solve for $F_D$

$F_T^2 = F_a^2 + F_D^2$

$F_T^2 - F_a^2 = F_D^2$

$\sqrt{F_T^2 - F_a^2} = F_D$

7. $x^2 + (4)^2 = (12)^2$. Then,

$x^2 + 16 = 144$

$x^2 = 144 - 16$

$x^2 = 128$

$x = \sqrt{128} \approx 11.31$

9. $x^2 - 10x + 16 = 0$

$x^2 - 10x = -16$

$x^2 - 10x + 25 = -16 + 25$

$x^2 - 10x + 25 = 9$

$(x - 5)^2 = 9$

$x - 5 = \pm 3$

$x = 5 \pm 3$

$x = 5 + 3$ or $x = 5 - 3$

$x = 8 \quad\quad x = 2$

11. $x^2 - 14x + 13 = 0$

$x^2 - 14x = -13$

$x^2 - 14x + 49 = -13 + 49$

$(x - 7)^2 = 36$

$x - 7 = \pm 6$

$x = 7 \pm 6$

$x = 7 + 6$ or $x = 7 - 6$

$x = 13 \quad\quad x = 1$

13. $x^2 - 3x - 54 = 0$

$x^2 - 3x = 54$

$x^2 - 3x + \frac{9}{4} = \frac{216}{4} + \frac{9}{4}$

$(x - \frac{3}{2})^2 = \frac{225}{4}$

$x - \frac{3}{2} = \pm \frac{15}{2}$

$x = \frac{3}{2} \pm \frac{15}{2}$

$x = \frac{3}{2} + \frac{15}{2}$ or $x = \frac{3}{2} - \frac{15}{2}$

$x = \frac{18}{2} = 9 \quad\quad x = \frac{-12}{2} = -6$

15. $x^2 + 2x - 5 = 0$

$x^2 + 2x = 5$

$x^2 + 2x + 1 = 5 + 1$

$(x + 1)^2 = 6$

$x + 1 = \pm \sqrt{6}$

$x = -1 \pm \sqrt{6}$

$x = -1 + \sqrt{6}$ or $x = -1 - \sqrt{6}$

17.
$$2x^2 - 8x = -64$$
$$x^2 - 4x = -32$$
$$x^2 - 4x + 4 = -32 + 4$$
$$(x - 2)^2 = -28$$
$$x - 2 = \pm \sqrt{-28}$$
$$x = 2 \pm \sqrt{4}\sqrt{7}\sqrt{-1}$$
$$x = 2 \pm 2i\sqrt{7}$$
$$x = 2 + 2i\sqrt{7} \text{ or } x = 2 - 2i\sqrt{7}$$

19.
$$4x^2 - 2x + 12 = 0$$
$$x^2 - \frac{1}{2}x = -3$$
$$x^2 - \frac{1}{2}x + \frac{1}{16} = -\frac{48}{16} + \frac{1}{16}$$
$$(x - \frac{1}{4})^2 = \frac{-47}{16}$$
$$x - \frac{1}{4} = \pm \frac{\sqrt{-47}}{4}$$
$$x = \frac{1}{4} \pm \frac{\sqrt{47}\sqrt{-1}}{4}$$
$$x = \frac{1 + i\sqrt{47}}{4} \text{ or } x = \frac{1 - i\sqrt{47}}{4}$$

21.
$$3x^2 - 4x - 20 = 0$$
$$a = 3,\ b = -4,\ c = -20$$
$$b^2 - 4ac = (-4)^2 - 4(3)(-20)$$
$$= 16 + 240$$
$$= 256$$
Since the discriminant is positive, this equation has two distinct real solutions.

23.
$$2x^2 + 6x + 7 = 0$$
$$a = 2,\ b = 6,\ c = 7$$
$$b^2 - 4(a)(c) = b^2 - 4(2)(7)$$
$$= 36 - 56$$
$$= -20$$
Since the discriminant is negative, this equation has no real solutions.

25.
$$x^2 - 12x = -36$$
$$x^2 - 12x + 36 = 0$$
$$a = 1,\ b = -12,\ c = 36$$
$$b^2 - 4ac = (-12)^2 - 4(1)(36)$$
$$= 144 - 144$$
$$= 0$$
Since the discriminant is 0, the equation has one real solution.

27.
$$-3x^2 - 4x + 8 = 0$$
$$a = -3,\ b = -4,\ c = 8$$
$$b^2 - 4ac = (-4)^2 - 4(-3)(8)$$
$$= 16 + 96$$
$$= 112$$
Since the discriminant is positive, this equation has two real solutions.

29.
$$x^2 - 9x + 14 = 0$$
$$a = 1,\ b = -9,\ c = 14$$
$$x = \frac{-b \pm \sqrt{b^2 - 4ac}}{2a}$$
$$x = \frac{-(-9) \pm \sqrt{(-9)^2 - 4(1)(14)}}{2(1)}$$
$$x = \frac{9 \pm \sqrt{81 - 56}}{2}$$
$$x = \frac{9 \pm \sqrt{25}}{2}$$
$$x = \frac{9 \pm 5}{2}$$
$$x = \frac{9 + 5}{2} \quad \text{or} \quad x = \frac{9 - 5}{2}$$
$$x = \frac{14}{2} = 7 \qquad x = \frac{4}{2} = 2$$

**31.** $x^2 = 7x - 10$

$x^2 - 7x + 10 = 0$

$a = 1, b = -7, c = 10$

$$x = \frac{-b \pm \sqrt{b^2 - 4ac}}{2a}$$

$$x = \frac{-(-7) \pm \sqrt{(-7)^2 - 4(1)(10)}}{2(1)}$$

$$x = \frac{7 \pm \sqrt{49 - 40}}{2}$$

$$x = \frac{7 \pm \sqrt{9}}{2}$$

$$x = \frac{7 \pm 3}{2}$$

$$x = \frac{7 + 3}{2} = \frac{10}{2} = 5 \quad \text{or}$$

$$x = \frac{7 - 3}{2} = \frac{4}{2} = 2$$

**33.** $x^2 - 18 = 7x$

$x^2 - 7x - 18 = 0$

$a = 1, b = -7, c = -18$

$$x = \frac{-b \pm \sqrt{b^2 - 4ac}}{2a}$$

$$x = \frac{-(-7) \pm \sqrt{(-7)^2 - 4(1)(-18)}}{2(1)}$$

$$x = \frac{7 \pm \sqrt{49 + 72}}{2}$$

$$x = \frac{7 \pm \sqrt{121}}{2}$$

$$x = \frac{7 \pm 11}{2}$$

$$x = \frac{7 + 11}{2} = \frac{18}{2} = 9 \quad \text{or}$$

$$x = \frac{7 - 11}{2} = \frac{-4}{2} = -2$$

**35.** $6x^2 + x - 15 = 0$

$a = 6, b = 1, c = -15$

$$x = \frac{-b \pm \sqrt{b^2 - 4ac}}{2a}$$

$$x = \frac{-1 \pm \sqrt{1^2 - 4(6)(-15)}}{2(6)}$$

$$x = \frac{-1 \pm \sqrt{1 + 360}}{12}$$

$$x = \frac{-1 \pm \sqrt{361}}{12}$$

$$x = \frac{-1 \pm 19}{12}$$

$$x = \frac{-1 + 19}{12} \quad \text{or } x = \frac{-1 - 19}{12}$$

$$x = \frac{18}{12} = \frac{3}{2} \qquad x = \frac{-20}{12} = \frac{-5}{3}$$

**37.** $-2x^2 + 3x + 6 = 0$

$a = -2, b = 3, c = 6$

$$x = \frac{-b \pm \sqrt{b^2 - 4ac}}{2a}$$

$$x = \frac{-3 \pm \sqrt{3^2 - 4(-2)(6)}}{2(-2)}$$

$$x = \frac{-3 \pm \sqrt{9 + 48}}{-4}$$

$$x = \frac{3 \pm \sqrt{57}}{4}$$

$$x = \frac{3 + \sqrt{57}}{4} \text{ or } x = \frac{3 - \sqrt{57}}{4}$$

**39.** $3x^2 - 4x + 6 = 0$

$a = 3,\ b = -4,\ c = 6$

$$x = \frac{-b \pm \sqrt{b^2 - 4ac}}{2a}$$

$$x = \frac{-(-4) \pm \sqrt{(-4)^2 - 4(3)(6)}}{2(3)}$$

$$x = \frac{4 \pm \sqrt{16 - 72}}{6}$$

$$x = \frac{4 \pm \sqrt{-56}}{6}$$

$$x = \frac{4 \pm \sqrt{4}\sqrt{14}\sqrt{-1}}{6}$$

$$x = \frac{4 \pm 2i\sqrt{14}}{6}$$

$$x = \frac{\overset{1}{\cancel{2}}(2 \pm i\sqrt{14})}{\underset{3}{\cancel{6}}}$$

$$x = \frac{2 \pm i\sqrt{14}}{3}$$

$$x = \frac{2 + i\sqrt{14}}{3} \text{ or } x = \frac{2 - i\sqrt{14}}{3}$$

**41.** $2x^2 + 3x = 0$

$a = 2,\ b = 3,\ c = 0$

$$x = \frac{-b \pm \sqrt{b^2 - 4ac}}{2a}$$

$$x = \frac{-3 \pm \sqrt{3^2 - 4(2)(0)}}{2(2)}$$

$$x = \frac{-3 \pm \sqrt{9}}{4}$$

$$x = \frac{-3 \pm 3}{4}$$

$$x = \frac{-3 + 3}{4} \text{ or } x = \frac{-3 - 3}{4}$$

$$x = \frac{0}{4} = 0 \qquad x = \frac{-6}{4} = \frac{-3}{2}$$

**43.** $x^2 - 11x + 24 = 0$

$(x - 8)(x - 3) = 0$

$x - 8 = 0$ or $x - 3 = 0$

$x = 8 \qquad x = 3$

**45.** $x^2 = -3x + 40$

$x^2 + 3x - 40 = 0$

$(x + 8)(x - 5) = 0$

$x + 8 = 0$ or $x - 5 = 0$

$x = -8 \qquad x = 5$

**47.** $x^2 - 4x - 60 = 0$

$(x - 10)(x + 6) = 0$

$x - 10 = 0$ or $x + 6 = 0$

$x = 10 \qquad x = -6$

**49.** $x^2 + 11x + 12 = 0$

$a = 1,\ b = 11,\ c = 12$

$$x = \frac{-b \pm \sqrt{b^2 - 4ac}}{2a}$$

$$x = \frac{-11 \pm \sqrt{(11)^2 - 4(1)(12)}}{2(1)}$$

$$x = \frac{-11 \pm \sqrt{121 - 48}}{2}$$

$$x = \frac{-11 \pm \sqrt{73}}{2}$$

$$x = \frac{-11 + \sqrt{73}}{2} \text{ or } x = \frac{-11 - \sqrt{73}}{2}$$

**51.** $x^2 + 6x = 0$

$x(x + 6) = 0$

$x = 0$ or $x + 6 = 0$

$x = 0 \qquad x = -6$

**53.** $3x^2 = 9x - 10$

$3x^2 - 9x + 10 = 0$

$a = 3,\ b = -9,\ c = 10$

$$x = \frac{-b \pm \sqrt{b^2 - 4ac}}{2a}$$

$$x = \frac{-(-9) \pm \sqrt{(-9)^2 - 4(3)(10)}}{2(3)}$$

$$x = \frac{9 \pm \sqrt{81 - 120}}{6}$$

$$x = \frac{9 \pm \sqrt{39}}{6}$$

$$x = \frac{9 \pm \sqrt{-39}\ \sqrt{-1}}{6}$$

$$x = \frac{9 \pm i\sqrt{39}}{6}$$

$$x = \frac{9 + i\sqrt{39}}{6} \text{ or } x = \frac{9 - i\sqrt{39}}{6}$$

**55.** $x^2 + 3x - 6 = 0$

$x + 3x = 6$

$$x + 3x + \frac{9}{4} = \frac{24}{4} + \frac{9}{4}$$

$$\left(x + \frac{3}{2}\right)^2 = \frac{33}{4}$$

$$x + \frac{3}{2} = \pm\sqrt{\frac{33}{4}}$$

$$x = -\frac{-3}{2} \pm \frac{\sqrt{33}}{2}$$

$$x = \frac{-3 + \sqrt{33}}{2}$$

or $$x = \frac{-3 - \sqrt{33}}{2}$$

**57.** $-3x^2 - 5x + 8 = 0$

$3x^2 + 5x - 8 = 0$

$(3x + 8)(x - 1) = 0$

$3x + 8 = 0$ or $x - 1 = 0$

$3x = 8 \qquad x = 1$

$$x = \frac{-8}{3}$$

**59.** $2x^2 - 5x = 0$

$x(2x - 5) = 0$

$x = 0$ or $2x - 5 = 0$

$x = 0 \qquad 2x = 5$

$$x = \frac{5}{2}$$

**61.** $$x^2 + \frac{5x}{4} = \frac{3}{8}$$

$$x^2 + \frac{5x}{4} + \frac{25}{64} = \frac{24}{64} + \frac{25}{64}$$

$$\left(x + \frac{5}{8}\right)^2 = \frac{49}{64}$$

$$x + \frac{5}{8} = \pm\frac{7}{8}$$

$$x = -\frac{5}{8} \pm \frac{7}{8}$$

$$x = -\frac{5}{8} + \frac{7}{8} = \frac{2}{8} = \frac{1}{4}$$

or

$$x = -\frac{5}{8} - \frac{7}{8} = \frac{-12}{8} = \frac{-3}{2}$$

**63.** $x^2 + 6x + 5 \geq 0$

$(x + 1)(x + 5) \geq 0$

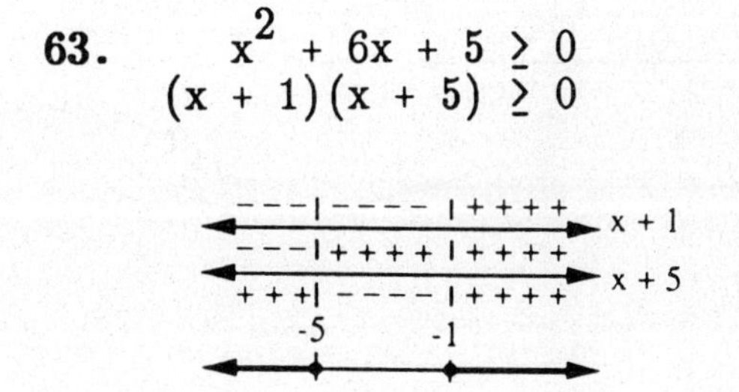

65. $x^2 \le 11x - 30$

$$x^2 - 11x + 30 \le 0$$
$$(x - 5)(x - 6) \le 0$$

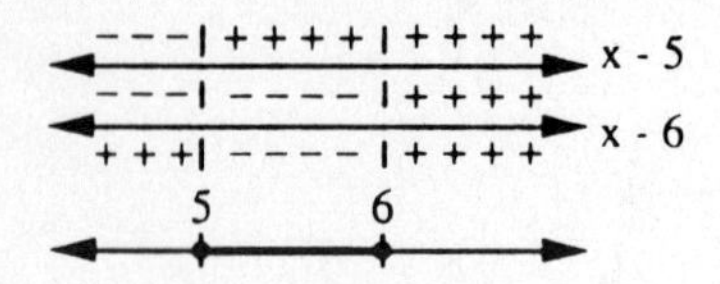


67. $3x^2 + 8x > 16$

$$3x^2 + 8x - 16 > 0$$
$$(3x - 4)(x + 4) > 0$$

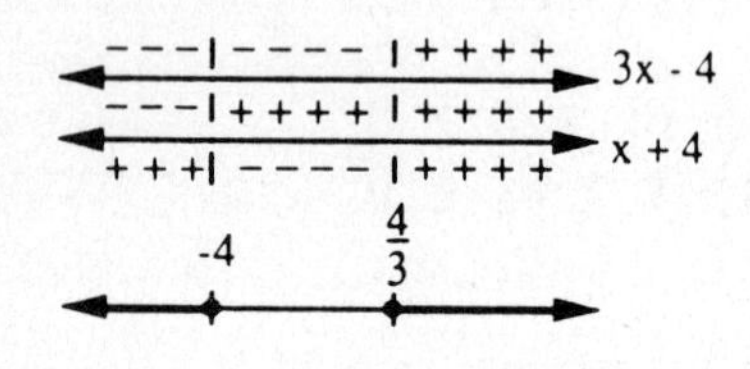


69. $5x^2 - 25 > 0$

$$5(x^2 - 5) > 0$$
$$x^2 = 5$$
$$x = \pm\sqrt{5}$$

71. $\dfrac{x + 2}{x - 3} > 0$

$$x \ne 3$$

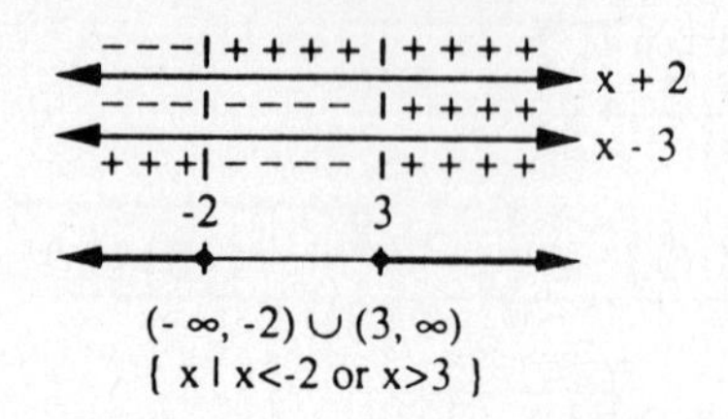


73. $\dfrac{2x - 4}{x + 1} \ge 0 \quad D: \ \{x | x \ne -1\}$

$$\frac{2(x - 2)}{x + 1} \ge 0$$

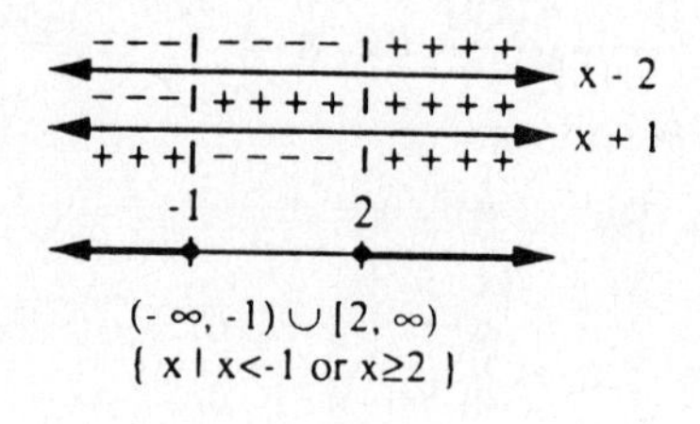


75. $(x + 3)(x + 1)(x - 2) > 0$

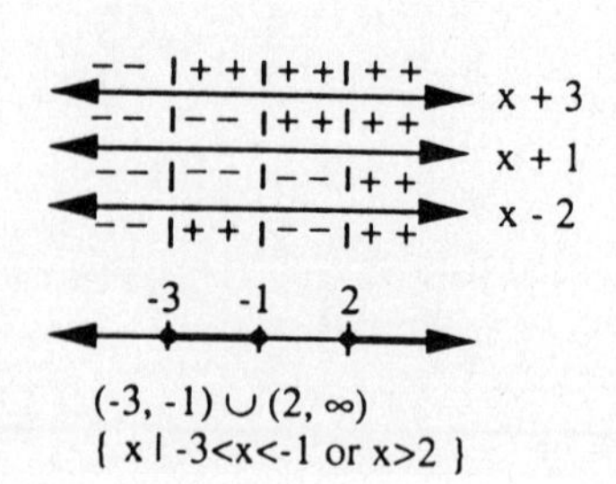

**77.** $(x + 4)(x - 1)(x - 3) \ge 0$

-- | + + | + + | + +   x + 4
-- | -- | + + | + +   x - 1
-- | -- | -- | + +   x - 3
-- | + + | -- | + +

-4   1   3

$[-4, 1] \cup [3, \infty)$
$\{ x \mid -4 \le x \le 1 \text{ or } x \ge 3 \}$

**83.** $$\frac{3}{x + 4} \ge -1$$

$$\frac{3}{x + 4} + 1 \ge 0$$

$$\frac{3 + 1(x + 4)}{x + 4} \ge 0$$

$$\frac{3 + x + 4}{x - 4} \ge 0$$

$$\frac{x+1}{x+4} \ge 0 \text{ x}$$

D: $\{x|x \neq -4\}$

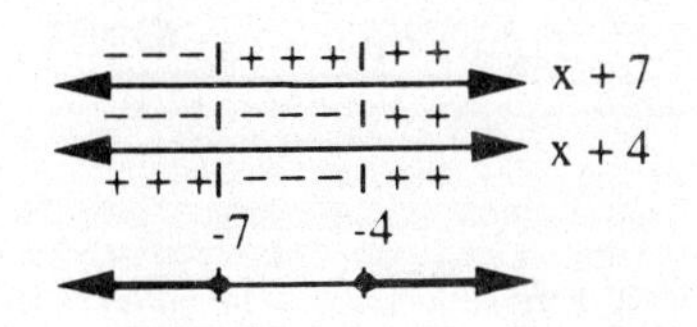


**79.** $$\frac{x(x - 4)}{x + 2} > 0$$

$$x \neq -2$$

-- | --- | + + + | + +   x
-- | --- | --- | + +   x - 4
-- | + + + | + + + | + +   x + 2
-- | + + + | --- | + +

-2   0   4

$(-2, 0) \cup (4, \infty)$
$\{ x \mid -2 < x < 0 \text{ or } x > 4 \}$

**85.** $$\frac{2x + 3}{x - 5} < 2$$

$$\frac{2x + 3}{x - 5} - 2 < 0$$

$$\frac{2x + 3 - 2(x - 5)}{x - 5} < 0$$

$$\frac{2x + 3 - 2x + 10}{x - 5} < 0$$

$$\frac{13}{x - 5} < 0$$

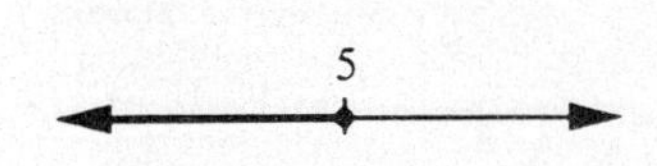


**81.** $$\frac{x - 3}{(x + 2)(x - 5)} \ge 0$$

D: $\{x|x \cdot 2, 5\}$

-- | -- | + + | + +   x - 3
-- | + + | + + | + +   x + 2
-- | -- | -- | + +   x - 5
-- | + + | -- | + +

-2   3   5

$(-2, 3] \cup (5, \infty)$
$\{ x \mid -2 < x \le 3 \text{ or } x > 5 \}$

**87.** $y = x^2 + 6x$

(a) Since a = 1 the parabola opens upward.

(b) $y = 0^2 + 6(0) = 0$

Number 87 (Continued)

(c) $x = \frac{-b}{2a}$ $y = \frac{4ac - b^2}{4a}$

$= \frac{-6}{2(1)}$ $= \frac{4(1)(0) - 6^2}{4(1)}$

$= -3$ $= \frac{-36}{4} = -9.$

The vertex is at (-3, -9).

(d) $0 = x^2 + 6x$

$= x(x + 6)$

$0 = x$ or $0 = x + 6$

$x = 0$ $\quad x = -6$

(e)

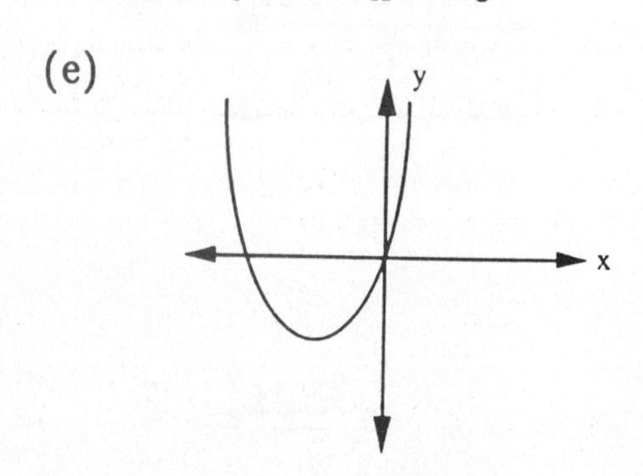


**89.** $y = 2x^2 + 4x - 16$

(a) Since a = 2 the parabola opens upward.

(b) $y = 2(0)^2 + 4(0) - 16 = -16$

(c) $x = \frac{-b}{2a}$ $y = \frac{4ac - b^2}{4a}$

$= \frac{-4}{2(2)}$ $= \frac{4(2)(-16) - 4^2}{4(2)}$

$= -1$ $= \frac{-128 - 16}{8}$

$= \frac{-144}{8} = -18.$

The vertex is at (-1, -18).

(d) $0 = 2x^2 + 4x - 16$

$= x^2 + 2x - 8$

$= (x + 4)(x - 2)$

$0 = x + 4$ or $0 = x - 2$

$-4 = x$ $\quad 2 = x$

Number 89 (Continued)

(e)

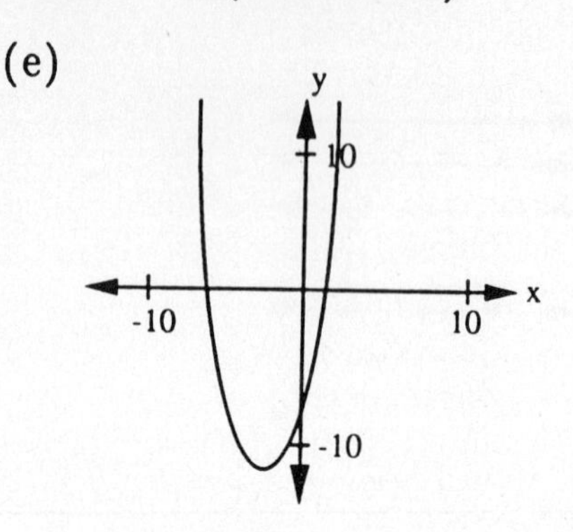


**91.** $y = -2x^2 - x + 15$

(a) Since a = -2 the parabola opens downward.

(b) $y = -2(0)^2 - 0 + 15 = 15$

(c) $x = \frac{-b}{2a}$ $y = \frac{4ac - b^2}{4a}$

$= \frac{-(-1)}{2(-2)}$ $= \frac{4(-2)(15) - (-1)^2}{4(-2)}$

$= \frac{-1}{4}$ $= \frac{121}{8}.$

The vertex is at $(-\frac{1}{4}, \frac{121}{8})$.

(d) $0 = -1(2x^2 + x - 15)$

$= -1(2x - 5)(x + 3)$

$0 = -1(2x - 5)$ or $0 = x + 3$

$5 = 2x$

$\frac{5}{2} = x$ or $-3 = x$

(e)

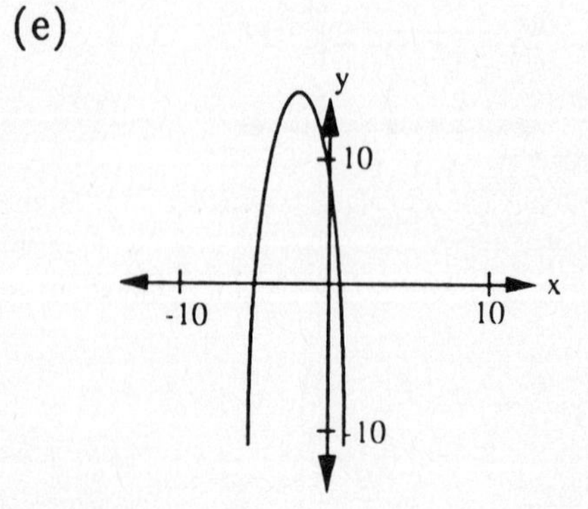

**93.** roots 3, - 2

$y = (x - 3)(x + 2)$

$y = x^2 + 2x - 3x - 6$

$y = x^2 - x - 6$

**95.** roots -3, -3

$y = (x + 3)(x + 3)$

$y = x^2 + 3x + 3x + 9$

$y = x^2 + 6x + 9$

**97.** first positive integer x, next consecutive integer x + 1

$x(x + 1) = 90$

$x^2 + x - 90 = 0$

$(x - 9)(x + 10) = 0$

$x - 9 = 0$ or $x + 10 = 0$

$x = 9$ $\quad x = -10$

$x + 1 = 10$

The two numbers are 9 and 10.

**99.** width x
length 2x - 1

$A = 1 \cdot w$

$66 = (2x - 1)(x)$

$66 = 2x^2 - x$

$0 = 2x^2 - x - 66$

$0 = (2x + 11)(x - 6)$

$0 = 2x + 11$ or $0 = x - 6$

$-11 = 2x$ $\quad 6 = x$

$\frac{-11}{2} = x$

width x = 6 inches

length $2x - 1 = 2(6) - 1$

$= 12 - 1$

$= 11$ inches

**101.** $d = -16t^2 + 1800 \quad t = 3$

$d = -16(3)^2 + 1800$

$d = -16(9) + 1800$

$d = -144 + 1800$

$d = 1656$ feet

**103.** $L = .0004t^2 + .16t + 20$

(a) $L = .0004(100)^2 + .16(100) + 20$

$= 40$ ml

(b) $53 = .0004t^2 + .16t + 20$

$0 = .0004t^2 + .16t - 33$

$$t = \frac{-.16 \pm \sqrt{(.16)^2 - 4(.0004)(-33)}}{2(.0004)}$$

$$= \frac{-.16 \pm \sqrt{.0784}}{.0008}$$

$$t = \frac{-.16 + \sqrt{.0784}}{.0008} \text{ or}$$

$$t = \frac{-.16 - \sqrt{.0784}}{.0008}$$

$t = 150^0C \quad t = -1100°C.$

The operating temperature is $150^0C$.

**105.** (a) The time, t, at which the cannon attains its maximum height is:

$t = \frac{-b}{2a}$

from $h = 5t^2 + 26t + 8$,

$t = \frac{-26}{2(-5)}$

$t = \frac{-26}{-10}$

$t = 2.6$ seconds

Number 105 (Continued)
(b) The maximum height at $t = 2.6$:

$$h = \frac{4ac - b^2}{4a}$$

$$h = \frac{4(-5)(8) - 26^2}{4(-5)}$$

$$h = \frac{-160 - 676}{-20}$$

$$h = \frac{-836}{-20}$$

$$h = 41.8 \text{ feet}$$

**107.** $(f + g)(x) = (x^2 - 3x + 4) + (2x - 5)$
$= x^2 - x - 1$

**109.** $(g - f)(x) = (2x - 5) - (x^2 - 3x + 4)$
$= 2x - 5 - x^2 + 3x - 4$
$= -x^2 + 5x - 9$

**111.** $(fg)x = (x^2 - 3x + 4)(2x - 5)$
$= 2x^3 - 11x^2 + 23x - 20$

**113.** $(\frac{f}{g})(x) = \frac{x^2 - 3x + 4}{2x - 5}$

**115.** $(f \circ g)(x) = (2x - 5)^2 - 3(2x - 5) + 4$
$= 4x^2 - 20x + 25 - 6x + 15 + 4$
$= 4x^2 - 26x + 44$

**117.** $(g \circ f)(x) = 2(x^2 - 3x + 4) - 5$
$= 2x^2 - 6x + 8 - 5$
$= 2x^2 - 6x + 3$

**119.** $(f + g)(x) = (3x + 2) + \sqrt{x - 4}$

**121.** $(fg)(x) = (3x + 2)(\sqrt{x - 4})$

**123.** $(f \circ g)(x) = 3\sqrt{x - 4} + 2$

**125.** Since the function passes the horizontal line test, it is one-to-one.

**127.** The function fails the horizontal line test and is, therefore, not one-to-one.

**129.** The function is one-to-one.

**131.** $f(x)$:
Domain: $\{-4, 0, 5, 6\}$
Range: $\{-3, 2, 3, 7\}$

$f^{-1}(x)$
Domain: $\{-3, 2, 3, 7\}$
Range: $\{-4, 0, 5, 6\}$

**133.** $y = f(x) = 4x - 2$
$f^{-1}(x) = \frac{x + 2}{4}$

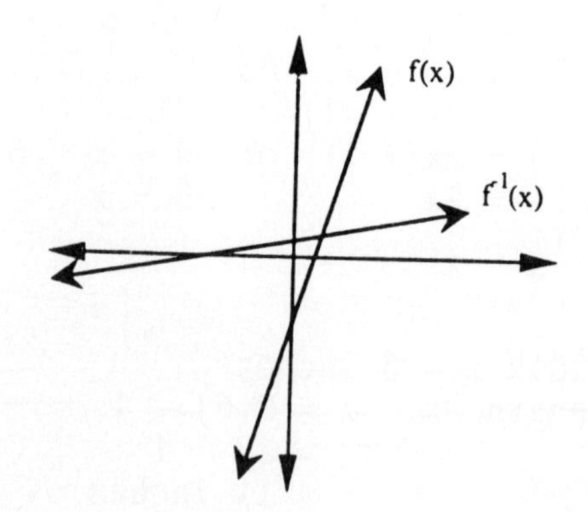

135. $y = f(x) = 2x + 5$

$$f^{-1}(x) = \frac{3x - 5}{2}$$

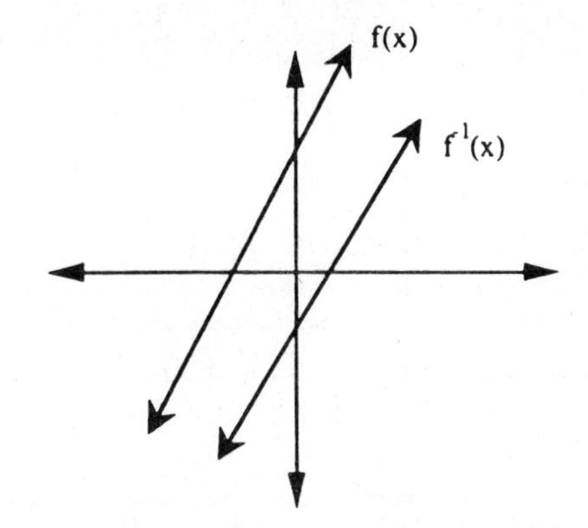


137. $$(f \circ f^{-1})(x) = \frac{5\left[\frac{2x + 4}{5}\right] - 4}{2}$$

$$= \frac{2x + 4 - 4}{2}$$

$$= \frac{2x}{2} = x$$

$$(f^{-1} \circ f)(x) = \frac{2\left[\frac{5x - 4}{2}\right] + 4}{5}$$

$$= \frac{5x - 4 + 4}{5}$$

$$= \frac{5x}{5} = x.$$

**PRACTICE TEST - CHAPTER 9**

1. $x^2 = -x + 12$

$x^2 + x = 12$

$x^2 + x + \frac{1}{4} = \frac{48}{4} + \frac{1}{4}$

$(x + \frac{1}{2})^2 = \frac{49}{4}$

$x = \frac{1}{2} = \pm \frac{7}{2}$

$x = -\frac{1}{2} \pm \frac{7}{2}$

$x = -\frac{1}{2} + \frac{7}{2}$ or $x = -\frac{1}{2} - \frac{7}{2}$

$x = \frac{6}{2} = 3$ $\qquad x = -\frac{8}{2} = -4$

2. $4x^2 + 8x = -12$

$\frac{4x^2}{4} + \frac{8x}{4} = \frac{-12}{4}$

$x^2 + 2x = -3$

$x^2 + 2x + 1 = -3 + 1$

$(x + 1)^2 = -2$

$\sqrt{(x + 1)^2} \pm \sqrt{-2}$

$x + 1 = i\sqrt{2}$ or $x + 1 = -i\sqrt{2}$

$x = -1 + i\sqrt{2}$ $\qquad x = -1 - i\sqrt{2}$

3. $x^2 - 5x - 6 = 0$

$a = 1, b = -5, c = -6$

$x = \frac{-b \pm \sqrt{b^2 - 4ac}}{2a}$

$= \frac{-(-5) \pm \sqrt{(-5)^2 - 4(1)(-6)}}{2(1)}$

$= \frac{5 \pm \sqrt{25 + 24}}{2}$

$= \frac{5 \pm \sqrt{49}}{2}$

Number 3 (Continued)

$x = \frac{5 + 7}{2}$ or $x = \frac{5 - 7}{2}$

$x = \frac{12}{2} = 6$ $\qquad x = \frac{-2}{2} = -1$

4. $x^2 + 5 = -8x$

$x^2 + 8x + 5 = 0$

$x = \frac{-8 \pm \sqrt{8^2 - 4(1)(5)}}{2}$

$x = \frac{-8 + \sqrt{44}}{2}$ or $x = \frac{-8 - \sqrt{44}}{2}$

$x = \frac{-8 + 2\sqrt{11}}{2}$ or $x = \frac{-8 - 2\sqrt{11}}{2}$

$x = -4 + \sqrt{11}$ $\qquad x = -4 - \sqrt{11}$

5. $3x^2 - 5x = 0$

$x(3x - 5) = 0$

$x = 0$ or $3x - 5 = 0$

$3x = 5$

$x = 0$ $\qquad x = \frac{5}{3}$

6. $-2x^2 = 9x - 5$

$0 = 2x^2 + 9x - 5$

$0 = (2x - 1)(x + 5)$

$2x - 1 = 0$ $\qquad x + 5 = 0$

$x = 1/2$ $\qquad x = -5$

7. Solve for b.

$P = 3a - b^2$

$P - 3a = -b^2$

$-1(P - 3a) = -1(-b^2)$

$3a - P = b^2$

$\sqrt{3a - P} = b$

**8.** Begin by rewriting the equation in standard form.

$5x^2 - 4x - 2 = 0$

The discriminant

$= b^2 - 4ac = (-4)^2 - 4(5)(-2)$
$= 56.$

Since the discriminant is greater than 0, the quadratic equation has two distinct real solutions.

**9.** $x^2 - x \geq 42$

$x^2 - x - 42 \geq 0$

$(x - 7)(x + 6) \geq 0$

$x - 7 = 0 \qquad x + 6 = 0$

$x = 7 \qquad x = -6$

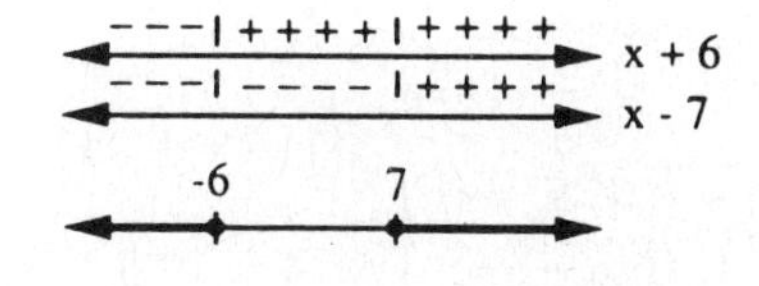


**10.** $\dfrac{(x + 3)(x - 4)}{x + 1} \geq 0$

Critical Points:

$x + 3 = 0 \quad x - 4 = 0 \quad x + 1 = 0$

$x = -3 \quad x = 4 \quad x = -1$

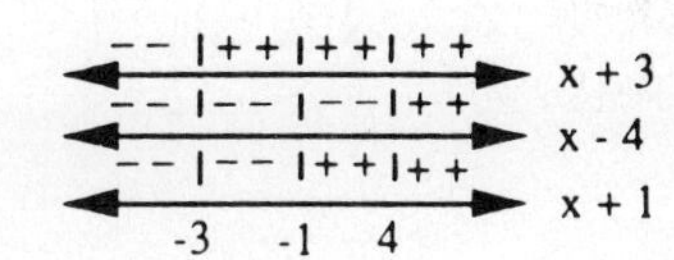


We want the quotient to be positive and this occurs where there is an even number of negative signs. Therefore, the solution is

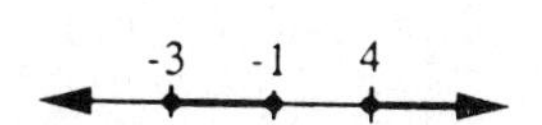


Note that -1 is excluded from the solution because x = -1 would yield a zero denominator.

**11.** $\dfrac{x + 3}{x - 3} \leq 1$

$\dfrac{x + 3}{x + 2} + 1 \leq 0$

$\dfrac{x + 3}{x + 2} + \dfrac{x + 2}{x + 2} \leq 0$

$\dfrac{2x + 5}{x + 2} \leq 0$

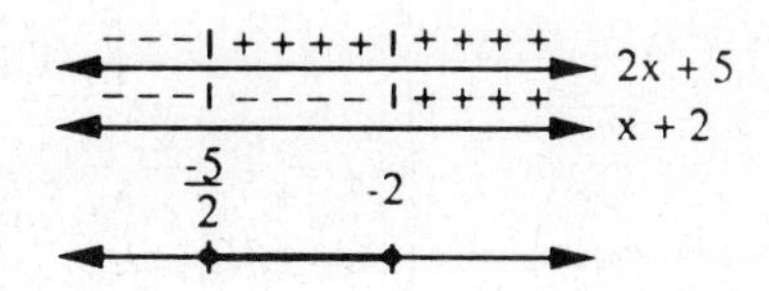


(a) $[-\frac{5}{2}, -2)$

(b) $\{x | -\frac{5}{2} \leq x < -2\}$

**12.** $y = x^2 - 2x - 8$

(a) Since $a = +1$, the parabola opens up.

(b) $y = 0^2 - 2(0) - 8$
$y = -8$
The y-intercept is $-8$.

(c) $x = \frac{-b}{2a} = \frac{-(-2)}{2(1)} = 1$
$y = 1^2 - 2(1) - 8$
$= -9$
Vertex: $(1, -9)$

(d) The x-intercepts occur where $y = 0$.
$0 = x^2 - 2x - 8$
$0 = (x - 4)(x + 2)$
$x - 4 = 0 \quad x + 2 = 0$
$x = 4 \quad x = -2$

The x-intercepts occur at and $-2$.

(e)

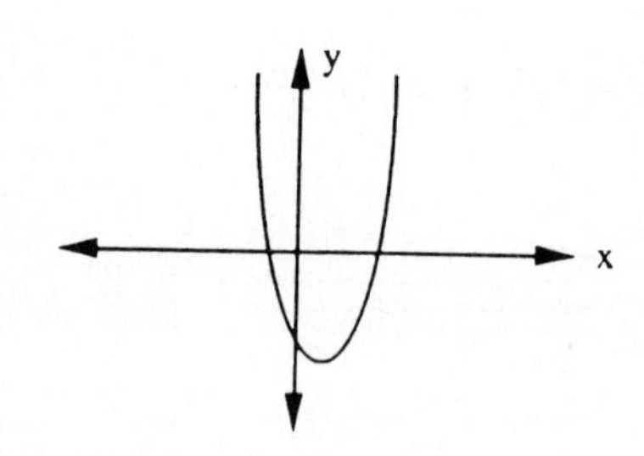


**13.** $y = -2x^2 - 3x + 9$

(a) Since $a = -2$, the parabola opens down.

(b) $y = -2(0)^2 - 3(0) + 9$
$= 9$
The y-intercept is 9.

(c) $x = \frac{-b}{2a} = \frac{-(-3)}{2(-2)} = \frac{-3}{4}$
$y = -2(\frac{-3}{4})^2 - 3(\frac{-3}{4}) + 9$
$= \frac{81}{8}$
Vertex: $(-\frac{3}{4}, \frac{81}{8})$

(d) $0 = -2x^2 - 3x + 9$
$0 = 2x^2 + 3x - 9$
$0 = 2x^2 - 3x + 6x - 9$
$0 = x(2x - 3) + 3(2x - 3)$
$0 = (x + 3)(2x - 3)$

$x + 3 = 0 \quad 2x - 3 = 0$
$x = -3 \quad x = +3/2$

x-intercepts are $-3$ and $+3/2$.

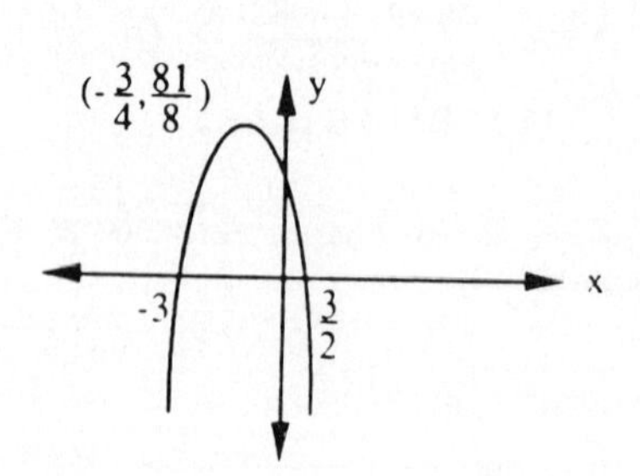

**14.** Since $-6$ and $1/2$ are the x-intercepts the factors of the equation are

$(x-(-6))$ and $(2x - 1)$
$(x + 6)(2x - 1) = 0$

$2x^2 - x + 12x - 6 = 0$

$2x^2 + 11x - 6 = 0$

is the equation.

**15.** Let $x =$ the width of the rectangle.
An $2x + 4$ = the length

$A = LW$
$48 = x(2x + 4)$
$48 = 2x^2 + 4x$
$0 = 2x^2 + 4x - 48$
$0 = x^2 + 2x - 24$
$0 = (x + 6)(x - 4)$

$x + 6 = 0$    $x - 4 = 0$
$x = -6$    $x = 4$

The width of the rectangle is 4 ft while the length is $2x + 4 = 12$ ft.

**16.** (a) $d = -16t^2 + 64t + 80.$

The maximum height occurs at the vertex of the parabola. Therefore,

$$t = \frac{-b}{2c} = \frac{-64}{2(-16)} = 2 \text{ sec.}$$

The maximum height occurs at 2 sec.

Number 16 (Continued)

(b) The maximum height occurs when $t = 2$. Therefore,

$a = -16(2)^2 + 64(2) + 80$
$= 144$ ft.

**17.** $(g - f)(x) = (2x - 4) - (x^2 - x + 8)$
$= 2x - 4 - x^2 + x - 8$
$= -x^2 + 3x - 12$

**18.** $(f \circ g)(x) = (2x - 4)^2 - (2x - 4) + 8$
$= 4x^2 - 16x + 16 - 2x + 4 + 8$
$= 4x^2 - 18x + 28$

**19.** $y = f(x) = 2x + 4$
$y = 2x + 4$
$y - 4 = 2x$
$\frac{y - 4}{2} = x$

$f^{-1}(x) = \frac{x - 4}{2}$

## EXERCISE SET 10.1

1. The center is (0,0).
Therefore h = 0 and k = 0.
The radius is 3.

$$(x - h)^2 + (y - k)^2 = r^2$$
$$(x - 0)^2 + (y - 0)^2 = 3^2$$
$$x^2 + y^2 = 9$$

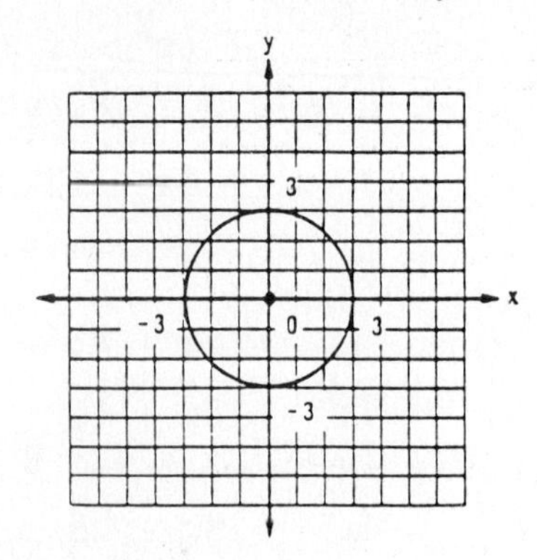


3. The center is (3,0).
Therefore, h = 3 and k = 0.
The radius is 1.

$$(x - h)^2 + (y - k)^2 = r2$$
$$(x - 3)^2 + (y - 0)^2 = 1^2$$
$$(x - 3)^2 + y^2 = 1$$

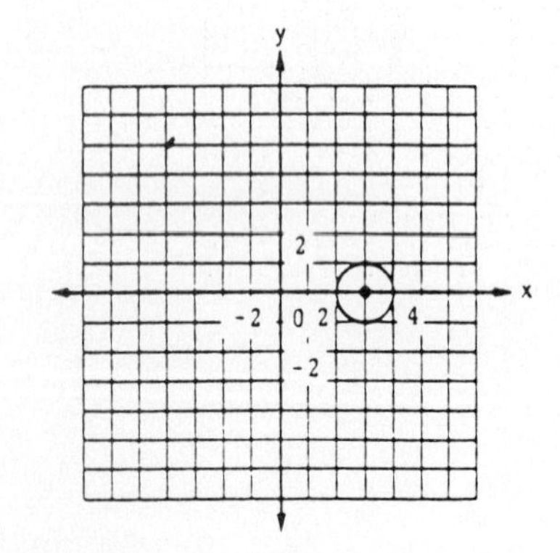


5. The center is (-6, 5).
Therefore, h = - 6 and k = 5.
The radius is 5.

$$(x - h)^2 + (y - k)^2 = r^2$$
$$(x - (-6))^2 + (y - 5)^2 = 5^2$$
$$(x + 6)^2 + (y - 5)^2 = 25$$

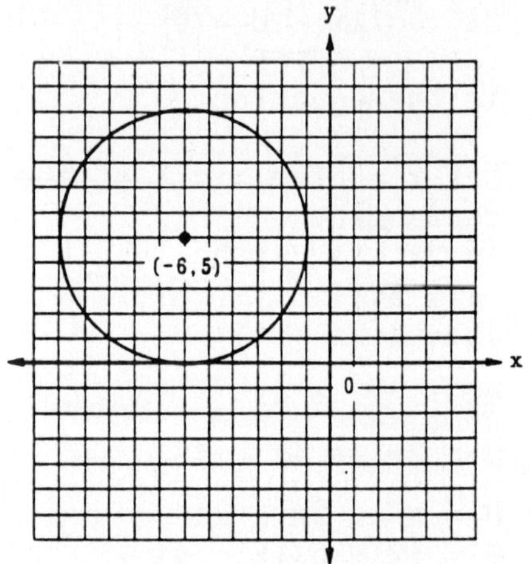


7. The center is (4, 7)
Therefore, h = 4 and k = 7.

The radius is $\sqrt{8}$.

$$(x - h)^2 + (y - k)^2 = r^2$$
$$(x - 4)^2 + (y - 7)^2 = (\sqrt{8})^2$$
$$(x - 4)^2 + (y - 7)^2 = 8$$

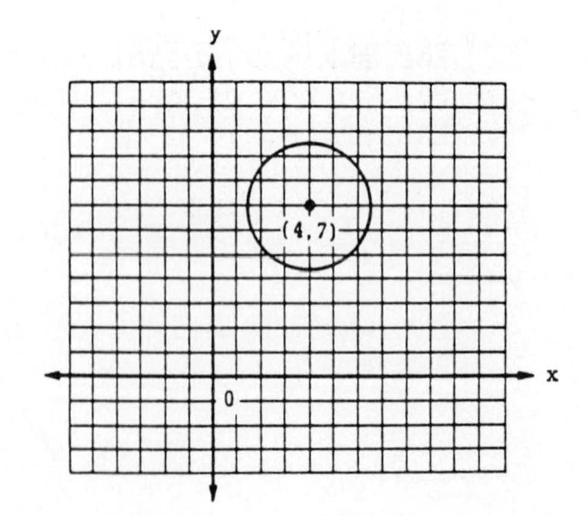

**9.** Rewrite the equation as

$$x^2 + y^2 = 4^2.$$

Center (0, 0) radius is 4.

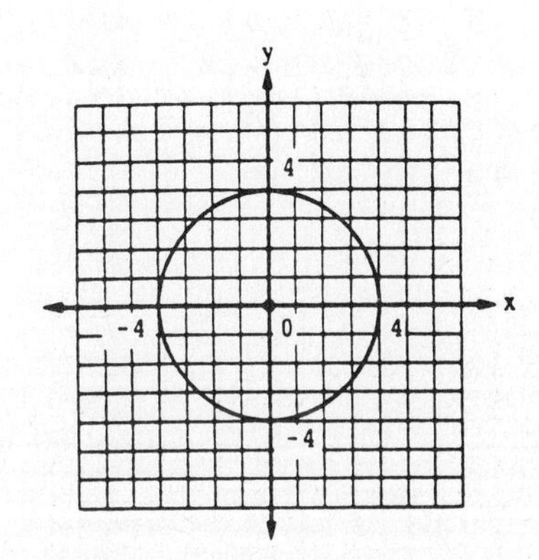


**11.** Rewrite the equation as

$$x^2 + y^2 = (\sqrt{3})^2.$$

Center (0, 0) radius $\sqrt{3}$.

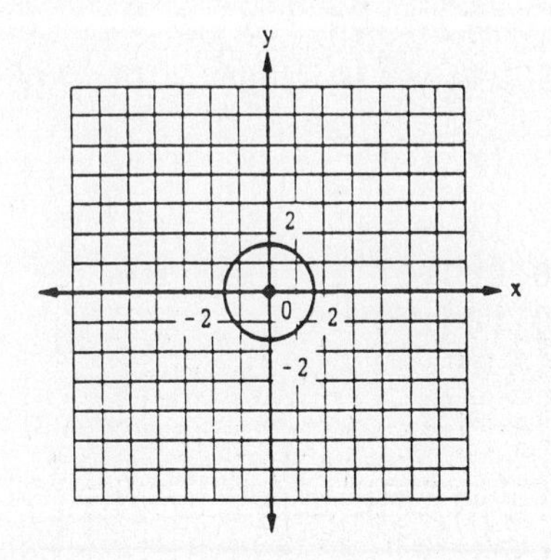


**13.** Rewrite the equation as

$$x^2 + (y - 3)^2 = 2^2.$$

Center (0,3) radius 2.

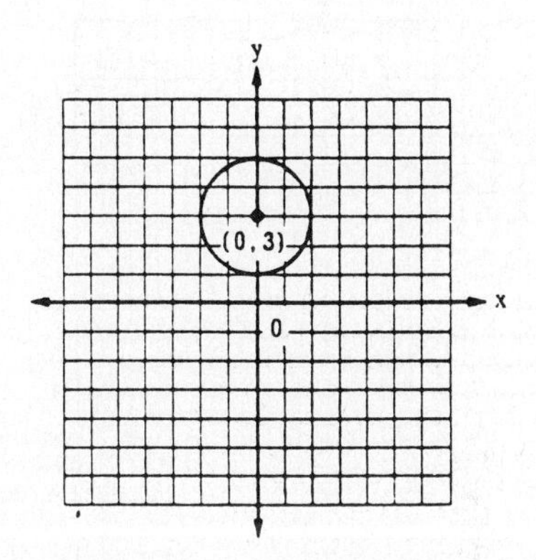


**15.** Rewrite the equation as

$$(x - 2)^2 + (y - (-3))^2 = 4^2.$$

Center (2, -3) radius 4.

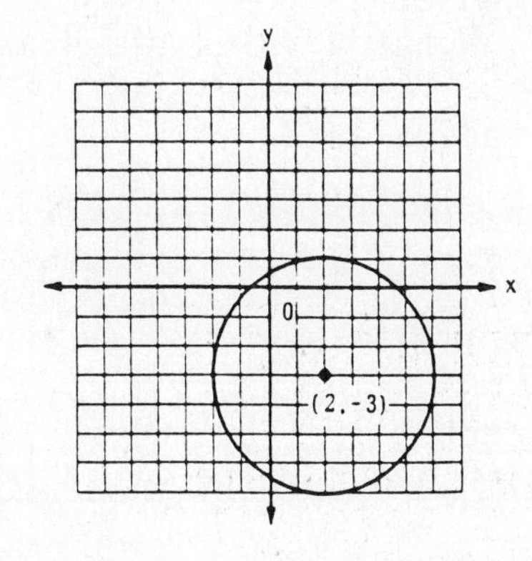

**17.** Rewrite the equation as

$(x - (-1))^2 + (y - 4)^2 = 6^2.$

Center (-1, 4) radius is 6.

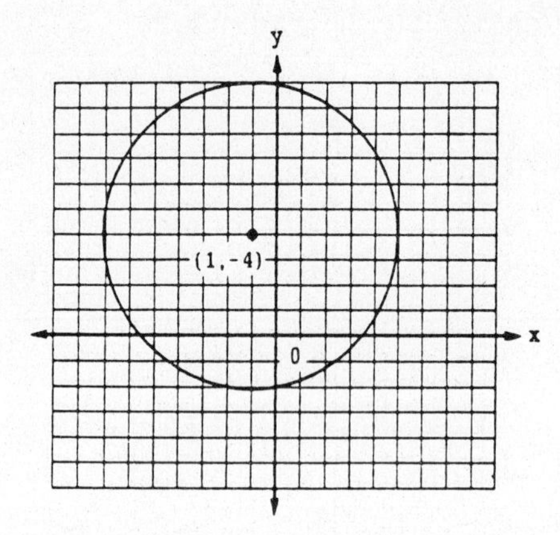


**19.** The center is (0, 0).
Thus, h has a value of 0 and k is 0.
Thc radius is 1.

$$(x - h)^2 + (y - k)^2 = r^2$$
$$(x - 0)^2 + (y - 0)^2 = 1^2$$
$$x^2 + y^2 = 1$$

**21.** The center is (-4, 6).
Thus, h has a value of -4 and k is 6.
The radius is 2.

$$(x - h)^2 + (y - k)^2 = r^2$$
$$(x - (-4))^2 + (y - 6)^2 = 2^2$$
$$(x + 4)^2 + (y - 6)^2 = 4$$

**23.** The center is (-5, -3).
Thus, h has a value of -5 and k is -3.
The radius is 2.

$$(x - h)^2 + (y - k)^2 = r^2$$
$$(x - (-5))^2 (y - (-3))^2 = 2^2$$
$$(x + 5)^2 + (y + 3)^2 = 4$$

**25.**

$$x^2 + y^2 + 10y - 75 = 0$$
$$x^2 + y^2 + 10y = 75$$
$$x^2 + y^2 + 10y + 25 = 75 + 25$$
$$x^2 + (y + 5)^2 = 100$$
$$x^2 + (y + 5)^2 = 10^2$$

The center is (0, -5).
The radius is 10.

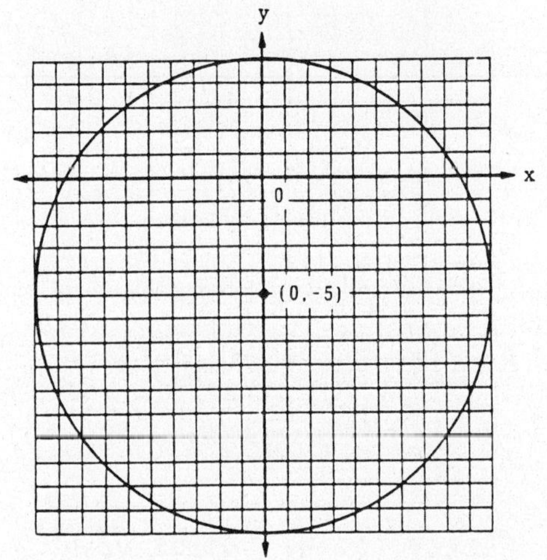


**27.**

$$x^2 + 8x - 9 + y^2 = 0$$
$$x^2 + 8x + y^2 = 9$$
$$x^2 + 8x + 16 + y^2 = 9 + 16$$
$$(x + 4)^2 + y^2 = 25$$
$$(x + 4)^2 + y^2 = 5^2$$

The center is (-4, 0).
The radius is 5.

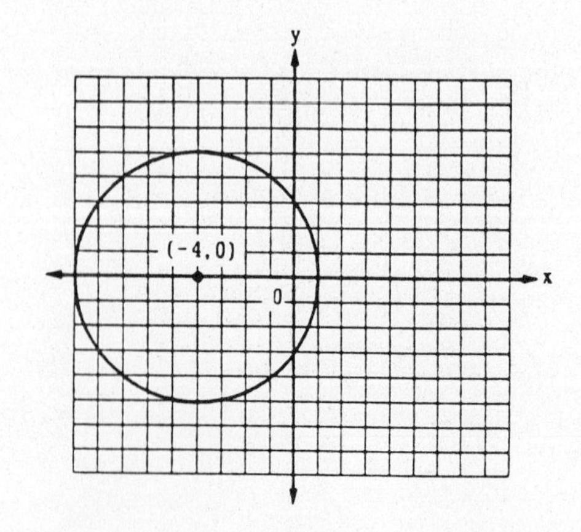

**29.**

$$x^2 + y^2 + 2x - 4y - 4 = 0$$
$$x^2 + 2x + y^2 - 4y = 4$$
$$x^2 + 2x + 1 + y^2 - 4y + 4 = 4+1+4$$
$$(x + 1)^2 + (y - 2)^2 = 9$$
$$(x + 1)^2 + (y - 2)^2 = 3^2$$

The center is (-1, 2).
The radius is 3.

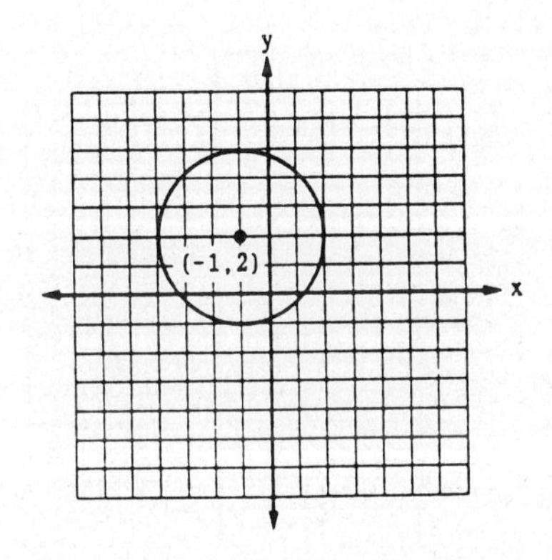


**31.**

$$x^2 + y^2 + 6x - 2y + 6 = 0$$
$$x^2 + 6x + y^2 - 2y = -6$$
$$x^2+6x + 9 + y^2-2y+1 = -6 + 9 + 1$$
$$(x + 3)^2 + (y - 1)^2 = 4$$
$$(x + 3)^2 + (y - 1)^2 = 2^2$$

The center is (-3, 1).
The radius is 2.

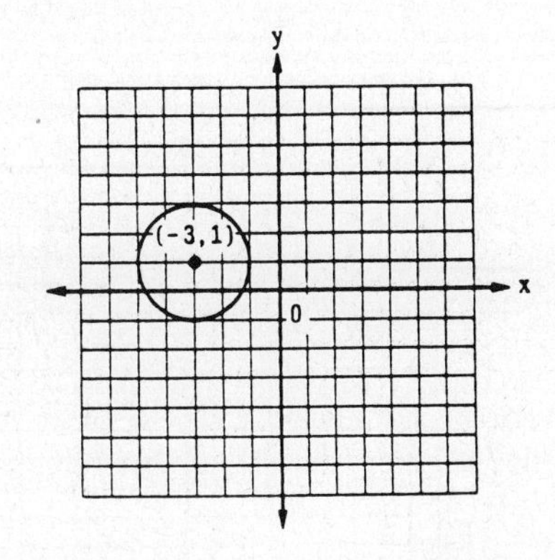


**33.**

$$x^2 + y^2 - 8x + 2y + 13 = 0$$
$$x^2 - 8x + y^2 + 2y = -13$$
$$x^2-8x + 16 + y^2 + 2y + 1 = -13 + 16 + 1$$
$$(x - 4)^2 + (y + 1)^2 = 4$$
$$(x - 4)^2 + (y + 1)^2 = 2^2$$

The center is (4, -1).
The radius is 2.

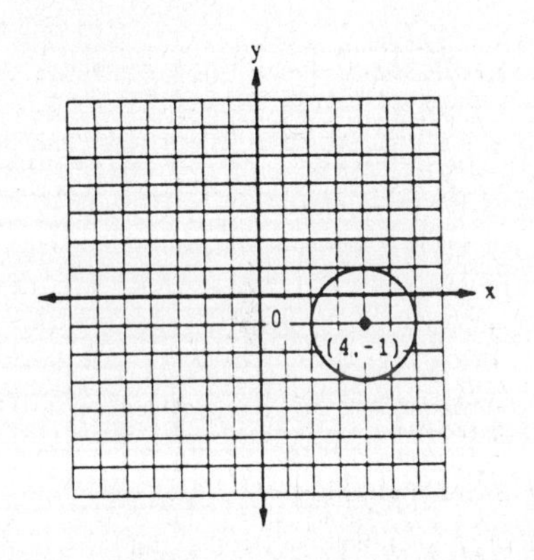


**35.**

$$8x^3 - 64 = 8(x^3 - 8)$$
$$= 8(x-2)(x^2+2x+4)$$

**37.** Solve for P

$$E = \frac{V+P^2}{2}$$
$$2E = V + P^2$$
$$2E - V = P^2$$
$$\sqrt{2E-V} = P$$

## EXERCISE SET 10.2

1. Rewrite the equation as:

$$\frac{x^2}{2^2} + \frac{y^2}{1^2} = 1$$

Thus, a = 2 and the x intercepts are ± 2.

Since b = 1, the y intercepts are ± 1.

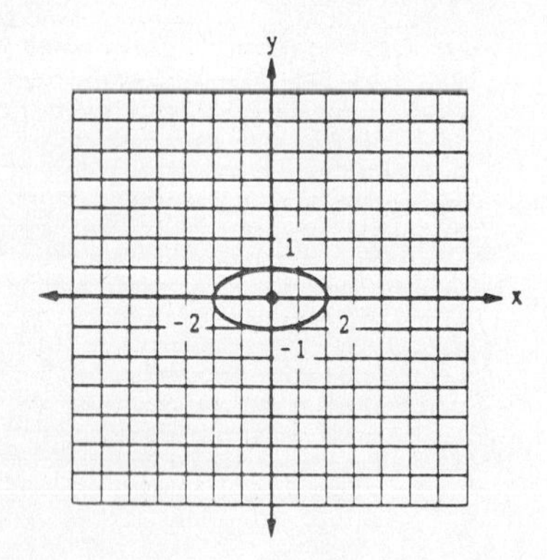


3. Rewrite the equation as:

$$\frac{x^2}{2^2} + \frac{y^2}{3^2} = 1$$

Thus, a = 2 and the x intercepts are ± 2.

Since b = 3, the y intercepts are ± 3.

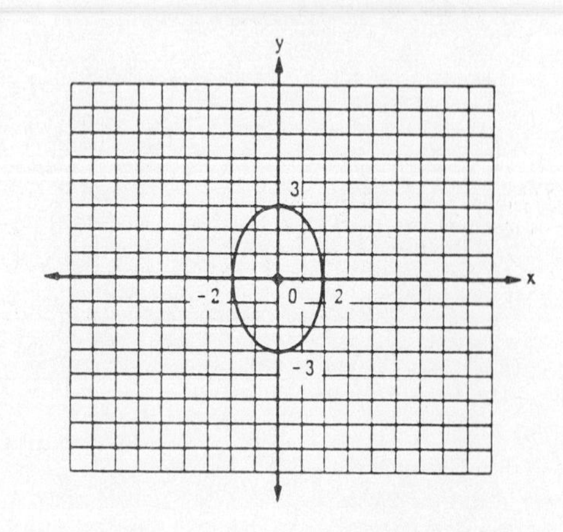


5. Rewrite the equation as:

$$\frac{x^2}{4^2} + \frac{y^2}{5^2} = 1$$

Thus, a = 4 and the x intercepts are ± 4.

Since b = 5, the y intercepts are ± 5.

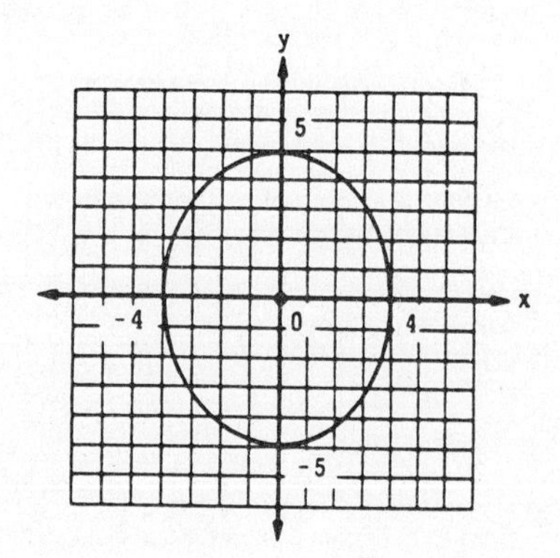


7. $$\frac{9x^2 + 12y^2}{108} = \frac{108}{108}$$

$$\frac{x^2}{12} + \frac{y^2}{9} = 1$$

Since $a^2 = 12$, $a = \sqrt{12}$ and the x intercepts are $\pm \sqrt{12}$.

We know that $b^2 = 9$, thus b = 3 and the y intercepts are ± 3.

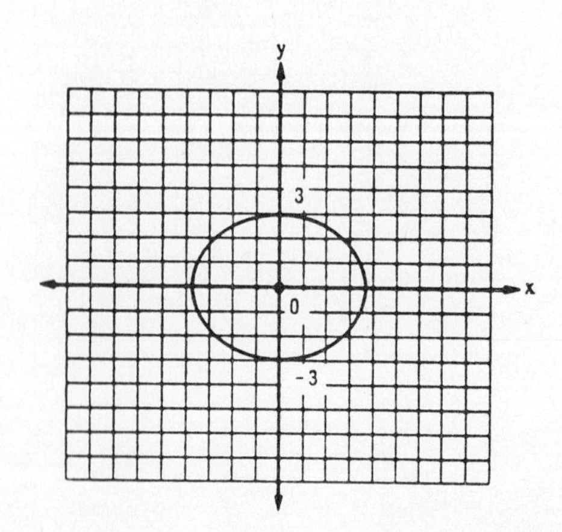

**9.** $25x^2 + 16y^2 = 400.$
Divide each term by 400.

$$\frac{25x^2 + 16y^2}{400} = \frac{400}{400}$$

$$\frac{x^2}{16} + \frac{y^2}{25} = 1$$

Since $a^2 = 16$, $a = 4$ and the x intercepts are $\pm$ 4.

We know that $b^2 = 25$, thus $b = 5$ and the y intercepts are $\pm$ 5.

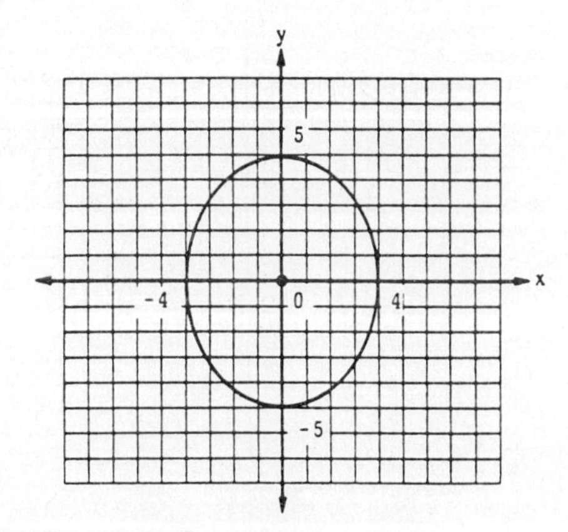


**11.** $9x^2 + 25y^2 = 225$
Divide each term by 225.

$$\frac{9x^2 + 25y^2}{225} = \frac{225}{225}$$

$$\frac{x^2}{25} + \frac{y^2}{9} = 1$$

Since $a^2 = 25$, $a = 5$ and the x intercepts are $\pm$ 5.

Number 11 (Continued)

We know that $b^2 = 9$, thus $b = \pm 3$ and the y intercepts are $\pm$ 3.

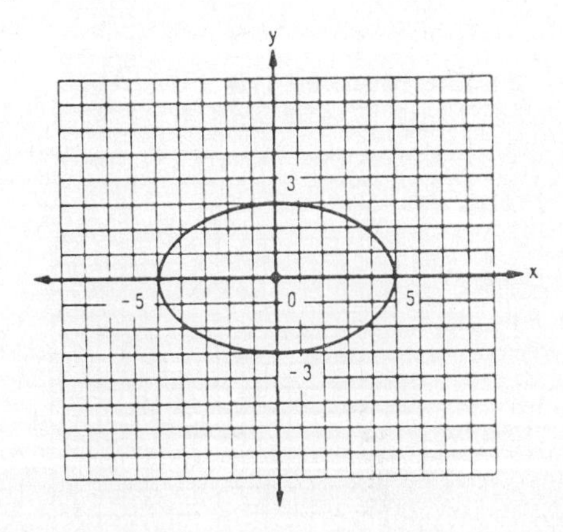


**13.** The set of points in a plane, the sum of whose distance from two fixed points is constant.

**15.** The graph becomes more circular as b approaches the value of a. When a = b, the graph is a circle.

**17.** $|2x-4| \leq 8$. This gives

$$\begin{aligned} 2x - 4 &= 8 \text{ or } & 2x - 4 &= -8 \\ 2x &= 12 \text{ or } & 2x &= -4 \\ x &= 6 \text{ or } & x &= -2 \end{aligned}$$

The graph is:

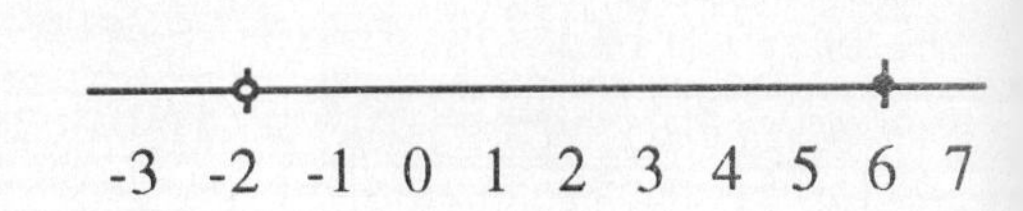


**19.** $|2x-4| > 8$

$$\begin{aligned} 2x - 4 &> 8 \text{ or } & 2x - 4 &< -8 \\ 2x &> 12 \text{ or } & 2x &< -4 \\ x &> 6 \text{ or } & x &< -2 \end{aligned}$$

## JUST FOR FUN 10.2

1. Rewrite the equation as

$$\frac{x^2}{4^2} + \frac{(y - 2)^2}{3^2} = 1$$

The center is (0, 2).
Thus $a = 4$ and $b = 3$.

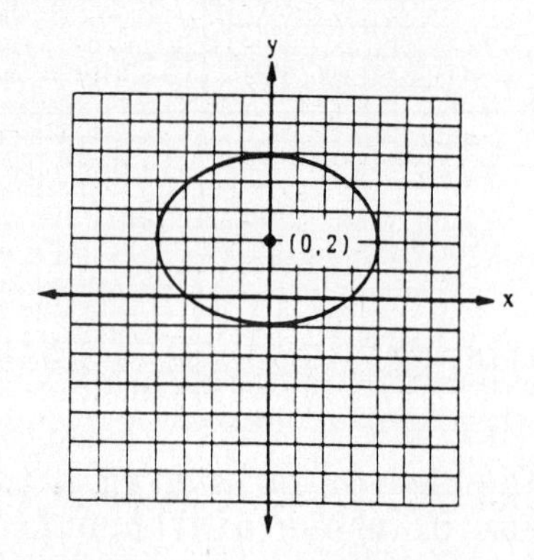


3. $x^2 + 4y^2 - 4x - 8y - 92 = 0$

$$x^2 - 4x + 4y^2 - 8y = 9^2$$

Complete the square for x:

$$x^2 - 4x + 4 + 4y^2 - 8y = 92 + 4$$
$$(x - 2)^2 + 4y^2 - 8y = 96$$
$$(x - 2)^2 + 4(y^2 - 2y) = 96$$

Complete the square for y:

$$(x - 2)^2 + 4(y^2 - 2y + 1) = 96 + 4$$
$$(x - 2)^2 + 4(y - 1)^2 = 100$$

Number 3 (Continued)

Divide each term by 100:

$$\frac{(x - 2)^2}{100} + \frac{(y - 1)^2}{25} = 1$$

Rewrite the equation as

$$\frac{(x - 2)^2}{10^2} + \frac{(y - 1)^2}{5^2} = 1$$

Center (2, 1) $a = 10$ and $b = 5$.

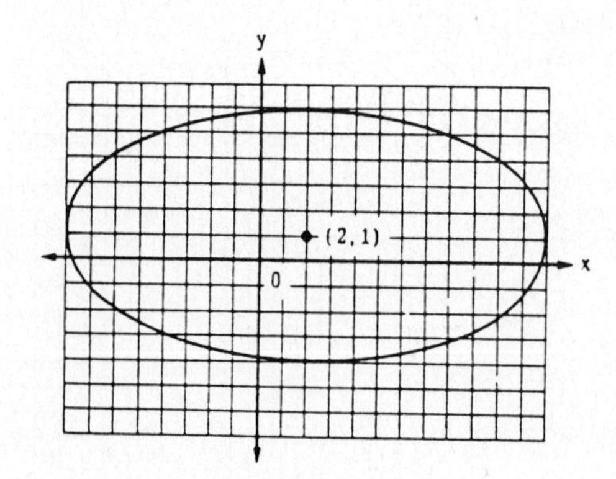

**EXERCISE SET 10.3**

1. The graph opens upward.

   The vertex is at (2, 3).

   When $x = 0$, the y intercept is 7.

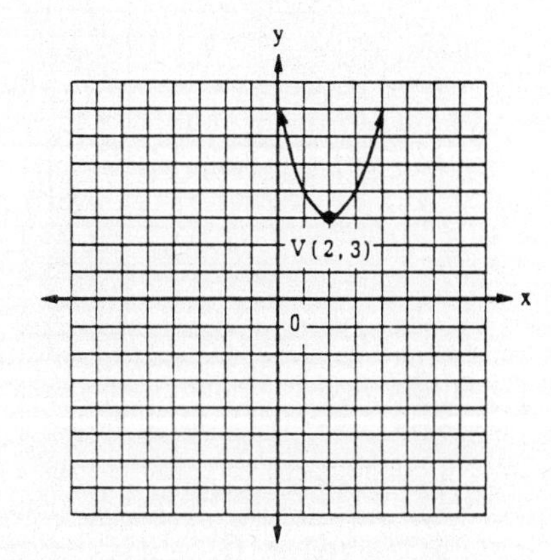


3. The graph opens downward since a = -1.

   The vertex is at (3, -6).

   When $x = 0$, the y intercept is -15.

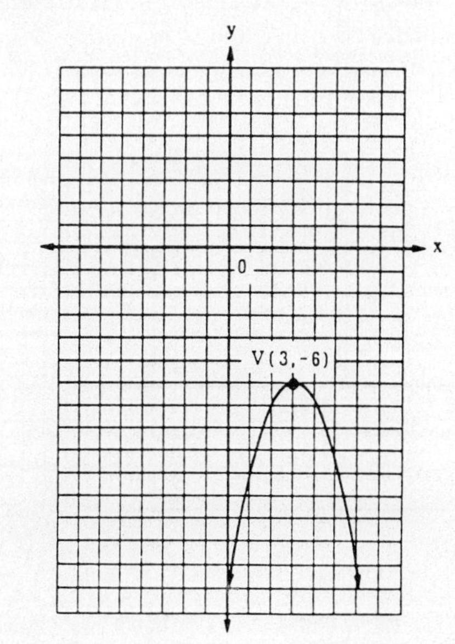


5. The graph opens to the right.

   The vertex is at (-3, 4).

   When $y = 0$, the x intercept is 13.

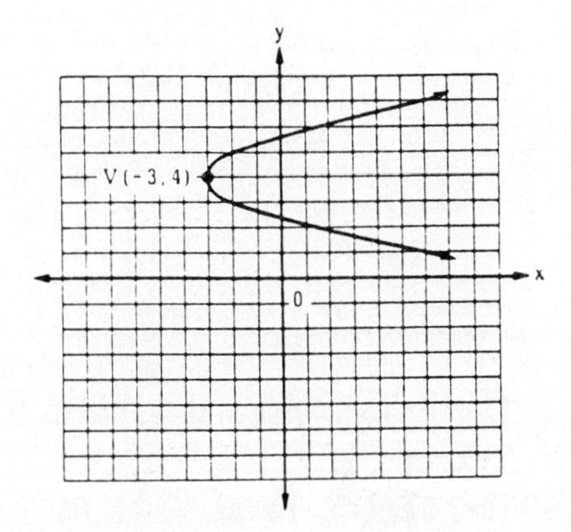


7. The graph opens upward.

   The vertex is at (-6, -4).

   When $x = 0$, the y intercept is 68.

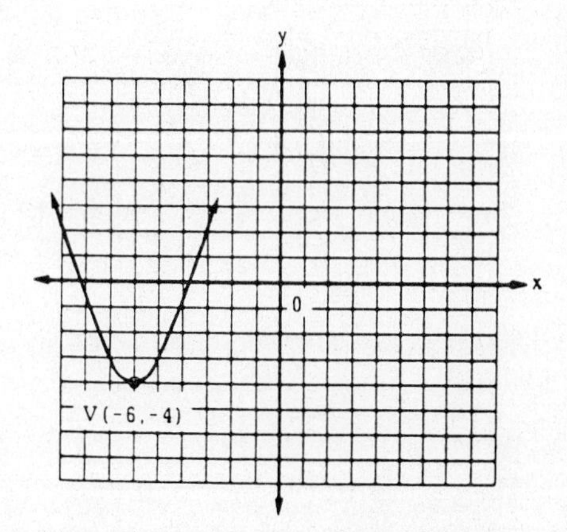


9. The graph opens to the left.

   The vertex is at (-6, -3).

Number 9 (Continued)

When $y = 0$, the x intercept is $-51$.

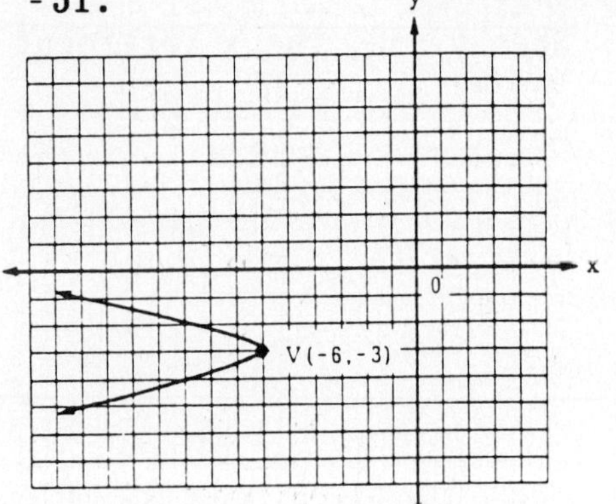


**11.** The graph opens downward.

The vertex is at $(-\frac{1}{2}, 6)$.

When $x = 0$, the y intercept is $11/2$.

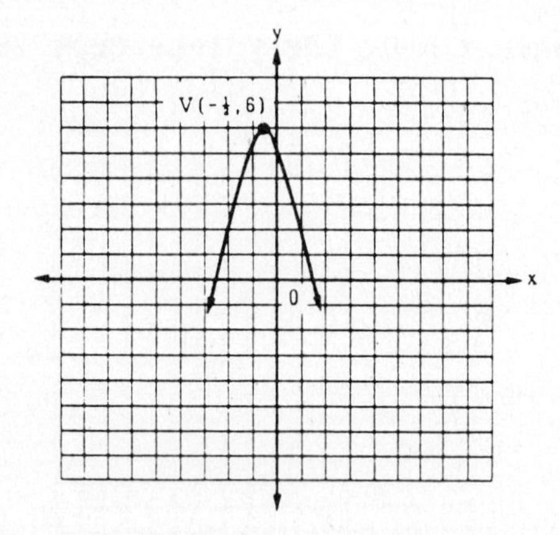


**13.** $y = x^2 + 2x$

Complete the square.

$y = x^2 + 2x + 1 - 1$

$y = (x + 1)^2 - 1$

Number 13 (Continued)

The graph opens upward.
The vertex is at $(-1, -1)$.
When $x = 0$, the y intercept is 0.

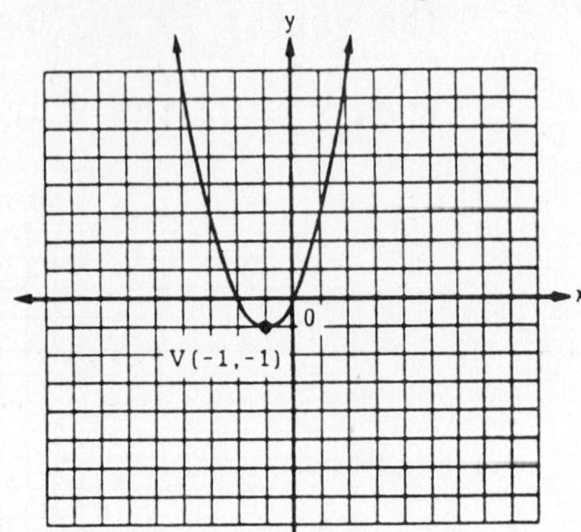


**15.** $x = y^2 + 6y$

Complete the square.

$x = y^2 + 6y + 9 - 9$

$x = (y + 3)^2 - 9$

The graph opens to the right.

The vertex is at $(-9, -3)$.

When $x = 0$, the y intercepts are 0 and $-6$.

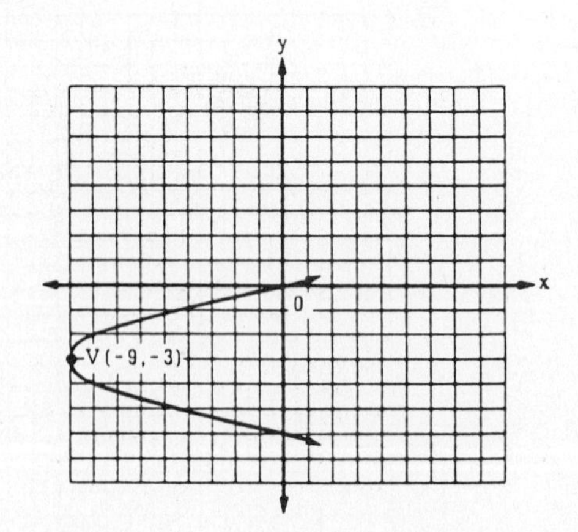

**17.** $y = x^2 + 2x - 15$

$y = x^2 + 2x + 1 - 1 - 15$

$y = (x + 1)^2 - 16$

The graph opens upward.
The vertex is at (-1, -16.)
When x = 0, the y intercept is -15.
When y = 0, the x intercepts are 3 and -5.

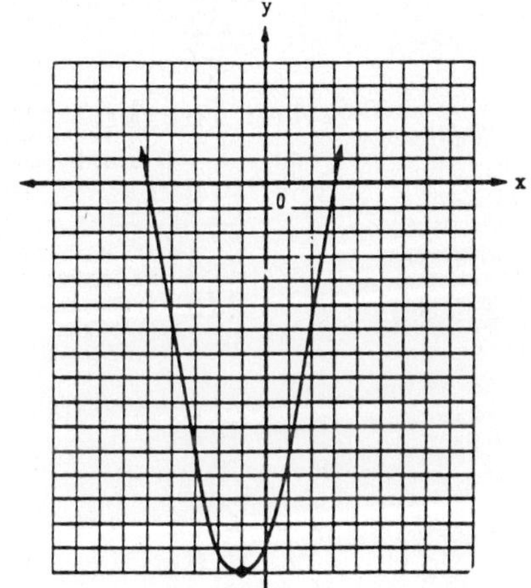


**19.** $x = -y^2 + 6y - 9$

$x = -(y^2 - 6y + 9)$

$x = -(y - 3)^2$

The graph opens to the left, since a = -1.

The vertex is at (0, 3).

When y = 0, the x intercept is -9.

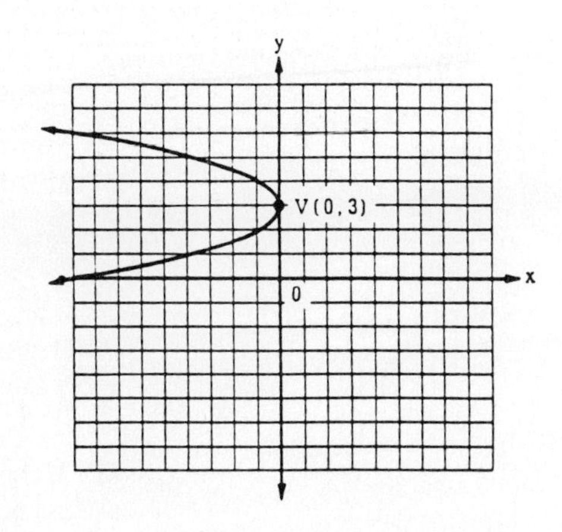


**21.** $y = x^2 + 7x + 10$

$y = x^2 + 7x + \frac{49}{4} - \frac{49}{4} + 10$

$y = (x + \frac{7}{2})^2 - \frac{9}{4}$

The graph opens upward.

The vertex is at $(-\frac{7}{2}, -\frac{9}{4})$.

When x = 0, the y intercept is 10.
When y = 0, the x intercepts are -2 and -5.

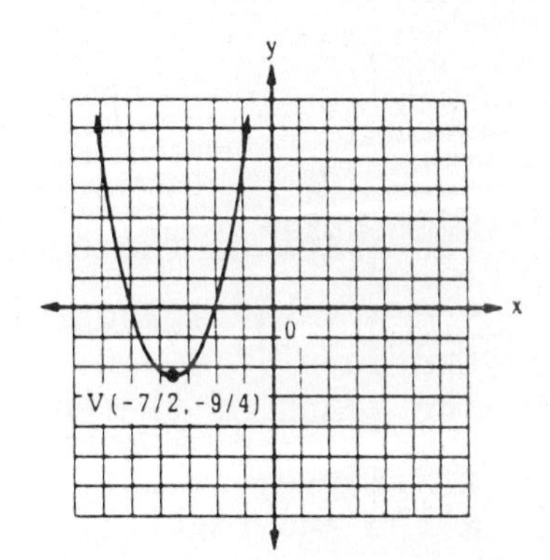


**23.** $y = 2x^2 - 4x - 4$

$y = 2(x^2 - 2x - 2)$

$= 2(x^2 - 2x + 1 - 1 - 2)$

$= 2(x^2 - 2x + 1 - 3)$

$= 2(x - 1)^2 - 2(3)$

$= 2(x - 1)^2 - 6$

The graph opens upward.
The vertex is (1,-6).
When x = 0, y = -4. Therefore, the y intercept is -4.

Number 23 (Continued)

When $y = 0$, the x-intercepts are at approximately 2.7 and -.7.

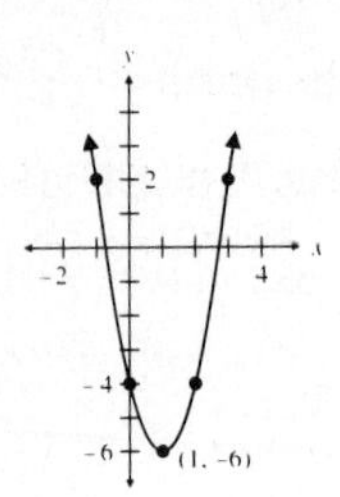


25. The graph of the equation $y = a(x-h)^2 + k$ opens up if $a > 0$ and down if $a < 0$.
The graph of the equation $x = a(y-h)^2 + h$ opens left if $a < 0$ and right if $a > 0$.

27. All parabolas of the form $y = a(x-h)^2 + k$, will be functions since they will pass the vertical line test.

29. The graphs will share the same vertex but $y = 2(x-3)^2 + 4$ will open up and $y = -2(x-3)^2 + 4$ will open down.

31. $\begin{vmatrix} 4 & 0 & 3 \\ 5 & 2 & -1 \\ 3 & 6 & 4 \end{vmatrix}$ Expansion along row 1:

$$= 4\begin{vmatrix} 2 & -1 \\ 6 & 4 \end{vmatrix} + 3\begin{vmatrix} 5 & 2 \\ 3 & 6 \end{vmatrix}$$

$$= 4(8+6) + 3(30-6)$$
$$= 128$$

33. $T = \dfrac{km_1m_2}{R^2}$

Solve for k:

$$\frac{3}{2} = \frac{k(6)(4)}{4^2}$$

$$\frac{3}{2} = \frac{24h}{16}$$

$$48 = 48k$$

$$1 = k$$

$$T = \frac{m_1m_2}{R^2}$$

$$T = \frac{6(10)}{2^2}$$

$$= 15$$

**EXERCISE SET 10.4**

1. The value of $a^2$ is 4; the positive root of 4 is 2.

   The value of $b^2$ is 1; the positive root of 1 is 1.

   Asymptotes are: $y = \frac{1}{2}x$

   and $y = -\frac{1}{2}x$.

   The graph intersects the x axis and the vertices are at 2 and -2.

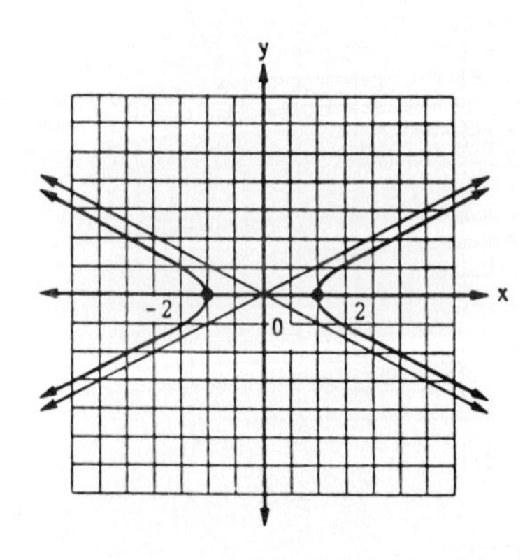


3. The value of $b^2$ is 9; the positive root of 9 is 3.

   The value of $a^2$ is 16; the positive root of 16 is 4.

   Asymptotes are $y = \frac{3}{4}x$ and $y = \frac{-3}{4}x$.

   The graph intersects the y axis and the vertices are at 3 and -3.

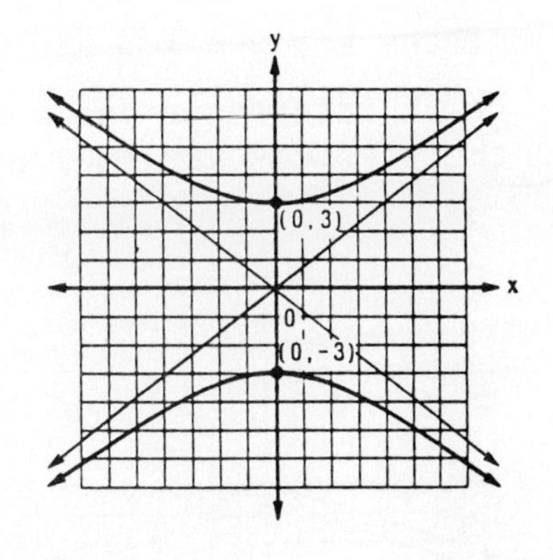


5. The value of $b^2$ is 25; the positive root of 25 is 5.

   The value of $a^2$ is 36; the positive root of 36 is 6.

   Asymptotes are $y = \frac{5}{6}x$ and $y = \frac{-5}{6}x$.

   The graph intersects the y axis and the vertices are at 5 and -5.

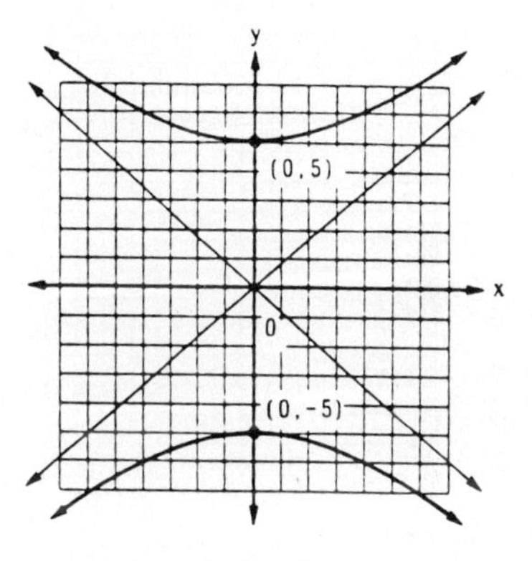


7. The value of $a^2$ is 4; therefore a is 2.

   The value of $b^2$ is 4; therefore b is 2.

   Asymptotes are $y = \pm x$.

   The graph intersects the x axis and the vertices are at 2 and -2.

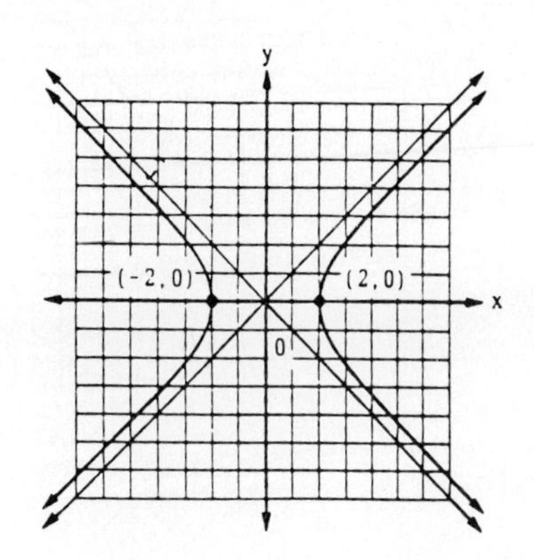

9. The value of $b^2$ is 16; therefore b is 4.

The value of $a^2$ is 81; therefore a is 9.

Asymptotes are $y = \pm \frac{4}{9}x$.

The graph intersects the y axis and the vertices are at 4 and -4.

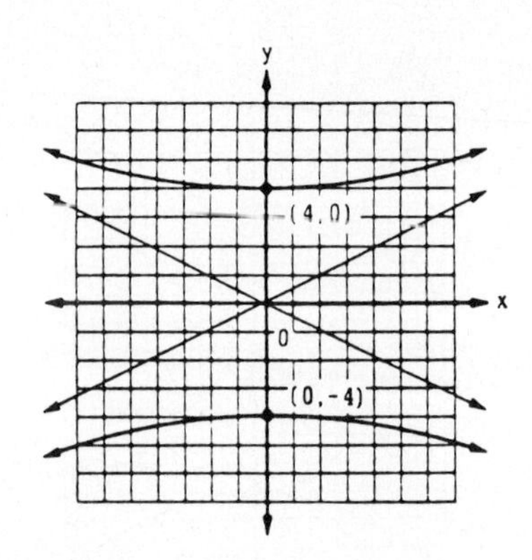


11. The value of $b^2$ is 25; therefore b is 5.

The value of $a^2$ is 16; therefore a is 4.

Asymptotes are $y = \pm \frac{5x}{4}$.

The graph intersects the y axis and the vertices are at 5 and -5.

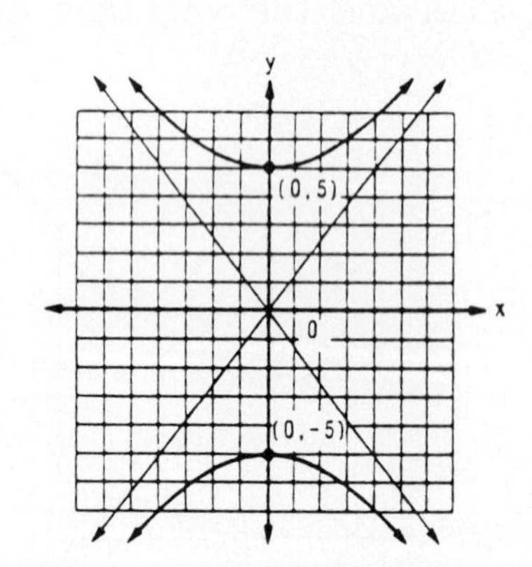


13. $16x^2 - 4y^2 = 64$

Divide each term of the equation by 64.

$$\frac{16x^2 - 4y^2}{64} = \frac{64}{64}$$

$$\frac{x^2}{4} - \frac{y^2}{16} = 1$$

The value of a is 2.
The value of b is 4.
Asymptotes are $y = \pm 2x$.
The graph intersects the x axis and the vertices are at 2 and -2.

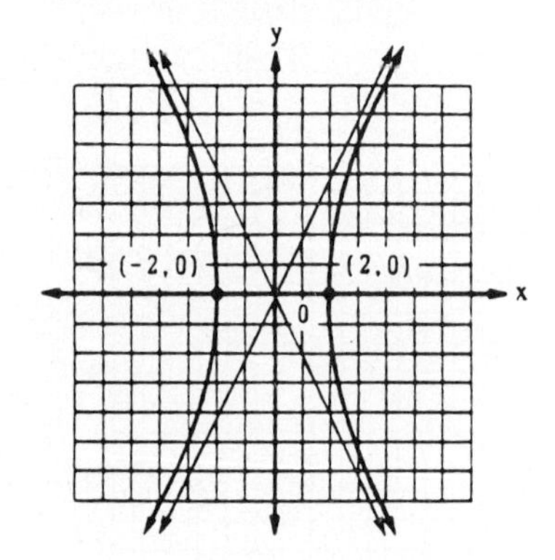


15. $9y^2 - x^2 = 9$

Divide each term of the equation by 9.

$$\frac{9y^2 - x^2}{9} = \frac{9}{9}$$

$$\frac{y^2}{1} - \frac{x^2}{9} = 1$$

The value of b is 1.
The value of a is 3.

Asymptotes are $y = \pm \frac{1}{3}x$.

The graph intersects the y axis and the vertices are at 1 and -1.

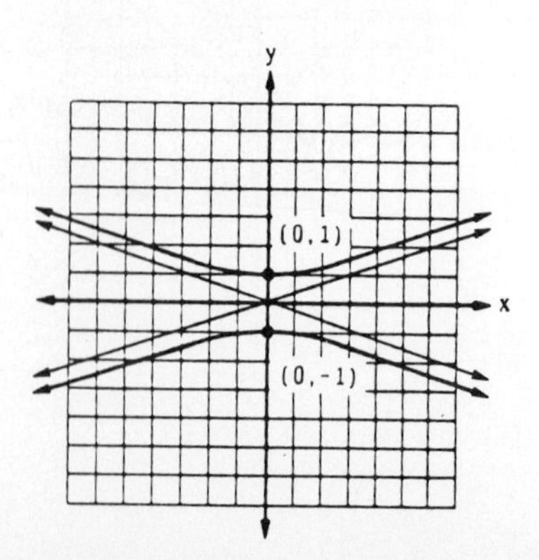

**17.** $4y^2 - 36x^2 = 144$
Divide each term of the equation by 144.

$$\frac{4y^2 - 36x^2}{144} = \frac{144}{144}$$

$$\frac{y^2}{36} - \frac{x^2}{4} = 1$$

The value of b is 6.
The value of a is 2.
Asymptotes are $y = \pm 3x$.
The graph intersects the y axis and the vertices are at 6 and -6.

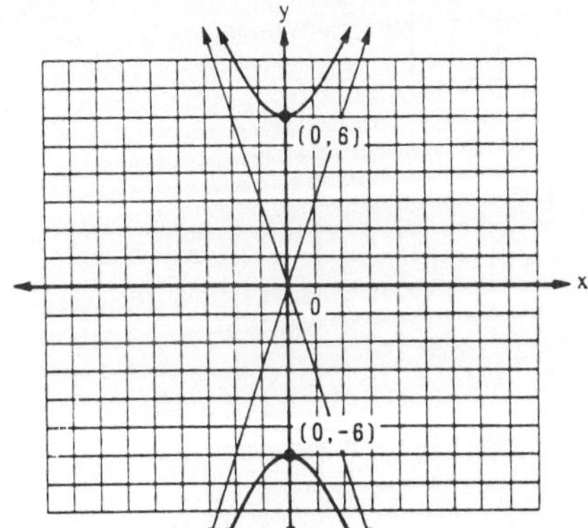


**19.** $25x^2 - 4y^2 = 100$
Divide each term of the equation by 100

$$\frac{25x^2 - 4y^2}{100} = \frac{100}{100}$$

$$\frac{x^2}{4} - \frac{y^2}{25} = 1$$

The value of a is 2.
The value of b is 5.
Asymptotes are $y = \pm \frac{5x}{2}$.
The graph intersects the x axis and the vertices are at 2 and -2.

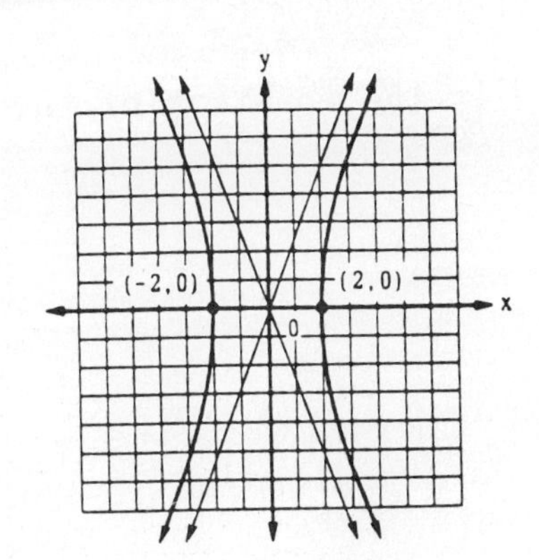


**21.** $81x^2 - 9y^2 = 729$
Divide each term of the equation by 729.

$$\frac{81x^2 - 9y^2}{729} = \frac{729}{729}$$

$$\frac{x^2}{9} - \frac{y^2}{81} = 1$$

The value of a is 3.
The value of b is 9.
Asymptotes are $y = \pm 3x$.
The graph intersects the x axis and the vertices are at 3 and -3.

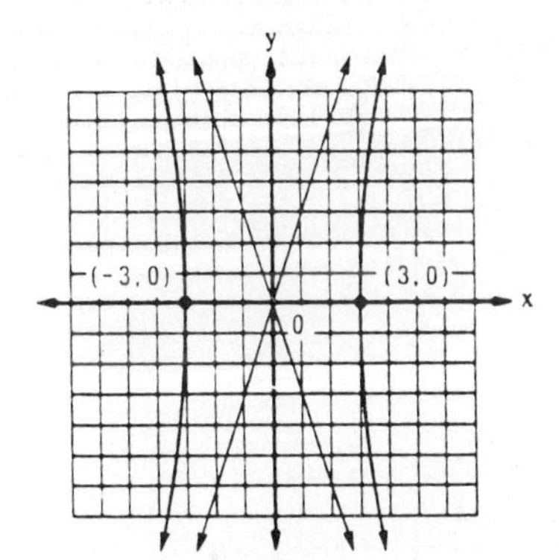


**23.** $xy = 10$ or $y = \frac{10}{x}$

| x | y |
|---|---|
| -4 | -5/2 |
| -3 | -10/3 |
| -2 | -5 |
| -1 | -10 |
| 1 | 10 |
| 2 | 5 |
| 3 | 10/3 |
| 4 | 5/2 |

Number 23 (Continued)

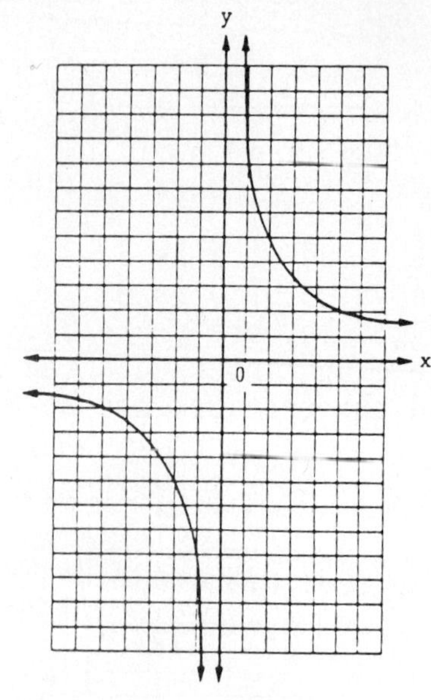


**25.** $xy = -8$ or $y = \frac{-8}{x}$

| x | y |
|---|---|
| -4 | 2 |
| -3 | 8/3 |
| -2 | 4 |
| -1 | 8 |
| 1 | -8 |
| 2 | -4 |
| 3 | -8/3 |
| 4 | -2 |

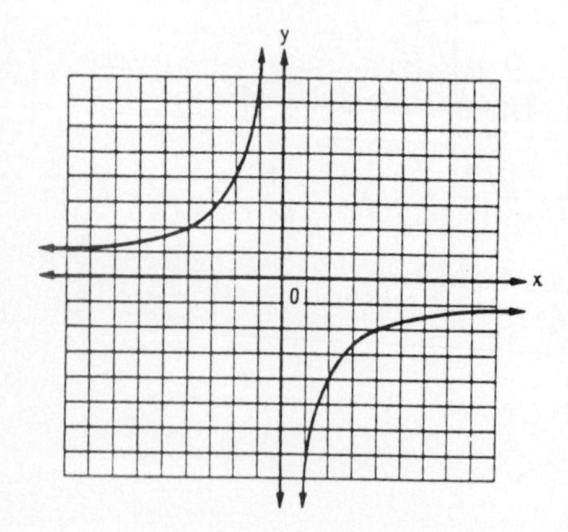


**27.** $y = \frac{12}{x}$

| x | y |
|---|---|
| -4 | -3 |
| -3 | -4 |
| -2 | -6 |
| -1 | -12 |
| 1 | 12 |
| 2 | 6 |
| 3 | 4 |
| 4 | 3 |

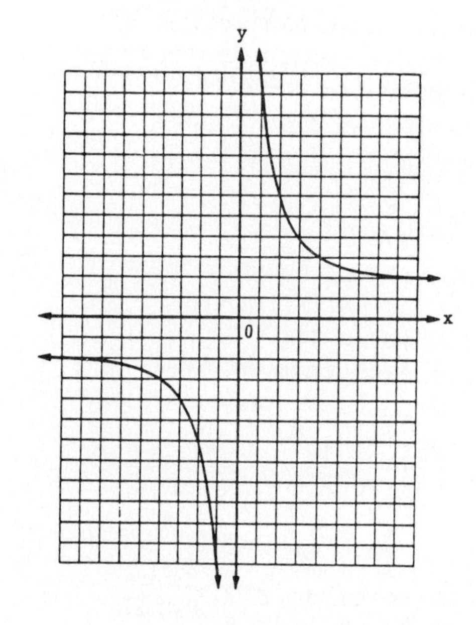


**29.** Asymptotes are lines that the hyperbola approaches.

$$y = \pm \frac{b}{a}x$$

is the equation of the hyperbola.

**31.** The graph of

$$\frac{x^2}{a^2} - \frac{y^2}{b^2} = 1$$

is a hyperbola with vertices at (a,0) and (-a,0). The graph has a transverse axis along the x-axis and the asymptotes are described by the equations $y = \pm \frac{b}{a} x$.

**33.** The transverse axis of both graphs would be along the x-axis. The vertices of the second graph will be closer to the origin, and the hyperbola of the second graph will open wider than that of the first graph.

**35.** $y = 6x^2 + 4x + 3$

$y = 6\ (x^2 + \frac{2}{3}x + \frac{1}{2})$

$y = 6\ (x^2 + \frac{2}{3}x + \frac{1}{9}) - \frac{2}{3} + 3$

$y = 6\ (x + \frac{1}{3})^2 + \frac{7}{3} \rightarrow$ parabola.

The graph opens upward.

The vertex is at $(-\frac{1}{3}, \frac{7}{3})$.

When x = 0, the y intercept is 3.

**37.** $5x^2 - 5y^2 = 25$

Divide each term by 25.

$$\frac{5x^2 - 5y^2}{25} = \frac{25}{25}$$

$$\frac{x^2}{5} - \frac{y^2}{5} = 1 \rightarrow \text{hyperbola}$$

The value of a is $\sqrt{5}$.

The value of b is $\sqrt{5}$.

Asymptotes are $y = \pm x$.

The graph intersects the x axis and the vertices are at $\sqrt{5}$ and $-\sqrt{5}$.

**39.** $x = y^2 + 6y - 7$

$x = y^2 + 6y + 9 - 9 - 7$

$x = (y + 3)^2 - 16 \rightarrow$ parabola.

The graph opens to the right.

The vertex is at (-16, -3).

When y = 0, the x interecept is -7.

**41.** $-2x^2 + 4y^2 = 16$

$$\frac{4y^2 - 2x^2}{16} = \frac{16}{16}$$

$$\frac{y^2}{4} - \frac{x^2}{8} = 1 \rightarrow \text{hyperbola}$$

The value of b is 2.

The value of a is $\sqrt{8}$.

Asymptotes are $y = \pm \frac{\sqrt{2}}{2}x$.

The graph intersects the y axis and the vertices are at 2 and -2.

**43.** $5x^2 + 10y^2 = 12$

Divide each term by 12.

$$\frac{5x^2 + 10y^2}{12} = \frac{12}{12}$$

$$\frac{5x^2}{12} + \frac{5y^2}{6} = 1$$

$$\frac{x^2}{\frac{12}{5}} + \frac{y^2}{\frac{6}{5}} = 1 \rightarrow \text{ellipse}$$

Thus, a $\sqrt{\frac{12}{5}} = \frac{2\sqrt{15}}{5}$ and the x intercepts are $\pm \frac{2\sqrt{15}}{5}$. Since $b = \sqrt{\frac{6}{5}} = \frac{\sqrt{30}}{5}$ the y intercepts are $\pm \frac{\sqrt{30}}{5}$.

**45.** $x = 3y^2 - y + 7$

$x = 3(y^2 - \frac{1}{3}y + \frac{7}{3})$

$x = 3(y^2 - \frac{1}{3}y + \frac{1}{36}) - \frac{1}{12} + 7$

$x = 3(y - \frac{1}{6})^2 + \frac{83}{12} \rightarrow$ parabola.

The graph opens to the right.
The vertex is at $(\frac{83}{12}, \frac{1}{6})$.
When $y = 0$, the x intercept is 7.

**47.** $6x^2 + 6y^2 = 36$
Divide each term by 6.

$\frac{6x^2 + 6y^2}{6} = \frac{36}{6}$

$x^2 + y^2 = 6 \rightarrow$ Circle Center (0,0).

Radius is $\sqrt{6}$.

**49.** $-3x^2 - 3y^2 = -27$
Divide each term by -3.

$\frac{-3x^2 - 3y^2}{-3} = \frac{-27}{-3}$

$x^2 + y^2 = 9 \rightarrow$ Circle Center (0,0).

Radius is 3.

**51.** $-6y^2 + x = -9$
Divide each term by -9.

$\frac{-6y^2 + x^2}{-9} = \frac{-9}{-9}$

$\frac{2y^2}{3} - \frac{x^2}{9} = 1$

$\frac{y^2}{\frac{3}{2}} - \frac{x^2}{9} = 1 \rightarrow$ hyperbola

Number 51 (Continued)

The value of b is $\sqrt{\frac{3}{2}} = \frac{\sqrt{6}}{2}$.
The value of a is 3.

Asymptotes are $y = \pm \frac{\sqrt{6}}{6}x$.
The graph intersects the y axis and the vertices are at $\frac{\sqrt{6}}{2}$ and $-\frac{\sqrt{6}}{2}$.

**55.** The domain is the set of values that can be used for the independent variable. The range is the set of values that are obtained for the dependent variables.

**57.** $-2x^2 + 6x - 5 = 0$

$2x^2 - 6x + 5 = 0$

$x = \frac{-(-6) \pm \sqrt{(-6)^2 - 4(2)(5)}}{2(2)}$

$x = \frac{6 + \sqrt{-4}}{4}$ or $x = \frac{6 - \sqrt{-4}}{4}$

$x = \frac{6 + 2i}{4}$ or $x = \frac{6 - 2i}{4}$

$= \frac{3 + i}{2}$ or $= \frac{3 - i}{2}$

**JUST FOR FUN**

1. The value of a is 3.
The value of b is 2.
The center is at (3, -2).

Number 1 (Continued)

Asymptotes are:

$$y - k = \pm \frac{b}{a}(x - h)$$

$$y - (-2) = \pm \frac{2}{3}(x - 3)$$

$$y + 2 = \pm \frac{2}{3}(x - 3)$$

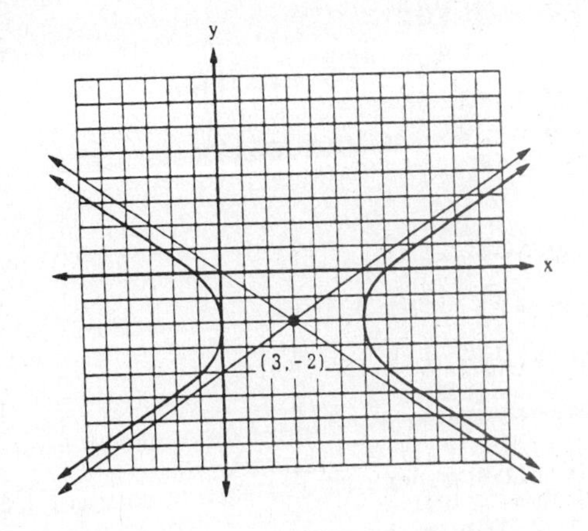


3.

$$y^2 - 4x^2 + 2y + 8x - 7 = 0$$

$$y^2 + 2y - 4(x^2 - 2x) = 7$$

$$y^2 + 2y + 1 - 4(x^2 - 2x + 1) = 7 + 1 - 4$$

$$\frac{(y + 1)^2 - 4(x - 1)^2}{4} = \frac{4}{4}$$

$$\frac{(y + 1)^2}{4} - \frac{(x - 1)^2}{1} = 1$$

The value of b is 2.
The value of a is 1.
The center is at (1, -1).
Asymptotes are:

$$y - k = \pm \frac{b}{a}(x - h)$$

$$y - (-1) = \pm 2(x - 1)$$

$$y + 1 = \pm 2(x - 1)$$

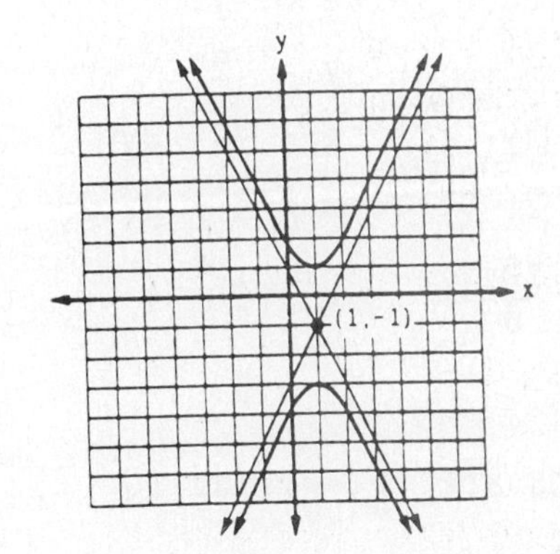

## Exercise Set 10.5

1. Solve $x - 2y = 4$ for x.

$x = 4 + 2y$

Substitute $4 + 2y$ for x in the equation $x^2 + y^2 = 4$ and solve for y.

$$x^2 + y^2 = 4$$
$$(4 + 2y)^2 + y^2 = 4$$
$$(4 + 2y)(4 + 2y) + y^2 = 4$$
$$16 + 8y + 8y + 4y^2 + y^2 = 4$$
$$16 + 16y + 5y^2 = 4$$
$$5y^2 + 16y + 12 = 0$$
$$(5y + 6)(y + 2) = 0$$
$$5y + 6 = 0 \quad \text{or} \quad y + 2 = 0$$
$$5y = -6 \qquad y = -2$$
$$y = \frac{-6}{5}$$

Find the corresponding value of x and for each value of y.

| | |
|---|---|
| $y = \frac{-6}{5}$ | $y = -2$ |
| $x = 4 + 2y$ | $x = 4 + 2y$ |
| $x = 4 + 2(-\frac{6}{5})$ | $x = 4 + 2(-2)$ |
| $x = 4 - \frac{12}{5}$ | $x = 4 - 4$ |
| $x = \frac{8}{5}$ | $x = 0$ |

Solutions are $(\frac{8}{5}, \frac{-6}{5})$ and $(0,-2)$.

3. Substitute $y = x^2 - 5$ in the equation $3x + 2y = 10$ and solve for x.

$$3x + 2y = 10$$
$$3x + 2(x^2 - 5) = 10$$
$$3x + 2x^2 - 10 = 10$$
$$3x + 2x^2 - 20 = 0$$
$$2x^2 + 3x - 20 = 0$$
$$(2x - 5)(x + 4) = 0$$
$$2x - 5 = 0 \quad \text{or} \quad x + 4 = 0$$
$$2x = 5 \qquad x = -4$$
$$x = \frac{5}{2}$$

Number 3 (Continued)

Find the corresponding value of y for each value of x.

| | |
|---|---|
| $x = \frac{5}{2}$ | $x = -4$ |
| $y = (\frac{5}{2})^2 - 5$ | $y = (-4)^2 - 5$ |
| $y = \frac{25}{4} - 5$ | $y = 16 - 5$ |
| $y = \frac{5}{2}$ | $y = 11$ |

Solutions are $(\frac{5}{2}, \frac{5}{4})$ and $(-4,11)$.

5. Solve $x - y = 6$ for x.

$x = 6 + y$.

Substitute $6 + y$ for x in the equation $2x^2 - y^2 = -8$ and solve for y.

$$2x^2 - y^2 = -8$$
$$2(6 + y)^2 - y^2 = -8$$
$$2(6 + y)(6 + y) - y^2 = -8$$
$$2(36 + 6y + 6y + y^2) - y^2 = -8$$
$$2(36 + 12y + y^2) - y^2 = -8$$
$$72 + 24y + 2y^2 - y^2 = -8$$
$$72 + 24y + y^2 = -8$$
$$80 + 24y + y^2 = 0$$
$$y^2 + 24y + 80 = 0$$
$$(y + 4)(y + 20) = 0$$
$$y + 4 = 0 \quad \text{or} \quad y + 20 = 0$$
$$y = -4 \qquad y = -20$$

Find the corresponding value of x for each value of y.

| | | |
|---|---|---|
| $y = -4$ | or | $y = -20$ |
| $x = 6 + y$ | | $x = 6 + y$ |
| $x = 6 - 4$ | | $x = 6 - 20$ |
| $x = 2$ | | $x = -14$ |

Solutions are $(2,-4)$ and $(-14,-20)$.

7. Solve $x^2 + y^2 = 1$ for $x^2$.

$$x^2 = 1 - y^2.$$

Substitute $1 - y^2$ for $x^2$ in the equation $x^2 - 4y^2 = 16$ and solve for y.

$$x^2 - 4y^2 = 16$$
$$1 - y^2 - 4y^2 = 16$$
$$1 - 5y^2 = 16$$
$$- 5^2 = 15$$
$$- 5y^2 = 15$$
$$y = \pm \sqrt{-3}$$
$$y = \pm 3i$$

This system of equations has no real solution.

9. Solve $6x - y = 5$ for y.

$$y = -5 + 6x.$$

Substitute $-5 + 6x$ for y in the equation $xy = 4$ and solve for x.

$$xy = 4$$
$$x(-5 + 6x) = 4$$
$$-5x + 6x^2 = 4$$
$$6x^2 - 5x - 4 = 0$$
$$(3x-4)(2x+1) = 0$$

$3x - 4 = 0$ or $2x + 1 = 0$

$3x = 4$ $\quad 2x = -1$

$x = \frac{4}{3}$ $\quad x = -\frac{1}{2}$

Find the corresponding value of y for each value of x.

| | |
|---|---|
| $x = \frac{4}{3}$ | $x = -\frac{1}{2}$ |
| $y = -5 + 6x$ | $y = -5 + 6x$ |
| $y = -5 + 6(\frac{4}{3})$ | $y = -5 + 6(-\frac{1}{2})$ |
| $y = -5 + 8$ | $y = -5 - 3$ |
| $y = 3$ | $y = -8$ |

Solutions are $(\frac{4}{3}, 3)$ and $(\frac{-1}{2}, -8)$.

11. Solve $y = x^2 - 3$ for $x^2$.

$$x^2 = y + 3$$

Substitute $y + 3$ for $x^2$ in the equation $x^2 + y^2 = 9$ and solve for y.

$$x^2 + y^2 = 9$$
$$y + 3 + y^2 = 9$$
$$y^2 + y - 6 = 0$$
$$(y + 3)(y - 2) = 0$$

$y + 3 = 0$ or $y - 2 = 0$

$y = -3$ $\quad y = 2$

Find the corresponding value of x for each value of y.

| | |
|---|---|
| $y = -3$ | $y = 2$ |
| $x^2 = y + 3$ | $x^2 = y + 3$ |
| $x^2 = -3 + 3$ | $x^2 = 2 + 3$ |
| $x^2 = 0$ | $x^2 = 5$ |
| $x = 0$ | $x = \pm \sqrt{5}$ |

Solutions are $(0, -3)$ $(\sqrt{5}, 2)$ and $(-\sqrt{5}, 2)$.

13.
$$x^2 - y^2 = 4$$
$$x^2 + y^2 = 4$$
$$2x^2 = 8$$
$$x^2 = 4$$
$$x = \pm \sqrt{4}$$
$$x = \pm 2$$

Now solve for y.

| | |
|---|---|
| $x = 2$ | $x = -2$ |
| $x^2 - y^2 = 4$ | $x^2 - y^2 = 4$ |
| $(2) - y^2 = 4$ | $(-2)^2 - y^2 = 4$ |
| $4 - y^2 = 4$ | $4 - y^2 = 4$ |
| $- y^2 = 0$ | $- y^2 = 0$ |
| $y^2 = 0$ | $y^2 = 0$ |
| $y = 0$ | $y = 0$ |

Solutions are $(2,0)$ and $(-2,0)$

15. (1) $x^2 + y^2 = 13$

(2) $2x^2 + 3y^2 = 30$

Multiply equation (1) by -2 and add the result to equation 2 to eliminate x and solve for y.

(1) $-2x^2 - y^2 = -26$

(2) $2x^2 + 3y^2 = 30$

$y^2 = 4$

$y = \pm 2$

If $y = 2$ then

$x^2 + (2)^2 = 13$

$x^2 = 9$

$x = \pm 3$

If $y = -2$ then

$x^2 + (-2)^2 = 13$

$x^2 = 9$

$x = \pm 3$

The solutions are:
(3,2), (-3,2), (3,-2), (-3,-2)

17. (1) $4x^2 + 9y^2 = 36$

(2) $2x^2 - 9y^2 = 18$

Add equation (1) and (2) to eliminate y and solve for x.

(1) $4x^2 + 9y^2 = 36$

(2) $2x^2 - 9y^2 = 18$

$6x^2 = 54$

$x^2 = 9$

$x = \pm 3$

If $x = 3$ then

$4x^2 + 9y^2 = 36$ becomes

$4(3)^2 + 9y^2 = 36$ or

$36 + 9y^2 = 36$ or $9y^2 = 0$

or $y = 0$.

Number 17 (Continued)

If $x = -3$ then

$4x^2 + 9y^2 = 36$ becomes

$4(-3)^2 + 9y^2 = 36$ or

$36 + 9y^2 = 36$ or $9y^2 = 0$

or $y = 0$

Solutions: (3,0), (-3,0)

19. (1) $2x^2 + 3y^2 = 21$

(2) $x^2 + 2y^2 = 12$

Multiply equation (2) by -2 and add the result to equation (1)

(1) $2x^2 + 3y^2 = 21$

(2) $-2x^2 - 4y^2 = -24$

$-y^2 = -3$

$y = \pm \sqrt{3}$

If $y = \sqrt{3}$ then

$x^2 + 2(\sqrt{3})^2 = 12$

$x^2 = 6$

$x = \pm \sqrt{6}$

If $y = -\sqrt{3}$ then

$x^2 + 2(-\sqrt{3})^2 = 12$

$x^2 = 6$

$x = \pm \sqrt{6}$

Solutions $(\sqrt{6}, \sqrt{3})$, $(-\sqrt{6}, \sqrt{3})$, $(\sqrt{6}, -\sqrt{3})$ $(-\sqrt{6}, -\sqrt{3})$

21. (1) $-x^2 - 2y^2 = 6$

(2) $5x^2 + 25y^2 = 20$

Multiply equation (1) by 5. Add the result to equation (2) to eliminate x and solve for y.

(1) $-5x^2 - 10y^2 = 30$

(2) $5x^2 + 15y^2 = 20$

$5y^2 = 50$

$y^2 = 10$

$y = \pm \sqrt{10}$

Number 21 (Continued)

If $y = \sqrt{10}$, then

$-x^2 - 2(\sqrt{10})^2 = 6$

$-x^2 - 20 = 6$

$-x^2 = 26$

$x = \pm \sqrt{-26}$

If $y = -\sqrt{10}$ then

$-x^2 - 2(-\sqrt{10})^2 = 6$

$-x^2 - 20 = 6$

$-x^2 = 26$

$x = \pm \sqrt{-26}$

There are no real solutions to this system of equations.

**23.** (1) $x^2 + 4^2 = 9$

(2) $16x^2 - 4y^2 = 64$

Multiply equation (1) by -16. Add the result to equation (2)

(1) $-16x^2 - 16y^2 = -144$

(2) $16x^2 - 4y^2 = 64$

$-20y^2 = -80$

$y^2 = 4$

$y = \pm 2$

If $y = 2$ then

$x^2 + (2)^2 = 9$

$x^2 = 5$

$x = \pm \sqrt{5}$

If $y = -2$ then

$x^2 + (-2)^2 = 9$

$x^2 = 5$

$x = \pm \sqrt{5}$

Solutions: $(\sqrt{5}, 2)$, $(-\sqrt{5},2)$, $(\sqrt{5},-2)$ $(-\sqrt{5},2)$.

**25.** (1) $x + y = 12$

(2) $x^2 + y^2 = 74$

Solve equation (1) for x and substitute into equation (2).

(1) $x = 12 - y$

(2) $(12-y)^2 + y^2 = 74$

$144 - 24y + y^2 + y^2 = 74$

$2y^2 - 24y + 70 = 0$

$y^2 - 12y + 35 = 0$

$(y - 5)(y - 7) = 0$

$y - 5 = 0$ or $y - 7 = 0$

$y = 5$ or $y = 7$

Substitute these values into equation (1) to solve for x.

$x = 12 - 5$ or $x = 12 - 7$

$x = 7$ or $x = 5$

The two numbers are 7 and 5.

**27.** (1) $x^2 + y^2 = 34$

(2) $x^2 - y^2 = 16$

$2x^2 = 50$

$x^2 = 25$

$x = +5$ or $x = -5$

Substitute these values of x into equation (1) or (2) to solve for y.

If $x = 5$, then

(1) $(5)^2 + y^2 = 34$

$y^2 = 9$

$y = \pm 3$

If $x = -5$, then

(1) $(-5)^2 + y^2 = 34$

$y^2 = 9$

$y = \pm 3$

The pairs of x and y values which would satisfy this problem are (5,-3), (5,3), (-5,3), and (-5,-3)

**29.** (1) $x^2 = y^2 - 11$
or

$x^2 - y^2 = -11$

(2) $x^2 + y^2 = 16$

$2x^2 = 50$

$x^2 = 25$

$x = 5$ or $x = -5$

Substitute these values of x into equation (1) or (2) to solve for y. If $x = 5$ then

(2) $(5)^2 + y^2 = 61$

$y^2 = 36$

$y = \pm 6$

If $x = -5$ then

(2) $(5)^2 + y^2 = 61$

$y^2 = 36$

$y = \pm 6$

The pairs of x and y values which would satisfy this problem are: (5,6), (5,-6), (-5,6) and (-5,-6).

**31.** (1) $Lw = 84$
(2) $2L + 2w = 38$
Solve equation (1) for L and substitute this expression into equation (2)

(1) $L = \frac{84}{w}$

(2) $2\left(\frac{84}{w}\right) + 2w = 38$

$\frac{168}{w} + 2w = 38$

$w\left(\frac{168}{w} + 2w = 38\right)$

$168 + 2w^2 = 38w$

$2w^2 - 38w + 168 = 0$

$w^2 - 19w + 84 = 0$

$(w - 7)(w - 12) = 0$

$w - 7 = 0 \quad w - 12 = 0$

$w = 7$ or $w = 12$

Number 31 (Continued)

Substitute these values into equation (1) to solve for h.

(1) $h = \frac{84}{7}$ or $L = \frac{84}{12}$

$h = 12$ or $L = 7$

The length -= 12 and the width 7 ft or the width - 12 ft and the length = 7 ft.

**33.** (1) $L_1 + L_2 + 13 = 30$ or

$L_1 + L_2 = 17$

(2) $L_1^2 + L_2^2 = 13^2$ or

$L_1^2 + L_2^2 = 169$

Solve equation 1 for $L_1$.
Substitute this expression into equation (2) to solve for $L_2$.

(1) $L_1 + L_2 = 17$

$L_1 = 17 - L_2$

(2) $(17 - L_2)^2 + L_2^2 = 169$

$289 - 34L_2 + L_2^2 + L_2^2 = 169$

$2L_2^2 - 34L_2 + 120 = 0$

$L_2^2 - 17L_2 + 60 = 0$

$(L_2 - 15)(L_2 - 2) = 0$

$L_2 - 15 = 0$ or $L_2 - 2 = 0$

$L_2 = 15 \qquad L_2 = 2$

Substitute $L_2$ into equation (1) to give for $L_1$.

$L_1 = 17 - 15$ or $L_2 = 17 - 2$

$L_1 = 2 \qquad L_2 = 15$

The figures of the triangle should be 15 ft and 2 ft.

**35.** (1) $Lw = 300$
(2) $L = w + 5$
Substitute $L = w + 5$ into equation (1) and solve for w.

$$(w + 5)\ w = 300$$
$$w^2 + 5w = 300$$
$$w^2 + 5w - 300 = 0$$
$$(w + 20)(w - 15) = 0$$

$w + 20 = 0$ or $w - 15 = 0$
$w = -20$ $w = 15$
Since width cannot be negative, $w = 15$. Substitute $w = 15$ into equation 2 and solve for L.
(2) $L = 15 + 5$
$L = 20$
The rectangle is 15 ft by 20 ft.

**37.** (1) $2x + y = 20$
(2) $xy = 48$
Solve equation (1) for y. Substitute this expression into equation (2) to solve for x.
(1) $y = 20 - 2x$
(2) $x(20 - 2x) = 48$

$$20x - 2x^2 = 48$$
$$0 = 2x^2 - 20x + 48$$
$$0 = x^2 - 10x + 24$$
$$0 = (x-6)(x-4)$$

$x - 6 = 0$ or $x - 4 = 0$
$x = 6$ or $x = 4$
Substitute these values into equation (1) to solve for y
If $x = 6$ then
$2(5) + y = 20$
$y = 8$
The field could be 6 ft by 8 ft
If $x = 4$ then
$2(4) + y = 20$
$y = 12$
The field could be 4 ft by 12 ft.

**39.** Since the distance above the ground for both the snowball and the quarter will be the same,
$-16t^2 - 10t + 2000 = -16t^2 + 800t + 100$
$-10t + 200 = 800t + 100$
$1900 = 810t$
$2.35$ sec. $= t$

**41.** (1) $pr = 72$
(2) $(p + 120)(r - .02) = 72$
Solve equation (2) for P. Substitute this into equation 1 and solve for r.

(2) $p + 120 = \dfrac{72}{r-.02}$

$$p = \frac{72}{r-.02} - 120$$

(1) $\left[\dfrac{72}{r-.02} - 120\right] r = 72$

$$(r-.02)\left[\frac{72r}{r-.02} - 120r = 72\right]$$
$$72r - 120r^2 + 2.4r = 72r - 1.44$$
$$74.4r - 120r^2 = 72r - 1.44$$
$$0 = 120r^2 - 2.4r - 1.44$$
$$r = \frac{-(-2.4) \pm \sqrt{(-2.4)^2 - 4(120)(-1.44)}}{2(120)}$$
$$= \frac{2.4 \pm \sqrt{696.96}}{240}$$

$r = \dfrac{2.4+26.4}{240}$ or $r = \dfrac{2.4-26.4}{240}$
$= .12$ $r = -.1$
Since rate cannot be negative, r = 12.
(1) $p(.12) = 72$
$p = \$600$
\$600 is the amount of principle and 12% is the rate.

43. $80x + 900 = 120x - .2x^2$

$.2x^2 - 40x + 900 = 0$

$x^2 - 200x + 4500 = 0$

$$x = -\frac{(-200) \pm \sqrt{(200)^2 - 4(1)(4500)}}{2}$$

$$= \frac{200 \pm \sqrt{22000}}{2}$$

$$x = \frac{200+148.3}{2} \text{ or } x = \frac{200-148}{2}$$

$x = 174$ or $x = 26$

45. $.6x^2 + 9 = 12x - .2x^2$

$.8x^2 - 12x + 9 = 0$

$$x = \frac{-(-12) \pm \sqrt{(-12)^2 - 4(.8)(9)}}{2(.8)}$$

$$x = \frac{12 + \sqrt{115.2}}{1.6} \text{ or } x = \frac{12 - \sqrt{115.2}}{1.6}$$

$x \approx 14$ or $x \approx 1$

47. Answers vary from student to student.

49. The maximum number points of integration is 8.

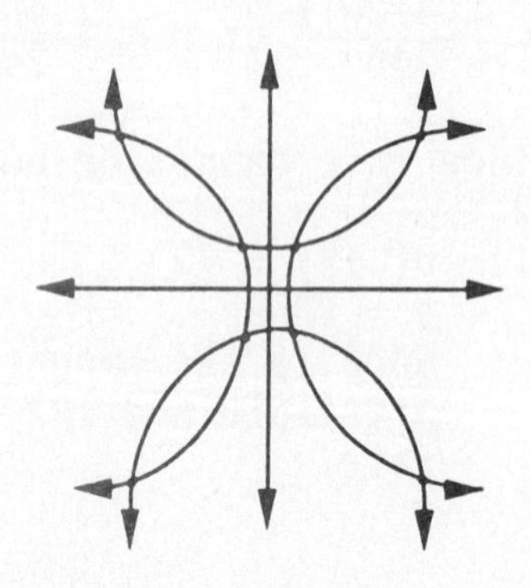

51. $A = 5000\ (1 + \frac{08}{2})^{2(2)}$

$= 5000\ (1.16985856)$

$= \$5849.29$

53. $\frac{3-4y}{3} \geq \frac{2y-6}{4} - \frac{7}{6}$

$12[\frac{3-4y}{3} \geq \frac{2y-6}{4} - \frac{7}{6}]$

$4(3-4y) \geq 3(2y-6) - 2(7)$

$12 - 16y \geq 6y - 18 - 14$

$44 \geq 22y$

$y \leq 2$

**JUST FOR FUN**

1. (1) $\frac{1}{2} L_1L_2 = 30$

or $L_1L_2 = 60$

(2) $L_1^2 + L_2^2 = 13^2$

Solve equation (1) for $L_1$ and substitute the resulting expression into equation 2.

(1) $L_1 = \frac{60}{L_2}$

(2) $(\frac{60}{L_2})^2 + L_2^2 = 169$

$\frac{3600}{L_2^2} + L_2^2 = 169$

$L_2^2[\frac{3600}{L_2^2} + L_2^2 = 169]$

$3600 + L_2^4 = 169L_2^2$

$L_2^4 - 169L_2^2 + 3600 = 0$

$(L_2^2 - 25)(L_2^2 - 144) = 0$

$L_2^2 - 25 = 0 \quad L_2^2 = 144 = 0$

$L_2^2 = 25 \quad L_2^2 = 144$

$L_2 = \pm 5$ or $L_2 = \pm 12$

Number 1 (Continued)

Since length cannot be negative

$L_2 = 5$ or $12$

If $L_2 = 5$ then

$$L_1 L_2 = 60$$

$$L_1(5) = 60$$

$$L_1 = 12$$

If $L_2 = 12$ then

$$L_1 L_2 = 60$$

$$L_1(12) = 60$$

$$L_1 = 5$$

The tenths of the time lags are 12 and 5 meters

**Exercise Set 10.6**

1. $x^2 + y^2 = 16 \rightarrow$ Center Circle (0,0). Radius is 4.

Select a test point: (0,0)

$x^2 + y^2 \geq 16$

$0 + 0 \geq 16$

$0 \geq 16$ False

The point (0,0) is not a solution, therefore shade outside the circle.

$y + x = 5 \rightarrow$ straight line

The x intercept is 5 and the y intercept is 5.

Select a test point: (0,0)

$y + x < 5$

$0 + 0 < 5$

$0 < 5$ True

The point (0,0) is a solution

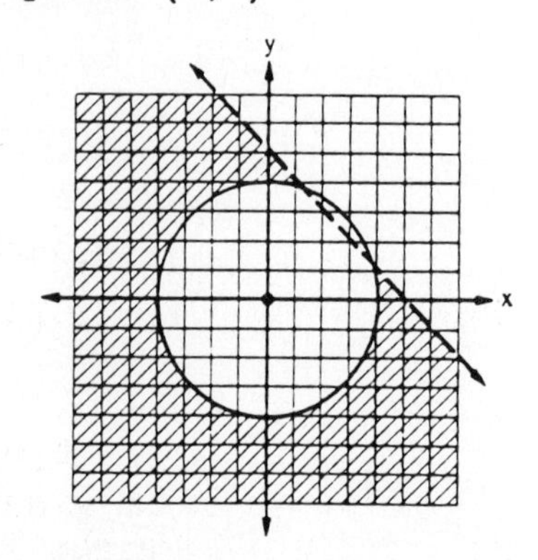


3. $4x^2 + y^2 > 16$

Divide each term by 16

$$\frac{4x^2 + y^2}{16} = \frac{16}{16}$$

$$\frac{x^2}{4} + \frac{y^2}{16} = 1 \rightarrow \text{ellipse}$$

$a = 2,\ b = 4$

Therefore, the x intercepts are $\pm$ 2 and the y intercepts are $\pm$ 4.

Number 3 (Continued)

Select a test point: (0,0)

$4x^2 + y^2 > 1$

$4(0) + 0 > 1$

$0 > 1$ False

The point (0,0) is not a solution. Therefore, shade all points outside the ellipse.

$y = 2x + 2 \rightarrow$ straight line

The x intercept is -1 and the y intercept is 2.

Select a test point (0,0)

$y \geq 2x + 2$

$0 \geq 2(0) + 2$

$0 \geq 2$ False

The point (0,0) is not a solution.

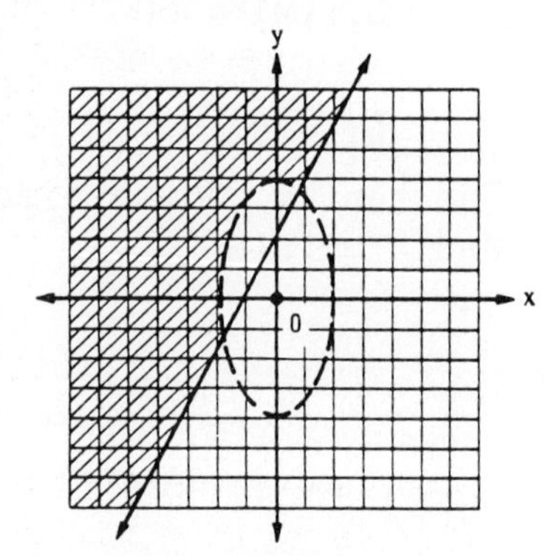


5. $x^2 + y^2 = 36 \rightarrow$ Circle Center (0,0). Radius is 6.

Select a test point: (0,0)

$x^2 + y^2 \leq 36$

$0 + 0 \leq 36$

$0 \leq 36$ True

The point (0,0) is a solution. Therefore, shade all points inside the circle.

$y = (x + 1)^2 - 5 \rightarrow$ parabola

The graph opens upward and the vertex is (-1, -5).

Number 5 (Continued)

Select a test point: (0,0)

$$y < (x + 1)^2 - 5$$
$$0 < (0 + 1)^2 - 5$$
$$0 < -4 \qquad \text{False}$$

The point (0,0) is not a solution. Thus all points outside the parabola will solve the inequality.

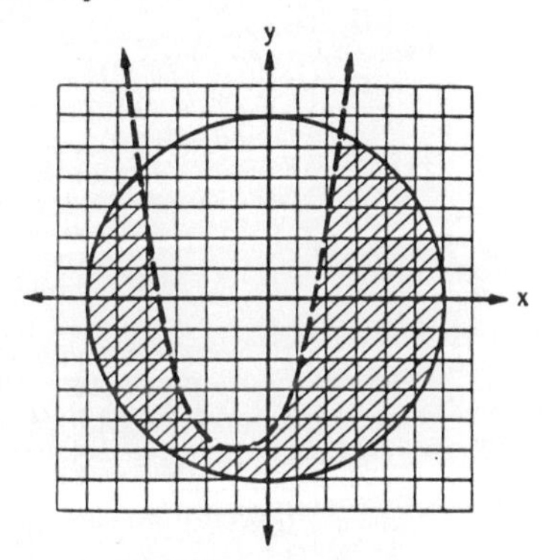


7. $\frac{y^2}{16} - \frac{x^2}{4} = 1 \rightarrow$ hyperbola

The graph intersects the y axis and the vertices are at ± 4.

Select a test point: (0,0)

$$\frac{y^2}{16} - \frac{x^2}{4} \geq 1$$
$$\frac{0}{16} - \frac{0}{4} \geq 1$$
$$0 \geq 1 \qquad \text{False}$$

The point (0,0) is not a solution.

$\frac{x^2}{4} - \frac{y^2}{1} = 1 \rightarrow$ hyperbola

The graph intersects the x axis and the vertices are at ± 2.

Number 7 (Continued)

Select a test point: (0,0)

$$\frac{x^2}{4} - \frac{y^2}{1} < 1$$
$$\frac{0}{4} - \frac{0}{1} < 1$$
$$0 < 1 \qquad \text{True}$$

The point (0,0) is a solution.

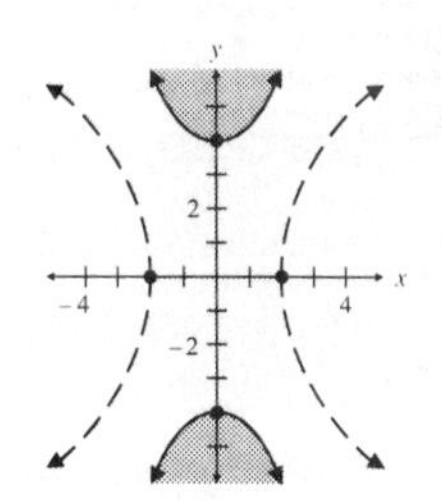


9. $xy = 6$ or $y = \frac{6}{x} \rightarrow$ hyperbola

Select a test point: (0,0)

$$xy \leq 6$$
$$(1)\ (0) \leq 6$$
$$0 \leq 6 \qquad \text{True}$$

The point (0,0) is a solution.

$2x - y = 8 \rightarrow$ straight line.

The x intercept is 4 and the y intercept is -8.

Select a test point: (0,0)

$$2x - y \leq 8$$
$$2(0) - 0 \leq 8$$
$$0 \leq 8 \qquad \text{True}$$

The point (0,0) is a solution.

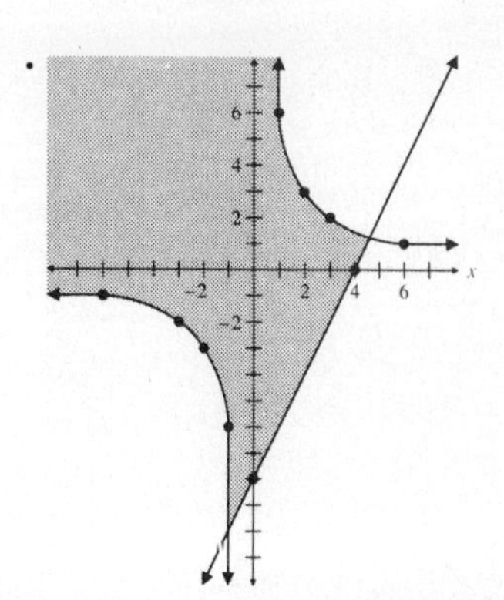

11. $(x - 3)^2 + (y + 2)^2 = 16 \rightarrow$ Circle Center (3,-2). Radius = 4.

Select a test point: (0,0)

$(x - 3)^2 + (y + 2)^2 \geq 16$

$(0 - 3)^2 + (0 + 2)^2 \geq 16$

$9 + 4 \geq 16$

$13 \geq 16$

False

The point (0,0) is not a solution. Therefore shade all points outside the circle.

$y = 4x - 2 \rightarrow$ straight line

The x intercept is 1/2 and the y intercept is -2.

Select a test point: (0,0)

$y \leq 4x - 2$

$0 \leq 4(0) - 2$

$0 \leq -2$ False

The point (0,0) is not a solution.

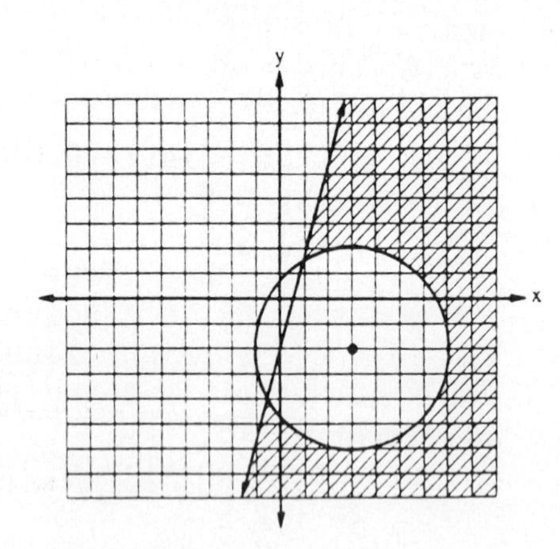


13. $\left(\frac{4x^{-2}y^3}{2xy^{-4}}\right)^2 \left(\frac{3xy^{-1}}{6x^4y^{-3}}\right)^{-2}$

$= [2x^{(-2-1)}y^{3-(-4)}]^2$

$\left[\frac{x^{(1-4)}y^{(-1(-3))}}{2}\right]^2$

$= [2x^{-3}y^7]^2 \left[\frac{x^{-3}y^2}{2}\right]^{-2}$

$= 4x^{-6}y^{14}\ \frac{x^6y^{-4}}{2^{-2}}$

$= 4(2)^2\ y^{(14-4)} = 16y^{10}$

15. $x^2 - 2x - 4 = 0$

$x = \frac{-(-2) \pm \sqrt{(-2)^2 - 4(1)(-4)}}{2}$

$x = \frac{2 + \sqrt{20}}{2}$ or $x = \frac{2 - \sqrt{20}}{2}$

$x = \frac{2 + 2\sqrt{5}}{2}$ or $x = \frac{2 - 2\sqrt{5}}{2}$

$x = 1 + \sqrt{5}$ or $x = 1 - \sqrt{5}$

**JUST FOR FUN**

1. $y = 4x - 6 \rightarrow$ straight line.
   The x intercept is 3/2 and the y intercept is -6.
   Select a test point: (0,0)

   $y > 4x - 6$

   $0 > 4(0) - 6$

   $0 > -6$ True

   The point (0,0) is a solution.

   $x^2 + y^2 = 36 \rightarrow$ Circle Center (0,0). Radius is 6.

Number 1 (Continued)

Select a test point: (0,0)

$$x^2 + y^2 \geq 36$$
$$0 + 0 \geq 36$$
$$0 \geq 36 \quad \text{False}$$

The point (0,0) is not a solution. Shade all points outside the circle.

$$2x + y = 8 \rightarrow \text{straight line.}$$

The x intercept is 4 and the y intercept is 8.

Select a test point: (0,0)

$$2x + y \leq 8$$
$$2(0) + 0 \leq 8$$
$$0 \leq 8 \quad \text{True}$$

The point (0,0) is a solution.

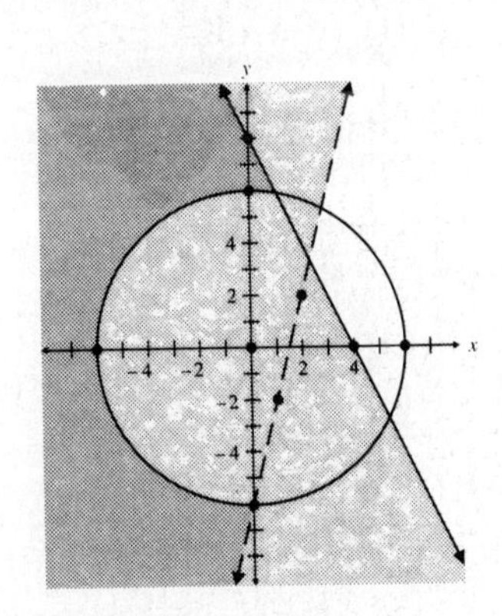

## REVIEW EXERCISES CHAPTER 10

1. The center is (0,0).
   Therefore, $h = 0$ and $k = 0$.
   The radius is 5.

$$(x - h)^2 + (y - k)^2 = r^2$$
$$(x - 0)^2 + (y - 0)^2 = 5^2$$
$$x^2 + y^2 = 5^2$$

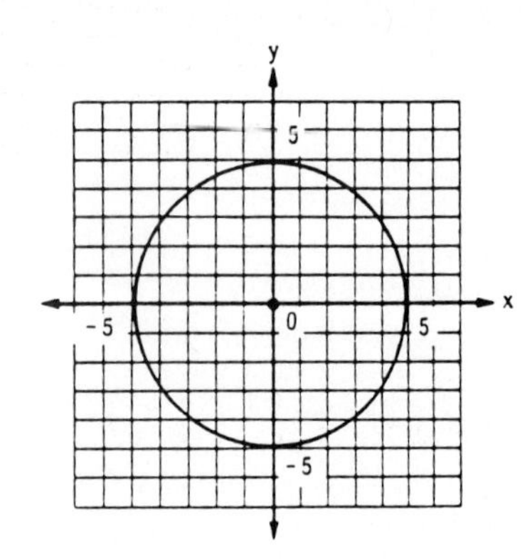


3. The center is (4,2).
   Therefore, $h = 4$ and $k = 2$.
   The radius is $\sqrt{8}$.

$$(x - h)^2 + (y - k)^2 = r^2$$
$$(x - 4)^2 + (y - 2)^2 = (\sqrt{8})^2$$

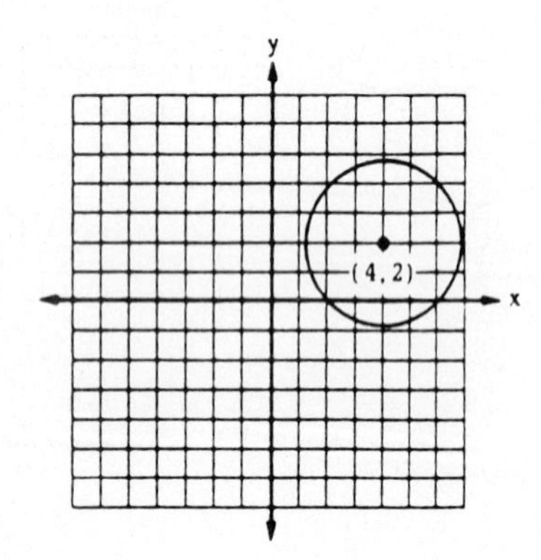


5. Rewrite the equation as

$$(x - 2)^2 + (y - (-3))^2 = 5^2.$$

Number 5 (Continued)

Center (2, -3), Radius is 5.

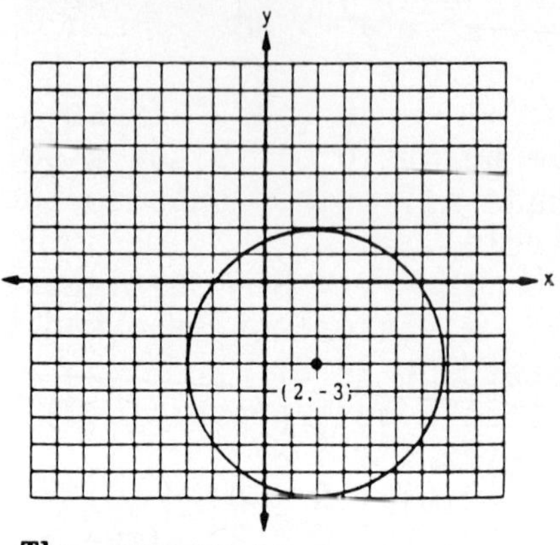


7. The center is (-1, 1).

   Thus, h has a value of -1 and k is 1.

   The radius is 2.

$$(x - h)^2 + (y - k)^2 = r^2$$
$$(x - (-1))^2 + (y - 1)^2 = 2^2$$
$$(x + 1)^2 + (y - 1)^2 = 4$$

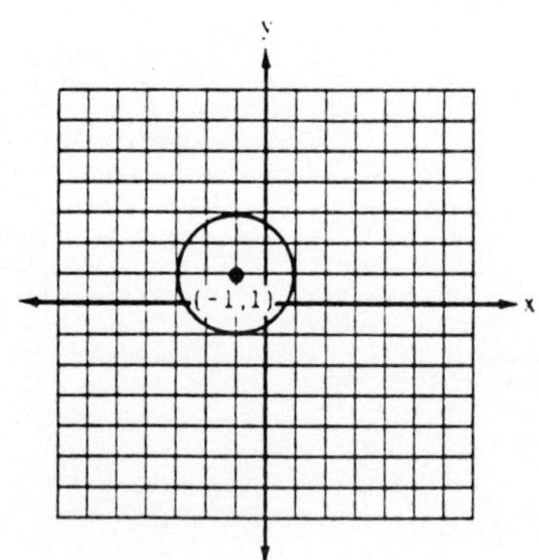


9. $x^2 + y^2 - 4y = 0$

$$x^2 + y^2 - 4y + 4 = 4$$
$$x^2 + (y - 2)^2 = 2^2$$

The center is (0,2).

25. $y = x^2 - 2x - 3$

$y = x^2 - 2x + 1 - 1 - 3$

$y = (x - 1)^2 - 4$

The graph opens upward.

The vertex is at (1, -4).

When x = 0, the y intercept is -3.
When y = 0, the x intercepts are -1 and 3.

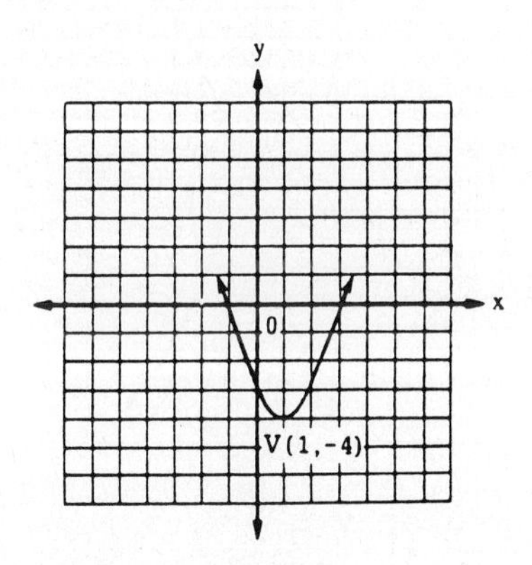


27. $x = y^2 + 5y + 4$

$x = y^2 + 5y + \frac{25}{4} - \frac{25}{4} + 4$

$x = (y + \frac{5}{2})^2 - \frac{9}{4}$

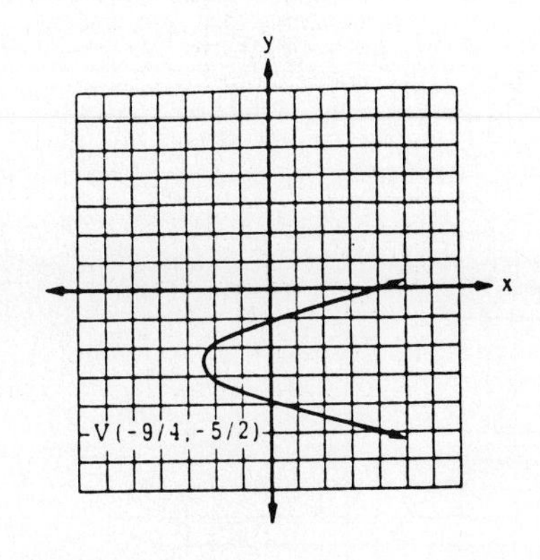


29. $\frac{x^2}{4} - \frac{y^2}{9} = 1$

The value is a is 2.
The value of b is 3.

Asymptotes are $y = \pm \frac{3}{2}x$.

The graph intersects the x axis and the vertices are at 2 and -2.

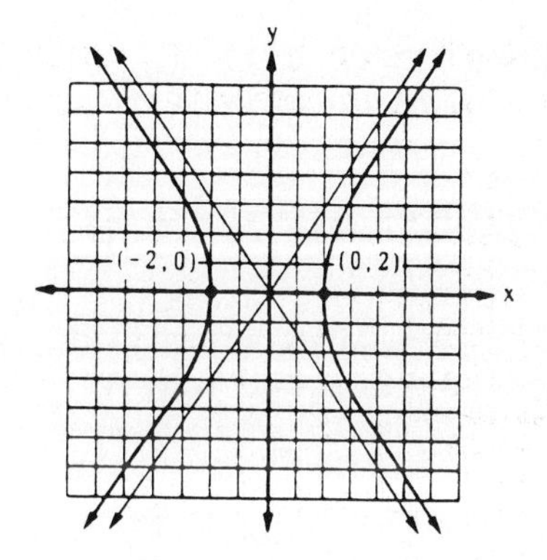


31. $\frac{y^2}{9} - \frac{x^2}{25} = 1$

The value of b is 3.
The value of a is 5.

Asymptotes are $y = \pm \frac{3}{5}x$.

The graph intersects the y axis and the vertices are at 3 and -3.

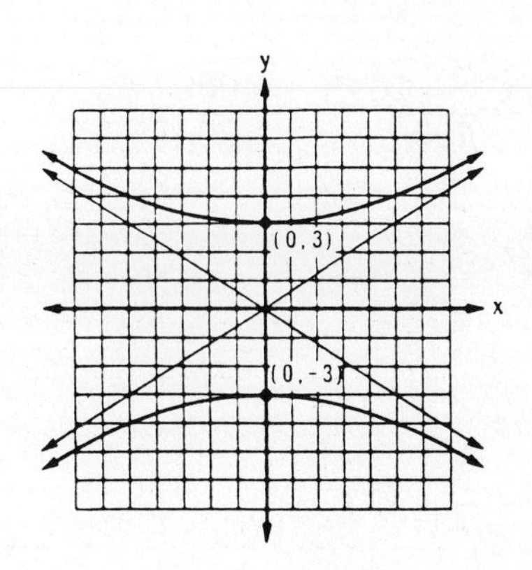

**33.** $9y^2 - 4x^2 = 36$

Divide each term by 36

$$\frac{9y^2 - 4x^2}{36} = \frac{36}{36}$$

$$\frac{y^2}{4} - \frac{x^2}{9} = 1$$

The value of b is 2.
The value of a is 3.

Asymptotes are $y = \pm \frac{2}{3}x$.

The graph intersects the y axis and the vertices are at 2 and -2.

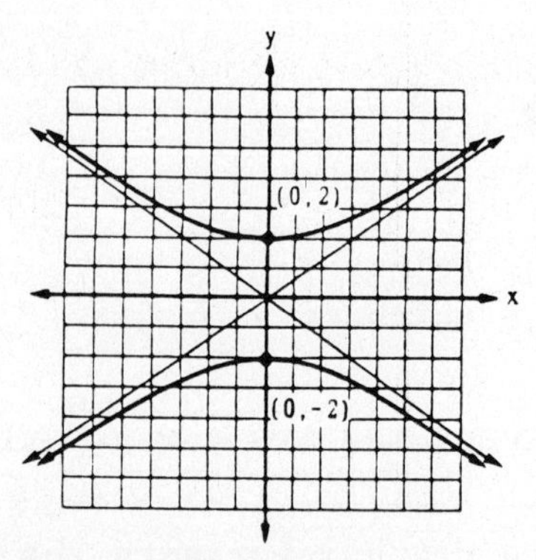


**35.** $25x^2 - 16y^2 = 400$
Divide each term by 400

$$\frac{25x^2 - 16y^2}{400} = \frac{400}{400}$$

$$\frac{x^2}{16} - \frac{y^2}{25} = 1$$

Number 35 (Continued)

The value of a is 4.
The value of b is 5.

Asymptotes are $y = \pm \frac{5}{4}x$.

The graph intersects the x axis and the vertices are at 4 and -4.

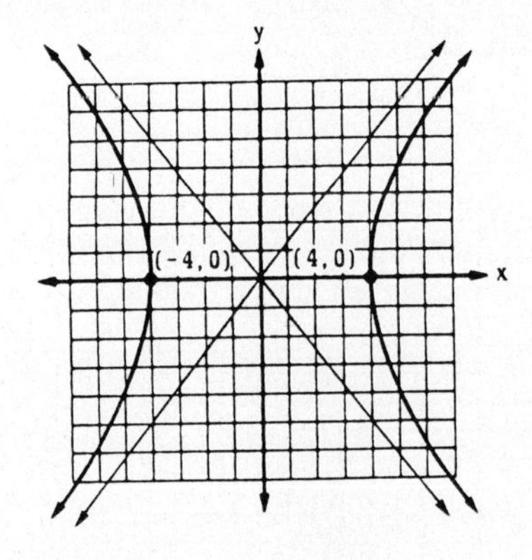


**37.** $xy = 6$ or $y = \frac{6}{x}$

| x | y |
|---|---|
| -4 | -3/2 |
| -3 | -2 |
| -2 | -3 |
| -1 | -6 |
| 1 | 6 |
| 2 | 3 |
| 3 | 2 |
| 4 | 3/2 |

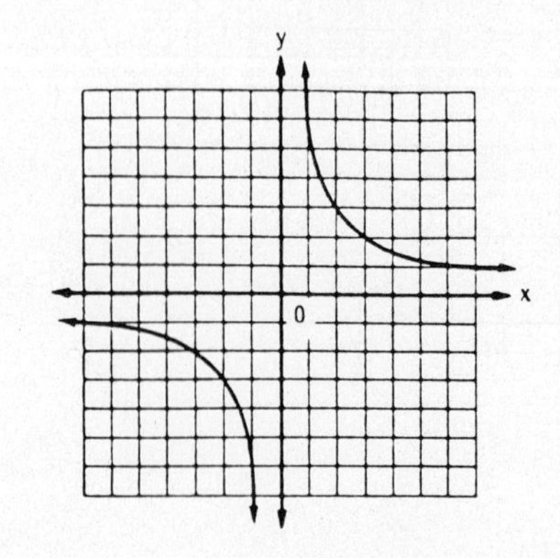

**39.** $y = \frac{3}{x}$

| x | y |
|---|---|
| -4 | -3/4 |
| -3 | -1 |
| -2 | -3/2 |
| -1 | -3 |
| 1 | 3 |
| 2 | 3/2 |
| 3 | 1 |
| 4 | 3/4 |

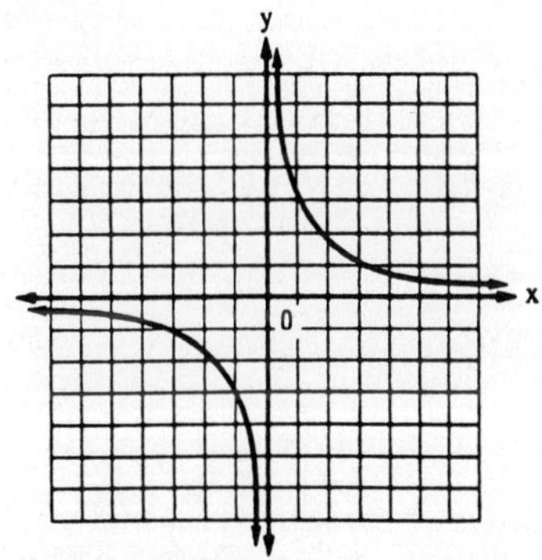


**41.** $\frac{x^2}{4} - \frac{y^2}{16} = 1 \rightarrow$ hyperbola

The value of $a$ is 2.
The value of $b$ is 4.

Asymptotes are $y = \pm 2x$.

The graph crosses the $x$ axis and the vertices are at 2 and -2.

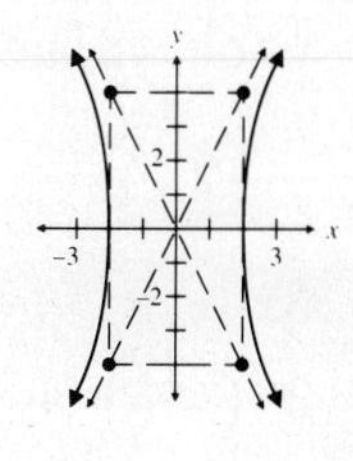

**43.** $(x - 4)^2 + (y + 2)^2 = 16$

$(x - 4)^2 + (y + 2)^2 = 4^2 \rightarrow$ circle

The center is (4, -2).
The radius is 4.

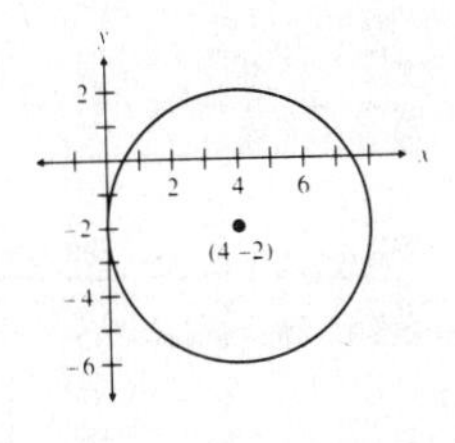


**45.** $\frac{x^2}{64} + \frac{y^2}{9} = 1 \rightarrow$ ellispse

Thus, $a = 8$ and the $x$ intercepts are $\pm$ 8.

Since $b = 3$, the $y$ intercepts are $\pm$ 3.

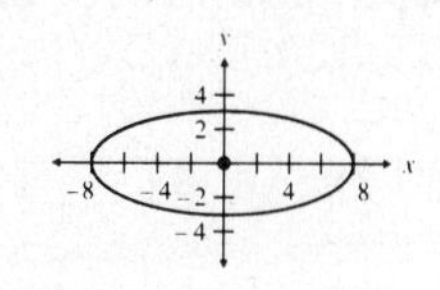

**47.** $4x^2 + 9y^2 = 36$

Divide each term by 36.

$$\frac{4x^2 + 9y^2}{36} = \frac{36}{36}$$

$$\frac{x^2}{9} + \frac{y^2}{4} = 1 \rightarrow \text{ellipse.}$$

Thus, $a = 3$ and the $x$ intercepts are $\pm 3$.

Since $b = 2$, the $y$ intercepts are $\pm 2$.

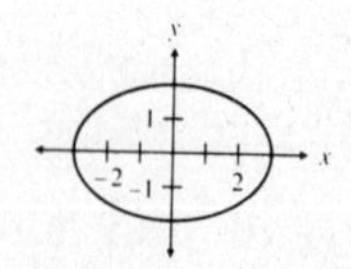

**49.** $x = -2(y + 3)^2 - 1 \rightarrow$ parabola.

The graph opens to the left since $a = -2$.

The vertex is at $(-1, -3)$.

When $y = 0$, the $x$ intercept is $-19$.

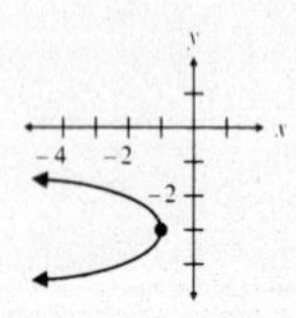

**51.** $x^2 - 4x + y^2 + 6y = 4 \rightarrow$ circle

$$(x^2 - 4x + 4) + (4^2 + 6y + 9) = -4 + 4 + 9$$

$$(x - 2)^2 + (y + 3)^2 = 9$$

Center: $(2, -3)$ Radius 3

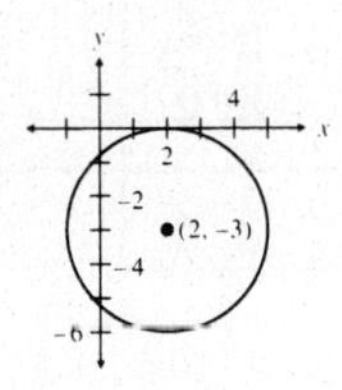


**53.** Substitute $y = 3x + 9$ in the equation.

$$x^2 + y^2 = 9$$
$$x^2 + (3x + 9)^2 = 9$$
$$x^2 + (3x + 9)(3x + 9) = 9$$
$$x^2 + 9x^2 + 27x + 27x + 81 = 9$$
$$10x^2 + 54x + 81 = 9$$
$$10x^2 + 54x + 72 = 0$$
$$5x^2 + 27x + 36 = 0$$
$$(5x + 12)(x + 3) = 0$$

$5x + 12 = 0$ or $x + 3 = 0$

$5x = -12$ $\qquad x = -3$

$x = -\frac{12}{5}$

Find the corresponding value of $y$ for each value of $x$.

| | |
|---|---|
| $x = -\frac{12}{5}$ | $x = -3$ |
| $x = 3x + 9$ | $y = 3x + 9$ |
| $y = 3\left(\frac{-12}{5}\right) + 9$ | $y = 3(-3) + 9$ |
| $y = \frac{9}{5}$ | $y = 0$ |

Solutions are $\left(\frac{-12}{5}, \frac{9}{5}\right)$ and $(-3, 0)$.

**55.** Solve $x^2 + y^2 = 4$ for $x^2$.

$$x^2 = 4 - y^2$$

Substitute $4 - y^2$ for $x^2$ in the equation $x^2 - y^2 = 4$ and solve for y.

$$x^2 - y^2 = 4$$
$$4 - y^2 - y^2 = 4$$
$$4 - 2y^2 = 4$$
$$-2y^2 = 0$$
$$y^2 = 0$$
$$y = 0$$

Number 55 (Continued)

Find the corresponding values of x.

$$y = 0$$
$$x^2 + y^2 = 4$$
$$x^2 + 0 = 4$$
$$x^2 = 4$$
$$x = \pm\sqrt{4}$$
$$x = \pm 2$$

Solutions are (2,0) and (-2,0).

**57.**
$$x^2 + y^2 = 16$$
$$\underline{x^2 - y^2 = 16}$$
$$2x^2 = 32$$
$$x^2 = 16$$
$$x = \pm\sqrt{16}$$
$$x = \pm 4$$

Now solve for y.

| $x = 4$ | $x = -4$ |
|---|---|
| $x^2 + y^2 = 16$ | $x^2 + y^2 = 16$ |
| $(4)^2 + y^2 = 16$ | $(-4)^2 + y^2 = 16$ |
| $16 + y^2 = 16$ | $16 + y^2 = 16$ |
| $y^2 = 0$ | $y^2 = 0$ |
| $y = 0$ | $y = 0$ |

Solutions are (4,0) and (-4,0).

**59.**
$$-4x^2 + y^2 = -12$$
$$8x^2 + 2y^2 = -8$$
$$-2[-4x^2 + y^2 = -12]$$
$$8x^2 + 2y^2 = -8$$

gives

$$8x^2 - 2y^2 = 24$$
$$\underline{8x^2 + 2y^2 = -8}$$
$$16x^2 = 16$$
$$x^2 = 1$$
$$x = \pm\sqrt{1}$$
$$x = \pm 1$$

Now solve for y.

| $x = 1$ | $x = -1$ |
|---|---|
| $8x^2 + 2y^2 = -8$ | $8x^2 + 2y^2 = -8$ |
| $8(1)^2 + 2y^2 = -8$ | $8(-1)^2 + 2y^2 = -8$ |
| $8 + 2y^2 = -8$ | $8 + 2y^2 = -8$ |
| $2y^2 = -16$ | $2y^2 = -16$ |
| $y^2 = -8$ | $y^2 = -8$ |
| $y = \pm\sqrt{-8}$ | $y = \pm\sqrt{-8}$ |
| $y = \pm\sqrt{4}\sqrt{2}\sqrt{-1}$ | $y = \pm\sqrt{4}\sqrt{2}\sqrt{-1}$ |
| $y = \pm 2i\sqrt{2}$ | $y = \pm 2i\sqrt{2}$ |

Since y is an imaginary number for both values of x, this system of equations has no real solution.

61. (1) $x^2 = y^2 - 9$

or $x^2 - y^2 = -9$

(2) $\underline{x^2 + y^2 = 41}$

$2x^2 = 32$

$x^2 = 16$

$x = \pm 4$

If $x = 4$ then

$4^2 = y^2 - 9$

$25 = y^2$

$y = \pm 5$

If $x = -4$ then

$(-4)^2 = y^2 - 9$

$25 = y^2$

$y = \pm 5$

The pairs of numbers are (4,5) (4,-5) (-4,5) and (-4,-5)

63. $x^2 + y^2 = 9 \rightarrow$ Circle (0,0)
Radius is 3.

Select a test point: (0,0)

$x^2 + y^2 < 9$

$0 + 0 < 9$

$0 < 9$ True

The point (0,0) is a solution, therefore shade inside the circle.

$2x + y = 6 \rightarrow$ straight line

The x intercept is 3 and the y intercept is 6.

Select a test point: (0,0)

$2x + y \geq 6$

$2(0) + 0 \geq 6$

$0 \geq 6$ False

The point (0,0) is not a solution.

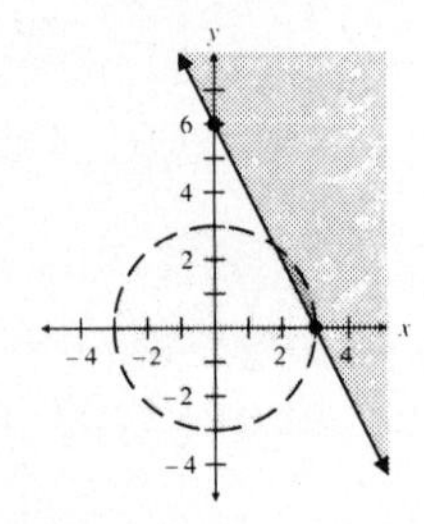


The overlapping region is that small part in the upper right side of the circle.

65. $4x^2 + 9y^2 = 36$

$\frac{4x^2 + 9y^2}{36} = \frac{36}{36}$

$\frac{x^2}{9} + \frac{y^2}{4} = 1 \rightarrow$ ellipse

$a = 3, b = 2$

Thus, the x intercepts are $\pm 3$ and the b intercepts are at $\pm 2$.

Number 65 (Continued)

Select a test point: (0,0)

$$\frac{x^2}{9} + \frac{y^2}{4} \leq 1$$

$$\frac{0}{9} + \frac{0}{4} \leq 1$$

$$0 \leq 1 \qquad \text{True}$$

The point (0,0) is a solution, therefore shade all points inside the ellipse.

$x^2 + y^2 > 25 \rightarrow$ Circle Center (0,0). Radius is 5.

Select a test point (0,0)

$$x^2 + y^2 > 25$$

$$0 + 0 > 25$$

$$0 > 25 \qquad \text{False}$$

The answer is NO real solution.

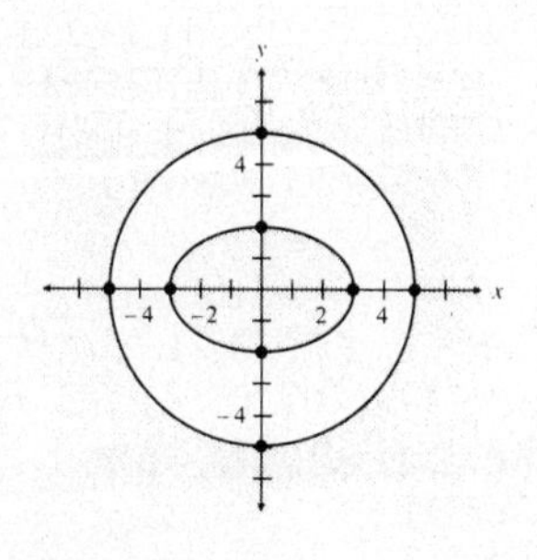

## PRACTICE TEST

1. The center is (-3, -1).

   Therefore, $h = -3$ and $k = -1$.

   The radius is 9.

$$(x - h)^2 + (y - k)^2 = r^2$$
$$(x - (-3))^2 + (y - (-1))^2 = 9^2$$
$$(x + 3)^2 + (y + 1)^2 = 9^2$$

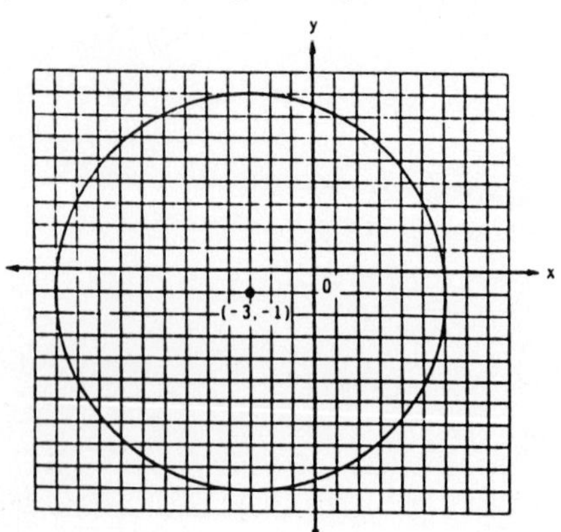


2. $x^2 + y^2 - 2x - 6y + 1 = 0$

$$(x^2 - 2x) + (y^2 - 6y) = -1$$
$$(x^2 - 2x + 1) + (y^2 - 6y + 9) = -1 + 1 + 9$$
$$(x - 1)^2 + (y - 3)^2 = 9$$
$$(x - 1)^2 + (y - 3)^2 = 3^2$$

Center: (1,3)
Radius: (3)

$$(x-1)^2 \div (y - 3)^2 = 3^2$$

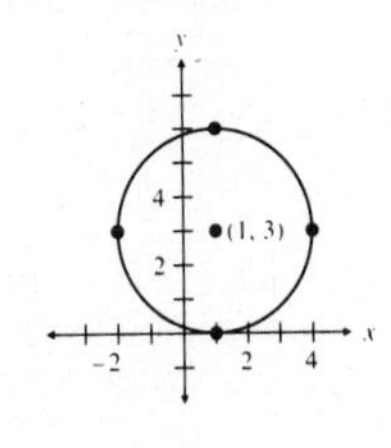


3. $9x^2 + 16y^2 = 144$

$$\frac{9x^2 + 16y^2}{144} = \frac{144}{144}$$
$$\frac{x^2}{16} + \frac{y^2}{9} = 1$$

Since $a^2 = 16$, $a = 4$ and the x intercepts are ± 4.

We know that $b^2 = 9$; thus $b = \pm 3$ and the y intercepts are ± 3.

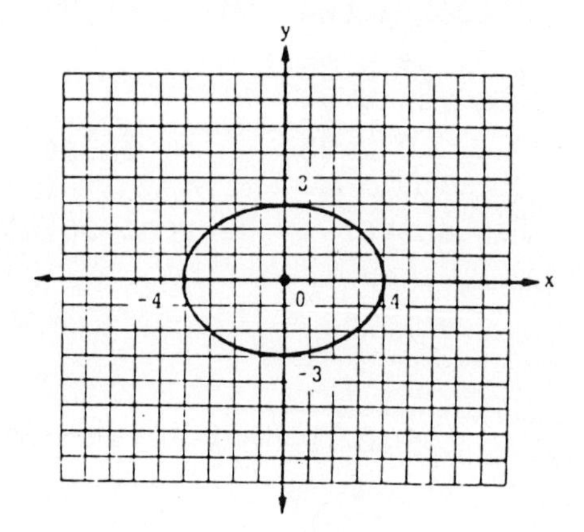


4. $y = -2\ (x - 3)^2 - 9$

   The parabola opens down since $a = -2$.

   The vertex is (3, -9).

Number 4 (Continued)

When $x = 0$, $y = -27$, so $-27$ is the $y$ - intercept.

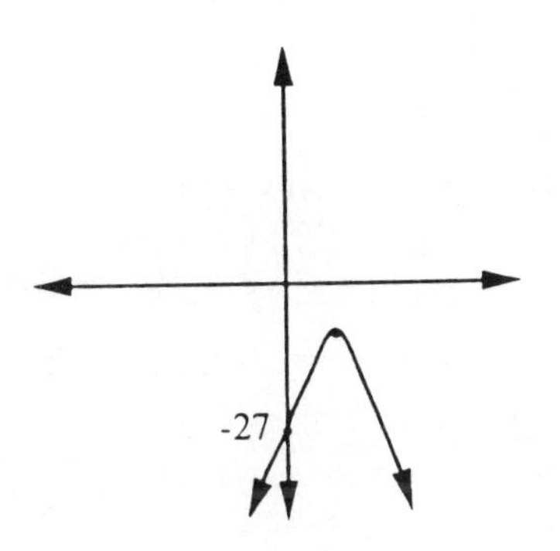


**6.** $y = \frac{8}{x}$

| x | y |
|---|---|
| -4 | -2 |
| -2 | -4 |
| -1 | -8 |
| 0 | undefined |
| 1 | 8 |
| 2 | 4 |
| 4 | 2 |

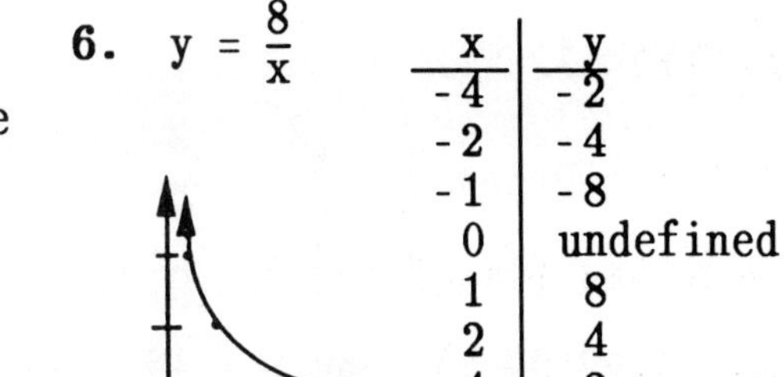

**5.** 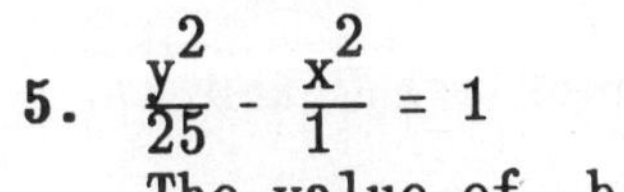

$$\frac{y^2}{25} - \frac{x^2}{1} = 1$$

The value of $b$ is 5.

The value of $a$ is 1.

Asymptotes are $y = \pm\ 5x$.

The graph intersects the $y$ axis and the vertices are at 5 and -5.

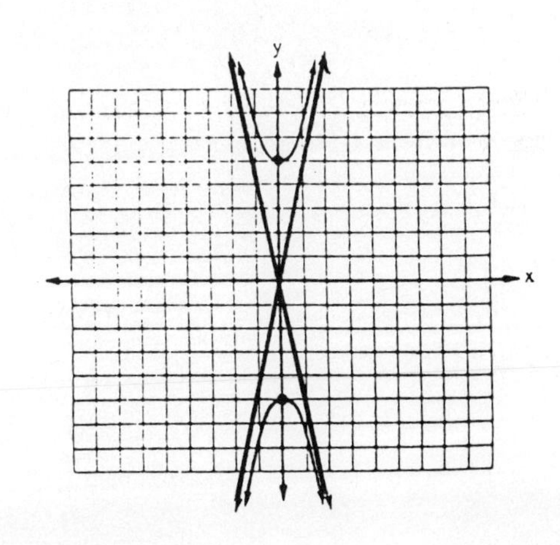


**7.** (1) $x^2 + y^2 = 16$

(2) $2x^2 - y^2 = 2$

Add equations (1) and (2)

(1) $x^2 + y^2 = 16$

(2) $2x^2 - y^2 = 2$

$3x^2 = 18$

$x^2 = 6$

$x = \pm\ \sqrt{6}$

If $x = \sqrt{6}$ then

$(\sqrt{6})^2 + y^2 = 16$

$y^2 = 10$

$y = \pm\ \sqrt{10}$

If $x = -\ \sqrt{6}$ then

$(-\sqrt{6})^2 + y^2 = 16$

$y^2 = 10$

$y = \pm\ \sqrt{10}$

Solutions: $(\sqrt{6}, \sqrt{10})$, $(\sqrt{6}, -\sqrt{10})$, $(-\sqrt{6}, \sqrt{10})$, $(-\sqrt{6}, -\sqrt{10})$

8. $\frac{x^2}{9} - \frac{y^2}{25} < 1$

Graph the hyperbola $\frac{x^2}{3^2} - \frac{y^2}{5^2} = 1$

Use dashed lines. Using (0,0) as a check point, $\frac{0}{9} - \frac{0}{25} < 1$ is true, therefore shade the region containing (0,0)

$$x^2 + y^2 \leq 4$$

Graph the circle $x^2 + y^2 = 2^2$ using a solid line. Using (0,0) as a check point,

$$0^2 + 0^2 \leq 4$$

is true. Therefore, shade the region insdie the circle which contains the point (0,0).

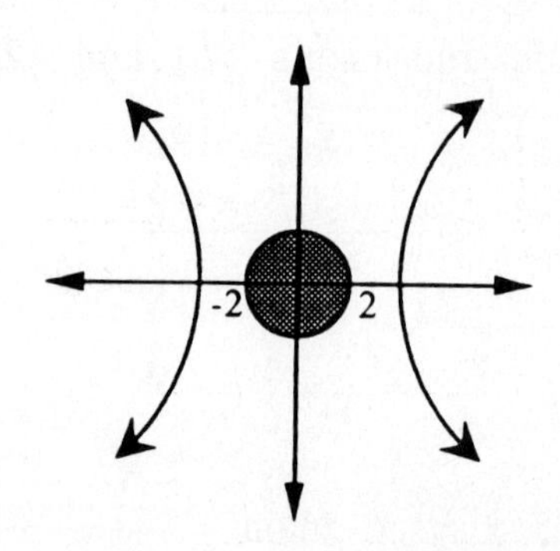


The solution lies where the shaded regions overlap. In this case, the solution is all points inside and including the circle $x^2 + y^2 = 4$.

9. (1) $xy = 54$
(2) $x + y = 29$
Solve equation (2) for x.
(2) $x = 29 - y$
(1) $(29 - y)(y) = 54$

$$29y - y^2 = 54$$
$$0 = y^2 - 29y + 54$$
$$0 = (y - 27)(y - 2)$$

$y - 27 = 0$ or $y - 2 = 0$

$y = 27$ $\qquad$ $y = 2$

If $y = 27$ then
$x(27) = 54$
$x = 2$

If $y = 2$ then
$x(2) = 54$
$x = 27$

The two numbers are 2 and 27.

**CUMULATIVE REVIEW TEST**

1. $2x + 3y = 12$

| x | y |
|---|---|
| 0 | 4 |
| 6 | 0 |

$x + 3y = 9$

| x | y |
|---|---|
| 0 | 3 |
| 9 | 0 |

Solution: (3,2)

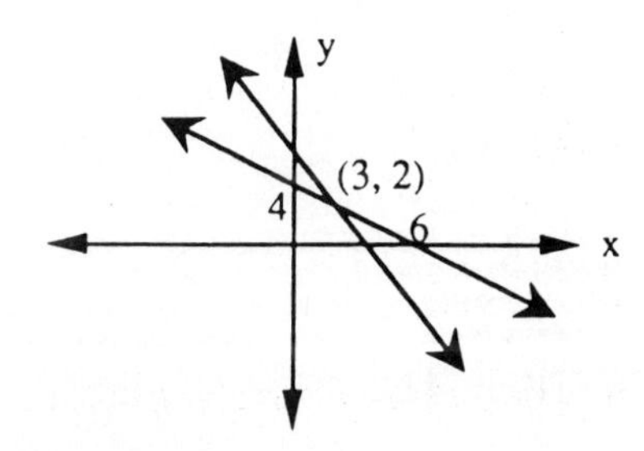


2. (1) $2x - y = 6$
(2) $3x + 2y = 4$
Multiply equation (1) by 2 and add the result to equation (2)
(1) $4x - 2y = 12$
(2) $3x + 2y = 4$

$7x = 16$
$x = 16/7$

Substitute $x = 16/7$ into equation (1) and (2) and solve for y.

$2(\frac{16}{7}) - y = 6$

$\frac{32}{7} - y = 6$

$32 - 7y = 42$

$-7y = 10$

$y = -10/7$

Solution: $(\frac{16}{7}, \frac{-10}{7})$

3. $(\frac{2x^{1/2}y^{2/3}}{x^2})^2 \; (\frac{3x^{1/3}y^{1/2}}{y^{-3/2}})$

$= [2x^{(1/2-2)}y^{2/3}]^2$

$[3x^{1/3}y^{1/2-(-3/2)}]$

$= [2x^{-3/2}y^{2/3}]^2 \; [3x^{1/3}y^2]$

$= (4x^{-3}y^{4/3}) \cdot (3x^{1/3}y^2)$

$= 12x^{-3+1/3} \; y^{4/3+2}$

$= 12x^{-8/3} \; y^{10/3} = \frac{12y^{10/3}}{x^{8/3}}$

4. $f(x) = x^2 + 2x + 5$

$f(a+3) = (a+3)^2 + 2(a+3) + 5$

$= a^2 + 6a + 9 + 2a + 6 + 5$

$= a^2 + 8a + 20$

5. $\frac{6x^2 + 5x - 4}{2x^2 - 3x + 1} \cdot \frac{4x^2 - 1}{8x^3 + 1}$

$= \frac{(3x + 4)(2x - 1)}{(x - 1)(2x - 1)} \cdot$

$\frac{(2x-1)(2x+1)}{(2x+1)(4x^2-2x+1)}$

$= \frac{(3x + 4)(2x - 1)}{(x - 1)(4x^2 - 2x + 1)}$

6. $\dfrac{x}{x + 3} - \dfrac{x + 1}{2x^2 - 2x - 24}$

$= \dfrac{x}{x - 3} - \dfrac{x + 1}{2(x + 3)(x - 4)}$

$= \dfrac{2(x - 4)\ x}{2(x + 3)(x - 4)} - \dfrac{x + 1}{2(x + 3)(x - 1)}$

$= \dfrac{2x^2 - 8x - x - 1}{2(x + 3)(x - 4)} =$

$\dfrac{2x^2 - 9x - 1}{2(x + 3)(x - 4)}$

7. $\dfrac{y + 1}{y + 3} + \dfrac{y - 3}{y - 2} = \dfrac{2y^2 - 15}{y^2 + y - 6}$

$\dfrac{y + 1}{y + 3} + \dfrac{y - 3}{y - 2} = \dfrac{2y^2 - 15}{(y + 3)(y - 2)}$

$\dfrac{(y+3)(y-2)(y+1)}{y+3} + \dfrac{(y-3)(y+3)(y-2)}{y-2}$

$= \dfrac{2y^2 - 15}{(y+5)(y-2)} \cdot (y+3)(y-2)$

$$\begin{aligned}
(y-2)(y+1) + (y-3)(y+3) &= 2y^2 - 15 \\
y^2 - y - 2 + y^2 - 9 &= 2y^2 - 15 \\
2y^2 - y - 11 &= 2y^2 - 15 \\
-y - 11 &= -15 \\
-y &= -4 \\
y &= 4
\end{aligned}$$

8. $\sqrt{\dfrac{12x5y^3}{8z}} = \sqrt{\dfrac{3x5y^3}{2z}} = \sqrt{\dfrac{3x5y^3}{\sqrt{2z}}\ \dfrac{\sqrt{2z}}{\sqrt{2z}}}$

$= \sqrt{\dfrac{6x5y^3z}{2z}} = \dfrac{x^2y\ \sqrt{6xyz}}{2z}$

9. $\dfrac{6(\sqrt{3} + \sqrt{5})}{(\sqrt{3} - \sqrt{5})(\sqrt{3} + \sqrt{5})}$

$= \dfrac{6\sqrt{3} + 6\sqrt{5}}{3 - 5} = \dfrac{6\sqrt{3} + 6\sqrt{5}}{-2}$

$= -3\sqrt{3} - 3\sqrt{5}$

10. $(3\ \sqrt[3]{2x + 2})^3 = (\sqrt[3]{80x} - 24)^3$

$$\begin{aligned}
27(2x + 2) &= 80x - 24 \\
54x + 54 &= 80x - 24 \\
78 &= 26x \\
3 &= x
\end{aligned}$$

11. $(5 - 4i)(5 + 4i)$

$= 25 - 16i^2$

$= 25 + 16 = 41$

12. $(x - 3)^2 = 28$

$$\begin{aligned}
\sqrt{(x - 3)^2} &= \sqrt{28} \\
x - 3 &= \pm 2\sqrt{7} \\
x &= 3 \pm 2\sqrt{7}
\end{aligned}$$

13. $3x^2 - 4x - 8 = 0$

$x = \dfrac{-(-4) \pm \sqrt{(-4)^2 - 4(3)(-8)}}{2(3)}$

$x = \dfrac{4 \pm \sqrt{112}}{6}$

$x = \dfrac{4 \pm 4\sqrt{7}}{6}$

$x = \dfrac{2(2 \pm 2\sqrt{7})}{2 \cdot 3}$

$x = \dfrac{2 \pm 2\sqrt{7}}{3}$

14. $\frac{3x - 2}{x + 4} \geq 0$

$3x - 2 = 0$ $\quad$ $x + 4 = 0$

$x = 2/3$ $\quad$ $x = -4$

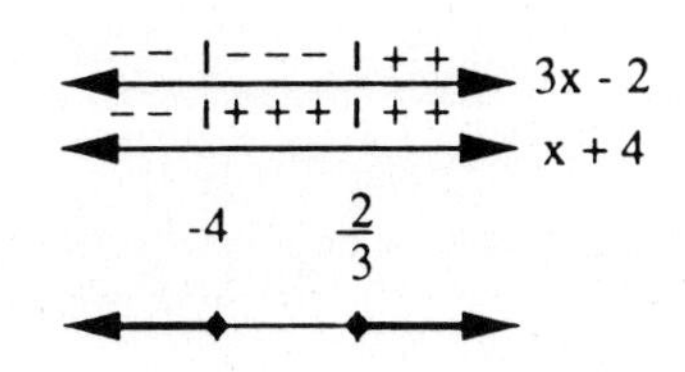


15. $y = x^2 - 4x + 4$

(a) Since a = 1, the parabola opens up.

(b) The y-intercept occurs when x = 0

$y = 0^2 - 4(0) + 4$

$y = 4$

(c) $x = \frac{-b}{2a} = \frac{-(-4)}{2(1)} = 2$

$y = 2^2 - 4(2) + 4$

$= 0$

Vertex: (2,0)

(d) The x-intercept is (2,0)

(e)

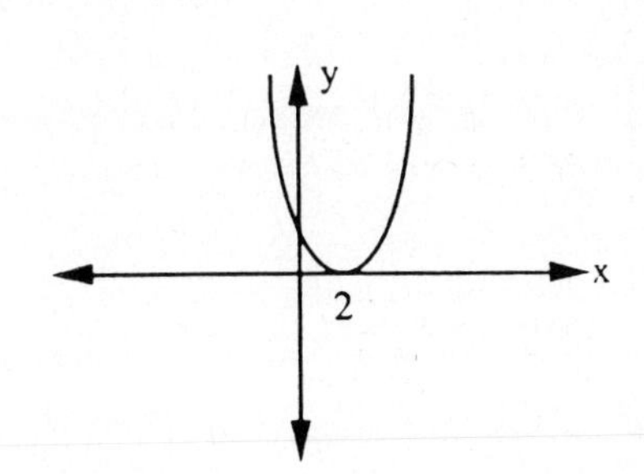


16. (a) $(f \circ g)(x) = (2x-3)^2 + 6(2x-3)$

$= 4x^2 - 12x + 9 + 12x - 18$

$= 4x^2 - 9$

(b) $(g \circ f)(x) = 2(x^2+6x) - 3$

$= 2x^2 + 12x - 3$

17. $9x^2 + 4y^2 = 36$

$\frac{9x^2 + 4y^2}{36} = \frac{36}{36}$

$\frac{x^2}{4} + \frac{y^2}{9} = 1$

$\frac{x^2}{(2)^2} + \frac{y^2}{(3)^2} = 1$

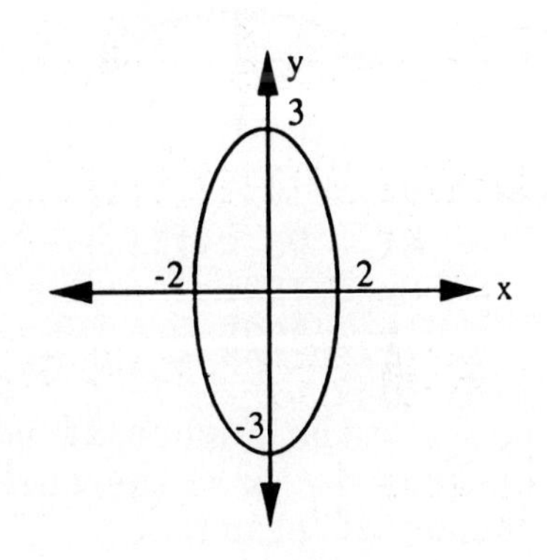

**18.** $\frac{y^2}{25} - \frac{x^2}{16} = 1$

$\frac{y^2}{5^2} - \frac{x^2}{4^2} = 1$

(a) Hyperbola - opens vertically

(b) Asymptotes: $y = \pm \frac{5}{4}x$

(c) y - intercepts: $\pm 5$

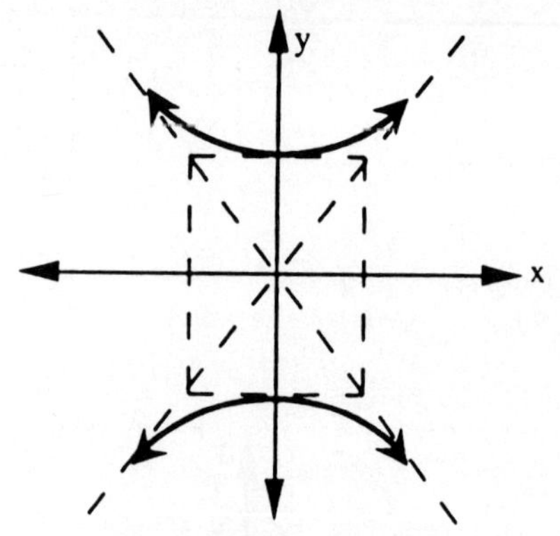


**19.** $x - .2x - [25] = 155$
$.8x - 25 = 155$
$.8x = 180$
$x = \$225$

**20.** Let x = the number of pounds of cashews 4 - x = the number of pounds of peanuts.

$7x + 5(4 - x) = 25$
$7x + 20 - 5x = 25$
$2x + 20 = 25$
$2x = 5$
$x = 2\ 1/2$
lbs of cashews

$4 - x = 1\frac{1}{2}$ lbs or 3/2 lbs of peanuts

## EXERCISE SET 11.1

1. $y = 2^x$

| x | -2 | -1 | 0 | 1 | 2 |
|---|---|---|---|---|---|
| y | 1/4 | 1/2 | 1 | 2 | 4 |

D: $\mathbb{R}$
R: $\{y|y > 0\}$

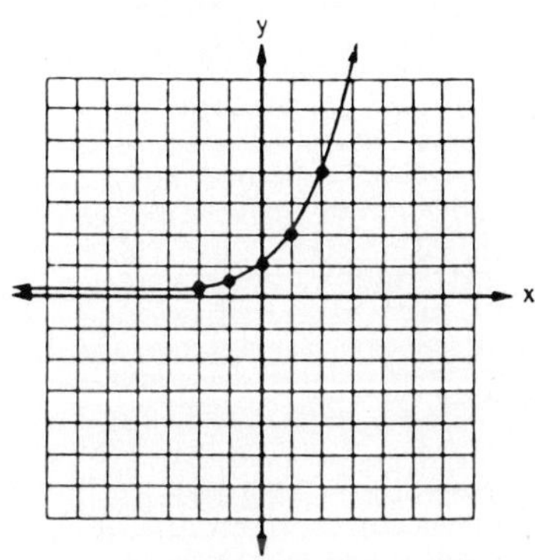


3. $y = (1/2)^x$

| x | -2 | -1 | 0 | 1 | 2 |
|---|---|---|---|---|---|
| y | 4 | 2 | 1 | 1/2 | 1/4 |

D: $\mathbb{R}$
R: $\{y|y > 0\}$

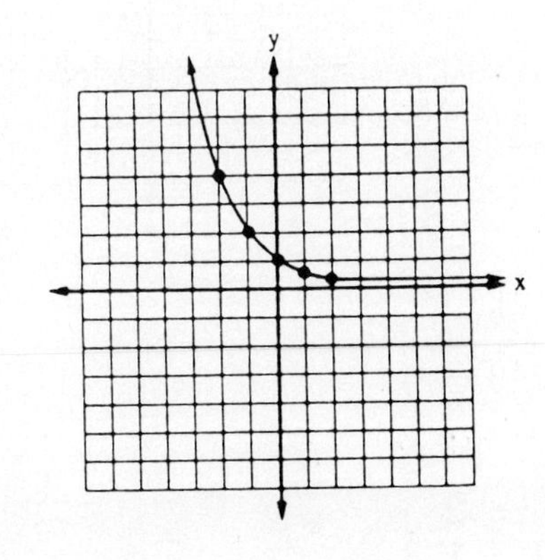


5. $y = 4^x$

| x | -2 | -1 | 0 | 1 | 2 |
|---|---|---|---|---|---|
| y | 1/16 | 1/4 | 1 | 4 | 16 |

D: $\mathbb{R}$
R: $\{y|y > 0\}$

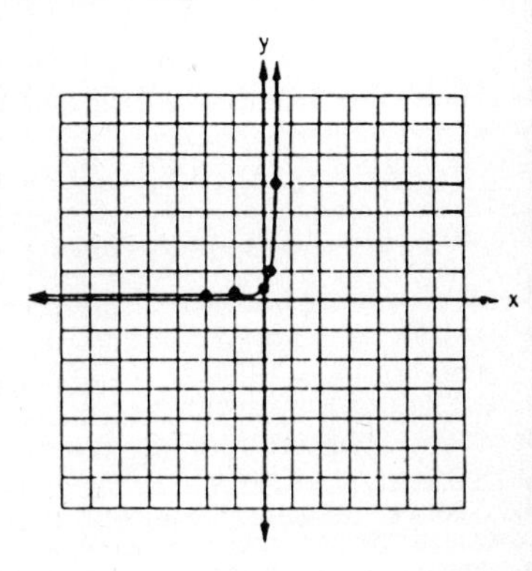


7. $y = \log_2 x$

$x = 2^y$

| x | 1/4 | 1/2 | 1 | 2 | 4 |
|---|---|---|---|---|---|
| y | -2 | -1 | 0 | 1 | 2 |

D: $(x|x > 0)$
R: $\mathbb{R}$

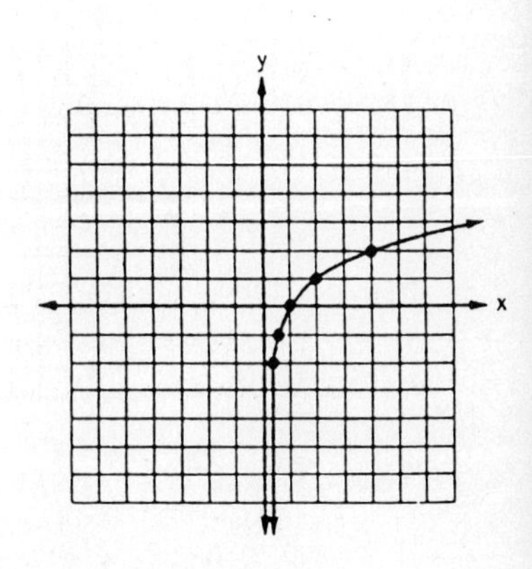


9. $y = \log_{1/2} x$

$x = (1/2)^y$

| x | 4 | 2 | 0 | 1/2 | 1/4 |
|---|---|---|---|---|---|
| y | -2 | -1 | 0 | 1 | 2 |

Number 9 (Continued)

D: $\{x|x > 0\}$
R: $\mathbb{R}$

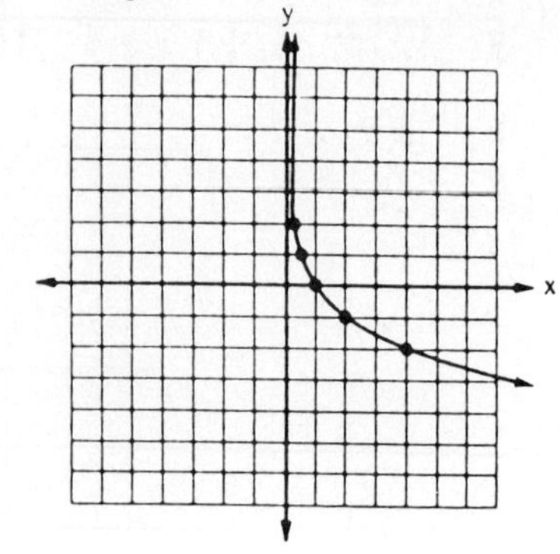


**11.** $y = \log_5 x$

$x = 5^y$

| x | 1/25 | 1/5 | 1 | 5 | 25 |
|---|---|---|---|---|---|
| y | -2 | -1 | 0 | 1 | 2 |

D: $\{x|x > 0\}$
R: $\mathbb{R}$

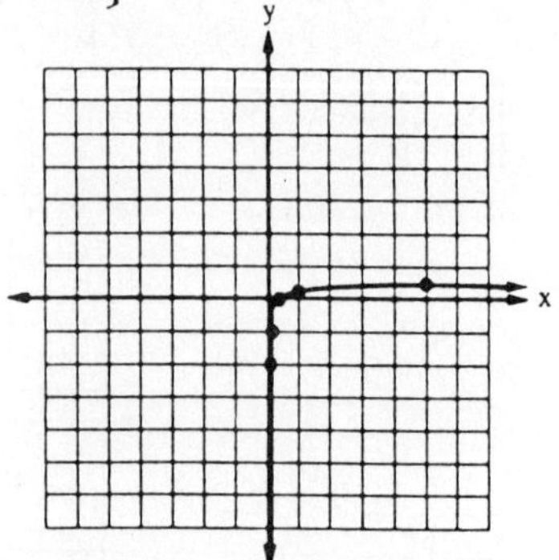


**13.** $y = 2^x$

| x | -2 | -1 | 0 | 1 | 2 |
|---|---|---|---|---|---|
| y | 1/4 | 1/2 | 1 | 2 | 4 |

$y = \log_{1/2} x$ can be written as $x = (1/2)^y$

| x | 4 | 2 | 1 | 1/2 | 1/4 |
|---|---|---|---|---|---|
| y | -2 | -1 | 0 | 1 | 2 |

Number 13 (Continued)

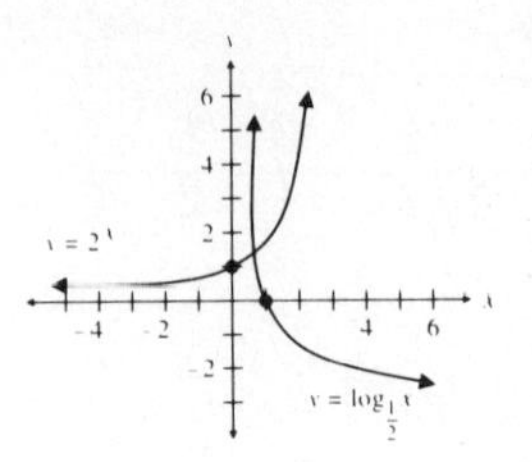

**15.** $y = 2^x$

| x | -2 | -1 | 0 | 1 | 2 |
|---|---|---|---|---|---|
| y | 1/4 | 1/2 | 1 | 2 | 4 |

$y = \log_2 x$

$x = 2^y$

| x | 1/4 | 1/2 | 1 | 2 | 4 |
|---|---|---|---|---|---|
| y | -2 | -1 | 0 | 1 | 2 |

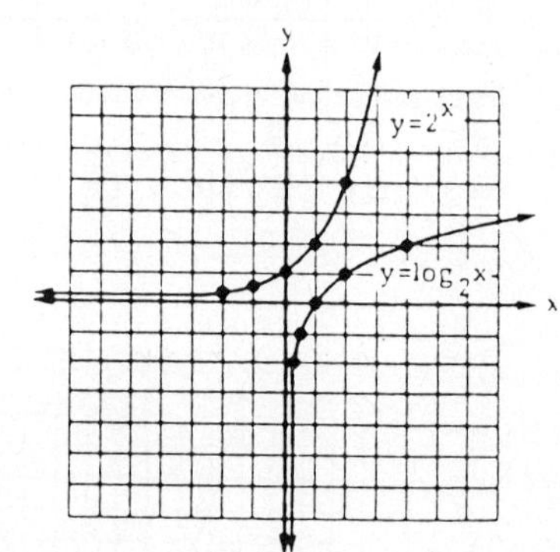


**17.** $2^3 = 8$
$\log_2 8 = 3$

**19.** $3^5 = 243$
$\log_3 243 = 5$

**21.** $8^{1/3} = 2$
$\log_8 2 = 1/3$

**23.** $(1/4)^2 = 1/16$
$\log_{1/4} 1/16 = 2$

**25.** $5^{-2} = 1/25$
$\log_5(1/25) = -2$

**27.** $4^{-3} = 1/64$
$\log_4 (1/64) = -3$

**29.** $16^{-1/2} = 1/4$
$\log_{16}(1/4) = -1/2$

**31.** $8^{-1/3} = 1/2$
$\log_8(1/2) = -1/3$

**33.** $\log_2 8 = 3$
$2^3 = 8$

**35.** $\log_4 64 = 3$
$4^3 = 64$

**37.** $\log_{1/2} (1/16) = 4$
$(1/2)^4 = 1/16$

**39.** $\log_5(1/125) = -3$
$5^{-3} = 1/125$

**41.** $\log_{125} 5 = 1/3$
$125^{1/3} = 5$

**43.** $\log_{27}(1/3) = -1/3$
$27^{-1/3} = 1/3$

**45.** $\log_{10} 1000 = 3$
$10^3 = 1000$

**47.** $y = \log_6 36$
$6^y = 36$
Since $6^2 = 36$, $y = 2$.

**49.** $3 = \log_2 x$
$2^3 = x$
Since $2^3 = 8$, $x = 8$.

**51.** $y = \log_{64} 8$
$64^y = 8$
Since $64^{1/2} = 8$, $y = 1/2$.

**53.** $3 = \log_a 64$
$a^3 = 64$
Since $4^3 = 64$, $a = 4$.

**55.** $4 = \log_{1/2} x$
$(1/2)^4 = x$
Since $(1/2)^4 = 1/16$, $x = 1/16$.

**57.** $5 = \log_a 32$
$a^5 = 32$
Since $2^5 = 32$, $a = 2$.

**59.** $1/3 = \log_8 x$
$8^{1/3} = x$
Since $8^{1/3} = 2$, $x = 2$.

**61.** $g = 2^n \quad n = 8$
$g = 2^8$
$g = 256$ gametes

63. $A = p(1 + r)^n$

$p = \$8000$

$r = \frac{12\%}{4} = 3\%$, and

$n = 5.4 = 20$

$A = 8000(11 + .03)^{20}$

$A = 8000(1.03)^{20}$

$A = 8000(1.8061112)$

$A = \$14448.89$

65. $A = A_0 \cdot 2^{-t/5600}$

$A_0 = 60$, $t = 10,000$

$A = 60 \cdot 2^{-10,000/5600}$

$A = 60 \cdot 2^{-1.7857143}$

$A = 60(.2900323)$

$A \approx 17.4$ grams

67. $y = 80(2)^{-0.4t}$

(a) $t = 10$

$y = 80(2)^{-.4(10)}$

$y = 80(2)^{-4} - 80(1/16)$

$y = 5$ grams

(b) $t = 100$

$y = 80(2)^{-.4(100)}$

$y = 80(2)^{-40}$

$y = 80(9.094947 \times 10^{-13})$

$y \approx 7.28 \times 10^{-11}$ grams

69. $R = \log_{10} I \quad R = 7$

$7 = \log_{10} I$

$10^7 = I$

$10,000,000 = I$

71. $y = \log_a x$

(a) The base a must be positive and must not be equal to one.

(b) The argument x represents a number that is greater than 0.

(c) $\mathbb{R}$

73. (a) The graph looks like a horizontal line through y = 1.

(b) Yes. A horizontal line will pass the vertical line test.

(c) No. f(x) is not 1-1 and therefore does not have an inverse function.

75. The functions $y = a^x$ and $y = \log_a x$ for $a \neq 1$ are inverses of each other and are symmetric with respect to the line $y = x$. For each ordered pair (x,y) on the graph of $y = a^x$, the ordered pair (y,x) is on the graph of $y = \log_a x$.

77. $y = a^x + k$ will have the same shape as the graph $y = a^x$. However, $y = a^x + k$ will be k units higher or lower than $y = a^x$. If $k > 0$ the graph will be shifted k units up and if $k < 0$, the graph will be lowered k units. The y-intercept of $y = a^x + k$ will be $(0, 1 + k)$.

79. $24x^2 - 6xy + 16\,xy - 4y^2$

$= 2[12x^2 - 3xy + 8xy - 2y^2]$

$= 2[3x(4x-y) + 2y(4x-y)]$

$= 2(3x+2y)(4x-y)$

81. $4x^4 - 36x^2$

$= 4x^2(x^2 - 9)$

$= 4x^2(x+3)(x-3)$

## EXERCISE SET 11.2

1. $\log_3 7 \cdot 12$
$= \log_3 7 + \log_3 12$

3. $\log_8 7(x + 3)$
$= \log_8 7 + \log_8 (x + 3)$

5. $\log_4 15/7$
$= \log_4 15 - \log_4 7$

7. $\log_{10} (\frac{\sqrt{x}}{x - 3})$
$= \log_{10} (\frac{x^{1/2}}{x - 3})$
$\log_{10} x^{1/2} - \log_{10} (x - 3)$
$= 1/2 \log_{10} x - \log_{10} (x - 3)$

9. $\log_8 x^4 = 4 \log_8 x$

11. $\log_{10} 3(8^2)$
$= \log_{10} 3 + \log_{10} 8^2$
$= \log_{10} 3 + 2 \log_{10} 8$

13. $\log_4 \sqrt{\frac{x^5}{x + 4}}$
$= \log_4 (\frac{x^5}{x + 4})^{1/2}$
$= 1/2 [\log_4 (\frac{x^5}{x + 4})]$
$= 1/2 [\log_4 x^5 - \log_4(x + 4)]$
$= 1/2 [5 \log_4 x - \log_4(x + 4)]$

15. $\log_{10}[\frac{x^4}{(x + 2)^3}] = \log_{10} x^4$
$- \log_{10} (x + 2)^3$
$= 4\log_{10} x - 3\log_{10} (x + 2)$

17. $\log_8[\frac{x(x - 6)}{x^3}] = \log_8 x +$
$\log_8 (x - 6) - \log_8 x^3$
$= \log_8 x + \log_8 (x - 6) - 3 \log_8 x$
$= -2\log_8 x + \log_8 (x - 6)$

19. $\log_{10} (\frac{2x}{3}) = \log_{10} 2x - \log_{10} 3$
$= \log_{10} 2 + \log_{10} x - \log_{10} 3$
$= \log_{10} 2 + 2\log_{10} x - \log_{10} 3$

21. $2 \log_{10} x - \log_{10}(x - 2)$
$= \log_{10} x^2 - \log_{10} (x - 2)$
$= \log_{10}[\frac{x^2}{x - 2}]$

23. $2(\log_5 x - \log_5 4) = 2\log_5 (\frac{x}{4})$
$= \log_5 (\frac{x}{4})^2$

25. $[\log_{10} x + \log_{10}(x - 4)] - \log_{10}(x + 1)$
$= \log_{10}[x(x - 4)] - \log_{10}(x + 1)$
$= \log_{10} [\frac{x(x - 4)}{x + 1}]$

27. $1/2[\log_7 (x - 2) - \log_7 x]$
$= 1/2 \log_7 [\frac{x - 2}{x}]$
$= \log_7 [\frac{x - 2}{x}]^{1/2}$

29. $(2\log_9 5 - 4\log_9 6) + \log_9 3$

$= (\log_9 5^2 - \log_9 6^4)$
$+ \log_9 3$

$= \log_9 \left(\frac{5^2}{6^4}\right) + \log_9 3$

$= \log_9 \left[\left(\frac{5^2}{6^4}\right) \cdot (3)\right]$

31. $4\log_6 3 - [2\log_6 (x+3)$
$+ 4\log_6 x]$

$= \log_6 3^4 - [\log_6 (x+3)^2 + \log_6 x^4]$

$= \log_6 \left[\frac{3^4}{(x+3)^2 \cdot x^4}\right]$

33. $\log_{10} 10 = \log_{10} (2)(5)$
$= \log_{10} 2 + \log_{10}(5)$
$= .3010 + 6990$
$= 1$

35. $\log_{10} 2.5 = \log_{10} \left(\frac{5}{2}\right)$
$= \log_{10} 5 - \log_{10} 2$
$= .6990 - .3010$
$= .398$

37. $\log_{10} 25 = \log_{10} 5^2$
$= 2(\log_{10} 5)$
$= 2(.6990)$
$= 1.3980$

39. False
$\log_{10} 3 + \log_{10} 4 = \log_{10} 3(4)$
$= \log_{10} 12$

41. True

43. False
$\frac{\log_5 8}{\log_5 2} = \log_2 8$
(Change of base formula)

45. False
$2\log_{10} 10 = \log_{10}(10)^2$

47. Varies from student to student

49. Varies from student to student

51. $\frac{2x+5}{x^2 - 7x + 12} - \frac{x-4}{2x^2 - x - 15}$

$= \frac{2x+5}{(x-4)(x-3)} - \frac{(x-4)}{(2x+5)(x-3)}$

$= \frac{(2x+5)(2x+5)}{(x-4)(x-3)(2x+5)} -$

$\frac{(x-4)(x-4)}{(2x+5)(x-3)(x-4)}$

$= \frac{(4x^2 + 20x + 25) - (x^2 - 8x + 16)}{(x-4)(x-3)(2x+5)}$

$= \frac{4x^2 + 20x + 25 - x^2 + 8x - 16}{(x-4)(x-3)(2x+5)}$

$= \frac{3x^2 + 28x + 9}{(x-4)(x-3)(2x+5)}$

53. $\sqrt[3]{4x^4y^7} \cdot \sqrt[3]{12x^7y^{10}}$

$= \sqrt[3]{48x^{11}y^{17}}$

$= 2x^3y^5 \sqrt{6x^2y^2}$

**EXERCISE SET 11.3**

**1.** $870 = 8.7 \times 10^2$

$\log 870 = \log 8.7 + \log 10^2$
$= .9395 + 2$
$= 2.9395$

**3.** $8 = 8.0 \times 10^0$

$\log 8 = \log 8.0 + \log 10^0$
$= .9031 + 0$
$= 0.9031$

**5.** $1000 = 1.0 \times 10^3$

$\log 1000 = \log 1.0 + \log 10^3$
$= 0.0000 + 3$
$= 3.0000$

**7.** $0.0000857 = 8.57 \times 10^{-5}$

$\log 0.0000857 = \log 8.57 + \log 10^{-5}$
$= .9330 - 5$
$= 5.9330 - 4.0670$

**9.** $100 = 1.0 \times 10^2$

$\log 100 = \log 1.0 + \log 10^2$
$= 0.0000 + 2$
$= 2.0000$

**11.** $1.74 = 1.74 \times 10^0$

$\log 1.74 = \log 1.74 + \log 10^0$
$= .2405 + 0$
$= 0.2405$

**13.** $.375 = 3.75 \times 10^{-1}$

$\log .375 = \log 3.75 + \log 10^{-1}$
$= .5740 - 1$
$= 9.5740 - .4260$

**15.** $.00872 = 8.72 \times 10^{-3}$

$\log .00872 = \log 8.72 + \log 10^{-3}$
$= .9405 - 3$
$= 7.9405 - 2.0595$

**17.** antilog 0.5416 = 3.48

**19.** antilog 2.3201 = 209

**21.** antilog (-1.0585) = .0874

**23.** antilog 0.0000 = 1.00

**25.** antilog 2.5011 = 317

**27.** antilog (-.1543) = .701

**29.** $\log N = 2.0000$
$N = \text{antilog } 2.0000$
$N = 100$

**31.** $\log N = -3.104$
$N = \text{antilog } (-3.104)$
$N = .000787$

**33.** $\log N = 3.8202$
$N = \text{antilog } 3.8202$
$N = 6610$

**35.** $\log N = -1.06$
$N = \text{antilog } (-1.06)$
$N = .0871$

**37.** $\log N = -.3686$
$N = \text{antilog } (-.3686)$
$N = 0.428$

**39.** $\log N = -.3936$
$N = \text{antilog } (-.3936)$
$N = .404$

**41.** $\log 2370 = 3.3747$
Therefore, $10^{3.3747} \approx 2370$

**43.** $\log .0410 = -1.3872$
Therefore $10^{-1.3872} \approx .0410$.

**45.** $\log 102 = 2.0086.$
Therefore $10^{2.0086} \approx 102$

**47.** $\log .00128 = -2.8928.$
Therefore $10^{-2.8928} \approx .00128.$

**49.** $10^{2.5866} = 386.0$

**51.** $10^{-.158} = 0.695$

**53.** $10^{-1.6091} = .0246$

**55.** $10^{1.1903} = 15.5$

**57.**
$\log 1 = x$
$10^x = 1$
$x = 0$
So, $\log 1 = 0$

**59.**
$\log 0.1 = x$
$10^x = .1$
$10^x = \frac{1}{10}$
$10^x = 10^{-1}$
$x = -1.$
Therefore $\log 0.1 = -1$

**61.**
$\log .01 = x$
$10^x = .01$
$10^x = \frac{1}{100}$
$10^x = 10^{-2}$
$x = -2$
Therefore $\log .01 = -2$

**63.**
$\log .001 = x$
$10^x = .001$
$10^x = \frac{1}{1000}$
$10^x = 10^{-3}$
$x = -3.$
Therefore, $\log .001 = -3.$

**65.** $R = \log I$
(a) $R = \log 12000 = 4.08$
(b) $4.29 = \log I$
$10^{4.29} = I$
$I = 19{,}500$

**67.**
$\log E = 11.8 + 1.5\, m_s$
(a) $\log E = 11.8 + 1.5\,(6)$
$\log E = 20.8$
$10^{20.8} = E$
$E = 6.31 \times 10^{20}$
(b) $\log (1.2 \times 10^{15}) = 11.8 + 1.5\, m_s$
$15.07918125 = 11.8 + 1.5\, m_s$
$3.27918125 = 1.5\, m_s$
$m_s = 2.19$

**69.**
$pH = -\log[H_3O^+]$
$pH = -\log[2.8 \times 10^{-3}]$
$pH = -\log[.0028]$
$pH = -(-2.55)$
$pH = 2.55$

**71.** No. $10^3 = 1000$ so $\log_{10} 1000 = 3.$
Therefore log 6250 must be greater than 3 since 6250 is greater than 1000.

**73.** No. $10^{-2} = .01$ and $\log_{10} .01 = -2.$
$10^{-3} = .001$ and $\log_{10} .001 = -3.$
Since .0024 falls between .01 and .001 the log .0024 must fall between -2 and -3.

**75.** $-3x^2 - 4x - 8 = 0$

$$x = \frac{-(-4) \pm \sqrt{(-4)^2 - 4(-3)(-8)}}{2(-3)}$$

$$x = \frac{4 \pm \sqrt{-80}}{-6}$$

$$x = \frac{-2 \pm 2i\sqrt{5}}{3}$$

**77.** $\dfrac{2x - 3}{5x + 10} < 0.$

Find the critical values.

$2x - 3 = 0 \quad 5x + 10 = 0$

$x = 3/2 \qquad x = -2$

-- | --- | ++   2x - 3

-- | +++ | ++   5x + 10

-2   $\frac{3}{2}$

The solution lies where there is an odd number of negative signs.

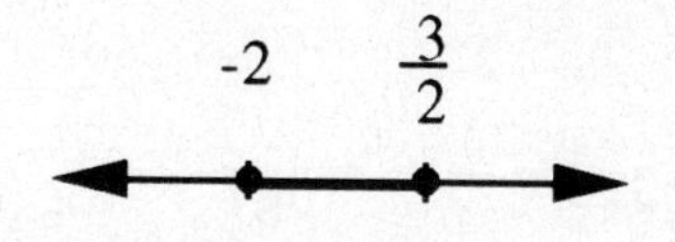

## EXERCISE SET 11.4

**1.** $3^x = 243$
$3^x = 3^5$
$x = 5$

**3.** $5^x = 125$
$5^x = 5^3$
$x = 3$

**5.** $1.05^x = 15$
$\log (1.05)^x = \log 15$
$x \log 1.05 = \log 15$
$x = \frac{\log 15}{\log 1.05}$

$x = \frac{1.1761}{0.0212}$
$x \approx 55.50$

**7.** $1.63^{x+1} = 25$
$\log (1.63)^{x+1} = \log 25$
$(x + 1) \log 1.63 = \log 25$
$x + 1 = \frac{\log 25}{\log 1.63}$
$x + 1 = \frac{1.3979}{.2122}$
$x + 1 = 6.59$
$x = 5.59$

**9.** $\log_4 (x + 1)^3 = 3$
$(x + 1)^3 = 4^3$
$x + 1 = 4$
$x = 3$

**11.** $\log_2 (x + 4)^2 = 4$
$(x + 4)^2 = 2^4$
$x^2 + 8x + 16 = 16$
$x^2 + 8x = 0$
$x(x + 8) = 0$
$x = 0$ or $x + 8 = 0$
$x = -8$

**13.** $\log(2x - 3)^3 = 3$
$3 \log(2x - 3) = 3$
$\log(2x - 3) = 1$
$2x - 3 = \text{antilog } 1$
$2x - 3 = 10$
$2x = 13$
$x = \frac{13}{2}$

**15.** $\log(x + 2) = \log(3x - 1)$
$\log(x + 2) - \log(3x - 1) = 0$
$\log[\frac{x + 2}{3x - 1}] = 0$
$\frac{x + 2}{3x - 1} = \text{antilog } 0$
$\frac{x + 2}{3x - 1} = 1$
$x + 2 = 3x - 1$
$3 = 2x$
$x = \frac{3}{2}$

**17.** $\log(3x - 1) + \log 4 = \log(9x + 2)$
$\log(3x - 1) + \log 4 - \log(9x + 2) = 0$
$\log [\frac{4(3x - 1)}{9x + 2}] = 0$
$\frac{4(3x - 1)}{9x + 2} = \text{antilog } 0$
$\frac{4(3x - 1)}{9x + 2} = 1$
$12x - 4 = 9x + 2$
$3x = 6$
$x = 2$

**19.** $\log x + \log(3x - 5) = \log 2$

$$\log x + \log(3x - 5) - \log 2 = 0$$
$$\log\left[\frac{x(3x - 5)}{2}\right] = 0$$
$$\frac{x(3x - 5)}{2} = \text{antilog } 0$$
$$\frac{x(3x - 5)}{2} = 1$$
$$3x^2 - 5x = 2$$
$$3x^2 - 5x - 2 = 0$$
$$(3x + 1)(x - 2) = 0$$

$3x + 1 = 0$ or $x - 2 = 0$

$3x = -1$ $\quad$ $x = 2$

$x = \frac{-1}{3}$ $\quad$ True

False

Check: $x = \frac{-1}{3}$

$$\log x + \log(3x - 5) = \log 2$$
$$\log(\tfrac{-1}{3}) + \log[3\,(\tfrac{-1}{3}) - 5] = \log 2$$

Logarithms of negative numbers are not real numbers.

Check: $x = 2$

$$\log x + \log(3x - 5) = \log 2$$
$$\log 2 + \log[3(2) - 5] = \log 2$$
$$\log 2 + \log 1 = \log 2$$
$$\log(2 \cdot 1) = \log 2$$
$$\log 2 = \log 2$$

2 is the only solution.
-1/3 is an extraneous solution.

**21.** $\log x + \log 4 = .56$

$$\log 4x = .56$$
$$4x = \text{antilog } .56$$
$$4x = 3.36$$
$$x \approx .91$$

**23.** $2 \log x - \log 4 = 2$

$$\log x^2 - \log 4 = 2$$
$$\log\left(\frac{x^2}{4}\right) = 2$$
$$\frac{x^2}{4} = \text{antilog } 2$$
$$\frac{x^2}{4} = 100$$
$$x^2 = 400$$
$$x^2 - 400 = 0$$
$$(x + 20)(x - 20) = 0$$

$x + 20 = 0$ or $x - 20 = 0$

$x = -20$ $\quad$ $x = 20$

False $\quad$ True

Check: $x = -20$

$$2\log x - \log 4 = 2$$
$$2\log(-20) - \log 4 = 2$$

Logarithms of negative numbers are not real numbers

Check: $x = 20$

$$2 \log x - \log 4 = 2$$
$$2 \log 20 - \log 4 = 2$$
$$\log\left(\frac{400}{4}\right) = 2$$
$$\log 100 = 2$$
$$100 = \text{antilog } 2$$
$$100 = 100$$

Thus, 20 is the only solution.
-20 is an extraneous solution.

**25.** $\log x + \log(x - 3) = 1$

$$\log x(x - 3) = 1$$
$$x(x - 3) = \text{antilog } 1$$
$$x(x - 3) = 10$$
$$x^2 - 3x = 10$$
$$x^2 - 3x - 10 = 0$$
$$(x - 5)(x + 2) = 0$$

$x - 5 = 0$ or $x + 2 = 0$

$x = 5$ or $x = -2$

True $\quad$ False

Thus, 5 is the only solution.
-2 is an extraneous solution.

**27.** $\log x = \frac{1}{3} \log 27$

$\log x = \log 27^{1/3}$
$\log x = \log 3$
Therefore, $x = 3$

**29.** $\log_8 x = 3 \log_8 2 - \log_8 4$

$\log_8 x = \log_8 2^3 - \log_8 4$

$\log_8 x = \log_8 \frac{8}{4}$

$\log_8 x = \log_8 2$

Therefore $x = 2$.

**31.** $A = P(1 + r)^n$

$A = 1200 (1 + .1)^5$
$= 1200 (1.61051)$
$= \$1932.61$

**33.** Let x = the initial amount of bacteria present initially.

$2224 = x(2)^4$

$\log 2224 = \log x(2)^4$

$\log 2224 = \log x + \log 2^4$

$0 = \log x + \log 2^4 - \log 2224$

$0 = \log \frac{2^4 x}{2224}$

$\text{antilog } 0 = \frac{16x}{2224}$

$1 = \frac{x}{139}$

$x = 139$
(bacteria present initially)

**35.** Let $c = \$50000$, $n = 12$, $r = .15$

$S = c(1 - r)^n$

$S = 50000(1 - .15)^{12}$

$\log S = \log 50000(1 - .15)^{12}$
$= \log 50000 + \log(1 - .15)^{12}$
$= \log 50000 + 12 \log(.85)$
$= 4.6990 + 12(0.9294 - 1)$
$= 4.6990 + 11.1528 - 12$
$= 3.8518$
$S = \text{antilog } 3.8518$

Calculator: $S \approx \$7112.10$

**37.** Let $P_{out} = 12.6$ watts,
$P_{in} = 0.146$ watts

$p = \log[10(\frac{12.6}{0.146})]$
$= \log 10 + \log 12.6 - \log 0.146$
$= 1 + 1.1004 - (9.1644 - 10)$
$= 2.1004 - 9.1664 + 10$

Calculator: $P \approx 2.94$

**39.** (a) Let $d = 120$
$d = 10 \log I$
$120 = 10 \log I$
$12 = \log I$
$I = \text{antilog}$
$I = 1 \times 10^{12}$
$I = 1{,}000{,}000{,}000{,}000$

(b) Let $d = 120 - 70 = 50$
$d = 10 \log I$
$50 = 10 \log I$
$5 = \log I$
$I = \text{antilog } 5$
$I = 1 \times 10^5$
$I = 100{,}000$

**41.** $\frac{x - 4}{2} - \frac{2x - 5}{5} > 3$

$10[\frac{x - 4}{2} - \frac{2x - 5}{5} > 3]$

$5(x - 4) - 2(2x - 5) > 30$

$5x - 20 - 4x + 10 > 30$

$x - 10 > 30$

$x > 40$

(a) $\{ x \mid x > 40\}$

(b) $(40, \infty)$

**43.**

$$\begin{array}{r} x^2 - 4x + 3 \\ 2x - 3 \\ \hline -3x^2 + 12x - 9 \\ 2x^3 - 8x^2 + 6x \phantom{{}-9} \\ \hline 2x^3 - 11x^2 + 18x - 9 \end{array}$$

**EXERCISE SET 11.5**

1. $\ln 50 = 3.9120$

3. $\ln 302 = 5.7104$

5. $\ln N = 1.6$
$e^{1.6} = N$
$N = 4.95$

7. $\ln N = -2.63$
$e^{-2.63} = N$
$N = .0721$

9. $\ln 40 = \frac{\log 40}{\log e}$
$= 3.6889$

11. $\ln .046 = \frac{\log .046}{\log e}$
$= -3.0791$

13. $\log_3 25 = \frac{\log 25}{\log 3}$
$= 2.9300$

15. $\log_2 20 = \frac{\log 20}{\log 2}$
$= 4.3219$

17. $\ln 2700 = \frac{\log 2700}{\log e}$

$= 7.9010$

19. $\log_3 .0049 = \frac{\log .0049}{\log 3}$
$= -4.8411$

21. $\ln x + \ln(x - 1) = \ln 12$
$\ln x(x - 1) = \ln 12$
$e^{\ln (x(x - 1))} = e^{\ln 12}$
$x (x - 1) = 12$
$x^2 - x - 12 = 0$
$(x - 4)(x + 3) = 0$

$x - 4 = 0$ or $x + 3 = 0$
$x = 4$ $\quad$ $x = -3$

Since it is not possible to take the log of a negative number, $x = 4$.

23. $\ln x = 5 \ln 2 - \ln 8$
$\ln x = \ln 2^5 - \ln 8$
$\ln x = \ln \frac{32}{8}$
$\ln x = \ln 4$
$e^{\ln x} = e^{\ln 4}$
$x = 4$

25. $\ln(x^2 - 4) - \ln(x + 2) = \ln 1$
$\ln(x^2 - 4) - \ln(x + 2) = 0$
$\ln(x^2 - 4) = \ln(x + 2)$
$e^{\ln(x^2 - 4)} = e^{\ln(x + 2)}$
$x^2 - 4 = x + 2$
$x^2 - x - 6 = 0$
$(x - 3)(x + 2) = 0$

$x - 3 = 0$ or $x + 2 = 0$
$x = 3$ $\quad$ $x = -2$

Since it is not possible to take the log of 0 and $\ln(-2 + 2) = \ln 0$, $x = 3$.

27. $P = 500\ e^{(1.6)(1.2)}$
$= 500\ e^{1.92}$
$= 3410.48$

**29.** $50 = P_0\ e^{-.05(3)}$

$50 = P_0 e^{-.15}$

$50 = P_0(.860708)$

$P_0 = \frac{50}{.860708} = 58.09$

**31.** $90 = 30e^{1.4t}$

$\frac{90}{30} = \frac{30}{30}\ e^{1.4t}$

$3 = e^{1.4t}$

$\ln 3 = \ln e^{1.4t}$

$\ln 3 = 1.4t$

$t = \frac{\ln 3}{1.4} = .7847$

**33.** $100 = 50\ e^{k(3)}$

$\frac{100}{50} = \frac{50}{50}\ e^{3k}$

$2 = e^{3k}$

$\ln 2 = \ln e^{3k}$

$\ln 2 = 3k$

$k = \frac{\ln 2}{3} = .2310$

**35.** $20 = 40e^{k(2.4)}$

$.5 = e^{2.4k}$

$\ln .5 = \ln e^{2.4k}$

$\ln .5 = 2.4k$

$k = \frac{\ln .5}{2.4} = -.2888$

**37.** $A = 6000\ e^{-.08(3)}$

$= 6000\ e^{-.24}$

$= 6000\ (.7866278611)$

$= 4719.77$

**39.** $V = V_0 e^{kt}$ Solve for $V_0$

$\frac{V}{e^{kt}} = V_0$

**41.** $P = 150\ e^{4t}$ Solve for t

$\frac{P}{150} = e^{4t}$

$\ln(\frac{P}{150}) = \ln e^{4t}$

$\ln(\frac{P}{150}) = 4t$

$\frac{\ln P - \ln 150}{4} = t$

**43.** $A = A_0\ e^{kt}$ Solve for k

$\frac{A}{A_0} = e^{kt}$

$\ln(\frac{A}{A_0}) = \ln e^{kt}$

$\ln A - \ln A_0 = kt$

$\frac{\ln A - \ln A_0}{t} = k$

**45.** $\ln y - \ln x = 2.3$ Solve for y

$\ln\ (\frac{y}{x}) = 2.3$

$e^{2.3} = \frac{y}{x}$

$x(e^{2.3}) = y$

**47.** $\ln y - \ln(x + 3) = 6.$ Solve for y

$\ln(\frac{y}{x + 3}) = 6$

$e^6 = \frac{y}{x + 3}$

$e^6(x + 3) = y$

**49.** $x = k(\ln I_0 - \ln I)$

Solve for $I_0$

$$\frac{x}{k} = \ln\left(\frac{I_0}{I}\right)$$

$$e^{\frac{x}{k}} = e^{\ln\left(\frac{I_0}{I}\right)}$$

$$e^{\frac{x}{k}} = \frac{I_o}{I}$$

$$Ie^{\frac{x}{k}} = I_0$$

**51.** $\ln M = \ln Q - \ln(1 - q)$

$$M = \frac{Q}{1 - Q}$$

$$M - MQ = Q$$

$$M = Q + MQ$$

$$M = Q(1 + M)$$

$$\frac{M}{1 + M} = Q$$

**53.** $A = P_0 e^{kt}$

(a) $= 5000\, e^{.08(2)}$
$= 5000\, (e^{.16})$
$= 5000\, (1.1735)$
$= \$5867.55$

Number 53 (Continued)

(b) If the amount in the account is to double, then $A = 2(5000) = \$10,000$.

$10,000 = 5000\, e^{.08t}$
$2 = e^{.08t}$
$\ln 2 = \ln e^{.08t}$
$\ln 2 = .08t$
$\frac{\ln 2}{.08} = t$
$8.66$ years $= t$

**55.** $f(t) = 1 - e^{-.04t}$

(a) $f(t) = 1 - e^{-.04(50)}$
$= 1 - e^{-2}$
$= 1 - .135335$
$= .864700$ or 86.4%

(b) $.75 = 1 - e^{-.04t}$
$-.25 = -e^{-.04t}$
$.25 = e^{-.04t}$
$\ln .25 = \ln e^{-.04t}$
$\ln .25 = -.04t$
$t = \frac{\ln .25}{-.04}$
$t = 34.66$ days

**57.** $f(P) = .37 \ln P + 05$

(a) $f(P) = .37 \ln(972,000) + .05$
$= 5.10012311 + .05$
$= 5.15$ ft/sec

(b) $f(P) = .37 \ln(8,567,000) + .05$
$= 5.906 + .05$
$= 5.96$ ft/sec.

Number 57 (Continued)

(c)
$$5 = .37 \ln P + .05$$
$$4.95 = 37 \ln P$$
$$13.378378 = \ln P$$
$$e^{13.378378} = e^{\ln P}$$
$$P = e^{13.378378}$$
$$= 646{,}000$$

**59.** $P(t) = 5.026\, e^{.018t}$

(a) t = number of years since 1988. Therefore, 2000 - 1988 = 12 years elapse between 1988 and 2000.

$$P(t) = 5.026\, e^{.018(12)}$$
$$= 5.026\, e^{.216}$$
$$= 6.24 \text{ billion}$$

(b) $P(t) = 2(5.026)$. Therefore,

$$2(5.026) = 5.026\, e^{.018t}$$
$$\frac{2(5.026)}{5.026} = \frac{5.026}{5.026}\, e^{.018t}$$
$$2 = e^{.018t}$$
$$\ln 2 = \ln e^{.018t}$$
$$\ln 2 = .018t$$
$$\frac{\ln 2}{.018} = t$$
$$38.51 \text{ years} = t.$$

**61.** $f(t) = V_0\, e^{-.0001205t}$

$$9 = 20\, e^{-.000125t}$$
$$.45 = e^{-.0001205t}$$
$$\ln .45 = -.0001205t$$
$$\ln .45 = -.0001205t$$
$$\frac{\ln .45}{-.0001205} = t$$
$$t = 6626.62 \text{ years}$$

Number 61 (Continued)

(b) Let x equal the surgical amount of carbon 14 and .5x equal the remaining amount.

$$.5x = xe^{-.0001205t}$$
$$\frac{.5\cancel{x}}{\cancel{x}} = \frac{\cancel{x}e^{-.0001205t}}{\cancel{x}}$$
$$.5 = e^{-.0001205t}$$
$$\ln .5 = \ln e^{-.0001205t}$$
$$\ln .5 = -.0001205t$$
$$\frac{\ln .5}{-.0001205} = t$$
$$t = 5752.26 \text{ years}$$

**63.** $e \approx 2.718$

**65.** $e^x = 184.93$

$$\ln e^x = \ln 184.93$$
$$x = \ln 184.93 = 5.2200$$

**67.** $e^{-1.73} = .1773$

Press [c] 1.73 [+/-] [inv] [ln]

**69.**
$$\frac{\frac{3}{x^2} - \frac{2}{x}}{\frac{x}{4}} = \frac{4x^2\left(\frac{3}{x^2}\right) - 4x^2\left(\frac{2}{x}\right)}{4x^2\left(\frac{x}{4}\right)}$$
$$= \frac{12 - 8x}{x^3}$$

**71.** $\sqrt[3]{128\, x^7 y^9 z^{13}} = 4x^2y^3z^4\, \sqrt[3]{2xz}$

## REVIEW EXERCISES

1. $y = 2^x$

| x | -2 | -2 | 0 | 1 | 2 | 3 |
|---|---|---|---|---|---|---|
| y | 1/4 | 1/2 | 1 | 2 | 4 | 8 |

Domain: $\mathbb{R}$
Range: $\{y \mid y > 0\}$

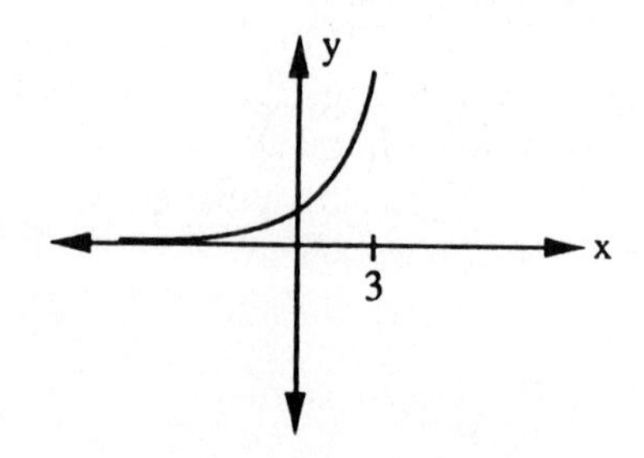


3. $y = \log_2 x$ means $2^y = x$

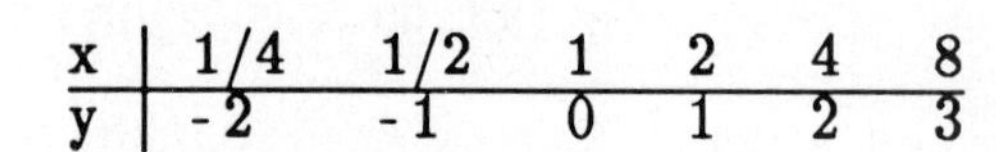

| x | 1/4 | 1/2 | 1 | 2 | 4 | 8 |
|---|---|---|---|---|---|---|
| y | -2 | -1 | 0 | 1 | 2 | 3 |

Domain: $\{x \mid x > 0\}$
Range: $\mathbb{R}$

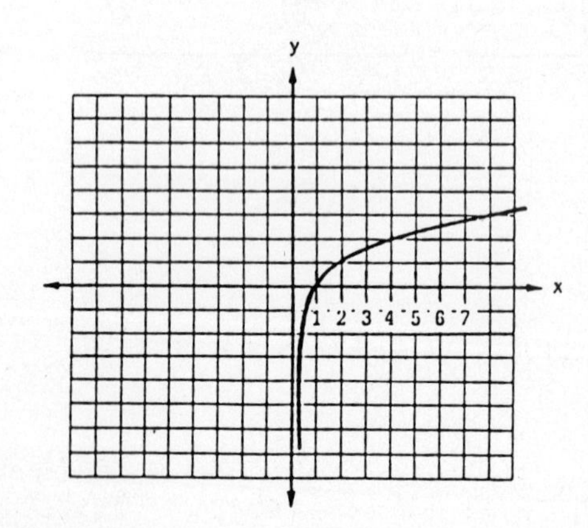


5. $y = 3^x$

| x | -2 | -1 | 0 | 1 | 2 |
|---|---|---|---|---|---|
| y | 1/9 | 1/3 | 1 | 3 | 9 |

Domain: $\mathbb{R}$
Range: $\{y \mid y > 0\}$
$y = \log_3 x$

| x | 1/9 | 1/3 | 1 | 3 | 9 |
|---|---|---|---|---|---|
| y | -2 | -1 | 0 | 1 | 2 |

Domain: $\{x \mid x > 0\}$
Range: $\mathbb{R}$

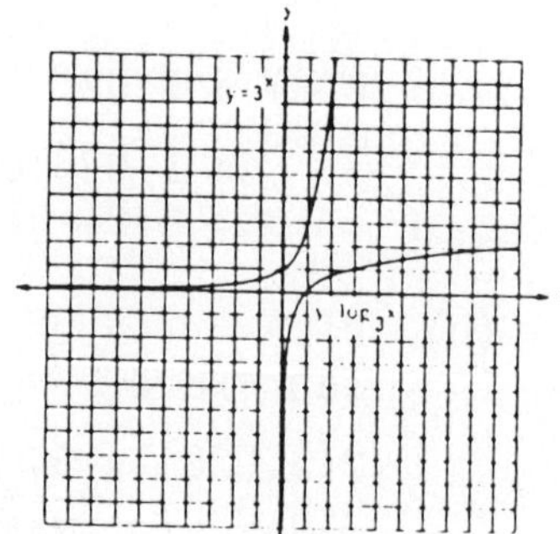

7. $8^{1/3} = 2$
$\log_8 2 = 1/3$

9. $25^{1/2} = 5$
$\log_{25} 5 = 1/2$

11. $\log_{1/3} (1/9) = 2$
$(1/3)^2 = 1/9$

13. $\log_2 32 = 5$
$2^5 = 32$

15. $2 = \log_4 x$
$x = 4^2 = 16$

17. $-3 = \log_{1/4} x$

$$x = (1/4)^{-3}$$
$$x = \frac{1}{(1/4)^3}$$
$$x = 64$$

19. $\log(x - 8)^5 = 5 \log(x - 8)$

21. $\log \frac{x^4}{39(2x + 8)} = \log x^4$

$- \log 39(2x + 8)$

$= 4 \log x -$

$[\log 39 + \log(2x - 8)]$

$= 4 \log x - \log 39 -$

$\log(2x + 8)$

23. $2 \log x - 3 \log (x + 1)$

$= \log x^2 - \log(x + 1)^3$

$= \log[x^2/(x + 1)^3]$

25. $[2 \log_8 (x + 3) + 4 \log_8 (x - 1)]$

$- 1/2 \log_8 x$

$= \log_8 (x + 3)^2 +$

$\log_8 (x - 1)^4 - \log_8 x^{1/2}$

$= \log_8 [(x + 3)^2 (x - 1)^4/\sqrt{x}]$

27. $3 \ln x + \frac{1}{2} \ln(x + 1)$

$- 3 \ln(x + 4)$

$= \ln x^3 + \ln (x + 1)^{1/2}$

$- \ln (x + 4)^3$

$= \ln \frac{x^3\sqrt{x + 1}}{(x + 4)^3}$

29. $\log 0.000716 = -3.1451$

31. $\log 17600 = 4.2455$

33. antilog $2.9186 \approx 829$

35. antilog $(-1.3747) = .0422$

37. $\log N = -1.2262$

$N =$ antilog $(-1.2262)$

$N = 0.0594$

39. $(3.2)^x = 187$

$$\log 3.2^x = \log 187$$
$$x \log 3.2 = \log 187$$
$$x = \frac{\log 187}{\log 3.2} \approx 4.498$$

41. $49^x = \frac{1}{7}$

$$\log 49^x = \log 1/7$$
$$x \log 49 = \log 1/7$$
$$x = \frac{\log 1/7}{\log 49} = -\frac{1}{2}$$

43. $\log(3x + 2) = \log 300$

$$10^{\log(3x + 2)} = 10^{\log 300}$$
$$3x + 2 = 300$$
$$3x = 298$$
$$x = \frac{298}{3}$$

45. $\log_3 x + \log_3 (2x + 1) = 1$

$\log_3 x(2x + 1) = 1$

$$3^1 = x(2x + 1)$$
$$0 = 2x^2 + x - 3$$
$$0 = (2x + 3)(x - 1)$$

$2x + 3 = 0$ or $x - 1 = 0$

$x = -3/2$ or $x = 1$

It is not possible to take the log of a negative number. Therefore $x = 1$.

**47.** $\ln(x + 1) - \ln(x - 2) = \ln 4$

$$\ln\left(\frac{x+1}{x-2}\right) = \ln 4$$
$$e^{\ln\left(\frac{x+1}{x-2}\right)} = e^{\ln 4}$$
$$\frac{x+1}{x-2} = 4$$
$$x + 1 = 4(x - 2)$$
$$x + 1 = 4x - 8$$
$$1 = 3x - 8$$
$$9 = 3x$$
$$x = 3$$

**49.** $P_0 = 80\ e^{-.02(10)}$

$$= 80\ e^{-.2}$$
$$= 65.50$$

**51.** $w = w_0\ e^{kt}$. Solve for $w_0$.

$$w_0 = \frac{w}{e^{kt}}.$$

**53.** $150 = 600\ e^{kt}$

Solve for t

$$\frac{150}{600} = e^{kt}$$
$$.25 = e^{kt}$$
$$\ln .25 = \ln e^{kt}$$
$$\ln .25 = kt$$
$$\frac{\ln .25}{t} = k$$

**55.** $\ln(y+3) - \ln(x+1) = \ln 5$

Solve for y

$$\ln\left(\frac{y+3}{x+1}\right) = \ln 5$$
$$e^{\ln\left(\frac{y+3}{x+1}\right)} = e^{\ln 5}$$
$$\frac{y+3}{x+1} = 5$$
$$y + 3 = 5 \cdot (x + 1)$$
$$y = 5(x + 1) - 3$$
$$y = 5x + 5 - 3$$
$$y = 5x + 2$$

**57.** $\ln 450 = 6.1092$

**59.** $\log_5 .0862 = \log .0862 \div \log 5 = -1.5229$

**61.** $R = 1000(.5)^{.00041t}$

$$= 1000(.5)^{.000041(20,000)}$$
$$= 1000(.5^{.82})$$
$$= 1000(.56644194)$$
$$= 566.5 \text{ mg}$$

**63.** $A = P_0(e^{kt})$

Let $2P_0$ = the amount in the account.

$$2P_0 = P_0 e^{.07t}$$
$$2 = e^{.07t}$$
$$\ln 2 = \ln e^{.07t}$$
$$\ln 2 = .07t$$
$$t = \frac{\ln 2}{.07}$$
$$t = 9.90 \text{ years}$$

## PRACTICE TEST

1. $y = 2^x$

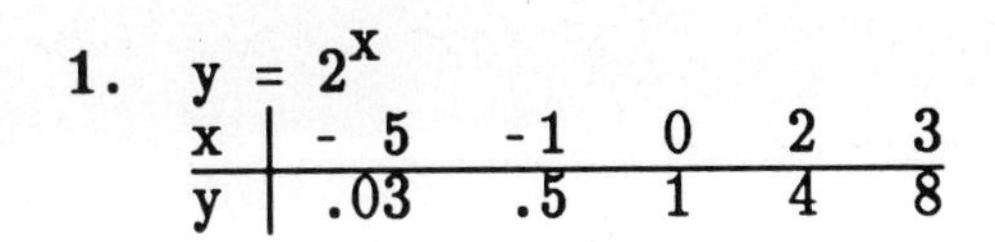

| x | - 5 | -1 | 0 | 2 | 3 |
|---|---|---|---|---|---|
| y | .03 | .5 | 1 | 4 | 8 |

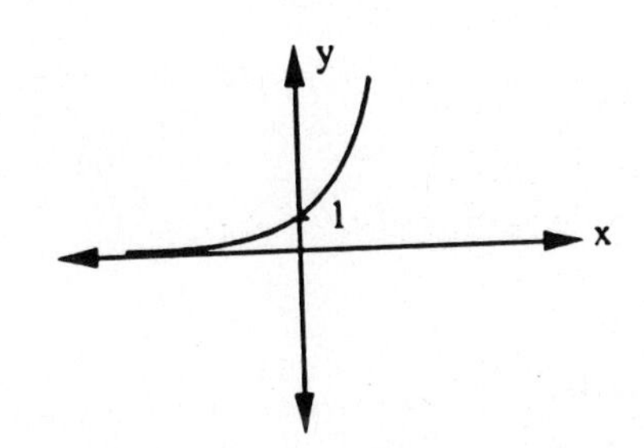


2. $y = \log_2 x$ means $2^y = x$

| x | 1/4 | 1/2 | 1 | 2 | 4 |
|---|---|---|---|---|---|
| y | -2 | -1 | 0 | 1 | 2 |

D: $\{x \mid x > 0\}$
R: $\mathbb{R}$

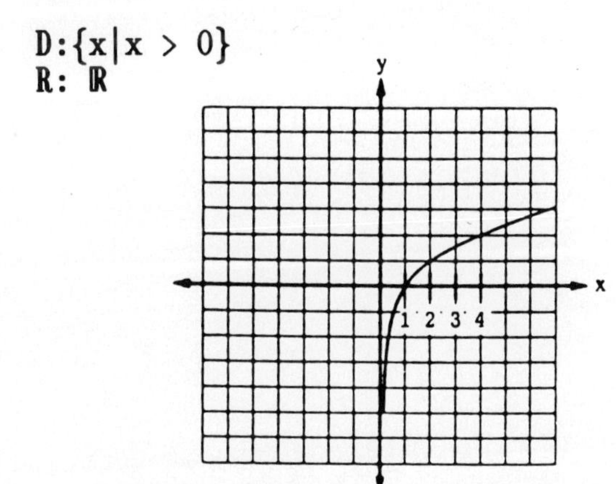


3. $4^{-3} = \frac{1}{64}$

$\log_4 1/64 = -3$

4. $\log_3 243 = 5$

$3^5 = 243$

5. $4 = \log_2 x$

$2^4 = x$
$16 = x$

6. $y = \log_{27} 3$

$27^y = 3$
$(3^3)^y = 3$
$3^{3y} = 3^1$
$3y = 1$
$y = \frac{1}{3}$

7. $\log_3 \frac{x(x - 4)}{x^2}$

$= \log_3 x(x - 4) - \log_3 x^2$
$= \log_3 x + \log_3(x - 4) - 2 \log_3 x$

8. $[3 \log_8(x - 4) + 2 \log_8 (x + 1)] - 1/2 \log_8 x$

$= \log_8(x - 4)^3 + \log_8(x + 1)^2 - \log_8 x^{1/2}$

$= \log_8[(x - 4)^3 (x + 1)^2/\sqrt{x}]$

9. $\log 4620 = 3.6646$

10. $\log .000638 = -3.1952$

11. $\log N = -2.3002$
$N = \text{antilog } (-2.3002)$
$N = .00501$

12. $3^x = 123$
$\log_3 123 = x$
$\frac{\log 123}{\log 3} = x$
$x = 4.38$

13. $\log 4x = \log(x + 3) + \log 2$
$\log 4x = \log 2(x + 3)$
$10^{\log 4x} = 10^{\log 2(x + 3)}$
$4x = 2(x + 3)$
$4x = 2x + 6$
$2x = 6$
$x = 3$

**14.** Let P = 1500, r = .12, n = 10

$$A = P(1 + r)^n$$
$$A = 1500(1 + .12)^{10}$$
$$\log A = \log(1500)(1.12)^{10}$$
$$= \log 1500 + 10 \log 1.12$$
$$= 3.1761 + 10(.0492)$$
$$= 3.1761 + .4922$$
$$= 3.6683$$
$$A = \text{antilog } 3.6683$$

Calculator: $A \approx \$4658.77$

**15.**

$$V = \nu_o e^{-.0001205t}$$
$$40 = 60e^{-.0001205t}$$
$$\frac{40}{60} = e^{-.0001205t}$$
$$\ln(\tfrac{2}{3}) = \ln e^{-.0001205t}$$
$$\ln(\tfrac{2}{3}) = -0001205t$$
$$\ln(\tfrac{2}{3}) = t$$
$$\frac{\ln(2/3)}{-.0001205} = t$$
$$3364.86 = t$$

## CHAPTER 12

### EXERCISE SET 12.1

1. $a_n = 2n$

$a_1 = 2(1) = 2$

$a_2 = 2(2) = 4$

$a_3 = 2(3) = 6$

$a_4 = 2(4) = 8$

$a_5 = 2(5) = 10$

2, 4, 6, 8, 10

3. $a_n = \frac{n + 5}{n}$

$a_1 = \frac{1 + 5}{1} = 6$

$a_2 = \frac{2 + 5}{2} = \frac{7}{2}$

$a_3 = \frac{3 + 5}{3} = \frac{8}{3}$

$a_4 = \frac{4 + 5}{4} = \frac{9}{4}$

$a_5 = \frac{5 + 5}{5} = \frac{10}{5} = 2$

$6, \frac{7}{2}, \frac{8}{3}, \frac{9}{4}, 2$

5. $a_n = \frac{1}{n}$

$a_1 = \frac{1}{1} = 1$

$a_2 = \frac{1}{2}$

$a_3 = \frac{1}{3}$

$a_4 = \frac{1}{4}$

$a_5 = \frac{1}{5}$

$1, \frac{1}{2}, \frac{1}{3}, \frac{1}{4}, \frac{1}{5}$

7. $a_n = \frac{n + 2}{n + 1}$

$a_1 = \frac{1 + 2}{1 + 1} = \frac{3}{2}$

$a_2 = \frac{2 + 2}{2 + 1} = \frac{4}{3}$

$a_3 = \frac{3 + 2}{3 + 1} = \frac{5}{4}$

$a_4 = \frac{4 + 2}{4 + 1} = \frac{6}{5}$

$a_5 = \frac{5 + 2}{5 + 1} = \frac{7}{6}$

$\frac{3}{2}, \frac{4}{3}, \frac{5}{4}, \frac{6}{5}, \frac{7}{6}$

9. $a_n = (-1)^n$

$a_1 = (-1)^1 = -1$

$a_2 = (-1)^2 = 1$

$a_3 = (-1)^3 = -1$

$a_4 = (-1)^4 = 1$

$a_5 = (-1)^5 = -1$

-1, 1, -1, 1, -1

11. $a_n = (-2)^{n+1}$

$a_1 = (-2)^{1+1} = (-2)^2 = 4$

$a_2 = (-2)^{2+1} = (-2)^3 = -8$

$a_3 = (-2)^{3+1} = (-2)^4 = 16$

$a_4 = (-2)^{4+1} = (-2)^5 = -32$

$a_5 = (-2)^{5+1} = (-2)^6 = 64$

4, -8, 16, -32, 64

13. $a_n = 2n + 3$

$a_{12} = 2(12) + 3 = 24 + 3 = 27$

15. $a_n = 2n - 4$
$a_5 = 2(5) - 4 = 10 - 4 = 6$

17. $a_n = (-2)^n$
$a_4 = (-2)^4 = 16$

19. $a_n = \frac{n^2}{2n + 1}$
$a_9 = \frac{9^2}{2(9) + 1} = \frac{81}{18 + 1} = \frac{81}{19}$

21. $a_n = 2n + 5$
$a_1 = 2(1) + 5 = 2 + 5 = 7$
$a_2 = 2(2) + 5 = 4 + 5 = 9$
$a_3 = 2(3) + 5 = 6 + 5 = 11$
$s_1 = a_1 = 7$
$s_3 = a_1 + a_2 + a_3$
$= 7 + 9 + 11$
$= 27$

23. $a_n = 2^n + 1$
$a_1 = 2^1 + 1 = 3$
$a_2 = 2^2 + 1 = 4 + 1 = 5$
$a_3 = 2^3 + 1 = 8 + 1 = 9$
$s_1 = a_1 = 3$
$s_3 = a_1 + a_2 + a_3$
$= 3 + 5 + 9$
$= 17$

25. $a_n = (-1)^{2n}$
$a_1 = (-1)^{2(1)} = (-1)^2 = 1$
$a_2 = (-1)^{2(2)} = (-1)^4 = 1$
$a_3 = (-1)^{2(3)} = (-1)^6 = 1$

$s_1 = a_1 = 1$
$s_3 = a_1 + a_2 + a_3$
$= 1 + 1 + 1$
$= 3$

27. $a_n = \frac{n^2}{2}$
$a_1 = \frac{1^2}{2} = \frac{1}{2}$
$a_2 = \frac{2^2}{2} = \frac{4}{2}$
$a_3 = \frac{3^2}{2} = \frac{9}{2}$
$s_1 = a_1 = \frac{1}{2}$
$s_3 = a_1 + a_2 + a_3$
$= \frac{1}{2} + \frac{4}{2} + \frac{9}{2}$
$= \frac{14}{2}$
$= 7$

29. Each term is twice the preceding term.

64, 128, 256

31. Each term is two more than the preceding term.

15, 17, 19

**33.** Each denominator is one more than the preceding one while the numerators always equal one.

$\frac{1}{6}, \frac{1}{7}, \frac{1}{8}$

**35.** Each term is -1 times the previous term.
1, -1, 1

**37.** Each denominator is three times the previous one while the numerators always equal one.

$\frac{1}{81}, \frac{1}{243}, \frac{1}{729}$

**39.** Each term is -1/2 times the preceding term.

$\frac{1}{16}, \frac{-1}{32}, \frac{1}{64}$

**41.** Each term is eight less than the previous term.

-25, -33, -41

**43.** $\sum_{n=1}^{5} (3n - 1)$

$[3(1)-1] + [3(2)-1] + [3(3)-1] + [3(4)-1] + [3(5) - 1]$

$= 2 + 5 + 8 + 11 + 14 = 40$

**45.** $\sum_{k=1}^{6} (2k^2 - 3)$

$= [2(1)^2 - 3] + [2(2)^2 - 3] + [2(3)^2 -3] + [2(4)^2 - 3] + [2(5)^2 - 3] + [2(6)^2 - 3]$
$= -1 + 5 + 15 + 29 + 47 + 69$
$= 164$

**47.** $\sum_{n=2}^{4} \frac{n^2 + n}{n + 1}$

$= [\frac{2^2 + 2}{2 + 1}] + [\frac{3^2 + 3}{3 + 1}] + [\frac{4^2 + 4}{4 + 1}] = 2 + 3 + 4 = 9$

**49.** $a_n = n + 3$

$\sum_{n=1}^{5} (n + 3)$

**51.** $a_n = \frac{n^2}{4}$

$\sum_{n=1}^{3} \frac{n^2}{4}$

**53.** $\sum_{n=1}^{5} x_i$

$= x_1 + x_2 + x_3 + x_4 + x_5$
$= 2 + 3 + 5 + (-1) + 4$
$= 13$

55. $\left(\sum_{i=1}^{5} x_i\right)^2$

$= [x_1 + x_2 + x_3 + x_4 + x_5]^2$

$= [2 + 3 + 5 + (-1) + 5]^2$

$= 13^2 = 169$

57. $\sum_{i=3}^{5} \frac{2x_i}{3}$

$= [\frac{2x_3}{3} + \frac{2x_4}{3} + \frac{2x_5}{3}]$

$= \frac{2(5)}{3} + \frac{2(-1)}{3} + \frac{2(4)}{3}$

$= \frac{10}{3} - \frac{2}{3} + \frac{8}{3} = \frac{16}{3}$

59. $\bar{x} = \frac{15 + 20 + 25 + 30 + 35}{5}$

$= 25$

61. $\bar{x} = \frac{72 + 83 + 4 + 60 + 18 + 20}{6}$

$= 42.83$

63. $\sum_{i=1}^{n} x_i = x_1 + x_2 + x_3 + x_4 + \cdots + x_n$

65. A sequence is a list of numbers arranged in a specific order.

67. The nth partial sum is the sum of the first consecutive n terms of a series.

69. $2x^2 + 15 = 13x$

$2x^2 - 13x + 15 = 0$

$(x - 5)(2x - 3) = 0$

$x - 5 = 0 \quad 2x - 3 = 0$

$x = 5 \quad 2x = 3$

$x = 3/2$

71. $\frac{x^2}{4} + \frac{y^2}{1} = 1$

$\frac{x^2}{2^2} + \frac{y^2}{1^2} = 1^2$

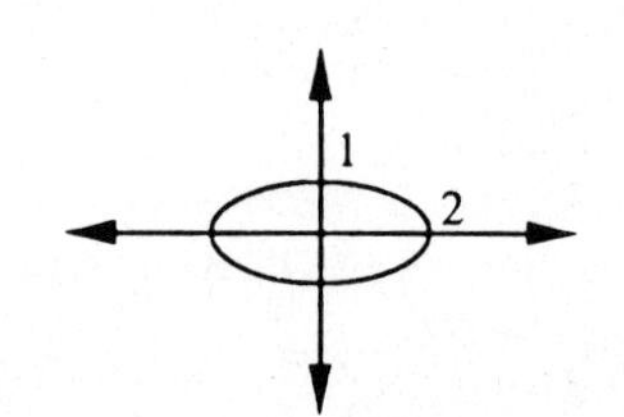


**JUST FOR FUN**

$\sum = \sqrt{\frac{N(\Sigma x^2 f) - (\Sigma xf)^2}{n(n-1)}}$

$\sum x^2 f = 1^2(3) + 3^2(4) + 5^2(5) + 7^2(0) + 9^2(2) = 231$

$S = \sqrt{\frac{14(231) - (58)^2}{14(14-1)}}$

$= \sqrt{\frac{4564 - 3364}{182}}$

$= \sqrt{6.5934} = 2.57$

## EXERCISE SET 12.2

1. $a_1 = 3$

$a_2 = 3 + (2 - 1)(4)$
$= 3 + 4$
$= 7$

$a_3 = 3 + (3 - 1)(4)$
$= 3 + 2(4)$
$= 3 + 8$
$= 11$

$a_4 = -5 + (4 - 1)(2)$
$= 3 + 3(4)$
$= 3 + 12$
$= 15$

$a_5 = 3 + (5 - 1)(4)$
$= 3 + 4(4)$
$= 3 + 16$
$= 19$

3, 7, 11, 15, 19

$a_n = 3 + 4(n - 1)$

3. $a_1 = -5$

$a_2 = -5 + (2 - 1)(2)$
$= -5 + 2$
$= -3$

$a_3 = -5 + (3 - 1)(2)$
$= -5 + 2(2)$
$= -5 + 4$
$= -1$

Number 3 (Continued)

$a_4 = -5 = (4 - 1)(2)$
$= -5 + 3(2)$
$= -5 + 6$
$= 1$

$a_5 = -5 + (5 - 1)(2)$
$= -5 + 4(2)$
$= -5 + 8$
$= 3$
$= -5, -3, -1, 1, 3$

$a_n = -5 + 2(n - 1)$

5. $a_1 = \frac{1}{2}$

$a_2 = \frac{1}{2} + (2 - 1)(\frac{3}{2})$
$= \frac{1}{2} + \frac{3}{2}$
$= \frac{4}{2}$
$= 2$

$a_3 = \frac{1}{2} + (3 - 1)(\frac{3}{2})$
$= \frac{1}{2} + 2(\frac{3}{2})$
$= \frac{1}{2} + \frac{6}{2}$
$= \frac{7}{2}$

Number 5 (Continued)

$$a_4 = \frac{1}{2} + (4 - 1)(\frac{3}{2})$$
$$= \frac{1}{2} + 3(\frac{3}{2})$$
$$= \frac{1}{2} + \frac{9}{2}$$
$$= \frac{10}{2}$$
$$= 5$$

$$a_5 = \frac{1}{2} + (5 - 1)(\frac{3}{2})$$
$$= \frac{1}{2} + 4(\frac{3}{2})$$
$$= \frac{1}{2} + \frac{12}{2}$$
$$= \frac{13}{2}$$

$\frac{1}{2}, 2, \frac{7}{2}, 5, \frac{13}{2}$

$$a_n = \frac{1}{2} + \frac{3}{2}(n - 1)$$

7. $a_1 = 100$

$$a_2 = 100 + (2 - 1)(-5)$$
$$= 100 - 5$$
$$= 100 - 5$$
$$= 95$$

$$a_3 = 100 + (3 - 1)(-5)$$
$$= 100 + 2(-5)$$
$$= 100 - 10$$
$$= 90$$

$$a_4 = 100 + (4 - 1)(-5)$$
$$= 100 + 3(-5)$$
$$= 100 - 15$$
$$= 85$$

Number 7 (Continued)

$$a_5 = 100 + (5 - 1)(-5)$$
$$= 100 + 4(-5)$$
$$= 100 - 20$$
$$= 80$$

100, 95, 90, 85, 80

$$a_n = 100 - 5(n - 1)$$

9. $a_7 = a_1 + (7 - 1)d$

$$a_7 = 4 + 6(3)$$
$$= 4 + 18$$
$$= 22$$

11. $a_{18} = a_1 + (18 - 1)d$

$$a_{18} = -6 + 17(-1)$$
$$= -6 - 17$$
$$= -23$$

13. $a_{10} - a_1 + (10 - 1)d$

$$a_{10} = -\frac{6}{3} + 9(\frac{5}{3})$$
$$= -\frac{6}{3} + \frac{45}{3}$$
$$= \frac{39}{3}$$
$$= 13$$

**15.** $a_9 = a_1 + (9 - 1)d$

$19 = 3 + 8d$

$16 = 8d$

$d = \frac{16}{8} = 2$

**17.** $a_n = 4 + (n - 1)(3)$

$28 = 4 + 3n - 3 = 1 + 3n$

$27 = 3n$

$n = \frac{27}{3} = 9$

**19.** $a_n = -\frac{7}{3} + (n - 1)(-\frac{2}{3})$

$-\frac{17}{3} = -\frac{7}{3} + (n - 1)(\frac{-2}{3})$

$-17 = -7 + (n - 1)(-2)$

$-17 = -7 - 2n + 2$

$= -5 - 2n$

$-12 = -2n$

$n = \frac{-12}{-2}$

$= 6$

**21.** $S_{10} = \frac{\overset{5}{\cancel{10}}(1 + 19)}{\cancel{2}} = 5(20) = 100$

$a_{10} = 1 + (10 - 1)d$

$19 = 1 + 9d$

$18 = 9d$

$d = \frac{18}{9} = 2$

**23.** $S_8 = \frac{8(\frac{3}{5} + 2)}{2} = \frac{8(\frac{13}{5})}{2}$

$= \overset{4}{\cancel{8}}(\frac{13}{5})(\frac{1}{\cancel{2}}) = 4(\frac{13}{5}) = \frac{52}{5}$

$a_8 = \frac{3}{5} + (8 - 1)d$

$2 = \frac{3}{5} + 7d$

$\frac{7}{5} = 7d$

$d = \frac{7}{5} \div 7$

$= \frac{7}{5} \cdot \frac{1}{\cancel{7}}$

$= \frac{1}{5}$

**25.** $S_5 = \frac{5(\frac{12}{5} + \frac{28}{5})}{2} = \frac{\cancel{5}(\frac{40}{\cancel{5}})}{2} = \frac{40}{2} = 20$

$a_5 = \frac{12}{5} + (5 - 1)d$

$\frac{28}{5} = \frac{12}{5} + 4d$

$\frac{16}{5} = 4d$

$d = \frac{16}{5} \div 4 = \frac{\overset{4}{\cancel{16}}}{5} \cdot \frac{1}{4} = \frac{4}{5}$

**27.** $S_{11} = \frac{11(7 + 67)}{2} = \frac{11(\overset{37}{\cancel{74}})}{\cancel{2}} = 407$

$a_{11} = 7 + (11 - 1)d$

$67 = 7 + 10d$

$60 = 10d$

$d = \frac{60}{10} = 6$

**29.** $a_1 = -4$

$a_2 = -4 + (2 - 1)(-2)$
$= -4 - 2$
$= -6$

$a_3 = -4 + (3 - 1)(-2)$
$= -4 + 2(-2)$
$= -4 - 4$
$= -8$

$a_4 = -4 + (4 - 1)(-2)$
$= -4 + 3(-2)$
$= -4 - 6$
$= -10$
-4, -6, -8, -10

$a_{10} = -4 + (10 - 1)(-2)$
$= -4 + 9(-2)$
$= -4 - 18$
$= -22$

$$S_{10} = \frac{\overset{5}{\cancel{10}}[-4 + (-22)]}{\cancel{2}} = 5(-26)$$
$= -130$

**31.** $a_1 = -8$

$a_2 = -8 + (2 - 1)(-5)$
$= -8 - 5$
$= -13$

$a_3 = -8 + (3 - 1)(-5)$
$= -8 + 2(-5)$
$= -8 - 10$
$= -18$

$a_4 = -8 + (4 - 1)(-5)$
$= -8 + 3(-5)$
$= -8 - 15$
$= -23$
-8, -13, -18, -23

Number 31 (Continued)

$a_{10} = -8 + (10 - 1)(-5)$
$= -8 + 9(-5)$
$= -8 - 45$
$= -53$

$$S_{10} = \frac{\overset{5}{\cancel{10}}[-8 + (-53)]}{\cancel{2}}$$
$= 5(-61)$
$= -305$

**33.** $a_1 = 100$

$a_2 = 100 + (2 - 1)(-7)$
$= 100 - 7$
$= 93$

$a_3 = 100 + (3 - 1)(-7)$
$= 100 + 2(-7)$
$= 100 - 14$
$= 86$

$a_4 = 100 + (4 - 1)(-7)$
$= 100 + 3(-7)$
$= 100 - 21$
$= 79$
100, 93, 86, 79

$a_{10} = 100 + (10 - 1)(-7)$
$= 100 + 9(-7)$
$= 100 - 63$
$= 37$

$$S_{10} = \frac{\overset{5}{\cancel{10}}(100 + 37)}{\cancel{2}} = 5(137) = 685$$

**35.** $a_1 = \frac{9}{5}$

$$a_2 = \frac{9}{5} + (2 - 1)(\tfrac{3}{5})$$
$$= \frac{9}{5} + \frac{3}{5}$$
$$= \frac{12}{5}$$

$$a_3 = \frac{9}{5} + (3 - 1)(\tfrac{3}{5})$$
$$= \frac{9}{5} + 2(\tfrac{3}{5})$$
$$= \frac{9}{5} + \frac{6}{15}$$
$$= \frac{15}{5}$$

$$a_4 = \frac{9}{5} + (4 - 1)\frac{3}{5}$$
$$= \frac{9}{5} + 3(\tfrac{3}{5})$$
$$= \frac{9}{5} + \frac{9}{5}$$
$$= \frac{18}{5}$$

$\frac{9}{5}, \frac{12}{5}, \frac{15}{5}, \frac{18}{5}$

$$a_{10} = \frac{9}{5} + (10 - 1)(\tfrac{3}{5})$$
$$= \frac{9}{5} + 9(\tfrac{3}{5})$$
$$= \frac{9}{5} + \frac{27}{5}$$
$$= \frac{36}{5}$$

$$S_{10} = \frac{10(\frac{9}{5} + \frac{36}{5})}{5} = \frac{10(\overset{9}{\cancel{\frac{45}{5}}})}{2} = \frac{90}{2} = 45$$

**37.** $d = -6 - (-8)$
$= -6 + 8$
$= 2$

$$a_n = -8 + (n - 1)(2)$$
$$= -8 + 2n - 2$$
$$= 2n - 10$$

$$42 = 2n - 10$$
$$52 = 2n$$
$$n = \frac{52}{2} = 26$$

$$S_{26} = \frac{\overset{13}{\cancel{26}}(-8 + 42)}{\cancel{2}} = 13(34) = 442$$

**39.** $d = \frac{2}{2} - \frac{1}{2} = \frac{1}{2}$

$$a_n = \frac{1}{2} + (n - 1)(\tfrac{1}{2})$$
$$\frac{17}{2} = \frac{1}{2} + (n - 1)(\tfrac{1}{2})$$
$$17 = 1 + (n - 1))(1)$$
$$= 1 + n - 1$$
$$= n$$
$$n = 17$$

$$S_{17} = \frac{17(\frac{1}{2} + \frac{17}{2})}{2} = \frac{17(\overset{9}{\cancel{\frac{18}{2}}})}{2} = \frac{153}{2}$$

**41.** $d = 14 - 7 = 7$

$$a_n = 7 + (n - 1)(7)$$
$$= 7 + 7n - 7$$
$$= 7n$$
$$63 = 7n$$
$$n = \frac{63}{7} = 9$$

$$S_9 = \frac{9(7 + 63)}{2} = \frac{9(\overset{35}{\cancel{70}})}{\cancel{2}} = 315$$

**43.** $d = 12 - 9 = 3$

$a_n = 9 + (n - 1)(3)$
$= 9 + 3n - 3$
$= 6 + 3n$

$93 = 6 + 3n$
$87 = 3n$
$n = \frac{87}{3} = 29$

$S_{29} = \frac{29(9 + 93)}{2} = \frac{29(\overset{51}{\cancel{102}})}{\cancel{2}} = 1479$

**45.** $a_1 = 1 \quad n = 1{,}000 \quad d = 1$

$a_{1000} = 1 + (1000 - 1)(1)$
$= 1 + 999$
$= 1000$

$S_{1000} = \frac{\overset{500}{\cancel{1000}}(1 + 1000)}{\cancel{2}}$
$= 500(1001)$
$= 500{,}500$

**47.** $\frac{1610}{6} = 268\frac{1}{3}$

So the largest number $\leq 1610$ which is divisible by 6 is $6(268) = 1608$.

$a_1 = 12 \quad a_n = 1608 \quad d = 6$

$1608 = 12 + (n - 1)(6)$
$= 12 + 6n - 6$
$= 6 + 6n$

$1602 = 6n \quad n = \frac{1602}{6} = 267$

**49.** $d = -6$ in. $= -.5$ ft.

$n = 11$
$a_1 = 6$
$a_{11} = 6 + (11 - 1)(-.5)$
$= 6 + 10(-.5)$
$= 6 - 5$
$= 1$ ft.

**51.** We need the sum of the first n layers, with $a_1 = 20$, and $a_n = 1$, $d = -1$.

$1 = 20 + (n - 1)(-1)$
$= 20 - n + 1$
$= 21 - n$

$-20 = -n$
$n = 20$

$S_{20} = \frac{\overset{10}{\cancel{20}}(1 + 20)}{\cancel{2}} = 10(21)$
$= 210$ logs.

**53.** An arithmetic series is the sum of the terms of an arithmetic sequence.

**55.** The system will have one solution since the slopes of the lines are different and the lines must intersect.

**57.** $|x - 2| < 4$

$-4 < x - 2 < 4$
$-2 < x < 6$

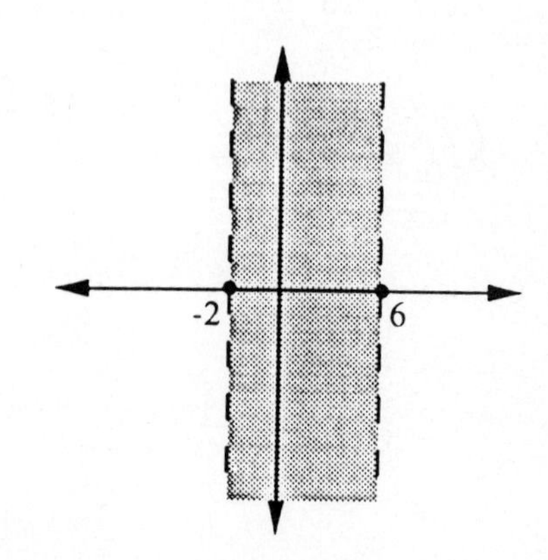

## EXERCISE SET 12.3

1. $a_1 = 5$

$a_2 = 5(3)^{2-1} = 5(3) = 15$

$a_3 = 5(3)^{3-1} = 5(3)^2 = 5(9) = 45$

$a_4 = 5(3)^{4-1} = 5(3)^3 = 5(27) = 135$

$a_5 = 5(3)^{5-1} = 5(3)^4 = 5(81) = 405$

5, 15, 45, 135, 405

3. $a_1 = 90$

$a_2 = 90(\frac{1}{3})^{2-1} = 90(\frac{1}{3}) = 30$

$a_3 = 90(\frac{1}{3})^{3-1} = 90(\frac{1}{3})^2 = 90(\frac{1}{9})$
$= 10$

$a_4 = 90(\frac{1}{3})^{4-1} = 90(\frac{1}{3})^3 = 90(\frac{1}{27}) = \frac{10}{3}$

$a_5 = 90(\frac{1}{3})^{5-1} = 90(\frac{1}{3})^4 = 90(\frac{1}{81}) = \frac{10}{9}$

$90, 30, 10, \frac{10}{3}, \frac{10}{9}$

5. $a_1 = -15$

$a_2 = -15(-2)^{2-1} = -15(-2)^1$
$= -15(-2)$
$= 30$

$a_3 = -15(-2)^{3-1} = -15(-2)^2$
$= -15(4)$
$= -60$

$a_4 = -15(-2)^{4-1} = -15(-2)^3$
$= -15(-8)$
$= 120$

$a_5 = -15(-2)^{5-1} = -15(-2)^4$
$= -15(16)$
$= -240$

-15, 30, -60, 120, -240

7. $a_1 = 3$

$a_2 = 3(\frac{3}{2})^{2-1} = 3(\frac{3}{2}) = \frac{9}{2}$

$a_3 = 3(\frac{3}{2})^{3-1} = 3(\frac{3}{2})^2 = 3(\frac{9}{4}) = \frac{27}{4}$

$a_4 = 3(\frac{3}{2})^{4-1} = 3(\frac{3}{2})^3 = 3(\frac{27}{8}) = \frac{81}{8}$

$a_5 = 3(\frac{3}{2})^{5-1} = 3(\frac{3}{2})^4 = 3(\frac{81}{16}) = \frac{243}{16}$

$3, \frac{9}{2}, \frac{27}{4}, \frac{81}{8}, \frac{243}{16}$

9. $a_6 = 5(2)^{6-1} = 5(2)^5 = 5(32)$
$= 160$

11. $a_7 = 18(3)^{7-1} = 18(3)^6 = 18(729)$
$= 13,122$

13. $a_8 = 2(\frac{1}{2})^{8-1} = 2(\frac{1}{2})^7 = 2(\frac{1}{128})$
$= \frac{1}{64}$

15. $a_{12} = -3(-2)^{12-1} = -3(-2)^{11}$
$= -3(-2048)$
$= 6144$

17. $S_5 = \frac{3(1 - 4^5)}{1 - 4} = \frac{3(1 - 1024)}{-3}$ (3 cancelled, leaving $-1$ in the denominator)

$= \frac{-1023}{-1}$
$= 1023$

19. $S_7 = \frac{80(1 - 2^7)}{1 - 2} = \frac{80(1 - 128)}{-1}$
$= \frac{80(-127)}{-1}$
$= 10,160$

**21.** $S_9 = \dfrac{-30[1 - (-\frac{1}{2})^9]}{1 - (-\frac{1}{2})}$

$= \dfrac{-30[1 - (-\frac{1}{512})]}{\frac{3}{2}}$

$= \dfrac{-30(1 + \frac{1}{512})}{\frac{3}{2}}$

$= \dfrac{-30(\frac{513}{512})}{\frac{3}{2}}$

$= -30(\frac{513}{512})(\frac{2}{3}) = -\frac{2565}{128}$ (cancelling: $-30 \to -10 \to -5$; $512 \to 256 \to 128$)

**23.** $S_5 = \dfrac{-9[1 - (\frac{2}{5})^5]}{1 - \frac{2}{5}} = \dfrac{-9(1 - \frac{32}{3125})}{\frac{3}{5}}$

$= \dfrac{-9(\frac{3093}{3125})}{\frac{3}{5}} = -9(\frac{3093}{3125})(\frac{5}{3})$ (cancelling: $-9 \to -3$; $3125 \to 625$)

$= -\frac{9279}{625}$

**25.** $r = \frac{5}{2} \div 5 = \frac{5}{2} \cdot \frac{1}{5} = \frac{1}{2}$

$a_n = 5(\frac{1}{2})^{n-1}$

**27.** $r = -6 \div 2 = -3$

$a_n = 2(-3)^{n-1}$

**29.** $r = -3 \div -1 = 3$

$a_n = -1(3)^{n-1}$

**31.** Consider the series 28, ___, 112, · · ·

$112 = 28r^{3-1} \quad r^2 = \frac{112}{28} = 4$

$r = 2 \quad$ or $\quad r = -2$

Then in the original series:

$112 = a_1(2)^{5-1}$ or $112 = a_1(-2)^{5-1}$

$112 = 16a_1$ or $112 = 16a_1$

$a_1 = \frac{112}{16} = 7$

**33.** Consider the series 15, ___, ___, 405, · · ·

$405 = 15r^{4-1}$

$r^3 = \frac{405}{15} = 27$

$r = 3$

Then in the original series:

$405 = a_1(3)^{5-1}$

$81a_1 = 405$

$a_1 = \frac{405}{81} = 5$

**35.** Each year the salary is 1.15 times the previous year's salary.

$a_1 = 20{,}000 \quad r = 1.15 \quad n = 25$

$a_n = 20{,}000(1.15)^{25-1}$

$\approx 20{,}000(28.625176)$

$= \$572{,}503.52$

**37.** $r = \frac{1}{2}$. Let $a_n$ be the amount left after the nth day.

After 1 day there are 300 (1/2) = 150 grams left, so $a_1 = 150$.

(a) $37.5 = 150(\frac{1}{2})^{n-1}$

$(\frac{1}{2})^{n-1} = \frac{37.5}{150} = \frac{1}{4} = (\frac{1}{2})^2$

$n - 1 = 2$

$n = 3$ days

(b) $a_8 = 150(\frac{1}{2})^{8-1} = 150(\frac{1}{128})$

$= 1.172$ grams.

**39.** Each year the population is 1.06 times the population in the previous year, so $r = 1.06$.

Let $a_n$ be the population after the nth year.

After the 1st year the population is 217.3(1.06) and equals 230.338 million, so $a_1 = 230.338$.

(a) $a_{12} = 230.338(1.06)^{12-1}$

$\approx 230.338(1.898299)$

$\approx 437.25$ million

(b) $a_n = 2(217.3) = 434.6$

$434.6 = 230.338(1.06)^{n-1}$

$(1.06)^{n-1} = \frac{434.6}{230.338}$

$\approx 1.886793$

Using the properties of logarithms

$(n - 1) \log(1.06) = \log(1.886793)$

$\approx \frac{0.275724}{0.025306}$

$\approx 10.896$

$n = 11.896 \approx 11.9$ years

**41.** $r = 1/2$

(a) After 1 m. there is 1/2 of the original light remaining, so $a_1 = 1/2$.

$a_1 = \frac{1}{2}$

$a_2 = \frac{1}{2}(\frac{1}{2})^{2-1} = (\frac{1}{2})(\frac{1}{2}) = \frac{1}{4}$

$a_3 = \frac{1}{2}(\frac{1}{2})^{3-1} = (\frac{1}{2})(\frac{1}{4}) = \frac{1}{8}$

$a_4 = \frac{1}{2}(\frac{1}{2})^{4-1} = (\frac{1}{2})(\frac{1}{8}) = \frac{1}{16}$

$a_5 = \frac{1}{2}(\frac{1}{2})^{5-1} = (\frac{1}{2})(\frac{1}{16}) = \frac{1}{32}$

(b) $a_n = \frac{1}{2}(\frac{1}{2})^{n-1} = (\frac{1}{2})^n$

(c) $a_7 = (\frac{1}{2})^7 = \frac{1}{128} \approx 0.0078$ or 0.78%

**43.** After 1 hour there is 2/3 of the original dye left, so $a_1 = 2/3$.

$r = \frac{2}{3}$ $n = 10$

$a_{10} = \frac{2}{3}(\frac{2}{3})^{10-1} = \frac{2}{3}(\frac{2}{3})^9 = (\frac{2}{3})^{10}$

$\approx 0.017$ or 1.7%

**45.** Let $a_n$ = value left after the nth year after the 1st year there is

$9800(\frac{4}{5}) = 7840$ of the value left,

so $a_1 = 7840$, $r = \frac{4}{5}$

$a_2 = 7840(\frac{4}{5})^{2-1} = 7840(\frac{4}{5}) = 6272$

$a_3 = 7840(\frac{4}{5})^{3-1} = 7840(\frac{16}{25})$

$= 5017.60$

Number 45 (Continued)

(a) \$7840, \$6272 \$5017.60

(b) $a_n = 7840(\frac{4}{5})^{n-1}$

(c) $a_5 = 7840(\frac{4}{5})^{5-1}$

$= \overset{1568}{\cancel{7840}}\ \frac{256}{\underset{125}{\cancel{625}}}$

$\approx 3211.26$

47. A geometric series is the sum of the terms of a geometric sequence.

49. Let x = the length of time it takes Mr. Donovan to load the truck and 2x = the length of time it takes Mrs. Donovan to load the truck.

$\frac{1}{x} + \frac{1}{2x} = \frac{1}{8}$

$8x(\frac{1}{x} + \frac{1}{2x} = \frac{1}{8})$

$8 + 4 = x$

$12 = x$

It will take Mr. Donovan 12 hrs to load the truck alone.

51. $\sqrt[3]{9x^2y}\ (\sqrt[3]{3x^4y^6} - \sqrt[3]{8xy^4})$

$= \sqrt[3]{27x^6y^7} - \sqrt[3]{72x^3y^5}$

$= 3x^2y^2\ \sqrt[3]{y} - 2xy\ \sqrt[3]{9y^2}$

**JUST FOR FUN**

1. (a) \$3211.26 (See #45, C)
   (b) C = 9800 r = 1/5 n = 5

$S = 9800(1 - \frac{1}{5})^5 = 9800(\frac{4}{5})^5$

$= \overset{392}{\cancel{9800}}\ \frac{1024}{\underset{125}{\cancel{3125}}} \approx 3211.26$

## EXERCISE SET 12.4

1. $r = 3 \div 6 = \frac{1}{2}$

$$S_\infty = \frac{6}{1 - \frac{1}{2}} = \frac{6}{\frac{1}{2}} = 6(\frac{2}{1}) = 12$$

3. $r = 2 \div 5 = \frac{2}{5}$

$$S_\infty = \frac{5}{1 - \frac{2}{5}} = \frac{5}{\frac{3}{5}} = 5(\frac{5}{3}) = \frac{25}{3}$$

5. $r = \frac{4}{15} \div \frac{1}{3} = \frac{4}{15}(\frac{3}{1}) = \frac{4}{5}$

$$S_\infty = \frac{\frac{1}{3}}{1 - \frac{4}{5}} = \frac{\frac{1}{3}}{\frac{1}{5}} = \frac{1}{3}(\frac{5}{1}) = \frac{5}{3}$$

7. $r = -1 \div 9 = \frac{-1}{9}$

$$S_\infty = \frac{9}{1 - (-\frac{1}{9})} = \frac{9}{1 + \frac{1}{9}} = \frac{9}{\frac{10}{9}}$$

$$= 9(\frac{9}{10}) = \frac{81}{10}$$

9. $r = \frac{1}{2} \div \frac{1}{2}$

$$S_\infty = \frac{1}{1 - \frac{1}{2}} = \frac{1}{\frac{1}{2}} = 1(\frac{2}{1}) = 2$$

11. $r = \frac{16}{3} \div 8 = \frac{\overset{2}{\cancel{16}}}{3}\cdot\frac{1}{\cancel{8}} = \frac{2}{3}$

$$S_\infty = \frac{8}{1 - \frac{2}{3}} = \frac{8}{\frac{1}{3}} = 8(\frac{3}{1}) = 24$$

13. $r = 20 \div -60 = \frac{20}{-60} = -\frac{1}{3}$

$$S_\infty = \frac{-60}{1 - (-\frac{1}{3})} = \frac{-60}{1 + \frac{1}{3}} = \frac{-60}{\frac{4}{3}}$$

$$= \overset{-15}{\cancel{-60}}\,\frac{3}{\cancel{4}} = -45$$

15. $r = -\frac{12}{5} \div 2 = -\frac{\cancel{12}}{5}(-\frac{1}{\cancel{12}}) = \frac{1}{5}$

$$S_\infty = \frac{-12}{1 - \frac{1}{5}} = \frac{12}{\frac{4}{5}} = \overset{-3}{\cancel{-12}}(\frac{5}{\cancel{4}}) = -15$$

17. $0.2727\ldots = 0.27 + 0.0027 + 0.000027 + \ldots$

$= 0.27 + 0.27(.01) + 0.27(.01)^2 + \ldots$

$r = .01$

$a_1 = 0.27$

$$S_\infty = \frac{0.27}{1 - .01} = \frac{0.27}{0.99} = \frac{27}{99}$$

$$= \frac{3}{11}$$

19. $0.5555\cdots = 0.5 + 0.05 + 0.005 + \ldots$

$= 0.5 + 0.5(.1) + 0.5(.1)^2 + \ldots$

$r = .1$

$a_1 = 0.5$

$$S_\infty = \frac{0.5}{1 - .1} = \frac{0.5}{0.9} = \frac{5}{9}$$

21. $0.515151 = 0.51 + 0.0051 + 0.000051 + \ldots$

$= 0.51 + 0.51(.01) + 0.51(.01)^2 + \ldots$

$r = .01$

$a_1 = 0.51$

$$S_\infty = \frac{0.51}{1 - .01} = \frac{0.51}{0.99}$$

$$= \frac{17}{33}$$

**23.** $r = .8$

$a_1 = 8$

$S_\infty = \frac{8}{1 - .8} = $ [illegible] $= \frac{80}{2} = 40$ ft.

**25.** When $|r| > 1$, the sum does not exist.

**27.**

$$\begin{array}{r|l} & 8x - 15 \\ \hline 2x+5 & 16x^2 + 0x - 18 \\ & \underline{16x^2 \; 40x} \\ & \quad 30x - 18 \\ & \quad \underline{30x - 75} \\ & \qquad\quad 57 \end{array}$$

Thus

$16x^2 +$ [illegible] $- 18 \div 2x + 5 =$

$8x -$ [illegible] $+ \frac{57}{2x + 5}$

**29.** $\sqrt{a^2 +\text{[illegible]} + 3} = -a$

$a^2$ [illegible] $a + 3 = a^2$

[illegible]$a + 3 = 0$

$a = -1/3$

**JUST F[illegible]UN**

1. [illegible] time the ball bounces it goes [illegible]nd then comes down the same [illegible]tance. Therefore, the total [illegible]tical distance will be twice [illegible] height it rises after each [illegible]unce plus the initial 10 feet [illegible]e height after each bounce form [illegible] infinite geometric sequence [illegible]ith $r = -.9$ and $a_1 = 9$.

$S_\infty = \frac{9}{1 - .9} = \frac{9}{.1} = 90$

Total distance $= 2S_\infty + 10$

$= 2(90) + 10$

$= 190$ feet

**EXERCISE SET 12.5**

**1.** $\binom{4}{2} = \frac{4!}{2!\cdot(4-2)!}$
$= \frac{4\cdot3\cdot2\cdot1}{(2\cdot1)(2\cdot1)}$
$= 6$

**3.** $\binom{5}{5} = 1$

**5.** $\binom{7}{0} = 1$

**7.** $\binom{8}{4} = \frac{8!}{4!(8-4)!}$
$= \frac{8\cdot7\cdot6\cdot5\cdot4\cdot3\cdot2\cdot1}{(4\cdot3\cdot2\cdot1)(4\cdot3\cdot2\cdot1)}$
$= 70$

**9.** $\binom{8}{2} = \frac{8!}{2!(8-2)!}$
$= \frac{40320}{2(720)}$
$= 28$

**11.** $(x + 4)^3 = x^3 + 3x^2(4) + \frac{3\cdot2x}{2\cdot1}(4)^2 + 4^3$
$= x^3 + 3x^2(4) + 3x(16) + 64$
$= x^3 + 12x^2 + 48x + 64$

**13.** $(a - b)^4 = a^4 + 4a^3(-b) + \frac{4\cdot3a^2}{2\cdot1}(-b)^2 + \frac{4\cdot3\cdot2a}{3\cdot2\cdot1}(-b)^3 + (-b)^4$
$= a^4 + 4a^3(-b) + 6a^2(b^2) + 4a(-b^3) + b^4$
$= a^4 - 4a^3b + 6a^2b^2 - 4ab^3 + b^4$

**15.** $(3a - b)^5 = (3a)^5 + 5(3a)^4(-b) + \frac{5\cdot4}{2\cdot1}(3a)^3(-b)^2 + \frac{5\cdot4\cdot3}{3\cdot2\cdot1}(3a)^2(-b)^3 + \frac{5\cdot4\cdot3\cdot2}{4\cdot3\cdot2\cdot1}(3a)(-b)^4 + (-b)^5$
$= 243a^5 + 5(81a^4)(-b) + 10(27a^3)(b^2) + 10(9a^2)(-b^3) + 5(3a)(b^4) - b^5$
$= 243a^5 - 405a^4b + 270a^3b^2 - 90a^2b^3 + 15ab^4 - b^5$

**17.** $(2x + \frac{1}{2})^4 = (2x)^4 + 4(2x)^3(\frac{1}{2}) + \frac{4\cdot3}{2\cdot1}(2x)^2(\frac{1}{2})^2 + \frac{4\cdot3\cdot2}{3\cdot2\cdot1}(2x)(\frac{1}{2})^3 + (\frac{1}{2})^4$
$= 16x^4 + 4(8x^3)(\frac{1}{2}) + 6(4x^2)(\frac{1}{4}) + 4(2x)(\frac{1}{8}) + \frac{1}{16}$
$= 16x^4 + 16x^3 + 6x^2 + x + \frac{1}{16}$

**19.** $$\left(\frac{x}{2} - 3\right)^4 = \left(\frac{x}{2}\right)^4 + 4\left(\frac{x}{2}\right)^3(-3) + \frac{4\cdot3}{2\cdot1}\left(\frac{x}{2}\right)^2(-3)^2 + \frac{4\cdot3\cdot2}{3\cdot2\cdot1}\left(\frac{x}{2}\right)(-3)^3 + (-3)^4$$
$$= \frac{x^4}{16} + 4\left(\frac{x^3}{8}\right)(-3) + 6\left(\frac{x^2}{4}\right)(9) + 4\left(\frac{x}{2}\right)(-27) + 81$$
$$= \frac{x^4}{16} - \frac{3x^3}{2} + \frac{27x^2}{2} - 54x + 81$$

**21.** $$(x + y)^{10} = x^{10} + 10x^9y + \frac{10\cdot9}{2\cdot1}x^8y^2 + \frac{10\cdot9\cdot8}{3\cdot2\cdot1}x^7y^3$$
$$= x^{10} + 10x^9y + 45x^8y^2 + 120x^7y^3 + \dots$$

**23.** $$(3x - y)^7 = (3x)^7 + 7(3x)^6(-y) + \frac{7\cdot6}{2\cdot1}(3x)^5(-y)^2 + \frac{7\cdot6\cdot5}{3\cdot2\cdot1}(3x)^4(-y)^3 + \cdots$$

Number 23 (Continued)

$$= 2187x^3(729x^6)(-y) + 21(243x^5)(y^2) + 35(81x^4)(-y^3) + 2187x^3+(729x^6)(-y)$$
$$= 2187x^7 - 5103x^6y + 5103x^5y^2 - 2835x^4y^3 + \cdots$$

**25.** $$(x^2 - 3y)^8 = (x^2)^8 + 8(x^2)^7(-3y) + \frac{8\cdot7}{2\cdot1}(x^2)^6(-3y)^2 + \frac{8\cdot7\cdot6}{3\cdot2\cdot1}(x^2)^5(-3y)^3 + \cdots$$
$$= x^{16} + 8(x^{14})(-3y) + 28(x^{12})(9y^2) + 56(x^{10})(-27y^3) + \cdots$$
$$= x^{16} - 24x^{14}y + 252x^{12}y^{12} - 1512x^{10}y^3 + \cdots$$

**27.** The first and last numbers in each row are 1 and the inner numbers are obtained by adding the two numbers in the row above.

```
    1
   1 1
  1 2 1
 1 3 3 1
1 4 6 4 1
```

**29.** Yes, $n! = n\cdot(n-1)!$
$$4! = 4\cdot3\cdot2\cdot1 = 4\cdot(3\cdot2\cdot1) = 4\cdot(3)! = 4\cdot(4-1)!$$

31. Yes, $(n-3)! = (n-3)(n-4)(n-5)!$ for $n \geq 5$.

$(7-3)! = (7-3)(7-4)(7-5)!$

$4! = (4)(3)(2!)$

$= (4)(3)(2)(1)$

$= 4!$

35. $16x^2 - 8x - 3$

$-\underline{12}x(\underline{4}) = -48$

$-\underline{12} + \underline{4} = -8$

$= 16x^2 - 12x + 4x - 3$

$= 4x(4x-3) + 1(4x-3)$

$= (4x+1)(4x-3)$

37. $S_n - S_n r = a_1 - a_1 r^n$

Solve for $S_n$

$S_n(1-r) = a_1 - a_1 r^n$

$$S_n = \frac{a_1 - a_1 r^n}{1 - r}$$

**JUST FOR FUN**

1. $(a + b)^n = \sum_{i=0}^{n} \binom{n}{i} a^{n-i} b^i$

**REVIEW EXERCISES**

1. $a_1 = 1 + 2 = 3$
   $a_2 = 2 + 2 = 4$
   $a_3 = 3 + 2 = 5$
   $a_4 = 4 + 2 = 6$
   $a_5 = 5 + 2 = 7$
   3, 4, 5, 6, 7

3. $a_1 = 1(1 + 1) = 1(2) = 2$
   $a_2 = 2(2 + 1) = 2(3) = 6$
   $a_3 = 3(3 + 1) = 3(4) = 12$
   $a_4 = 4(4 + 1) = 4(5) = 20$
   $a_5 = 5(5 + 1) = 5(6) = 30$
   2, 6, 12, 20, 30

5. $a_7 = 3(7) + 4 = 21 + 4 = 25$

7. $a_9 = \frac{9 + 7}{9^2} = \frac{16}{81}$

9. $a_1 = 3(1) + 2 = 3 + 2 = 5$
   $a_2 = 3(2) + 2 = 6 + 2 = 8$
   $a_3 = 3(3) + 2 = 9 + 2 = 11$
   $S_1 = a_1 = 5$
   $S_3 = a_1 + a_2 + a_3$
   $= 5 + 8 + 11$
   $= 24$

11. $a_1 = \frac{1 + 3}{1 + 2} = \frac{4}{3}$
   $a_2 = \frac{2 + 3}{2 + 2} = \frac{5}{4}$
   $a_3 = \frac{3 + 3}{3 + 2} = \frac{6}{5}$
   $S_1 = a_1 = \frac{4}{3}$
   $S_3 = a_1 + a_2 + a_3$
   $= \frac{4}{3} + \frac{5}{4} + \frac{6}{5}$
   $= \frac{80}{60} + \frac{75}{60} + \frac{72}{60}$
   $= \frac{227}{60}$

13. This is a geometric sequence.
   $r = 2 \div 1 = 2$
   $a_1 = 1$
   $a_5 = 1(2)^{5-1} = 2^4 = 16$
   $a_6 = 1(2)^{6-1} = 2^5 = 32$
   $a_7 = 1(2)^{7-1} = 2^6 = 64$
   16, 32, 64
   $a_n = 1(2)^{n-1} = 2^{n-1}$

15. This is a geometric sequence.
   $r = \frac{4}{3} \div \frac{2}{3} = \frac{\cancelto{2}{4}}{\cancel{3}}\left(\frac{\cancel{3}}{\cancel{2}}\right) = 2$
   $a_1 = \frac{2}{3}$
   $a_5 = \frac{2}{3}(2)^{5-1} = \frac{2}{3}(2^4) = \frac{2^5}{3} = \frac{32}{3}$
   $a_6 = \frac{2}{3}(2)^{6-1} = \frac{2^6}{3} = \frac{64}{3}$
   $a_7 = \frac{2}{3}(2)^{7-1} = \frac{2^7}{3} = \frac{128}{3}$
   $\frac{32}{3}, \frac{64}{3}, \frac{128}{3}$
   $a_n = \frac{2}{3}(2)^{n-1}$

**17.** This is a geometric sequence.

$r = 1 \div -1 = -1$

$a_1 = -1$

$a_5 = -1(-1)^{5-1} = -1(1) = -1$

$a_6 = -1(-1)^{6-1} = -1(-1) = 1$

$a_7 = -1(-1)^{7-1} = -1(1) = -1$

$-1, 1, -1$

$a_n = -1(-1)^{n-1} = (-1)^n$

**19.** $\sum_{n=1}^{3} (n^2+2) = (1^2+2)+(2^2+2) + (3^2+2)$

$= 3 + 6 + 11 = 20$

**21.** $\sum_{k=1}^{5} \frac{k^2}{3} = \frac{1^2}{3} + \frac{2^2}{3} + \frac{3^2}{3} + \frac{4^2}{3} + \frac{5^2}{3}$

$= \frac{1}{3} + \frac{4}{3} + \frac{9}{3} + \frac{16}{3} + \frac{25}{3} = \frac{55}{3}$

**23.** $\sum_{i=1}^{4} x_i = x_1 + x_2 + x_3 + x_4$

$= 3 + 9 + 5 + 10$

$= 27$

**25.** $\sum_{i=2}^{3} (x_i^2 + 1) = (x_2^2 + 1)$

$+ (x_3^2 + 1)$

$= (9^2 + 1) + (5^2 + 1)$

$= 108$

**27.** $a_1 = 5$

$a_2 = 5 + (2 - 1)(2) = 5 + 2 = 7$

$a_3 = 5 + (3 - 1)(2) = 5 + 2(2) = 9$

$a_4 = 5 + (4 - 1)(2) = 5 + 3(2) = 11$

$a_5 = 5 + (5 - 1)(2) = 5 + 4(2) = 13$

$5, 7, 9, 11, 13$

**29.** $a_1 = -12$

$a_2 = -12 + (2 - 1)(-1/2)$

$= -12 - 1/2$

$= -\frac{25}{2}$

$a_3 = -12 + (3 - 1)(-1/2)$

$= -12 + 2(-1/2)$

$= -12 - 1 = -13$

$a_4 = -12 + (4 - 1)(-1/2)$

$= -12 + 3(-1/2)$

$= -12 - \frac{3}{2} = \frac{-27}{2}$

$a_5 = -12 + (5 - 1)(-1/2)$

$= -12 + 4(-1/2)$

$= -12 - 2 = -14$

$-12, -\frac{25}{2}, -13, -\frac{27}{2}, -14$

**31.** $a_9 = 2 + (9 - 1)(3) = 2 + 8(3) = 26$

**33.** $34 = 50 + (5 - 1)d = 50 + 4d$

$-16 = 4d \qquad d = \frac{-16}{4} = -4$

**35.** $-13 = 12 + (n - 1)(-5)$

$= 12 - 5n + 5$

$= 17 - 5n$

$-30 = -5n \quad n = \frac{-30}{-5} = 6$

**37.** $21 = 7 + (8 - 1)d = 7 + 7d$

$14 = 7d \quad d = \frac{14}{7} = 2$

$S_8 = \frac{\overset{4}{\cancel{8}}(7 + 21)}{\cancel{2}} = 4(28) = 112$

**39.** $3 = \frac{3}{5} + (7 - 1)d = \frac{3}{5} + 6d$

$\frac{12}{5} = 6d \quad d = \frac{12}{5} \div 6 = \frac{\overset{2}{\cancel{12}}}{5}\left(\frac{1}{\cancel{6}}\right) = \frac{2}{5}$

$S_7 = \frac{7\left(\frac{3}{5} + 3\right)}{2} = \frac{7\left(\frac{18}{5}\right)}{2}$

$= 7\left(\frac{\overset{9}{\cancel{18}}}{5}\right)\left(\frac{1}{\cancel{2}}\right) = \frac{63}{5}$

**41.** $a_1 = 2$

$a_2 = 2 + (2 - 1)(4) = 2 + 4 = 6$

$a_3 = 2 + (3 - 1)(4) = 2 + 2(4) = 10$

$a_4 = 2 + (4 - 1)(4) = 2 + 3(4) = 14$

2, 6, 10, 14

$a_{10} = 22 + (10 - 1)(4)$

$= 2 + 9(4) = 38$

$S_{10} = \frac{\overset{5}{\cancel{10}}(2 + 38)}{\cancel{2}} = 5(40) = 200$

**43.** $a_1 = \frac{5}{6}$

$a_2 = \frac{5}{6} + (2 - 1)\left(\frac{2}{3}\right) = \frac{5}{6} + \frac{2}{3} = \frac{9}{6}$

$a_3 = \frac{5}{6} + (3 - 1)\left(\frac{2}{3}\right) = \frac{5}{6} + 2\left(\frac{2}{3}\right)$

$= \frac{5}{6} + \frac{4}{3} = \frac{13}{6}$

$a_4 = \frac{5}{6} + (4 - 1)\left(\frac{2}{3}\right) = \frac{5}{6} + 3\left(\frac{2}{3}\right)$

$= \frac{5}{6} + \frac{6}{3} = \frac{17}{6}$

$\frac{5}{6}, \frac{9}{6}, \frac{13}{6}, \frac{17}{6}$

$a_{10} = \frac{5}{6} + (10 - 1)\left(\frac{2}{3}\right) = \frac{5}{6} + 9\left(\frac{2}{3}\right)$

$= \frac{5}{6} + \frac{18}{3} = \frac{41}{6}$

$S_{10} = \frac{10\left(\frac{5}{6} + \frac{41}{6}\right)}{2} = \frac{10\left(\frac{46}{6}\right)}{2}$

$= \overset{5}{\cancel{10}}\left(\frac{\overset{23}{\cancel{46}}}{\underset{3}{\cancel{6}}}\right)\left(\frac{1}{\cancel{2}}\right) = \frac{115}{3}$

**45.** $d = 8 - 3 = 5$

$53 = 3 + (n - 1)(5) = 3 + 5n - 5$

$= 5n - 2$

$55 = 5n \quad n = \frac{55}{5} = 11$

$S_n = \frac{11(3 + 53)}{2} = \frac{11(\overset{28}{\cancel{56}})}{\cancel{2}} = 308$

**47.** $d = \frac{9}{10} - \frac{6}{10} = \frac{3}{10}$

$\frac{36}{10} = \frac{6}{10} + (n - 1)\left(\frac{3}{10}\right)$

$36 = 6 + (n - 1)(3) = 6 + 3n - 3$

$= 3 + 3n$

$33 = 3n \quad n = \frac{33}{3} = 11$

Number 47 (Continued)

$$S_{11} = \frac{11(\frac{6}{10} + \frac{36}{10})}{2} = \frac{11(\frac{42}{10})}{2}$$

$$= 11(\frac{\overset{21}{\cancel{42}}}{10})(\frac{1}{\cancel{2}}) = \frac{231}{10}$$

**49.** $a_1 = 5$

$a_2 = 5(2)^{2-1} = 5(2) = 10$

$a_3 = 5(2)^{3-1} = 5(2)^2 = 5(4) = 20$

$a_4 = 5(2)^{4-1} = 5(2)^3 = 5(8) = 40$

$a_5 = 5(2)^{5-1} = 5(2)^4 = 5(16) = 80$

5, 10, 20, 40, 80

**51.** $a_1 = 20$

$a_2 = 20(\frac{-2}{3})^{2-1} = 20(\frac{-2}{3}) = \frac{-40}{3}$

$a_3 = 20(\frac{-2}{3})^{3-1} = 20(\frac{-2}{3})^2$

$= 20(\frac{4}{9}) = \frac{80}{9}$

$a_4 = 20(\frac{-2}{3})^{4-1} = 20(\frac{-2}{3})^3$

$= 20(\frac{-8}{27}) = \frac{-160}{27}$

$a_5 = 20(\frac{-2}{3})^{5-1} = 20(\frac{-2}{3})^4 = 20(\frac{16}{81})$

$= \frac{320}{81}$

**53.** $a_7 = 12(\frac{1}{3})^{7-1} = \overset{4}{\cancel{12}}(\frac{1}{\underset{243}{\cancel{729}}}) = \frac{4}{243}$

**55.** $a_9 = -8(-2)^{9-1} = -8(256) = -2048$

**57.** $S_8 = \frac{12(1 - 2^8)}{1 - 2} = \frac{12(1 - 256)}{-1}$

$= \frac{12(-255)}{-1} = 3060$

**59.** $$S_5 = \frac{-84[1 - (-\frac{1}{4})^5]}{1 - (-\frac{1}{4})}$$

$$= \frac{-84[1 - (-\frac{1}{1024})}{1 + \frac{1}{4}}$$

$$= \frac{-84(1 + \frac{1}{1024})}{\frac{5}{4}} = \overset{-21}{\cancel{-84}}\ \frac{\overset{205}{\cancel{1025}}}{\underset{\underset{64}{\cancel{256}}}{\cancel{1024}}}\ \frac{4}{\cancel{5}}$$

$$= \frac{-4305}{64}$$

**61.** $r = 12 \div 6 = 2$

$a_n = 6(2)^{n-1}$

**63.** $r = -20 \div -4 = \frac{-20}{-4} = 5$

$a_n = -4(5)^{n-1}$

**65.** $r = \frac{7}{2} \div \frac{7}{2} \cdot \frac{1}{7} = \frac{1}{2}$

$$S_\infty = \frac{7}{1 - \frac{1}{2}} = \frac{7}{\frac{1}{2}} = 7(\frac{2}{1}) = 14$$

**67.** $r = -\frac{10}{3} \div -5 = (-\frac{10}{3})(-\frac{1}{5}) = \frac{2}{3}$

$S_\infty = \frac{-5}{1 - \frac{2}{3}} = \frac{-5}{\frac{1}{3}} = -5(\frac{3}{1}) = -15$

**69.** $r = 1 \div 2 = \frac{1}{2}$

$S_\infty = \frac{2}{1 - \frac{1}{2}} = \frac{2}{\frac{1}{2}} = 2(\frac{2}{1}) = 4$

**71.** $r = -\frac{24}{3} \div -12 = -\frac{24}{3}(\frac{-1}{12}) = \frac{2}{3}$

$S_\infty = \frac{-12}{1 - \frac{2}{3}} = \frac{-12}{\frac{1}{3}} = -12(\frac{3}{1}) = -36$

**73.** $0.5252\ldots = 0.52 + 0.0052 + 0.000052 + \cdots$
$= 0.52 + 0.52(.01) + 0.52(.01)^2 + \cdots$

$r = .01$

$S_\infty = \frac{0.52}{1 - .01} = \frac{0.52}{0.99} = \frac{52}{99}$

**75.** $(3x + y)^4 = (3x)^4 + 4(3x)^3y + \frac{4\cdot 3}{2\cdot 1}(3x)^2y^2 + \frac{4\cdot 3\cdot 2}{3\cdot 2\cdot 1}(3x)y^3 + y^4$

$= 81x^4 + 4(27x^3)y + 6(9x^2)y^2 + 4(3x)y^3 + y^4$

Number 75 (Continued)

$= 81x^4 + 108x^3y + 54x^2y^2 + 12xy^3 + y^4$

**77.** $(x - 2y)^9 = x^9 + 9x^8(-2y) + \frac{9\cdot 8}{2\cdot 1}x^7(-2y)^2 + \frac{9\cdot 8\cdot 7}{3\cdot 2\cdot 1}x^6(-2y)^3 + \cdots$

$= x^9 + 9x^8(-2y) + 36x^7(4y^2) + 84x^6(-8y^3) + \cdots$

$= x^9 - 18x^8y + 144x^7y^2 - 672^6y^3 + \cdots$

**79.** This is an arithmetic series with $d = 1$, $a_1 = 100, n = 200 - 99 = 101$, and $a_{101} = 200$.

$S_{101} = \frac{101(100 + 200)}{2} = \frac{101(300)}{2} = 15150$

**81.** This is a geometric sequence with $r = 2$ and $n = 10$ after the first doubling there is \$200, so $a_1 = 200$.

$a_{10} = 200(2)^{10-1} = 200(512) = \$102,400$.

**83.** This is an infinite geometric series with $r = .92$ and $a_1 = 8$.

$S_\infty = \frac{8}{1 - .92} = \frac{8}{0.08} = 100$ ft.